Texts in Applied Mathematics 20

T0156105

Editors
Stuart Antman
Philip Holmes
K.R. Sreenivasan

Advisors
G. Iooss
P. Holmes
D. Barkley
M. Dellnitz
P. Newton

For further volumes:
http://www.springer.com/series/1214

Mark H. Holmes

Introduction to Perturbation Methods

Second Edition

Mark H. Holmes
Department of Mathematical Sciences
Rensselaer Polytechnic Institute
110 8th Street
Troy, NY 12180-3590

ISSN 0939-2475
ISBN 978-1-4899-9613-8   ISBN 978-1-4614-5477-9 (eBook)
DOI 10.1007/978-1-4614-5477-9
Springer New York Heidelberg Dordrecht London

Mathematics Subject Classification (2010): 34-XX , 34D15, 34Exx, 35-XX, 35B25, 35B27, 76-XX, 76M45, 41-XX, 41A60

© Springer Science+Business Media New York 2013
Softcover reprint of the hardcover 2nd edition 2013
This work is subject to copyright. All rights are reserved by the Publisher, whether the whole or part of the material is concerned, specifically the rights of translation, reprinting, reuse of illustrations, recitation, broadcasting, reproduction on microfilms or in any other physical way, and transmission or information storage and retrieval, electronic adaptation, computer software, or by similar or dissimilar methodology now known or hereafter developed. Exempted from this legal reservation are brief excerpts in connection with reviews or scholarly analysis or material supplied specifically for the purpose of being entered and executed on a computer system, for exclusive use by the purchaser of the work. Duplication of this publication or parts thereof is permitted only under the provisions of the Copyright Law of the Publisher's location, in its current version, and permission for use must always be obtained from Springer. Permissions for use may be obtained through RightsLink at the Copyright Clearance Center. Violations are liable to prosecution under the respective Copyright Law.
The use of general descriptive names, registered names, trademarks, service marks, etc. in this publication does not imply, even in the absence of a specific statement, that such names are exempt from the relevant protective laws and regulations and therefore free for general use.
While the advice and information in this book are believed to be true and accurate at the date of publication, neither the authors nor the editors nor the publisher can accept any legal responsibility for any errors or omissions that may be made. The publisher makes no warranty, express or implied, with respect to the material contained herein.

Printed on acid-free paper

Springer is part of Springer Science+Business Media (www.springer.com)

To Colette, Matthew, and Marianna
A small family with big hearts

Preface

First, let me say hello and welcome to the subject of perturbation methods. For those who may be unfamiliar with the topic, the title can be confusing. The first time I became aware of this was during a family reunion when someone asked what I did as a mathematician. This is not an easy question to answer, but I started by describing how a certain segment of the applied mathematics community was interested in problems that arise from physical problems. Examples such as water waves, sound propagation, and the aerodynamics of airplanes were discussed. The difficulty of solving such problems was also described in exaggerated detail. Next came the part about how one generally ends up using a computer to actually find the solution. At this point I editorialized on the limitations of computer solutions and why it is important to derive, if at all possible, accurate approximations of the solution. This led naturally to the mentioning of asymptotics and perturbation methods. These terms ended the conservation because I was unprepared for people's reactions. They were not sure exactly what asymptotics meant, but they were quite perplexed about perturbation methods. I tried, unsuccessfully, to explain what it means, but it was not until sometime later that I realized the difficulty. For them, as in *Webster's Collegiate Dictionary*, the first two meanings for the word perturb are "to disturb greatly in mind (disquiet); to throw into confusion (disorder)." Although a cynic might suggest this is indeed appropriate for the subject, the intent is exactly the opposite. For a related comment, see Exercise 3.18(d).

In a nutshell, this book serves as an introduction to systematically constructing an approximation of the solution to a problem that is otherwise intractable. The methods all rely on there being a parameter in the problem that is relatively small. Such a situation is relatively common in applications, and this is one of the reasons that perturbation methods are a cornerstone of applied mathematics. One of the other cornerstones is scientific computing, and it is interesting that the two subjects have grown up together. However, this is not unexpected given their respective capabilities. Using a computer one can solve problems that are nonlinear, inhomogeneous, and multidimensional. Moreover, it is possible to achieve very high accuracy. The drawbacks are that computer solutions do not provide much insight into the physics of problems (particularly when one does not have access to the appropriate software or computer), and there is always the question as to whether or not the computed solution is correct. On the other hand, perturbation methods are also capable of dealing with nonlinear, inhomogeneous, and multidimensional problems (although not to the same extent as computer-generated solutions). The principal objective when using perturbation methods, at least as far as the author is concerned, is to provide a reasonably accurate expression for the solution. In doing this one can derive an understanding of the physics of a problem. Also, one can use the result, in conjunction with the original problem, to obtain more efficient numerical procedures for computing the solution.

The methods covered in the text vary widely in their applicability. The first chapter introduces the fundamental ideas underlying asymptotic approximations. This includes their use in constructing approximate solutions of transcendental equations as well as differential equations. In the second chapter, matched asymptotic expansions are used to analyze problems with layers. Chapter 3 describes a method for dealing with problems with more than one time scale. In Chap. 4, the WKB method for analyzing linear singular perturbation problems is developed, while in Chap. 5 a method for dealing with materials containing disparate spatial scales (e.g., microscopic vs. macroscopic) is discussed. The last chapter examines the topics of multiple solutions and stability.

The mathematical prerequisites for this text include a basic background in differential equations and advanced calculus. In terms of difficulty, the chapters are written so the first sections are either elementary or intermediate, while the later sections are somewhat more advanced. Also, the ideas developed in each chapter are applied to a spectrum of problems, including ordinary differential equations, partial differential equations, and difference equations. Scattered through the exercises are applications to integral equations, integrodifferential equations, differential-difference equations, and delay equations. What will not be found is an in-depth discussion of the theory underlying the methods. This aspect of the subject is important, and references to the more theoretical work in the area are given in each chapter.

The exercises in each section vary in their complexity. In addition to the more standard textbook problems, an attempt has been made to include problems from the research literature. The latter are intended to provide a window into the wide range of areas that use perturbation methods. Solutions to some of the exercises are available and can be obtained, at no charge, from the author's home page. Also included, in the same file, is an errata sheet. Readers who would like to make a contribution to this file or who have suggestions about the text can reach the author at holmes@rpi.edu.

I would like to express my gratitude to the many students who took my course in perturbation methods at Rensselaer. They helped me immeasurably in understanding the subject and provided much needed encouragement to write this book. It is a pleasure to acknowledge the suggestions of Jon Bell, Ash Kapila, and Bob O'Malley, who read early versions of the manuscript. I would also like to thank Julian Cole, who first introduced me to perturbation methods and is still, to this day, showing me what the subject is about.

Troy, NY   Mark H. Holmes

Preface to the Second Edition

It's interesting reading something you wrote 15 years earlier, not just because of what you *did* write but also because of what you did *not* write. You also realize how the subject has evolved, that certain topics should be rewritten and others included. It is for these reasons that this second edition was undertaken. As will be explained in the next paragraph, every section has been edited, many only in minor ways, while others have been completely revised; new material has also been added. This includes approximations for weakly coupled oscillators, analysis of problems that involve transcendentally small terms, an expanded discussion of Kummer functions, and metastability. Also, one of the core objectives of this book is to develop the ideas underlying perturbation methods and then demonstrate how they can be used in a wide variety of problems. To provide background on some of these areas, two appendices have been added, one on solving difference equations, the other on delay equations. Finally, a few things have also been removed, the most prominent of which is the appendix containing the numerical solution of boundary-value problems. The code given in the earlier edition is available at the author's home page and is also discussed at length in Holmes (2007). Finally, the references have been updated, new exercises added, and a few of the original exercises modified.

There is an interesting aside that is worth telling. The first edition was written using a commercial software program. At the time, Springer published their books using LaTeX, so they needed to use a translator program to convert the manuscript. With this, they had to redraw, by hand, all of the figures. They also had to write macros to produce some of the lettering used for balancing equations. The result was very nice. Unfortunately, none of this worked for this edition. The original files cannot be used with the current version of the commercial software program, and the converted LaTeX files exist only in fragments. Consequently, this edition was written practically from scratch, with every figure redrawn. With this effort come a few benefits. One is that the manuscript is now in LaTeX and should be useable for the foreseeable future. Second, many of the figures were redrawn using MATLAB, and the codes used for the figures are available at the author's home page. You might find them useful, either for teaching a class or else for the insights they provide into how to proceed when working out the homework problems. You might also enjoy the videos available, which show some of the solutions of the time-dependent problems solved in the book. Also, there is an errata page as well as answers to some of the exercises.

I would like to thank those who developed and have maintained TeXShop, a free and very good TeX previewer.

Troy, New York   Mark H. Holmes

Contents

Chapter 1
Introduction to Asymptotic Approximations

1.1 Introduction

We will be interested in this book in using what are known as asymptotic expansions to find approximate solutions of differential equations. Usually our efforts will be directed toward constructing the solution of a problem with only occasional regard for the physical situation it represents. However, to start things off, it is worth considering a typical physical problem to illustrate where the mathematical problems originate. A simple example comes from the motion of an object projected radially upward from the surface of the Earth. Letting $x(t)$ denote the height of the object, measured from the surface, from Newton's second law we obtain the following equation of motion:

$$\frac{\mathrm{d}^2 x}{\mathrm{d}t^2} = -\frac{gR^2}{(x+R)^2}, \quad \text{for } 0 < t, \tag{1.1}$$

where R is the radius of the Earth and g is the gravitational constant. We will assume the object starts from the surface with a given upward velocity, so $x(0) = 0$ and $x'(0) = v_0$, where v_0 is positive. The nonlinear nature of the preceding ordinary differential equation makes finding a closed-form solution difficult, and it is natural to try to see if there is some way to simplify the equation. For example, if the object does not get far from the surface, then one might try to argue that x is small compared to R and the denominator in (1.1) can be simplified to R^2. This is the type of argument often made in introductory physics and engineering texts. In this case, $x \approx x_0$, where $x_0'' = -g$ for $x_0(0) = 0$ and $x_0'(0) = v_0$. Solving this problem yields

$$x_0(t) = -\frac{1}{2}gt^2 + v_0 t. \tag{1.2}$$

One finds in this case that the object reaches a maximum height of $v_0^2/2g$ and comes back to Earth when $t = 2v_0/g$ (Fig. 1.1). The difficulty with

M.H. Holmes, *Introduction to Perturbation Methods*, Texts in Applied Mathematics 20, DOI 10.1007/978-1-4614-5477-9_1, © Springer Science+Business Media New York 2013

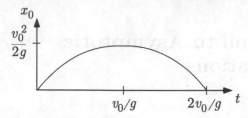

Figure 1.1 Schematic of solution $x_0(t)$ given in (1.2). This solution comes from the linearization of (1.1) and corresponds to the motion in a uniform gravitational field

this reduction is that it is unclear how to determine a correction to the approximate solution in (1.2). This is worth knowing since we would then be able to get a measure of the error made in using (1.2) as an approximation and it would also be possible to see just how the nonlinear nature of the original problem affects the motion of the object.

To make the reduction process more systematic, it is first necessary to scale the variables. To do this, let $\tau = t/t_c$ and $y(\tau) = x(t)/x_c$, where t_c is a characteristic time for the problem and x_c is a characteristic value for the solution. We have a lot of freedom in picking these constants, but they should be representative of the situation under consideration. Based on what is shown in Fig. 1.1, we take $t_c = v_0/g$ and $x_c = v_0^2/g$. Doing this the problem transforms into the following:

$$\frac{d^2 y}{d\tau^2} = -\frac{1}{(1+\varepsilon y)^2}, \quad \text{for } 0 < \tau, \tag{1.3}$$

where

$$y(0) = 0 \quad \text{and} \quad y'(0) = 1. \tag{1.4}$$

In (1.3), the parameter $\varepsilon = v_0^2/Rg$ is dimensionless and its value is important because it gives us a measure of how high the projectile gets in comparison to the radius of the Earth. In terms of the function $x_0(t)$ it can be seen from Fig. 1.1 that $\varepsilon/2$ is the ratio of the maximum height of the projectile to the radius of the Earth. Assuming $R = 4,000$ mi, then $\varepsilon \approx 1.5 \times 10^{-9} v_0^2 s^2/\text{ft}^2$. It would therefore appear that if v_0 is much less than 10^3 ft/s, then (1.2) is a reasonably good approximation to the solution. We can verify this assertion by reconstructing the first-order approximation in (1.2). This can be done by assuming that the dependence of the solution on ε can be determined using a Taylor series expansion about $\varepsilon = 0$. In other words, for small ε it is assumed that

$$y \sim y_0(\tau) + \varepsilon y_1(\tau) + \cdots.$$

The first term in this expansion is the scaled version of x_0, and this will be shown later (Sect. 1.6). What is important is that with this approach it is possible to estimate how well y_0 approximates the solution of (1.3), (1.4) by finding y_1. The method of deriving the first-order approximation (y_0) and its correction (y_1) is not difficult, but we first need to put the definition of an asymptotic approximation on a firm foundation. Readers interested in investigating the ideas underlying the nondimensionalization of a physical problem and some of the theory underlying dimensional analysis may consult Holmes (2009).

1.2 Taylor's Theorem and l'Hospital's Rule

As in the preceding example, we will typically end up expanding functions in powers of ε. Given a function $f(\varepsilon)$, one of the most important tools for doing this is Taylor's theorem. This is a well-known result, but for completeness it is stated below.

Theorem 1.1. *Given a function $f(\varepsilon)$, suppose its $(n+1)$st derivative $f^{(n+1)}$ is continuous for $\varepsilon_a < \varepsilon < \varepsilon_b$. In this case, if ε_0 and ε are points in the interval $(\varepsilon_a, \varepsilon_b)$, then*

$$f(\varepsilon) = f(\varepsilon_0) + (\varepsilon - \varepsilon_0)f'(\varepsilon_0) + \cdots + \frac{1}{n!}(\varepsilon - \varepsilon_0)^n f^{(n)}(\varepsilon_0) + R_{n+1}, \quad (1.5)$$

where

$$R_{n+1} = \frac{1}{(n+1)!}(\varepsilon - \varepsilon_0)^{n+1} f^{(n+1)}(\xi) \quad (1.6)$$

and ξ is a point between ε_0 and ε.

This result is useful because if the first $n+1$ terms from the Taylor series are used as an approximation of $f(\varepsilon)$, then it is possible to estimate the error using (1.6).

A short list of Taylor series expansions that will prove useful in this book is given in Appendix A. Examples using this list are given below.

Examples

1. Find the first three terms in the expansion of $f(\varepsilon) = \sin(e^\varepsilon)$ for $\varepsilon_0 = 0$.

   Given that $f' = e^\varepsilon \cos(e^\varepsilon)$ and $f'' = e^\varepsilon(\cos(e^\varepsilon) - \sin(e^\varepsilon))$, it follows that

   $$f(\varepsilon) = \sin(1) + \varepsilon\cos(1) + \frac{1}{2}\varepsilon^2(\cos(1) - \sin(1)) + \cdots . \quad \blacksquare$$

2. Find the first three terms in the Taylor expansion of $f(\varepsilon) = e^\varepsilon/(1-\varepsilon)$ for $\varepsilon_0 = 0$.

There are a couple of ways this can be done. One is the direct method using (1.5), the other is to multiply the series for e^ε with the series for $1/(1 - \varepsilon)$. To use the direct method, note that

$$f'(\varepsilon) = \frac{e^\varepsilon}{1 - \varepsilon} + \frac{e^\varepsilon}{(1 - \varepsilon)^2}$$

and

$$f''(\varepsilon) = \frac{e^\varepsilon}{1 - \varepsilon} + \frac{2e^\varepsilon}{(1 - \varepsilon)^2} + \frac{2e^\varepsilon}{(1 - \varepsilon)^3}.$$

Evaluating these at $\varepsilon = 0$ it follows from (1.5) that a three-term Taylor expansion is

$$f(\varepsilon) = 1 + 2\varepsilon + 5\varepsilon^2 + \cdots. \quad \blacksquare$$

Another useful result is l'Hospital's rule, which concerns the value of the limit of the ratio of two functions.

Theorem 1.2. *Suppose* $f(\varepsilon)$ *and* $\phi(\varepsilon)$ *are differentiable on the interval* $(\varepsilon_0, \varepsilon_b)$ *and* $\phi'(\varepsilon) \neq 0$ *in this interval. Also suppose*

$$\lim_{\varepsilon \downarrow \varepsilon_0} \frac{f'(\varepsilon)}{\phi'(\varepsilon)} = A,$$

where $-\infty \leq A \leq \infty$. *In this case,*

$$\lim_{\varepsilon \downarrow \varepsilon_0} \frac{f(\varepsilon)}{\phi(\varepsilon)} = A$$

if either one of the following conditions holds:

1. $f \to 0$ *and* $\phi \to 0$ *as* $\varepsilon \downarrow \varepsilon_0$, *or*

2. $\phi \to \infty$ *as* $\varepsilon \downarrow \varepsilon_0$.

The proofs of these two theorems and some of their consequences can be found in Rudin (1964).

1.3 Order Symbols

To define an asymptotic approximation, we first need to introduce order, or Landau, symbols.[1] The reason for this is that we will be interested in how functions behave as a parameter, typically ε, becomes small. For example,

---

[1] These symbols were first introduced by Bachmann (1894), and then Landau (1909) popularized their use. For this reason they are sometimes called Bachmann–Landau symbols.

the function $\phi(\varepsilon) = \varepsilon$ does not converge to zero as fast as $f(\varepsilon) = \varepsilon^2$ when $\varepsilon \to 0$, and we need a notation to denote this fact.

Definition 1.1.

1. $f = O(\phi)$ as $\varepsilon \downarrow \varepsilon_0$ means that there are constants k_0 and ε_1 (independent of ε) so that

$$|f(\varepsilon)| \leq k_0 |\phi(\varepsilon)| \quad \text{for } \varepsilon_0 < \varepsilon < \varepsilon_1.$$

We say that "f is big Oh of ϕ" as $\varepsilon \downarrow \varepsilon_0$.

2. $f = o(\phi)$ as $\varepsilon \downarrow \varepsilon_0$ means that for every positive δ there is an ε_2 (independent of ε) so that

$$|f(\varepsilon)| \leq \delta |\phi(\varepsilon)| \quad \text{for } \varepsilon_0 < \varepsilon < \varepsilon_2.$$

We say that "f is little oh of ϕ" as $\varepsilon \downarrow \varepsilon_0$.

These definitions may seem cumbersome, but they usually are not hard to apply. However, there are other ways to determine the correct order. Of particular interest is the case where ϕ is not zero near ε_0 (i.e., $\phi \neq 0$ if $\varepsilon_0 < \varepsilon < \varepsilon_\beta$ for some $\varepsilon_\beta > \varepsilon_0$). In this case we have that $f = O(\phi)$ if the ratio $|f/\phi|$ is bounded for ε near ε_0. Other, perhaps more useful, tests are identified in the next result.

Theorem 1.3.

1. If

$$\lim_{\varepsilon \downarrow \varepsilon_0} \frac{f(\varepsilon)}{\phi(\varepsilon)} = L, \tag{1.7}$$

where $-\infty < L < \infty$, then $f = O(\phi)$ as $\varepsilon \downarrow \varepsilon_0$.

2. If

$$\lim_{\varepsilon \downarrow \varepsilon_0} \frac{f(\varepsilon)}{\phi(\varepsilon)} = 0, \tag{1.8}$$

then $f = o(\phi)$ as $\varepsilon \downarrow \varepsilon_0$.

The proofs of these statements follow directly from the definition of a limit and are left to the reader.

Examples (for $\varepsilon \downarrow 0$)

1. Suppose $f = \varepsilon^2$. Also, let $\phi_1 = \varepsilon$ and $\phi_2 = -3\varepsilon^2 + 5\varepsilon^6$. In this case,

$$\lim_{\varepsilon \downarrow 0} \frac{f}{\phi_1} = 0 \quad \Rightarrow \quad f = o(\phi_1)$$

and

$$\lim_{\varepsilon \downarrow 0} \frac{f}{\phi_2} = -\frac{1}{3} \quad \Rightarrow \quad f = O(\phi_2). \quad \blacksquare$$

2. If $f = \varepsilon \sin(1 + 1/\varepsilon)$ and $\phi = \varepsilon$, then the limit in the preceding theorem does not exist. However, $|f/\phi| \leq 1$ for $0 < \varepsilon$, and so from the definition it follows that $f = O(\phi)$. $\blacksquare$

3. If $f(\varepsilon) = \sin(\varepsilon)$ then, using Taylor's theorem, $f = \varepsilon - \frac{1}{2}\varepsilon^2 \sin(\xi)$. Thus, $\lim_{\varepsilon \downarrow 0}(f/\varepsilon) = 1$, and from this it follows that $f = O(\varepsilon)$. $\blacksquare$

4. If $f = e^{-1/\varepsilon}$ then, using l'Hospital's rule, $f = o(\varepsilon^\alpha)$ for all values of α. We say in this case that f is *transcendentally small* with respect to the power functions ε^α. $\blacksquare$

Some properties of order symbols are examined in the exercises. Three that are worth mentioning are the following (the symbol $\Leftrightarrow$ stands for the statement "if and only if"):

(a) $f = O(1)$ as $\varepsilon \downarrow \varepsilon_0 \Leftrightarrow f$ is bounded as $\varepsilon \downarrow \varepsilon_0$.

(b) $f = o(1)$ as $\varepsilon \downarrow \varepsilon_0 \Leftrightarrow f \to 0$ as $\varepsilon \downarrow \varepsilon_0$.

(c) $f = o(\phi)$ as $\varepsilon \downarrow \varepsilon_0 \Rightarrow f = O(\phi)$ as $\varepsilon \downarrow \varepsilon_0$ (but not vice versa).

The proofs of these statements are straightforward and are left to the reader. Some of the other basic properties of these symbols are given in Exercises 1.2 and 1.3.

Two symbols we will use occasionally are $\ll$ and $\approx$. When we say that $f(\varepsilon) \ll \phi(\varepsilon)$, we mean that $f = o(\phi)$, and the statement that $\varepsilon \ll 1$, or that ε is small, means $\varepsilon \downarrow 0$. The symbol $\approx$ does not have a precise definition and it is used simply to designate an approximate numerical value. An example of this is the statement that $\pi \approx 3.14$.

Exercises

1.1.(a) What values of α, if any, yield $f = O(\varepsilon^\alpha)$ as $\varepsilon \downarrow 0$: (i) $f = \sqrt{1 + \varepsilon^2}$, (ii) $f = \varepsilon \sin(\varepsilon)$, (iii) $f = (1 - e^\varepsilon)^{-1}$, (iv) $f = \ln(1 + \varepsilon)$, (v) $f = \varepsilon \ln(\varepsilon)$, (vi) $f = \sin(1/\varepsilon)$, (vii) $f = \sqrt{x + \varepsilon}$, where $0 \leq x \leq 1$?
(b) For the functions listed in (a) what values of α, if any, yield $f = o(\varepsilon^\alpha)$ as $\varepsilon \downarrow 0$?

1.2. In this problem it is assumed that $\varepsilon \downarrow 0$.
(a) Show $f = O(\varepsilon^\alpha) \Rightarrow f = o(\varepsilon^\beta)$ for any $\beta < \alpha$.
(b) Show that if $f = O(g)$, then $f^\alpha = O(g^\alpha)$ for any positive α. Give an example to show that this result is not necessarily true if α is negative.

(c) Give an example to show that $f = O(g)$ does not necessarily mean that $e^f = O(e^g)$.

1.3. This problem establishes some of the basic properties of the order symbols, some of which are used extensively in this book. The limit assumed here is $\varepsilon \downarrow 0$.

(a) If $f = o(g)$ and $g = O(h)$, or if $f = O(g)$ and $g = o(h)$, then show that $f = o(h)$. Note that this result can be written as $o(O(h)) = O(o(h)) = o(h)$.

(b) Assuming $f = O(\phi_1)$ and $g = O(\phi_2)$, show that $f + g = O(|\phi_1| + |\phi_2|)$. Also, explain why the absolute signs are necessary. Note that this result can be written as $O(f) + O(g) = O(|f| + |g|)$.

(c) Assuming $f = O(\phi_1)$ and $g = O(\phi_2)$, show that $fg = O(\phi_1\phi_2)$. This result can be written as $O(f)O(g) = O(fg)$.

(d) Show that $O(O(f)) = O(f)$.

(e) Show that $O(f)o(g) = o(f)o(g) = o(fg)$.

1.4. Occasionally it is useful to state the order of a function more precisely. One way to do this is to say $f = O_s(\phi)$ as $\varepsilon \downarrow \varepsilon_0 \Leftrightarrow f = O(\phi)$ but $f \neq o(\phi)$ as $\varepsilon \downarrow \varepsilon_0$.

(a) What values of α, if any, yield $f = O_s(\varepsilon^\alpha)$ as $\varepsilon \downarrow 0$? (i) $f = \varepsilon \sin(\varepsilon)$, (ii) $f = (1 - e^\varepsilon)^{-1}$, (iii) $f = \ln(1 + \varepsilon)$, (iv) $f = \varepsilon \ln(\varepsilon)$, (v) $f = \sin(1/\varepsilon)$.

(b) Suppose the limit in (1.7) exists. In this case, show that $f = O_s(\phi)$ as $\varepsilon \downarrow \varepsilon_0 \Leftrightarrow 0 < \lim_{\varepsilon \downarrow \varepsilon_0} |f/\phi| < \infty$.

1.5. Suppose $f = o(\phi)$ for small ε, where f and ϕ are continuous.

(a) Give an example to show that it is not necessarily true that

$$\int_0^\varepsilon f d\varepsilon = o\left(\int_0^\varepsilon \phi d\varepsilon\right).$$

(b) Show that

$$\int_0^\varepsilon f d\varepsilon = o\left(\int_0^\varepsilon |\phi| d\varepsilon\right).$$

1.4 Asymptotic Approximations

Our objective is to construct approximations to the solutions of differential equations. It is therefore important that we state exactly what we mean by an approximation. To introduce this idea, suppose we are interested in finding an approximation of $f(\varepsilon) = \varepsilon^2 + \varepsilon^5$ for ε close to zero. Because $\varepsilon^5 \ll \varepsilon^2$, a reasonable approximation is $f(\varepsilon) \approx \varepsilon^2$. On the other hand, a lousy approximation is $f(\varepsilon) \approx \frac{2}{3}\varepsilon^2$. This is lousy even though the error $f(\varepsilon) - \frac{2}{3}\varepsilon^2$ goes to zero as $\varepsilon \downarrow 0$. The problem with this "lousy approximation" is that the error is of the same order as the function we are using to approximate $f(\varepsilon)$. This observation gives rise to the following definition.

Definition 1.2. Given $f(\varepsilon)$ and $\phi(\varepsilon)$, we say that $\phi(\varepsilon)$ is an *asymptotic approximation* to $f(\varepsilon)$ as $\varepsilon \downarrow \varepsilon_0$ whenever $f = \phi + o(\phi)$ as $\varepsilon \downarrow \varepsilon_0$. In this case we write $f \sim \phi$ as $\varepsilon \downarrow \varepsilon_0$.

As demonstrated in the preceding example, the idea underlying this definition is that $\phi(\varepsilon)$ serves as an asymptotic approximation of $f(\varepsilon)$, for ε close to ε_0, when the error is of higher order than the approximating function. In the case where $\phi(\varepsilon)$ is not zero near ε_0, we can make use of (1.8). In particular, we have that $f \sim \phi$ as $\varepsilon \downarrow \varepsilon_0$ if

$$\lim_{\varepsilon \downarrow \varepsilon_0} \frac{f(\varepsilon)}{\phi(\varepsilon)} = 1. \tag{1.9}$$

Examples

1. Suppose $f = \sin(\varepsilon)$ and $\varepsilon_0 = 0$. We can obtain an approximation of $f(\varepsilon)$ by expanding about $\varepsilon = 0$ using Taylor's theorem. This yields

$$f = \varepsilon - \frac{1}{6}\varepsilon^3 + \frac{1}{120}\varepsilon^5 \cos(\xi).$$

From this the following asymptotic approximations are obtained:

   (i) $f \sim \varepsilon$,

   (ii) $f \sim \varepsilon - \frac{1}{6}\varepsilon^3$.

However, it is not hard to show that the following expression is also an asymptotic approximation:

(iii) $f \sim \varepsilon + 2\varepsilon^2$.

In comparing these approximations with the Taylor series expansion it would appear that, for small ε, (ii) is the most accurate and (iii) is the least accurate. However, our definition of an asymptotic approximation says little about comparative accuracy. This weakness is corrected below by introducing asymptotic expansions. ∎

2. Suppose $f = x + e^{-x/\varepsilon}$, where $0 < x < 1$ is fixed. In this case $f \sim x$ for small ε. However, it is natural to ask how well this does in approximating the function for $0 < x < 1$. If we plot both together, we obtain the curves shown in Fig. 1.2. It is apparent that the approximation is quite good away from $x = 0$. It is also clear that we do not do so well near $x = 0$. This is true no matter what value of ε we choose since $f(0) = 1$. It should be remembered that with an asymptotic approximation, given a value of x, the approximation is a good one if ε is close to $\varepsilon_0 = 0$. What we are seeing in this example is that exactly how small ε must be can depend

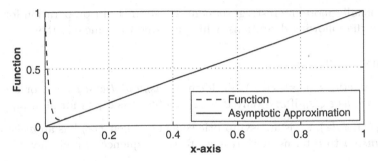

Figure 1.2 Comparison between the function $f = x + e^{-x/\varepsilon}$ and its asymptotic approximation $f \sim x$. Note that the two functions are essentially indistinguishable everywhere except near $x = 0$. In this plot, $\varepsilon = 10^{-2}$

on x (the closer we are to $x = 0$, the smaller ε must be). In later sections we will refer to this situation as a case where the approximation is not uniformly valid on the interval $0 < x < 1$. ■

3. Consider the function $f = \sin(\pi x) + \varepsilon^3$ for $0 \le x \le \frac{1}{2}$. For small ε it might seem reasonable to expect that $f \sim \sin(\pi x)$. For this to hold it is required that $f - \sin(\pi x) = o(\sin(\pi x))$ as $\varepsilon \downarrow 0$. If $x \ne 0$, then this is true since $\lim_{\varepsilon \downarrow 0}(\varepsilon^3 / \sin(\pi x)) = 0$. However, at $x = 0$ this requirement does not hold since $\sin(0) = 0$. Therefore, $\sin(\pi x)$ is not an asymptotic approximation of f over the entire interval $0 \le x \le \frac{1}{2}$. This problem of using an approximating function whose zeros do not agree with the original function is one that we will come across on numerous occasions. Usually it is a complication we will not worry a great deal about because the correction is relatively small. For instance, in this example the correction is $O(\varepsilon^3)$ while everywhere else the function is $O(1)$. This is not true for the correction that is needed to fix the approximation in the previous example. There the correction at, or very near, $x = 0$ is $O(1)$, and this is the same order as the value of the function through the rest of the interval. We are therefore not able to ignore the problem at $x = 0$; how to deal with this is the subject of Chap. 2. ■

1.4.1 Asymptotic Expansions

Two observations that come from the preceding examples are that an asymptotic approximation is not unique, and it also does not say much about the accuracy of the approximation. To address these shortcomings, we need to introduce more structure into the formulation. In the preceding examples, the sequence $1, \varepsilon, \varepsilon^2, \varepsilon^3, \ldots$ was used in the expansion of the function, but other, usually more interesting, sequences also arise. In preparation for this, we have the following definitions, which are due to Poincaré (1886).

Definition 1.3.

1. The functions $\phi_1(\varepsilon)$, $\phi_2(\varepsilon)$, ... form an *asymptotic sequence*, or are *well ordered*, as $\varepsilon \downarrow \varepsilon_0$ if and only if $\phi_{m+1} = o(\phi_m)$ as $\varepsilon \downarrow \varepsilon_0$ for all m.

2. If $\phi_1(\varepsilon)$, $\phi_2(\varepsilon)$, ... is an asymptotic sequence, then $f(\varepsilon)$ has an *asymptotic expansion* to n terms, with respect to this sequence, if and only if

$$f = \sum_{k=1}^{m} a_k \phi_k + o(\phi_m) \text{ for } m = 1, 2, \ldots, n \text{ as } \varepsilon \downarrow \varepsilon_0, \tag{1.10}$$

where the a_k are independent of ε. In this case we write

$$f \sim a_1 \phi_1(\varepsilon) + a_2 \phi_2(\varepsilon) + \cdots + a_n \phi_n(\varepsilon) \text{ as } \varepsilon \downarrow \varepsilon_0. \tag{1.11}$$

The ϕ_k are called the scale or gauge or basis functions.

To make use of this definition, we need to have some idea of what scale functions are available. We will run across a wide variety in this book, but a couple of our favorites will turn out to be the following ones:

1. $\phi_1 = (\varepsilon - \varepsilon_0)^\alpha$, $\phi_2 = (\varepsilon - \varepsilon_0)^\beta$, $\phi_3 = (\varepsilon - \varepsilon_0)^\gamma$, ..., where $\alpha < \beta < \gamma < \cdots$.

2. $\phi_1 = 1$, $\phi_2 = e^{-1/\varepsilon}$, $\phi_3 = e^{-2/\varepsilon}$, ...

The first of these is simply a generalization of the power series functions, and the second sequence is useful when we have to describe exponentially small functions. The verification that these do indeed form asymptotic sequences is left to the reader.

Now comes the harder question. Given a function $f(\varepsilon)$, how do we find an asymptotic expansion of it? The most commonly used methods include employing one of the following: (1) Taylor's theorem, (2) l'Hospital's rule, or (3) an educated guess. The last one usually relies on an intuitive understanding of the problem and, many times, on luck. The other two methods are more routine and are illustrated in the examples below.

Taylor's theorem is a particularly useful tool because if the function is smooth enough to let us expand about the point $\varepsilon = \varepsilon_0$, then any one of the resulting Taylor polynomials can be used as an asymptotic expansion (for $\varepsilon \downarrow \varepsilon_0$). Moreover, Taylor's theorem enables us to analyze the error very easily.

Examples (for $\varepsilon \ll 1$)

1. To find a three-term expansion of e^ε, we use Taylor's theorem to obtain

$$e^\varepsilon = 1 + \varepsilon + \frac{1}{2}\varepsilon^2 + \frac{1}{3}\varepsilon^3 + \cdots$$

$$\sim 1 + \varepsilon + \frac{1}{2}\varepsilon^2. \quad \blacksquare$$

2. Finding the first two terms in the expansion of $f(\varepsilon) = \cos(\varepsilon)/\varepsilon$ requires an extension of Taylor's theorem because the function is not defined at $\varepsilon = 0$. One way this can be done is to factor out the singular part of $f(\varepsilon)$ and apply Taylor's theorem to the remaining regular portion. The singular part here is $1/\varepsilon$, and the regular part is $\cos(\varepsilon)$. Using the expansion for $\cos(\varepsilon)$ in Appendix A we obtain

$$f(\varepsilon) \sim \frac{1}{\varepsilon}\left(1 - \frac{1}{2}\varepsilon^2 + \cdots\right)$$

$$\sim \frac{1}{\varepsilon} - \frac{1}{2}\varepsilon.$$

As expected, the expansion, like the function, is not defined at $\varepsilon = 0$. $\blacksquare$

3. To find a two-term expansion of

$$f(\varepsilon) = \frac{\sqrt{1+\varepsilon}}{\sin(\sqrt{\varepsilon})},$$

note that

$$\sqrt{1+\varepsilon} = 1 + \frac{1}{2}\varepsilon + \cdots$$

and

$$\sin(\sqrt{\varepsilon}) = \varepsilon^{1/2} - \frac{1}{6}\varepsilon^{3/2} + \cdots.$$

Consequently,

$$f(\varepsilon) \sim \frac{1 + \frac{1}{2}\varepsilon + \cdots}{\varepsilon^{1/2} - \frac{1}{6}\varepsilon^{3/2} + \cdots} = \frac{1}{\varepsilon^{1/2}}\frac{1 + \frac{1}{2}\varepsilon + \cdots}{1 - \frac{1}{6}\varepsilon + \cdots}$$

$$\sim \frac{1}{\varepsilon^{1/2}}\left(1 + \frac{1}{2}\varepsilon + \cdots\right)\left(1 + \frac{1}{6}\varepsilon + \cdots\right)$$

$$\sim \frac{1}{\varepsilon^{1/2}}\left(1 + \frac{1}{3}\varepsilon\right). \quad \blacksquare$$

A nice aspect about using Taylor's theorem is that the scale functions do not have to be specified ahead of time. This differs from the next procedure, which requires the specification of the scale functions before constructing the expansion.

To describe the second procedure for constructing an asymptotic expansion, suppose the scale functions ϕ_1, ϕ_2, ... are given, and the expansion of the function has the form $f \sim a_1\phi_1(\varepsilon) + a_2\phi_2(\varepsilon) + \cdots$. From the preceding definition, this means that $f = a_1\phi_1 + o(\phi_1)$. Assuming we can divide by ϕ_1, we get $\lim_{\varepsilon\downarrow\varepsilon_0}(f/\phi_1) = a_1$. This gives us the value of a_1, and with this information we can determine a_2 by noting $f = a_1\phi_1 + a_2\phi_2 + o(\phi_2)$. Thus, $\lim_{\varepsilon\downarrow\varepsilon_0}[(f - a_1\phi_1)/\phi_2] = a_2$. This idea can be used to calculate the other coefficients of the expansion, and one obtains the following formulas:

$$a_1 = \lim_{\varepsilon\downarrow\varepsilon_0} \frac{f}{\phi_1}, \tag{1.12}$$

$$a_2 = \lim_{\varepsilon\downarrow\varepsilon_0} \frac{f - a_1\phi_1}{\phi_2}, \tag{1.13}$$

$$a_3 = \lim_{\varepsilon\downarrow\varepsilon_0} \frac{f - a_1\phi_1 - a_2\phi_2}{\phi_3}. \tag{1.14}$$

This assumes that the scale functions are nonzero for ε near ε_0 and that each of the limits exists. If this is the case, then the preceding formulas for the a_k show that the asymptotic expansion is unique.

Example

Suppose $\phi_1 = 1$, $\phi_2 = \varepsilon$, $\phi_3 = \varepsilon^2$, ..., and

$$f(\varepsilon) = \frac{1}{1 + \varepsilon} + e^{-1/\varepsilon}.$$

From the preceding limit formulas we have that

$$a_1 = \lim_{\varepsilon\downarrow 0} \frac{f}{1} = 1,$$

$$a_2 = \lim_{\varepsilon\downarrow 0} \frac{f - 1}{\varepsilon} = \lim_{\varepsilon\downarrow 0} \left(\frac{-1}{1 + \varepsilon} + \frac{1}{\varepsilon}e^{-1/\varepsilon} \right) = -1,$$

$$a_3 = \lim_{\varepsilon\downarrow 0} \frac{f - 1 + \varepsilon}{\varepsilon^2} = \lim_{\varepsilon\downarrow 0} \left(\frac{1}{1 + \varepsilon} + \frac{1}{\varepsilon^2}e^{-1/\varepsilon} \right) = 1.$$

Thus, $f \sim 1 - \varepsilon + \varepsilon^2 + \cdots$. What is interesting here is that the exponential makes absolutely no contribution to this expansion. This is because $e^{-1/\varepsilon} = o(\varepsilon^\alpha)$ for all values of α; in other words, the exponential decays so quickly to zero that any power function considers it to be zero. In this case, we say that this exponential function is *transcendentally small* with respect to these scale functions. ∎

As seen in the previous example, two functions can have the same asymptotic expansion. In particular, using the power series functions $\phi_0 = 1$, $\phi_1 = \varepsilon$, $\phi_2 = \varepsilon^2, \ldots$ one obtains the same expansion as in the previous example for any of the following functions:

(i) $f_1 = \dfrac{\tanh(1/\varepsilon)}{1+\varepsilon}$,

(ii) $f_2 = \dfrac{1+e^{-1/\varepsilon}}{1+\varepsilon}$,

(iii) $f_3 = \dfrac{1}{1+\varepsilon} + \varepsilon^{100} \operatorname{sech}(-1/\varepsilon)$.

This observation brings up the idea of asymptotic equality, or asymptotic equivalence, with respect to a given sequence $\phi_1, \phi_2, \phi_3, \ldots$. We say that two functions f and g are *asymptotically equal to n terms* if $f - g = o(\phi_n)$ as $\varepsilon \downarrow \varepsilon_0$.

1.4.2 Accuracy Versus Convergence of an Asymptotic Series

It is not unusual to expect that to improve the accuracy of an asymptotic expansion, it is simply necessary to include more terms. This is what happens with Taylor series expansions, where in theory one should be able to obtain as accurate a result as desired by simply adding together enough terms. However, with an asymptotic expansion this is not necessarily true. The reason is that an asymptotic expansion only makes a statement about the series in the limit of $\varepsilon \downarrow \varepsilon_0$, whereas increasing the number of terms is saying something about the series as $n \to \infty$. In fact, an asymptotic expansion need not converge! Moreover, even if it converges, it does not have to converge to the function that was expanded! These two observations may seem to be major flaws in our definition of an asymptotic expansion, but they are, in fact, attributes that we will take great advantage of throughout this book.

A demonstration that a convergent asymptotic expansion need not converge to the function that was expanded can be found in the last example. It is not hard to show that the asymptotic series converges to the function $(1+\varepsilon)^{-1}$. This is clearly not equal to the original function since it is missing the exponential term.

A well-known example of a divergent asymptotic expansion arises with the Bessel function $J_0(z)$, which is defined as

$$J_0(z) = \sum_{k=0}^{\infty} \frac{(-1)^k z^{2k}}{2^{2k}(k!)^2}. \tag{1.15}$$

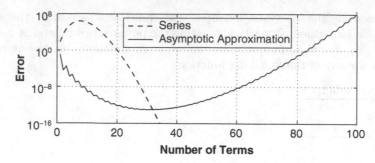

Figure 1.3 The error when using the convergent series (1.15) or the asymptotic series (1.16) to determine the value of the function $f(\varepsilon) = J_0(\varepsilon^{-1})$ for $\varepsilon = \frac{1}{15}$. The values are given as a function of the number of terms used

If we let $f(\varepsilon) = J_0(\frac{1}{\varepsilon})$, then it can be shown that an asymptotic expansion of $f(\varepsilon)$ for small ε is (Abramowitz and Stegun, 1972)

$$f \sim \sqrt{\frac{2\varepsilon}{\pi}}\left[\alpha \cos\left(\frac{1}{\varepsilon} - \frac{\pi}{4}\right) + \beta \sin\left(\frac{1}{\varepsilon} - \frac{\pi}{4}\right)\right], \qquad (1.16)$$

where

$$\alpha \sim 1 - \frac{1 \cdot 3^2 \varepsilon^2}{2! \, 8^2} + \frac{1 \cdot 3^2 \cdot 5^2 \cdot 7^2 \varepsilon^4}{4! \, 8^4} + \cdots \qquad (1.17)$$

and

$$\beta \sim \frac{\varepsilon}{8} - \frac{1 \cdot 3^2 \cdot 5^2 \varepsilon^3}{3! \, 8^3} + \cdots . \qquad (1.18)$$

It is not hard to show that the series expansions in (1.17) and (1.18) are divergent for all nonzero values of ε (Exercise 1.14). To see just how well a divergent series like (1.16) can approximate the function $f(\varepsilon)$, the errors in using (1.15) and (1.16)–(1.18) are shown in Fig. 1.3 for $\varepsilon = 1/15$. What is immediately apparent is that the asymptotic approximation does very well using only 1 or 2 terms, while the convergent series needs to include more than 20 terms to achieve the same error. If a smaller value of ε is used, then the situation is even more pronounced, eventually getting to the point that it is essentially impossible to calculate the value of the Bessel function using the series representation since so many terms are needed.

The preceding observations concerning convergence should always be kept in mind when using asymptotic expansions. Also, it should be pointed out that asymptotic approximations are most valuable when they can give insights into the structure of a solution or the physics of the problem under study. If decimal place accuracy is desired, then numerical methods, or perhaps numerical methods combined with an asymptotic solution, are probably worth considering.

The ideas underlying an asymptotic approximation are so natural that they are found in some of the earliest work in mathematics. The basis of the definitions given here can be traced back to at least Legendre (1825).

In his Traité des Fonctions Elliptiques he considered approximating a function with a series where the error committed using the first n terms is of the same order as the $(n + 1)$st term. He referred to this as a semiconvergent series. This is a poor choice of words since an asymptotic series need not converge. Somewhat earlier than this, Laplace (1812) had found use for asymptotic series to be able to evaluate certain special functions because the standard series expansions converged so slowly. He considered the fact that they did not converge to be "inconvenient," an interesting attitude that generally leads to disastrous results. The most significant contribution to the subject was made by Poincaré (1886). He was able to make sense of the numerous divergent series solutions of differential equations that had been found up to that time by introducing the concept of an asymptotic expansion. His ideas form the basis of our development. It is interesting that his paper actually marked the modern beginnings of two areas in mathematics: the asymptotic solution of a differential equation and the study of divergent series. The latter can also be called the theory of summability of a divergent series, and the books by Ford (1916) and Hardy (1954) are good references in this area. However, if the reader has not heard of this particular subject, it is not surprising because it has since died of natural causes. Readers interested in the early history of the development of asymptotic approximations may consult the reviews by McHugh (1971) and Schlissel (1977a).

1.4.3 Manipulating Asymptotic Expansions

It is not hard to show that two asymptotic expansions can be added together term by term, assuming that the same basis functions are used for both expansions. Multiplication is also relatively straightforward, albeit more tedious and limited to asymptotic sequences that can be ordered in particular ways (e.g., Exercise 1.12). What is not completely clear is whether or not an asymptotic approximation can be differentiated (assuming the functions involved are smooth). To be more specific, suppose we know that

$$f(x, \varepsilon) \sim \phi_1(x, \varepsilon) + \phi_2(x, \varepsilon) \quad \text{as } \varepsilon \downarrow \varepsilon_0. \tag{1.19}$$

What is of interest is whether or not it is true that

$$\frac{\mathrm{d}}{\mathrm{d}x} f(x, \varepsilon) \sim \frac{\mathrm{d}}{\mathrm{d}x} \phi_1(x, \varepsilon) + \frac{\mathrm{d}}{\mathrm{d}x} \phi_2(x, \varepsilon) \quad \text{as } \varepsilon \downarrow \varepsilon_0. \tag{1.20}$$

Based on the conjecture that things will go wrong if at all possible, the only possible answer to the preceding question is that, no, it is not always true. To give an example, let $f(x, \varepsilon) = e^{-x/\varepsilon} \sin(e^{x/\varepsilon})$. For small ε one finds that, for $0 < x < 1$,

$$f \sim 0 + 0 \cdot \varepsilon + 0 \cdot \varepsilon^2 + \cdots . \tag{1.21}$$

However, the function

$$\frac{\mathrm{d}}{\mathrm{d}x} f(x, \varepsilon) = -\frac{1}{\varepsilon} e^{-x/\varepsilon} \sin(e^{x/\varepsilon}) + \frac{1}{\varepsilon} \cos(e^{x/\varepsilon})$$

does not even have an expansion in terms of the scale functions used in (1.21).

Another problem is that even if ϕ_1 and ϕ_2 are well ordered, their derivatives may not be. For example, letting $\phi_1 = 1 + x$ and $\phi_2 = \varepsilon \sin(x/\varepsilon)$, for $0 \leq x \leq 1$, then $\phi_2 = o(\phi_1)$ for ε small, but $\frac{\mathrm{d}}{\mathrm{d}x}\phi_1$ and $\frac{\mathrm{d}}{\mathrm{d}x}\phi_2$ are not well ordered. Thus, the question that we need to address is when can we differentiate an expansion? Well, the answer to this is that if

$$f(x, \varepsilon) \sim a_1(x)\phi_1(\varepsilon) + a_2(x)\phi_2(\varepsilon) \quad \text{as } \varepsilon \downarrow \varepsilon_0, \qquad (1.22)$$

and if

$$\frac{\mathrm{d}}{\mathrm{d}x} f(x, \varepsilon) \sim b_1(x)\phi_1(\varepsilon) + b_2(x)\phi_2(\varepsilon) \quad \text{as } \varepsilon \downarrow \varepsilon_0, \qquad (1.23)$$

then $b_k = \frac{\mathrm{d}}{\mathrm{d}x} a_k$ [i.e., the expansion for $\frac{\mathrm{d}}{\mathrm{d}x} f$ can be obtained by differentiating the expansion of $f(x, \varepsilon)$ term by term]. Throughout this book, given (1.22), we will automatically assume that (1.23) also holds.

The other operation we will have use for is integration, and in this case the situation is better. If the expansion of a function is as given in (1.22) and all the functions involved are integrable, then

$$\int_a^b f(x, \varepsilon)\mathrm{d}x \sim \left(\int_a^b a_1(x)\mathrm{d}x \right)\phi_1(\varepsilon) + \left(\int_a^b a_2(x)\mathrm{d}x \right)\phi_2(\varepsilon) \quad \text{as } \varepsilon \downarrow \varepsilon_0.$$

$$(1.24)$$

It goes without saying here that the interval $a \leq x \leq b$ must be within the set of x where (1.22) holds.

Examples

1. Find the first two terms in the expansion of $f(\varepsilon) = \int_0^1 e^{\varepsilon x^2}\mathrm{d}x$.

   Given that $e^{\varepsilon x^2} \sim 1 + \varepsilon x^2 + \cdots$, and this holds for $0 \leq x \leq 1$, then

   $$f(\varepsilon) \sim \int_0^1 (1 + \varepsilon x^2 + \cdots)\mathrm{d}x$$

   $$= 1 + \frac{1}{3}\varepsilon + \cdots . \quad \blacksquare$$

2. Find the first two terms in the expansion of

   $$f(\varepsilon) = \int_0^1 \frac{\mathrm{d}x}{\varepsilon^2 + x^2} .$$

For $0 < x \leq 1$ the integrand has the expansion

$$\frac{1}{\varepsilon^2 + x^2} \sim \frac{1}{x^2} - \frac{\varepsilon^2}{x^4} + \cdots . \tag{1.25}$$

Unfortunately, the terms in this expansion are not integrable over the interval $0 \leq x \leq 1$. This can be avoided by taking the easy way out and simply calculating the integral and then expanding the result. This produces the following:

$$\begin{aligned} f(\varepsilon) &= \frac{1}{\varepsilon}\arctan\left(\frac{1}{\varepsilon}\right) \\ &= \frac{1}{\varepsilon}\left(\frac{\pi}{2} - \arctan(\varepsilon)\right) \\ &\sim \frac{\pi}{2\varepsilon} - 1 + \frac{1}{3}\varepsilon^2 + \cdots . \end{aligned} \tag{1.26}$$

For more complicated integrals it is not possible to carry out the integration, and this brings up the question of whether we can obtain (1.26) some other way. To answer this, note that for (1.25) to be well ordered we require $\varepsilon^2/x^4 \ll 1/x^2$, that is, $\varepsilon \ll x$. Therefore, to use (1.25), we write

$$f(\varepsilon) = \int_0^\delta \frac{dx}{\varepsilon^2 + x^2} + \int_\delta^1 \frac{dx}{\varepsilon^2 + x^2},$$

where $\varepsilon \ll \delta \ll 1$. The idea here is that the first integral is over a small interval containing the singularity, and the second integral is over the remainder of the interval where we can use (1.25). With this

$$\begin{aligned} \int_\delta^1 \frac{dx}{\varepsilon^2 + x^2} &\sim \int_\delta^1 \left(\frac{1}{x^2} - \frac{\varepsilon^2}{x^4} + \cdots\right) dx \\ &= -1 + \frac{1}{3}\varepsilon^2 + \frac{1}{\delta} - \frac{\varepsilon^3}{3\delta^3} + \cdots \end{aligned}$$

and

$$\begin{aligned} \int_0^\delta \frac{dx}{\varepsilon^2 + x^2} &= \frac{1}{\varepsilon}\arctan\left(\frac{\delta}{\varepsilon}\right) \\ &= \frac{\pi}{2\varepsilon} - \frac{1}{\delta} + \cdots . \end{aligned}$$

Adding these together we obtain (1.26). ∎

3. Find the first two terms in the expansion of

$$f(\varepsilon) = \int_0^{\pi/3} \frac{dx}{\varepsilon^2 + \sin x}.$$

Like the last example, this has a nonintegrable singularity at $x = 0$ when $\varepsilon = 0$. This is because for x near zero, $\sin x = x + O(x^3)$, and $1/x$ is not integrable when one of the endpoints is $x = 0$. Consequently, as before, we will split the interval. First, note that

$$\frac{1}{\varepsilon^2 + \sin x} \sim \frac{1}{\sin x}\left(1 - \frac{\varepsilon^2}{\sin x} + \cdots\right).$$

For this to be well ordered near $x = 0$ we require that $\varepsilon^2 / \sin x \ll 1$, that is, $\varepsilon^2 \ll x$. So, we write

$$f(\varepsilon) = \int_0^\delta \frac{dx}{\varepsilon^2 + \sin x} + \int_\delta^{\pi/3} \frac{dx}{\varepsilon^2 + \sin x},$$

where $\varepsilon^2 \ll \delta \ll 1$. With this

$$\int_\delta^{\pi/3} \frac{dx}{\varepsilon^2 + \sin x} \sim \int_\delta^{\pi/3} \frac{1}{\sin x}\left(1 - \frac{\varepsilon^2}{\sin x} + \cdots\right)dx$$

$$= \ln\left(\frac{1}{3}\sqrt{3}\right) - \ln\left(\tan\left(\frac{\delta}{2}\right)\right) + \varepsilon^2\left(\frac{1}{3}\sqrt{3} - \cot(\delta)\right) + \cdots$$

$$\sim \ln\left(\frac{2}{3}\sqrt{3}\right) + \frac{1}{3}\sqrt{3}\,\varepsilon^2 - \left(\ln(\delta) + \frac{\varepsilon^2}{\delta} + \frac{1}{12}\delta^2\right) + \cdots$$

and, setting $\eta = \delta/\varepsilon^2$,

$$\int_0^\delta \frac{dx}{\varepsilon^2 + \sin x} = \varepsilon^2 \int_0^\eta \frac{dr}{\varepsilon^2 + \sin(\varepsilon^2 r)}$$

$$\sim \varepsilon^2 \int_0^\eta \frac{dr}{\varepsilon^2 + \varepsilon^2 r - \frac{1}{6}\varepsilon^6 r^3 + \cdots}$$

$$\sim \int_0^\eta \left(\frac{1}{1+r} + \frac{\varepsilon^4 r^3}{6(1+r)^2} + \cdots\right)dr$$

$$= \ln(1+\eta) + \frac{1}{6}\varepsilon^4\left(\frac{1}{2}\eta^2 - 2\eta - 1 + 3\ln(1+\eta) + \frac{1}{1+\eta}\right) + \cdots$$

$$\sim \ln(\eta) + \frac{1}{\eta} + \cdots + \frac{1}{6}\varepsilon^4\left(\frac{1}{2}\eta^2 + \cdots\right) + \cdots.$$

Adding these we obtain

$$f(\varepsilon) \sim -2\ln(\varepsilon) + \ln\left(\frac{2}{3}\sqrt{3}\right). \quad \blacksquare$$

As a final comment, the last expansion contains a log term, which might not be expected given the original integral. The need for log scale functions

is not common, but it does occur. In the next chapter this will be mentioned
again in reference to what is known as the switchbacking problem.

Exercises

1.6. Are the following sequences well ordered (for $\varepsilon \downarrow 0$)? If not, arrange them
so they are or explain why it is not possible to do so.

(a) $\phi_n = (1 - e^{-\varepsilon})^n$ for $n = 0, 1, 2, 3, \ldots$.

(b) $\phi_n = [\sinh(\varepsilon/2)]^{2n}$ for $n = 0, 1, 2, 3, \ldots$.

(c) $\phi_1 = \varepsilon^5 e^{-3/\varepsilon}$, $\phi_2 = \varepsilon$, $\phi_3 = \varepsilon \ln(\varepsilon)$, $\phi_4 = e^{-\varepsilon}$, $\phi_5 = \frac{1}{\varepsilon} \sin(\varepsilon^3)$, $\phi_6 = \frac{1}{\ln(\varepsilon)}$.

(d) $\phi_1 = \ln(1+3\varepsilon^2)$, $\phi_2 = \arcsin(\varepsilon)$, $\phi_3 = \sqrt{1+\varepsilon}/\sin(\varepsilon)$, $\phi_4 = \varepsilon \ln[\sinh(1/\varepsilon)]$,
$\phi_5 = 1/(1 - \cos(\varepsilon))$.

(e) $\phi_1 = e^\varepsilon - 1 - \varepsilon$, $\phi_2 = e^\varepsilon - 1$, $\phi_3 = e^\varepsilon$, $\phi_4 = e^\varepsilon - 1 - \varepsilon - \frac{1}{2}\varepsilon^2$.

(f) $\phi_1 = 1$, $\phi_2 = \varepsilon$, $\phi_3 = \varepsilon^2$, $\phi_4 = \varepsilon \ln(\varepsilon)$, $\phi_5 = \varepsilon^2 \ln(\varepsilon)$, $\phi_6 = \varepsilon \ln^2(\varepsilon)$,
$\phi_7 = \varepsilon^2 \ln^2(\varepsilon)$.

(g) $\phi_k = \begin{cases} 1 & \text{if } k^{-1} \leq \varepsilon, \\ 0 & \text{if } 0 \leq \varepsilon < k^{-1}, \end{cases}$ where $k = 1, 2, 3, \ldots$.

(h) $\phi_k(\varepsilon) = g^k(\varepsilon)$ for $k = 1, 2, \ldots$, where $g(\varepsilon)$ is continuous and $g(0) = 0$.

(i) $\phi_k(\varepsilon) = g(\varepsilon^k)$ for $k = 1, 2, \ldots$, where $g(x) = x \sin(\frac{1}{x})$ for $k = 1, 2, \ldots$.

(j) $\phi_k(\varepsilon) = g(\varepsilon^k)$ for $k = 1, 2, \ldots$, where $g(x) = e^{-1/x}$.

1.7. Assuming $f \sim a_1 \varepsilon^\alpha + a_2 \varepsilon^\beta + \cdots$, find α, β (with $\alpha < \beta$) and nonzero
a_1, a_2 for the following functions:

(a) $f = \dfrac{1}{1 - e^\varepsilon}$.

(b) $f = \left[1 + \dfrac{1}{\cos(\varepsilon)}\right]^{3/2}$.

(c) $f = 1 + \varepsilon - 2\ln(1 + \varepsilon) - \dfrac{1}{1+\varepsilon}$.

(d) $f = \sinh(\sqrt{1 + \varepsilon x})$, for $0 < x < \infty$.

(e) $f = (1 + \varepsilon x)^{1/\varepsilon}$, for $0 < x < \infty$.

(f) $f = \displaystyle\int_0^\varepsilon \sin(x + \varepsilon x^2) dx$.

(g) $f = \displaystyle\sum_{n=1}^\infty (\tfrac{1}{2})^n \sin(\dfrac{\varepsilon}{n})$.

(h) $f = \prod_{k=0}^{n} (1 + \varepsilon k)$, where n is a positive integer.

(i) $f = \int_0^\pi \dfrac{\sin(x)}{\sqrt{1 + \varepsilon x}} dx$.

1.8. Find the first two terms in the expansion for the following functions:

(a) $f = \int_0^{\pi/4} \dfrac{dx}{\varepsilon^2 + \sin^2 x}$.

(b) $f = \int_0^1 \dfrac{\cos(\varepsilon x)}{\varepsilon + x} dx$.

(c) $f = \int_0^1 \dfrac{dx}{\varepsilon + x(x-1)} dx$.

1.9. This problem derives an asymptotic approximation for the Stieltjes function, defined as

$$S(\varepsilon) = \int_0^\infty \frac{e^{-t}}{1 + \varepsilon t} dt.$$

(a) Find the first three terms in the expansion of the integrand for small ε and explain why this requires that $t \ll 1/\varepsilon$.

(b) Split the integral into the sum of an integral over $0 < t < \delta$ and one over $\delta < t < \infty$, where $1 \ll \delta \ll 1/\varepsilon$. Explain why the second integral is bounded by $e^{-\delta}$, and use your expansion in part (a) to find an approximation for the first integral. From this derive the following approximation:

$$S(\varepsilon) \sim 1 - \varepsilon + 2\varepsilon^2 + \cdots.$$

1.10. Assume $f(\varepsilon)$ and $\phi(\varepsilon)$ are positive for $\varepsilon > 0$ and $f \sim \phi$ as $\varepsilon \downarrow 0$.
(a) Show $f^\alpha \sim \phi^\alpha$ as $\varepsilon \downarrow 0$.
(b) Give an example to show that it is not necessarily true that $e^f \sim e^\phi$ as $\varepsilon \downarrow 0$. What else must be assumed? Is it enough to impose the additional assumption that $\phi = O(1)$ as $\varepsilon \downarrow 0$?

1.11. In this problem, assume the functions are continuous and nonzero for ε near ε_0.
(a) Show that if $f \sim \phi$ as $\varepsilon \downarrow \varepsilon_0$, then $\phi \sim f$ as $\varepsilon \downarrow \varepsilon_0$.
(b) Suppose $f \sim \phi$ and $g \sim \varphi$ as $\varepsilon \downarrow \varepsilon_0$. Give an example to show that it is not necessarily true that $f + g \sim \phi + \varphi$ as $\varepsilon \downarrow \varepsilon_0$.

1.12. Suppose $f(\varepsilon) \sim a_0 \phi_0(\varepsilon) + a_1 \phi_1(\varepsilon) + \cdots$ and $g(\varepsilon) \sim b_0 \phi_0(\varepsilon) + b_1 \phi_1(\varepsilon) + \cdots$ as $\varepsilon \downarrow \varepsilon_0$, where $\phi_0, \phi_1, \phi_2, \ldots$ is an asymptotic sequence as $\varepsilon \downarrow \varepsilon_0$.
(a) Show that $f + g \sim (a_0 + b_0)\phi_0 + (a_1 + b_1)\phi_1 + \cdots$ as $\varepsilon \downarrow \varepsilon_0$.
(b) Assuming $a_0 b_0 \neq 0$, show that $fg \sim a_0 b_0 \phi_0^2$ as $\varepsilon \downarrow \varepsilon_0$. Also, discuss the possibilities for the next term in the expansion.

(c) Suppose that $\phi_i \phi_j = \phi_{i+j}$ for all i, j. In this case show that

$$fg \sim a_0 b_0 \phi_0 + (a_0 b_1 + a_1 b_0)\phi_1 + (a_0 b_2 + a_1 b_1 + a_2 b_0)\phi_2 + \cdots \text{ as } \varepsilon \downarrow \varepsilon_0.$$

(d) Under what conditions on the exponents will the following be an asymptotic sequence satisfying the condition in part (c): $\phi_0 = (\varepsilon - \varepsilon_0)^\alpha$, $\phi_1 = (\varepsilon - \varepsilon_0)^\beta$, $\phi_2 = (\varepsilon - \varepsilon_0)^\gamma$, ...?

1.13. This problem considers variations on the definitions given in this section. It is assumed that $\phi_1, \phi_2, \ldots$ form an asymptotic sequence.
(a) Explain why $f \sim a_1 \phi_1(\varepsilon) + a_2 \phi_2(\varepsilon) + \cdots + a_n \phi_n(\varepsilon)$, where $a_1 \neq 0$, implies that the series

$$\sum_{k=1}^{m} a_k \phi_k(\varepsilon)$$

is an asymptotic approximation of $f(\varepsilon)$ for $m = 1, \ldots, n$. Also, explain why the converse is not true.
(b) Assuming the a_m are nonzero, explain why the definition of an asymptotic expansion is equivalent to saying that $a_m \phi_m(\varepsilon)$ is an asymptotic approximation of

$$f - \sum_{k=1}^{m-1} a_k \phi_k(\varepsilon)$$

for $m = 1, \ldots, n$ (the sum is taken to be zero when $m = 1$).
(c) Explain why the definition of an asymptotic expansion (to n terms) is equivalent to saying that

$$f = \sum_{k=1}^{n} a_k \phi_k(\varepsilon) + o(\phi_n) \text{ as } \varepsilon \downarrow \varepsilon_0.$$

1.14. The absolute value of the ε^k term in (1.17), (1.18) is

$$\frac{[(2k-1)!]^2}{k2^{5k-2}[(k-1)!]^3} \quad \text{for } k = 1, 2, 3, \ldots.$$

With this show that each series in (1.17), (1.18) diverges for $\varepsilon > 0$.

1.15. The entropy jump $[\![S]\!]$ across a shock wave in a gas is given as (Cole and Cook, 1986)

$$[\![S]\!] = c_v \ln\left(\frac{1 + \frac{\gamma-1}{2}\varepsilon}{1 - \frac{\gamma+1}{2}\varepsilon}(1-\varepsilon)^\gamma\right),$$

where $c_v > 0$, $\gamma > 1$, and $\varepsilon > 0$ is the shock strength.
(a) Find a first-term expansion of the entropy jump for a weak shock (i.e., $\varepsilon \ll 1$).
(b) What is the order of the second term in the expansion?

1.16. This problem derives asymptotic approximations for the complete elliptic integral, defined as

$$K(x) = \int_0^{\pi/2} \frac{ds}{\sqrt{1 - x\sin^2 s}}.$$

It is assumed that $0 < x < 1$.
(a) Show that, for x close to zero, $K \sim \frac{\pi}{2}(1 + \frac{1}{4}x)$.
(b) Show that, for x close to one, $K \sim -\frac{1}{2}\ln(1 - x)$.
(c) Show that, for x close to one,

$$K \sim -\frac{1}{2}\left[1 + \frac{1}{4}(1 - x)\right]\ln(1 - x).$$

1.17. A well-studied problem in solid mechanics concerns the deformation of an elastic body when compressed by a rigid punch. This gives rise to having to evaluate the following integral (Gladwell, 1980):

$$I = \int_0^\infty N(x)\sin(\lambda x)dx,$$

where

$$N(x) = \frac{2\alpha\sinh(x) - 2x}{1 + \alpha^2 + x^2 + 2\alpha\cosh(x)} - 1$$

and α and λ are constants with $1 \le \alpha \le 3$. The infinite interval complicates finding a numerical value of I, and so we write

$$I = \int_0^{x_0} N(x)\sin(\lambda x)dx + R(x_0).$$

The objective is to take x_0 large enough that the error $R(x_0)$ is relatively small.
(a) Find a two-term expansion of $N(x)$ for large x.
(b) Use the first term in the expansion you found in part (a) to determine a value of x_0 so $|R(x_0)| \le 10^{-6}$.
(c) Based on the second term in the expansion you found in part (a), does it appear that the value of x_0 you found in part (b) is a reasonable choice?

1.5 Asymptotic Solution of Algebraic and Transcendental Equations

The examples to follow introduce, and extend, some of the basic ideas for using asymptotic expansions to find approximate solutions. They also illustrate another important point, which is that problems can vary considerably

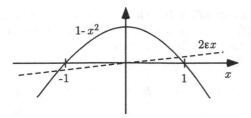

Figure 1.4 Sketch of functions appearing in quadratic equation in (1.28)

in type and complexity. The essence of using asymptotic expansions is adapting the method to a particular problem.

Example 1

One of the easiest ways to illustrate how an asymptotic expansion can be used to find an approximate solution is to consider algebraic equations. As a simple example consider the quadratic equation

$$x^2 + 0.002x - 1 = 0. \tag{1.27}$$

The fact that the coefficient of the linear term in x is much smaller than the other coefficients can be used to find an approximate solution. To do this, we start with the related equation

$$x^2 + 2\varepsilon x - 1 = 0, \tag{1.28}$$

where $\varepsilon \ll 1$.

As the first step we will assess how many solutions there are and their approximate location. With this in mind the functions involved in this equation are sketched in Fig. 1.4. It is evident that there are two real-valued solutions, one located slightly to the left of $x = 1$, the other slightly to the left of $x = -1$. In other words, the expansion for the solutions should not start out as $x \sim \varepsilon x_0 + \cdots$ because this would be assuming that the solution goes to zero as $\varepsilon \to 0$. Similarly, we should not assume $x \sim \frac{1}{\varepsilon}x_0 + \cdots$ as the solution does not become unbounded as $\varepsilon \to 0$. For this reason we will assume that the expansion has the form

$$x \sim x_0 + \varepsilon^\alpha x_1 + \cdots, \tag{1.29}$$

where $\alpha > 0$ (this inequality is imposed so the expansion is well ordered for small ε). It should be pointed out that (1.29) is nothing more than an educated guess. The motivation for making this assumption comes from the observation that in expanding functions one usually ends up using Taylor's theorem and (1.29) is simply a reflection of that type of expansion. The exponent α is to allow for a little flexibility.

Substituting (1.29) into (1.28) one finds that

$$x_0^2 + 2\varepsilon^\alpha x_0 x_1 + \cdots + 2\varepsilon(x_0 + \varepsilon^\alpha x_1 + \cdots) - 1 = 0. \qquad (1.30)$$

    ①   ②   ③   ④

The first problem to solve comes from the terms labeled ①, ④. This is because this equation is suppose to hold for small ε, and therefore we require it to hold as $\varepsilon \downarrow 0$. This gives us:

$O(1)$ $x_0^2 - 1 = 0$.

> The solutions are $x_0 = \pm 1$. An important observation here is that this problem has as many solutions as the original equation (i.e., both problems are quadratic).

With the preceding solution we now have a term in (1.30) that is $O(\varepsilon)$, namely, term ③. Since the right-hand side is zero, it must be that ③ balances with one of the other terms on the left-hand side of the equation. The only choice is ②, and this means $\alpha = 1$ (Exercise 1.23). With this we have the next order problem to solve:

$O(\varepsilon)$ $2x_0 x_1 + 2x_0 = 0$.

> The solution is $x_1 = -1$.

The process used to find x_0 and x_1 can be continued to systematically construct the other terms of the asymptotic expansion. We will not pursue this here but point out that the approximations we have calculated so far are $x \sim \pm 1 - \varepsilon$. To illustrate the accuracy of this result, one of the exact solutions is plotted in Fig. 1.5 along with its asymptotic approximation. It is clear that they are in very good agreement for small values of ε. Also, recall that the original problem was to construct approximations of the solutions of (1.27). In this case, $\varepsilon = 10^{-3}$, and so from the asymptotic approximation we get $x \approx -1.001, 0.999$, while the exact solution gives $x = -1.0010005 \cdots, 0.9990005 \cdots$.

It is appropriate to comment on the notation used in the preceding paragraphs. We referred to the equation that x_0 satisfies as the $O(1)$ equation. However, strictly speaking, based on the definition of the order symbol, every term in (1.30) is $O(1)$. So what we should have done was first introduce a new notation, such as $O_s(1)$ (Exercise 1.4). When stating that $f = O_s(1)$, we would mean that $f = O(1)$ but $f \neq o(1)$. We could then specify, without ambiguity, exactly what terms go into the x_0 equation. A few textbooks do

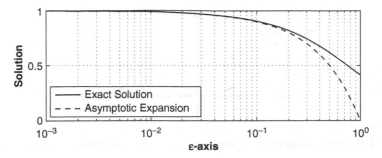

Figure 1.5 Comparison between positive root of (1.28) and the two-term asymptotic expansion $x \sim 1 + \varepsilon$. It is seen that for small ε the asymptotic approximation is very close to the exact value

exactly this to try to reduce the confusion for those who are first learning the subject. However, the truth is that few people actually use this symbolism. In other words, $O(\cdot)$ has two meanings, one connected with boundedness as expressed in the original definition and the other as an identifier of particular terms in an equation.

Example 2

As a second example consider the quadratic equation

$$\varepsilon x^2 + 2x - 1 = 0. \tag{1.31}$$

A sketch of the functions in this equation is given in Fig. 1.6. In this case, for small ε, one of the solutions is located slightly to the left of $x = 1/2$ while the second moves ever leftward as ε approaches zero. Another observation to make is that if $\varepsilon = 0$, then (1.31) becomes linear, that is, the order of the equation is reduced. This is significant because it fundamentally alters the nature of the equation and can give rise to what is known as a singular problem.

If we approach this problem in the same way as in the previous example, then the regular expansion given in (1.29) is used. Carrying out the calculations one finds that

$$x \sim \frac{1}{2} - \frac{\varepsilon}{8} + \cdots . \tag{1.32}$$

Not unexpectedly, we have produced an approximation for the solution near $x = \frac{1}{2}$. The criticism with this is that there are two solutions of (1.31) and the expansion has produced only one. One remedy is to use (1.32) to factor the quadratic Eq. (1.31) to find the second solution. However, there is a more direct way to find the solution that can be adapted to solving differential

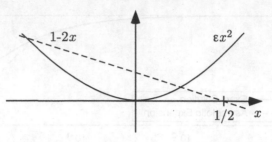

Figure 1.6 Sketch of functions appearing in quadratic equation in (1.31)

equations with a similar complication. To explain what this is, note that the problem is singular in the sense that if $\varepsilon = 0$, then the equation is linear rather than quadratic. To prevent this from happening we assume

$$x \sim \varepsilon^\gamma (x_0 + \varepsilon^\alpha x_1 + \cdots), \tag{1.33}$$

where $\alpha > 0$ (so the expansion is well ordered). Substituting this into (1.31) we have that

$$\underset{①}{\varepsilon^{1+2\gamma}(x_0^2 + 2\varepsilon^\alpha x_0 x_1 + \cdots)} + \underset{②}{2\varepsilon^\gamma (x_0 + \varepsilon^\alpha x_1 + \cdots)} - \underset{③}{1} = 0. \tag{1.34}$$

The terms on the left-hand side must balance to produce zero, and we need to determine the order of the problem that comes from this balancing. There are three possibilities. One of these occurs when $\gamma = 0$, and in this case the balance is between terms ② and ③. However, this leads to the expansion in (1.32), and we have introduced (1.33) to find the other solution of the problem. So we have the following other possibilities:

(i) ① $\sim$ ③ and ② is higher order.
   The condition ① $\sim$ ③ requires that $1 + 2\gamma = 0$, and so $\gamma = -\frac{1}{2}$. With this we have that $①, ③ = O(1)$ and $② = O(\varepsilon^{-1/2})$. This violates our assumption that ② is higher order (i.e., ② $\ll$ ①) so this case is not possible.

(ii) ① $\sim$ ② and ③ is higher order.
   The condition ① $\sim$ ② requires that $1 + 2\gamma = \gamma$, and so $\gamma = -1$. With this we have that $①, ② = O(\varepsilon^{-1})$ and $③ = O(1)$. In this case, the conclusion is consistent with the original assumptions, and so this is the balancing we are looking for.

With $\gamma = -1$ and (1.34), the problem we need to solve is

$$x_0^2 + 2\varepsilon^\alpha x_0 x_1 + \cdots + 2(x_0 + \varepsilon^\alpha x_1 + \cdots) - \varepsilon = 0.$$

From this we have

$O(1)$ $x_0^2 + 2x_0 = 0.$

We are interested in the nonzero solution, and so $x_0 = -2$. With this, from balancing it follows that $\alpha = 1$.

$O(\varepsilon)$ $2x_0x_1 + 2x_1 - 1 = 0$

The solution, for $x_0 = -2$, is $x_1 = -\frac{1}{2}$.

Thus, a two-term expansion of the second solution of (1.31) is

$$x \sim \frac{1}{\varepsilon}\left(-2 - \frac{\varepsilon}{2}\right).$$

As a final note, in the preceding derivation only the nonzero solution of the $O(1)$ equation was considered. One reason for this is that the zero solution ends up producing the solution given in (1.32).

Example 3

One of the comforts in solving algebraic equations is that we have a very good idea of how many solutions to expect. With transcendental equations this is harder to determine, and sketching the functions in the equation takes on a more important role in the derivation. To illustrate, consider the equation

$$x^2 + e^{\varepsilon x} = 5. \tag{1.35}$$

From the sketch in Fig. 1.7 it is apparent that there are two real-valued solutions. To find asymptotic expansions of them assume

$$x \sim x_0 + \varepsilon^\alpha x_1 + \cdots .$$

Substituting this into (1.35) and using Taylor's theorem on the exponential we obtain

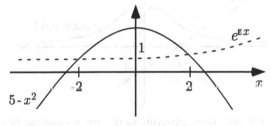

Figure 1.7 Sketch of functions appearing in transcendental equation in (1.35)

$$x_0^2 + 2x_0 x_1 \varepsilon^\alpha + \cdots + 1 + \varepsilon x_0 + \cdots = 5.$$

From the $O(1)$ equation we get that $x_0 = \pm 2$ and $\alpha = 1$, while the $O(\varepsilon)$ equation yields $x_1 = -1/2$. Hence, a two-term expansion of each solution is $x \sim \pm 2 - \varepsilon/2$.

Example 4

For the last example we investigate the equation

$$x + 1 + \varepsilon \operatorname{sech}\left(\frac{x}{\varepsilon}\right) = 0. \tag{1.36}$$

The functions in this equation are sketched in Fig. 1.8, and it is seen that there is one real-valued solution. To find an approximation of it, suppose we proceed in the usual manner and assume

$$x \sim x_0 + \varepsilon^\alpha x_1 + \cdots. \tag{1.37}$$

Substituting this into (1.36) and remembering $0 < \operatorname{sech}(z) \le 1$, it follows that $x_0 = -1$. The complication is that it is not possible to find a value of α so that the other terms in (1.36) balance. In other words, our assumption concerning the structure of the second term in (1.37) is incorrect. Given the character of the hyperbolic function, it is necessary to modify the expansion, and we now assume

$$x \sim -1 + \mu(\varepsilon), \tag{1.38}$$

where we are not certain what μ is other than $\mu \ll 1$ (so the expansion is well ordered). Substituting this into (1.36) we get

$$\mu + \varepsilon \operatorname{sech}(-\varepsilon^{-1} + \mu/\varepsilon) = 0. \tag{1.39}$$

Now, since $\operatorname{sech}(-\varepsilon^{-1} + \mu/\varepsilon) \sim \operatorname{sech}(-\varepsilon^{-1}) \sim 2\exp(-1/\varepsilon)$, we therefore have that $\mu = -2\varepsilon \exp(-1/\varepsilon)$. To construct the third term in the expansion, we would extend (1.38) and write

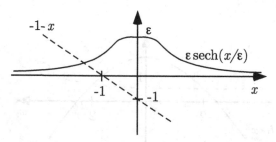

Figure 1.8 Sketch of functions appearing in quadratic equation in (1.36)

$$x \sim -1 - 2\varepsilon e^{-1/\varepsilon} + \nu(\varepsilon), \tag{1.40}$$

where $\nu \ll \varepsilon \exp(-1/\varepsilon)$. To find ν, we proceed as before, and the details are left as an exercise.

Exercises

1.18. Find a two-term asymptotic expansion, for small ε, of each solution x of the following equations:

(a) $x^2 + x - \varepsilon = 0$,

(b) $x^2 - (3 + \varepsilon)x + 1 + \varepsilon = 0$,

(c) $x^2 + (1 - \varepsilon - \varepsilon^2)x + \varepsilon - 2e^{\varepsilon^2} = 0$,

(d) $x^2 - 2x + (1 - \varepsilon^2)^{25} = 0$,

(e) $\varepsilon x^3 - 3x + 1 = 0$,

(f) $\varepsilon^2 x^3 - x + \varepsilon = 0$,

(g) $x^2 + \sqrt{1 + \varepsilon x} = e^{1/(2+\varepsilon)}$,

(h) $x^2 + \varepsilon\sqrt{2 + x} = \cos(\varepsilon)$,

(i) $x = \int_0^\pi e^{\varepsilon \sin(x+s)} ds$,

(j) $x^{2+\varepsilon} = \dfrac{1}{x + 2\varepsilon}$,

(k) $x^2 - 1 + \varepsilon \tanh(\frac{x}{\varepsilon}) = 0$,

(l) $\beta = x - (x + \beta)\dfrac{x - \alpha}{x + \alpha}e^{x/\varepsilon}$ where α and β are positive constants,

(m) $\varepsilon e^{x^2} = 1 + \dfrac{\varepsilon}{1 + x^2}$,

(n) $\left(\dfrac{1}{x} - 1\right)e^{\frac{1}{x}+\alpha} = \left(\dfrac{1}{\varepsilon} - 1\right)e^{\frac{1}{\varepsilon}}$ where $\alpha > 0$ is constant,

(o) $\varepsilon = \left(\frac{3}{2}\int_0^x \sqrt{rp(r)}dr\right)^{2/3}$ where $p(r)$ is smooth and positive and $p'(0) \neq 0$.

(p) $xe^{-x} = \varepsilon$.

1.19. This problem considers the equation $1 + \sqrt{x^2 + \varepsilon} = e^x$.
(a) Explain why there is one real root for small ε.
(b) Find a two-term expansion of the root.

1.20. In this problem you should sketch the functions in each equation and then use this to determine the number and approximate location of the real-valued solutions. With this, find a three-term asymptotic expansion, for small ε, of the nonzero solutions.

(a) $x = \tanh\left(\dfrac{x}{\varepsilon}\right)$,

(b) $x = \tan\left(\dfrac{x}{\varepsilon}\right)$.

1.21. To determine the natural frequencies of an elastic string, one is faced with solving the equation $\tan(\lambda) = \lambda$.
(a) After sketching the two functions in this equation on the same graph explain why there is an infinite number of solutions.
(b) To find an asymptotic expansion of the large solutions of the equation, assume that $\lambda \sim \varepsilon^{-\alpha}(\lambda_0 + \varepsilon^\beta \lambda_1)$. Find ε, α, β, λ_0, λ_1 (note λ_0 and λ_1 are nonzero and $\beta > 0$).

1.22. An important, but difficult, task in numerical linear algebra is to calculate the eigenvalues of a matrix. One of the reasons why it is difficult is that the roots of the characteristic equation are very sensitive to the values of the coefficients of the equation. A well-known example, due to Wilkinson (1964), illustrating this is the equation

$$x^{20} - (1 + \varepsilon)210x^{19} + 20,615x^{18} + \cdots + 20! = 0,$$

which can be rewritten as

$$(x - 1)(x - 2)\cdots(x - 20) = 210\varepsilon x^{19}.$$

(a) Find a two-term expansion for each root of this equation.
(b) The expansion in part (a) has the form $x \sim x_0 + \varepsilon^\alpha x_1$. Based on this result, how small does ε have to be so $|x - x_0| < 10^{-2}$ for every root? Does it seem fair to say that even a seemingly small error in the accuracy of the coefficients of the equation has a tremendous effect on the value of the roots?

1.23. Find a two-term asymptotic expansion, for small ε, of the following:
(a) The point $\mathbf{x}_m = (x_m, y_m)$ at which the function

$$f(\mathbf{x}) = x^2 + 2\varepsilon \sin(x + e^y) + y^2$$

attains its minimum value;
(b) The perimeter of the two-dimensional golf ball described as $r = 1 + \varepsilon \cos(20\theta)$, where $0 \le \theta \le 2\pi$. Explain why ε needs to be fairly small before this two-term expansion can be expected to produce an accurate approximation of the arc length. Also, explain why this happens in geometric terms.

1.24. To find the second term of the expansion when solving (1.30). we took $\alpha = 1$. Show that the choice $0 < \alpha < 1$ does not determine the next nonzero term in the expansion. Also, explain why the choice $1 < \alpha$ is not appropriate.

1.25. An important problem in celestial mechanics is to determine the position of an object given the geometric parameters of its orbit. For an elliptic orbit, one ends up having to solve Kepler's equation, which is

$$\varepsilon \sin(E) = E - M,$$

where $M = n(t - T)$ and ε is the eccentricity of the orbit. This equation determines E, the eccentric anomaly, in terms of the time variable t. After E is found, the radial and angular coordinates of the object are calculated using formulas from geometry. Note that in this equation, n, T, and ε are positive constants.

(a) After sketching the functions in Kepler's equation on the same graph, explain why there is at least one solution. Show that if M satisfies $j\pi \leq M \leq (j+1)\pi$, then there is exactly one solution and it satisfies $j\pi \leq E \leq (j+1)\pi$. To do this, you will need to use the fact that $0 \leq \varepsilon < 1$.

(b) The eccentricity for most of the planets in the Solar System is small (e.g., for the Earth $\varepsilon = 0.02$). Assuming $\varepsilon \ll 1$, find the first three terms in an asymptotic expansion for E.

(c) Show that your result agrees, through the third term, with the series solution (Bessel, 1824)

$$E = M + 2 \sum_{n=1}^{\infty} \frac{1}{n} J_n(n\varepsilon) \sin(nM).$$

It is interesting that Bessel first introduced the functions $J_n(x)$ when solving Kepler's equation. He found that he could solve the problem using a Fourier series, and this led him to an integral representation of $J_n(x)$. This is one of the reasons why these functions were once known as Bessel coefficients.

1.26. The Jacobian elliptic functions $\mathrm{sn}(x, k)$ and $\mathrm{cn}(x, k)$ are defined as follows: given x and k, they satisfy

$$x = \int_0^{\mathrm{sn}(x,k)} \frac{dt}{\sqrt{(1-t^2)(1-k^2 t^2)}}$$

and

$$x = \int_{\mathrm{cn}(x,k)}^{1} \frac{dt}{\sqrt{(1-t^2)[1+k^2(t^2-1)]}},$$

where $-1 \leq x \leq 1$ and $0 \leq k \leq 1$.

(a) Show that $\mathrm{sn}(x, 0) = \sin(x)$.

(b) For small values of k, the expansion for sn has the form $\mathrm{sn}(x, k) \sim s_0 + k^2 s_1 + \cdots$. Use the result from (a) to determine s_0 and then show that $s_1 = -\frac{1}{4}(x - \sin x \cos x) \cos x$.

(c) Show that, for small values of k, $\mathrm{cn}(x, k) \sim \cos(x) + k^2 c_1 + \cdots$, where $c_1 = \frac{1}{4}(x - \sin x \cos x) \sin x$.

1.27. In the study of porous media one comes across the problem of having to determine the permeability, $k(s)$, of the medium from experimental data (Holmes, 1986). Setting $k(s) = F'(s)$, this problem then reduces to solving the following two equations:

$$\int_0^1 F^{-1}(c - \varepsilon r)\mathrm{d}r = s,$$

$$F^{-1}(c) - F^{-1}(c - \varepsilon) = \beta,$$

where β is a given positive constant. The unknowns here are the constant c and the function $F(s)$, and they both depend on ε (also, s and β are independent of ε). It is assumed that $k(s)$ is smooth and positive.

(a) Find the first term in the expansion of the permeability for small ε.

(b) Show that the second term in the expansion in part (a) is $O(\varepsilon^3)$.

1.28. Consider functions $y(x)$ and $Y(x)$ defined, for $0 \le x \le 1$, through the equations

$$\int_y^2 \frac{\mathrm{d}r}{\mathrm{arctanh}(r - 1)} = 1 - x, \text{ and } Y(x) = 3e^{-x/\varepsilon}.$$

(a) Show that y is a positive, monotonically increasing function with $y \le 2$. Use this to show that in the interval $0 \le x \le 1$ there is a point of intersection x_s where $y(x_s) = Y(x_s)$.

(b) For $\varepsilon \ll 1$ find the first two terms in the expansion of x_s. It is worth pointing out that implicitly defined functions, like $y(x)$, arise frequently when one is using matched asymptotic expansions (the subject of Chap. 2).

The definitions of order symbols and asymptotic expansions for complex valued functions, vector-valued functions, and even matrix functions are obtained by simply replacing the absolute value with the appropriate norm. The following exercises examine how these extensions can be used with matrix perturbation problems.

1.29. Let $\mathbf{A}$ and $\mathbf{D}$ be (real) $n \times n$ matrices.

(a) Suppose $\mathbf{A}$ is symmetric and has n distinct eigenvalues. Find a two-term expansion of the eigenvalues of the perturbed matrix $\mathbf{A} + \varepsilon\mathbf{D}$. where $\mathbf{D}$ is positive definite. What you are finding is known as a Rayleigh–Schrödinger series for the eigenvalues.

(b) Suppose $\mathbf{A}$ is the identity and $\mathbf{D}$ is symmetric. Find a two-term expansion of the eigenvalues for the matrix $\mathbf{A} + \varepsilon\mathbf{D}$.

(c) Considering

$$\mathbf{A} = \begin{bmatrix} 0 & 1 \\ 0 & 0 \end{bmatrix} \quad \text{and} \quad \mathbf{D} = \begin{bmatrix} 0 & 0 \\ 1 & 0 \end{bmatrix},$$

show that a $O(\varepsilon)$ perturbation of a matrix need not result in a $O(\varepsilon)$ perturbation of the eigenvalues. This example also demonstrates that a smooth perturbation of a matrix need not result in a smooth perturbation of the eigenvalues. References for this material and its extensions to differential equations are Kato (1995) and Hubert and Sanchez-Palencia (1989).

1.30. Find a two-term expansion of $(\mathbf{A} + \varepsilon\mathbf{B})^{-1}$, where $\mathbf{A}$ is an invertible $n \times n$ matrix.

1.31. Let $\mathbf{C}$ be an $m \times n$ matrix with rank n. In this case the pseudoinverse of $\mathbf{C}$ is defined as $\mathbf{C}^{\dagger} \equiv (\mathbf{C}^T\mathbf{C})^{-1}\mathbf{C}^T$.
(a) Find a two-term expansion of $(\mathbf{A} + \varepsilon\mathbf{B})^{\dagger}$, where $\mathbf{A}$ is an $m \times n$ matrix with rank n.
(b) Show that the result in part (a) reduces to the expansion from Exercise 1.30 when $m = n$ and $\mathbf{A}$ is an invertible $n \times n$ matrix.
(c) A theorem due to Penrose (1955) states that the solutions of the least-squares problem of minimizing $||\mathbf{C} - \mathbf{B}||_2$ have the form $\mathbf{x} = \mathbf{C}^{\dagger}\mathbf{b} + (\mathbf{I} - \mathbf{C}^{\dagger}\mathbf{C})\mathbf{z}$, where $\mathbf{z}$ is an arbitrary n vector. If $\mathbf{C} = \mathbf{A} + \varepsilon\mathbf{B}$, where $\mathbf{A}$ is an $m \times n$ matrix with rank n, then use the result from part (a) to find a two-term asymptotic expansion of the solution. What does your expansion reduce to in the case where $m = n$ and $\mathbf{A} = \mathbf{I}$?

1.6 Introduction to the Asymptotic Solution of Differential Equations

The ideas used to construct asymptotic approximations of the solutions of algebraic and transcendental equations can also be applied to differential equations. What we consider in this section are called regular perturbation problems. Roughly speaking, this means that a regular Poincare expansion is sufficient to obtain an asymptotic approximation of the solution. In later chapters we will take up the study of nonregular, or singular, perturbation problems.

Example: The Projectile Problem

To illustrate how to find an asymptotic approximation of a solution of a differential equation, we return to the projectile example described in Sect. 1.1. Specifically, we consider the problem

$$\frac{d^2y}{d\tau^2} = -\frac{1}{(1+\varepsilon y)^2}, \quad \text{for } 0 < \tau, \tag{1.41}$$

where

$$y(0) = 0 \text{ and } y'(0) = 1. \tag{1.42}$$

The equation in (1.41) is weakly nonlinear because it is nonlinear, but it reduces to one that is linear when $\varepsilon = 0$. To construct an approximation of the solution, we start with a working hypothesis on what is the appropriate form of the asymptotic expansion. For the problem at hand we assume

$$y(\tau) \sim y_0(\tau) + \varepsilon^\alpha y_1(\tau) + \cdots, \tag{1.43}$$

where $\alpha > 0$. As in (1.29), the exponent α is included to allow for a little flexibility. We will also assume that the expansions for the derivatives of $y(\tau)$ can be obtained by differentiating (1.43).

To reduce the differential equation, recall that, for small z, $(1+z)^{-2} \sim 1 - 2z$. So substituting (1.43) into (1.41) yields

$$y_0''(\tau) + \varepsilon^\alpha y_1''(\tau) + \cdots = -\frac{1}{[1 + \varepsilon y_0(\tau) + \cdots]^2}$$
$$\sim -1 + 2\varepsilon y_0(\tau) + \cdots \tag{1.44}$$

and, from (1.42),

$$y_0(0) + \varepsilon^\alpha y_1(0) + \cdots = 0 \text{ and } y_0'(0) + \varepsilon^\alpha y_1'(0) + \cdots = 1. \tag{1.45}$$

The procedure for finding the terms in the expansion follows the argument used to solve algebraic equations. By equating like powers of ε, we obtain the following problems:

$O(1)$ $y_0''(\tau) = -1$,

  $y_0(0) = 0, y_0'(0) = 1.$

  The solution of this problem is $y_0(\tau) = -\frac{1}{2}\tau^2 + \tau$. Also, because of the second term on the right-hand side of (1.44), we must take $\alpha = 1$ in (1.43).

$O(\varepsilon)$ $y_1''(\tau) = 2y_0(\tau)$,

  $y_1(0) = 0, y_1'(0) = 0.$

  The solution here is $y_1(\tau) = \frac{1}{3}\tau^3 - \frac{1}{12}\tau^4$.

Therefore, we have that

$$y(\tau) \sim \tau\left(1 - \frac{1}{2}\tau\right) + \frac{1}{3}\varepsilon\tau^3\left(1 - \frac{1}{4}\tau\right). \tag{1.46}$$

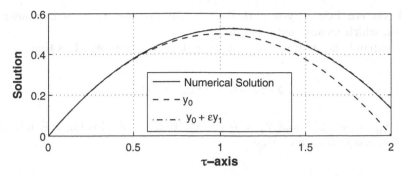

Figure 1.9 Comparison between the numerical solution of (1.41), a one-term asymptotic approximation of the solution, and a two-term asymptotic approximation of the solution. In these calculations $\varepsilon = 10^{-1}$. There is little difference between the numerical solution and the two-term expansion

This approximation applies for $0 \leq \tau \leq \tau_h$, where $\tau_h > 0$ is the point where $y(\tau_h) = 0$ (i.e., the time when the projectile returns to the surface of the Earth). A comparison between the asymptotic expansion and the numerical solution of the problem is given in Fig. 1.9. It is seen that the two-term expansion is essentially indistinguishable from the numerical solution, which indicates the accuracy of the approximation.

The $O(\varepsilon)$ term in (1.46) contains the contribution of the nonlinearity and the $O(1)$ term is the solution one obtains in a uniform gravitational field [it is the scaled version of the solution given in (1.2)]. Since $y_1' \geq 0$ for $0 \leq \tau \leq \tau_h$, it follows that the contribution of y_1 grows with time and increases the flight time (i.e., it makes τ_h larger). This is in agreement with the physics of the problem, which says that the gravitational force decreases with height, which should allow the projectile to stay up longer (Exercise 1.34). This is a fairly simple example, so there is not a lot that can be said about the solution. What this does illustrate, though, is that an asymptotic approximation is capable of furnishing insights into the properties of the solution, which allows us to develop an understanding of the situation that may not be possible otherwise.

Example: A Thermokinetic System

A thermokinetic model for the concentration $u(t)$ and temperature $q(t)$ of a mixture is (Gray and Scott, 1994)

$$\frac{d}{dt}\mathbf{y} = \mathbf{f}(\mathbf{y}, \varepsilon), \tag{1.47}$$

where $\mathbf{y} = (u, q)^{\mathrm{T}}$ and

$$\mathbf{f} = \begin{pmatrix} 1 - ue^{\varepsilon(q-1)} \\ ue^{\varepsilon(q-1)} - q \end{pmatrix}. \tag{1.48}$$

The initial condition is $\mathbf{y}(0) = \mathbf{0}$. We are assuming here that the nonlinearity is weak, which means that ε is small.

We expand the solution using our usual assumption, which is that

$$\mathbf{y} \sim \mathbf{y}_0(t) + \varepsilon \mathbf{y}_1(t) + \cdots,$$

where $\mathbf{y}_0 = (u_0, q_0)^{\mathrm{T}}$ and $\mathbf{y}_1 = (u_1, q_1)^{\mathrm{T}}$. Before substituting this into the differential equation, note that

$$e^{\varepsilon(q-1)} \sim 1 + \varepsilon(q-1) + \frac{1}{2}\varepsilon^2(q-1)^2 + \cdots$$

$$\sim 1 + \varepsilon(q_0 + \varepsilon q_1 + \cdots - 1) + \frac{1}{2}\varepsilon^2(q_0 + \varepsilon q_1 + \cdots - 1)^2 + \cdots$$

$$\sim 1 + \varepsilon(q_0 - 1) + \cdots$$

and

$$ue^{\varepsilon(q-1)} \sim (u_0 + \varepsilon u_1 + \cdots)[1 + \varepsilon(q_0 - 1) + \cdots]$$

$$\sim u_0 + \varepsilon[u_0(q_0 - 1) + u_1] + \cdots.$$

With this, (1.47) takes the form

$$\mathbf{y}_0' + \varepsilon \mathbf{y}_1' + \cdots = \begin{pmatrix} 1 - u_0 \\ u_0 - q_0 \end{pmatrix} + \varepsilon \begin{pmatrix} -u_1 - u_0(q_0 - 1) \\ -q_1 + u_1 + u_0(q_0 - 1) \end{pmatrix} + \cdots. \quad (1.49)$$

This leads to the following problems.

$$O(1) \quad \mathbf{y}_0' = \begin{pmatrix} 1 - u_0 \\ u_0 - q_0 \end{pmatrix}$$

The solution of this that satisfies the initial condition $\mathbf{y}_0(0) = \mathbf{0}$ is

$$\mathbf{y}_0 = \begin{pmatrix} 1 - e^{-t} \\ 1 - (1+t)e^{-t} \end{pmatrix}. \quad (1.50)$$

$$O(\varepsilon) \quad \mathbf{y}_1' = \begin{pmatrix} -u_1 - u_0(q_0 - 1) \\ -q_1 + u_1 + u_0(q_0 - 1) \end{pmatrix}$$

The initial condition is $\mathbf{y}_1(0) = \mathbf{0}$. The equation for u_1 is first order, and the solution can be found using an integrating factor. Once u_1 is determined, then the q_1 equation can be solved using an integrating factor. Carrying out the calculation one finds that

$$\mathbf{y}_1 = \begin{pmatrix} \dfrac{1}{2}(t^2 + 2t - 4)e^{-t} + (2 + t)e^{-2t} \\ \dfrac{1}{6}(t^3 - 18t + 30)e^{-t} - (2t + 5)e^{-2t} \end{pmatrix}. \qquad (1.51)$$

A comparison of the numerical solution for $q(t)$ and the preceding asymptotic approximation for $q(t)$ is shown in Fig. 1.10 for $\varepsilon = 0.1$. It is seen that even the one-term approximation, $q \sim 1 - (1 + t)e^{-t}$, produces a reasonably accurate approximation, while the two-term approximation is indistinguishable from the numerical solution. The approximations for $u(t)$, which are not shown, are also as accurate.

The previous example involved an implicit extension of the definition of an asymptotic expansion to include vector-valued functions. It is possible to do this because the extension is straightforward and does not require the introduction of any new ideas. For example, the vector version is simply Definition 1.3, but f and the a are vector functions. Alternatively, the vector version is obtained by expanding each component of $\mathbf{y}$ using Definition 1.3.

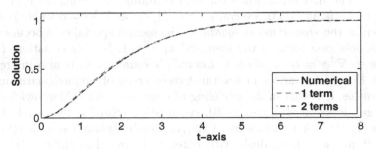

Figure 1.10 Comparison between the numerical solution for $q(t)$, and the one and two term approximations coming from (1.50) and (1.51). In the calculation, $\varepsilon = 0.1$

Example: A Nonlinear Potential Problem

The ideas developed in the preceding problem can also be applied to partial differential equations. An interesting example arises in the theory of the diffusion of ions through a solution containing charged molecules (i.e., a polyelectrolyte solution). If the solution occupies a domain Ω, then the electrostatic potential $\phi(\mathbf{x})$ in the solution satisfies the Poisson–Boltzmann equation

$$\nabla^2 \phi = -\sum_{i=1}^{k} \alpha_i z_i e^{-z_i \phi}, \text{ for } \mathbf{x} \in \Omega, \tag{1.52}$$

where the α_i are positive constants, z_i is the valence of the ionic species, and the sum is over all the ionic species that are present. For example, for a simple salt such as NaCl, in the sum $k = 2$ with $z_1 = -1$ and $z_2 = 1$. There is also an electroneutrality condition for the problem that states that the valences and constants must satisfy

$$\sum_{i=1}^{k} \alpha_i z_i = 0. \tag{1.53}$$

On the boundary $\partial\Omega$ a uniform charge is imposed, and this condition takes the form

$$\partial_n \phi = \varepsilon, \text{ on } \partial\Omega, \tag{1.54}$$

where $\mathbf{n}$ is the unit outward normal to $\partial\Omega$ (note $\partial_n \equiv \mathbf{n} \cdot \nabla$ is the normal derivative).

For cultural reasons it is worth pointing out two important special cases of (1.52). The first arises when the sum contains two terms, with $z_1 = -1$ and $z_2 = 1$. Using (1.53), we can rewrite (1.52) as $\nabla^2 \phi = \alpha \sinh(\phi)$, which is known as the *sinh–Poisson equation*. The second special case occurs when there is only one term in the sum and $z_1 = -1$. In this situation, (1.52) reduces to $\nabla^2 \phi = \alpha e^\phi$, which is *Liouville's equation*. It is also interesting to note that (1.52) arises in several diverse areas of applied mathematics. For example, it occurs in the modeling of semiconductors (Markowich et al., 1990), gel diffusion (Holmes, 1990), combustion (Kapila, 1983), relativistic field theory (D'Hoker and Jackiw, 1982), and hydrodynamic stability (Stuart, 1971). It has also been dealt with extensively by MacGillivray (1972) in deriving Manning's condensation theory.

The problem we must solve consists of the nonlinear partial differential equation in (1.52) subject to the boundary condition in (1.54). This is a very difficult problem, and a closed-form solution is beyond the range of current mathematical methods. To deal with this, in the Deybe–Hückel theory presented in most introductory physical chemistry textbooks, it is assumed that the potential is small enough that the Poisson–Boltzmann equation can be linearized. This reduction is usually done heuristically, but here we want to carry out the calculations more systematically. For our problem, a small potential means ε is small. Given the boundary condition in (1.54), the appropriate expansion of the potential, for small ε, is

$$\phi \sim \varepsilon(\phi_0(\mathbf{x}) + \varepsilon\phi_1(\mathbf{x}) + \cdots). \tag{1.55}$$

Substituting this into (1.52) we obtain

$$
\begin{aligned}
\varepsilon(\nabla^2\phi_0 &+ \varepsilon\nabla^2\phi_1 + \cdots) \\
&= -\sum \alpha_i z_i e^{-\varepsilon z_i(\phi_0 + \varepsilon\phi_1 + \cdots)} \\
&\sim \sum \alpha_i z_i \left[1 - \varepsilon z_i(\phi_0 + \varepsilon\phi_1 + \cdots) + \frac{1}{2}z_i^2\varepsilon^2\phi_0^2 + \cdots\right] \\
&\sim \varepsilon \sum \alpha_i z_i^2 \phi_0 + \varepsilon^2 \sum \alpha_i z_i^2 (\phi_1 + \frac{1}{2}z_i\phi_0^2) + \cdots.
\end{aligned}
\tag{1.56}
$$

The last step follows from the electroneutrality condition in (1.53). Now, setting

$$\kappa^2 = \sum_{i=1}^{k} \alpha_i z_i^2,$$

it follows from (1.56) that the $O(\varepsilon)$ problem is

$$\nabla^2\phi_0 = \kappa^2\phi_0, \quad \text{for } \mathbf{x} \in \Omega, \tag{1.57}$$

where, from (1.54),

$$\partial_n\phi_0 = 1 \text{ on } \partial\Omega. \tag{1.58}$$

To solve this problem, the domain Ω needs to be specified, which will be done in the example presented below.

The first-term approximation $\phi \sim \varepsilon\phi_0(\mathbf{x})$ is the basis of the classical Debye–Hückel theory in electrochemistry. One of the advantages of our approach is that it is relatively easy to find the correction to this approximation, and this is accomplished by solving the $O(\varepsilon^2)$ problem. From (1.56) one finds that

$$(\nabla^2 - \kappa^2)\phi_1 = -\lambda\phi_0^2, \quad \text{for } \mathbf{x} \in \Omega, \tag{1.59}$$

where $\lambda = \frac{1}{2}\sum \alpha_i z_i^3$ and, from (1.54),

$$\partial_n\phi_1 = 0, \quad \text{on } \partial\Omega. \tag{1.60}$$

It is assumed here that $\lambda \neq 0$.

Special Case

It remains to solve the problems we have derived for the first two terms in the expansion. In electrochemistry, an often-used domain is the region outside of the unit sphere (i.e., the region $1 < r < \infty$). Using spherical coordinates, the solution is independent of the angular variables and both (1.57) and (1.59) reduce to ordinary differential equations in the radial coordinate. In particular, (1.57) reduces to solving

$$\frac{1}{r^2}\frac{\mathrm{d}}{\mathrm{d}r}\left(r^2\frac{\mathrm{d}\phi_0}{\mathrm{d}r}\right) - \kappa^2\phi_0 = 0, \quad \text{for } 1 < r < \infty,$$

where $\phi_0'(1) = -1$. The bounded solution of this problem is

$$\phi_0 = \frac{1}{(1+\kappa)r}e^{\kappa(1-r)}. \tag{1.61}$$

With this, (1.59) becomes

$$\frac{1}{r^2}\frac{\mathrm{d}}{\mathrm{d}r}\left(r^2\frac{\mathrm{d}\phi_1}{\mathrm{d}r}\right) - \kappa^2\phi_1 = -\frac{\lambda}{r^2}e^{2\kappa(1-r)},$$

where $\phi_1'(1) = 0$. Given that $\exp(\pm\kappa r)/r$ are solutions of the associated homogeneous equation, using the method of variation of parameters one finds that

$$\phi_1 = \frac{\alpha}{r}e^{-\kappa r} + \frac{\gamma}{\kappa r}\left[e^{\kappa r}E_1(3\kappa r) - e^{-\kappa r}E_1(\kappa r)\right],$$

where $\gamma = \lambda e^{2\kappa}/(2\kappa(1+\kappa)^2)$, E_1 is the exponential integral, and

$$\alpha = \frac{\gamma}{\kappa(1+\kappa)}\left[(\kappa-1)e^{2\kappa}E_1(3\kappa) + (\kappa+1)E_1(\kappa)\right].$$

The accuracy of using just the first term in the approximation is illustrated in Fig. 1.11, which corresponds to the case where $z_1 = 1$, $z_2 = -2$, and $\alpha_2 = 1$. Evidently, the Deybe–Hückel approximation is rather good, at least for these parameter values. ∎

Our approach to regular perturbation expansions is constructive, and in the development little has been said about the theoretical foundation of the subject. For example, we have not made any attempt to prove rigorously that the expansions are asymptotic. This requires estimates of the error, and a nice introduction to this subject can be found in Murdock (1999).

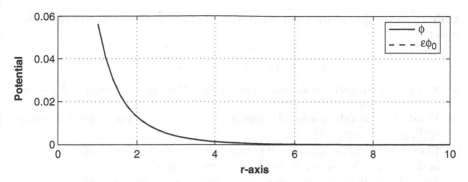

Figure 1.11 Comparison between the numerical solution of (1.52), (1.54), and the first-term approximation obtained from (1.61) for the region outside the unit sphere. In the calculation, $\varepsilon = 0.1$

Exercises

1.32. Find a two-term asymptotic expansion, for small ε, of the solution of the following problems:

(a) $y'' + \varepsilon y' - y = 1$, where $y(0) = 0$ and $y(1) = 1$.

(b) $y'' + f(\varepsilon y) = 0$, where $y(0) = y(1) = 0$ and $f(s)$ is a smooth positive function.

(c) $\nabla^2 u = 0$, for $x^2 + y^2 < 1$, where $u = e^{\varepsilon \cos(\theta)}$ when $x^2 + y^2 = 1$.

(d) $\partial_t u = \partial_x[D(\varepsilon u)\partial_x u]$, for $0 < x < 1$ and $0 < t$, where $u(0, t) = u(1, t) = 0$ and $u(x, 0) = \sin(4\pi x)$. Also, $D(s)$ is a smooth positive function.

1.33. Find a two-term asymptotic expansion, for small ε, of the solution of the following problems. Also, comment on how the boundary conditions help determine the form of the expansion.

(a) $y'' - y + \varepsilon y^3 = 0$, where $y(0) = 0$ and $y(1) = 1$.

(b) $y'' - y + y^3 = 0$, where $y(0) = 0$ and $y(1) = \varepsilon$.

(c) $y'' - y + \varepsilon y^3 = 0$, where $y(0) = 0$ and $y(1) = \varepsilon$.

(d) $y'' - y + \varepsilon y^3 = 0$, where $y(0) = \varepsilon$ and $y(1) = 1$.

1.34. This problem concerns aspects of the projectile problem.

(a) Assuming $\tau_h \sim \tau_0 + \varepsilon \tau_1$, find τ_0 and τ_1 from (1.46). Give a physical reason why τ_1 is positive.

(b) Find a two-term expansion for the time at which the projectile reaches its maximum height. How much higher does the projectile get in the nonuniform gravitational field? (You should find a first-term approximation for this.)

1.35. In the projectile problem, to account for air resistance, one obtains the equation

$$\frac{\mathrm{d}^2 x}{\mathrm{d}t^2} = -\frac{gR^2}{(x+R)^2} - \frac{k}{R+x}\frac{\mathrm{d}x}{\mathrm{d}t},$$

where k is a nonnegative constant. Assume here that $x(0) = 0$ and $x'(0) = v_0$.

(a) What is the nondimensional version of this problem if one uses the same scaling as in Sect. 1.1?

(b) Find a two-term asymptotic expansion of the solution for small ε. Assume in doing this that $\alpha = kv_0/(gR)$ is independent of ε.

(c) Does the addition of air resistance increase or decrease the flight time?

1.36. The eigenvalue problem for the vertical displacement, $y(x)$, of an elastic string with variable density is

$$y'' + \lambda^2 \rho(x, \varepsilon) y = 0, \quad \text{for } 0 < x < 1,$$

where $y(0) = y(1) = 0$. For small ε assume $\rho \sim 1 + \varepsilon\mu(x)$, where $\mu(x)$ is positive and continuous. In this case the solution $y(x)$ and eigenvalue λ depend on ε, and the appropriate expansions are $y \sim y_0(x) + \varepsilon y_1(x)$ and $\lambda \sim \lambda_0 + \varepsilon\lambda_1$ (better expansions will be discussed in Sect. 3.6).
(a) Find y_0 and λ_0.
(b) Find y_1 and λ_1.

1.37. An interesting problem that can be traced back to Bernoulli and Euler is that of a whirling elastic string that is held fixed at each end (something like a jump rope). The problem comes down to solving for a function $y(x)$ and a frequency of rotation ω that satisfy (Caughey, 1970)

$$y'' + \omega^2 \left(\frac{1}{\gamma\sqrt{1 + \varepsilon^2 y^2}} + \alpha^2 \right) y = 0, \quad \text{for } 0 < x < 1,$$

where $y'(0) = y'(1) = 0$. Here α is a constant that satisfies $0 < \alpha < 1$, and γ is a constant that depends on y through the equation

$$\gamma = \frac{1}{\alpha^2}\left(\frac{1}{1 - \alpha^2} - \int_0^1 \frac{\mathrm{d}x}{\sqrt{1 + \varepsilon^2 y^2}} \right).$$

There is one additional constraint imposed on the solution, and it is that

$$\int_0^1 y(x)\mathrm{d}x = 0.$$

What we have here is an eigenvalue problem where ω is the eigenvalue.
(a) Find a two-term expansion of ω for small ε. Note that to do this you will need to find the first term in the expansion for $y(x)$, but not necessarily the second term.

(b) Comment on the accuracy of the expansion for ω for $0 < \alpha < 1$.

(c) Caughey (1970) proves that whirling cannot occur at frequencies below $\omega = \pi$. Is your expansion in part (a) consistent with this result?

1.38. The equation for the displacement, $u(x)$, of a weakly nonlinear string, on an elastic foundation and with a uniform forcing, is

$$\frac{d}{dx}\left(\frac{u_x}{\sqrt{1 + \varepsilon(u_x)^2}}\right) - k^2 u = 1, \quad \text{for } 0 < x < 1,$$

where $u(0) = u(1) = 0$ and k is a positive constant. Find a two-term expansion of the solution for small ε. You should find, but do not need to solve, the problem for the second term.

1.39. The equation for the displacement, $u(x)$, of a nonlinear beam, on an elastic foundation and with a small periodic forcing, is

$$u'''' - \kappa u'' + k^2 u = \varepsilon F_0 \sin(\pi x), \quad \text{for } 0 < x < 1,$$

where $u(0) = u''(0) = u(1) = u''(1) = 0$, k is a positive constant, and F_0 is a nonzero constant. Also,

$$\kappa = \frac{1}{4}\int_0^1 (u_x)^2 dx.$$

Find a two-term expansion of the solution for small ε.

1.40. Consider the following eigenvalue problem:

$$\int_0^a K(x, s) y(s) ds = \lambda y(x), \quad \text{for } 0 < x < a.$$

This is a Fredholm integral equation, where the kernel $K(x, s)$ is known and is assumed to be smooth and positive. Also, the eigenfunction is taken to be positive and normalized so that

$$\int_0^a y^2 ds = a.$$

The eigenvalue λ and the eigenfunction $y(x)$ depend on the value of the parameter a. This exercise investigates how to find this dependence in the case where a is small.

(a) Find the first two terms in the expansions of λ and $y(x)$ for small a.

(b) By changing variables, transform the integral equation into (Knessl and Keller, 1991a)

$$\int_0^1 K(a\xi, ar)\phi(r)dr = \frac{\lambda}{a}\phi(\xi), \quad \text{for } 0 < \xi < 1. \tag{1.62}$$

What happens to the normalization?

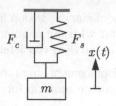

Figure 1.12 Figure for Exercise 1.42

(c) From part (b) find a two-term expansion of λ and $\phi(\xi)$ for small a.
(d) Explain why the expansions in parts (a) and (c) for λ are the same but those for $y(x)$ differ.

1.41. In quantum mechanics, the perturbation theory for bound states involves the time-independent (normalized) Schrödinger equation

$$\Psi'' - [V_0(x) + \varepsilon V_1(x)]\Psi = -E\Psi, \quad \text{for } -\infty < x < \infty,$$

where $\psi(-\infty) = \psi(\infty) = 0$. In this problem the eigenvalue E is the energy, V_1 is the perturbing potential, and ε is called the coupling constant. The potentials V_0 and V_1 are given continuous functions. This exercise examines what is known as a logarithmic perturbation expansion to find the corrections to the energy (Imbo and Sukhatme, 1984). To do this, it is assumed that the unperturbed ($\varepsilon = 0$) state is nonzero (more specifically, it is a nondegenerate ground state).
(a) Assuming $\psi \sim \psi_0(x) + \varepsilon\psi_1(x) + \varepsilon^2\psi_2(x)$ and $E \sim E_0 + \varepsilon E_1 + \varepsilon^2 E_2$, find what problem the first term in these expansions satisfies. In this problem assume

$$\int_{-\infty}^{\infty} \psi_0^2 dx = 1 \quad \text{and} \quad \int_{-\infty}^{\infty} |V_1(x)| dx < \infty.$$

(b) Letting $\psi = e^{\varphi(x)}$, find the problem $\varphi(x)$ satisfies.
(c) Expand $\varphi(x)$ for small ε, and from this find E_1 and E_2 in terms of ψ_0 and the perturbing potential.
(d) For a harmonic oscillator (thus, $V_0 = \lambda^2 x^2$ with $\lambda > 0$) with perturbing potential $V_1 = \alpha x e^{-\gamma x^2}$ (where α and γ are positive) show that

$$E \sim \lambda - \frac{1}{4}\left(\frac{\varepsilon\alpha}{\gamma + \lambda}\right)^2 \sqrt{\frac{\lambda}{\lambda + 2\gamma}}.$$

1.42. From Newton's second law the displacement, $x(t)$, of the mass in a mass–spring–dashpot system satisfies $mx'' = F_s + F_d + F_e$, where m is the mass, F_s the restoring force in the spring, F_d the damping force, and F_e the external force (Fig. 1.12). In this problem take $F_e = 0$, and assume $x(0) = 0$ and $x'(0) = v_0$.

(a) Suppose there is linear damping, $F_d = -cx'$, and the spring is linear, $F_s = -kx$. Nondimensionalize the problem using the time scale for the undamped ($c = 0$) system. Your answer should contain the parameter $\alpha = ck^{1/2}/m^{3/2}$; explain why α is the ratio of the period of the undamped oscillator to the damping time scale.

(b) Assume there is linear damping but the spring is nonlinear with $F_s = -kx - Kx^3$, where K is a positive constant. The equation in this case is known as Duffing's equation. Nondimensionalize the problem using the same scaling as in part (a). The problem should contain two parameters, α and $\varepsilon = Kv_0^2 k^2/m^3$. Explain why the latter is the ratio of the nonlinear to linear component of the spring.

(c) For the scaled problem in part (b) find a two-term expansion of the solution in the case where the spring is weakly nonlinear (thus, ε is small).

1.43. In the nonlinear potential example, suppose that $k = 2$, $z_1 = 1$, and $z_2 = -1$.

(a) Show that the equation takes the form

$$\nabla^2 \phi = \alpha \sinh(\phi) \text{ for } \mathbf{x} \in \Omega,$$

where α is a positive constant and the boundary condition is given in (1.54).

(b) Find the problems satisfied by the first and second terms in the expansion for ϕ.

1.44. Consider the second-order difference equation

$$y_{n+1} - 2y_n + y_{n-1} = (\alpha + \varepsilon f_n)y_n, \text{ for } n = 1, 2, 3, \ldots,$$

where, for this problem, α is a positive constant and $f_n = (1/2)^n$.

(a) Find a two-term expansion of the solution for small ε.

(b) Suppose $y_0 = 1$, $y_1 = 0$, and $\alpha = 2$. Calculate the exact solution, and the two-term expansion you found in part (a), for $n = 2, 3, \ldots, 10$. Comment on the accuracy of the approximation.

1.45. The theory of self-gravitating annuli orbiting a central mass is used to study planetary rings, rings around galaxies, and other such phenomena. A problem that arises in this theory is to find the mass density, $\rho(r)$, of a ring. This is determined from the following problem (Christodoulou and Narayan, 1992):

$$\frac{1}{r}\frac{d}{dr}\left(r\left(\frac{d}{dr}\rho^{1/n}\right)\right) + \alpha\rho = \frac{1}{r^3} - \frac{\beta}{r^{2q}}, \text{ for } 1 - \varepsilon < r < 1 + \varepsilon,$$

where $\rho(1 - \varepsilon) = \rho(1 + \varepsilon) = 0$. Also, $\alpha > 0$, $\beta > 1$, $q > 3/2$, and $n \geq 1$ are constants. This problem is concerned with what is known as a slender annulus approximation, that is, $\varepsilon \ll 1$.

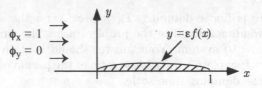

Figure 1.13 Figure for Exercise 1.46

(a) Rescale the independent variable so the interval is independent of ε.
(b) Assuming the density is bounded for small ε, find the first term in an expansion of the mass density.
(c) Find the second term in the expansion of the mass density.

1.46. To find the flow of air over an airplane wing, as illustrated in Fig. 1.13, one can reduce the problem to solving the following integral equation for $u(x, \varepsilon)$ (Geer and Keller, 1968):

$$\int_{\alpha(\varepsilon)}^{\beta(\varepsilon)} K(x, \zeta, \varepsilon) u(\zeta, \varepsilon) \, d\zeta = -2\pi f'(x), \quad \text{for } \alpha(\varepsilon) < x < \beta(\varepsilon),$$

where

$$K(x, \zeta, \varepsilon) = \frac{f(x) - (x - \zeta) f'(x)}{(x - \zeta)^2 + \varepsilon^2 f^2(x)}$$

and $0 \leq \alpha(\varepsilon) < \beta(\varepsilon) \leq 1$. This equation is used to determine $u(x, \varepsilon)$ as well as the endpoints $\alpha(\varepsilon)$ and $\beta(\varepsilon)$. The function $f(x)$ is known and is associated with the shape of the airplane wing (Fig. 1.13). For this reason, $f(x)$ is assumed to be smooth, positive for $0 < x < 1$, and $f(0) = f(1) = 0$. We will also assume $f'(x) \neq 0$ at $x = 0, 1$.
(a) For small ε, explain why it is not appropriate to expand K using the binomial expansion.
(b) In the case where the wing can be described by the ellipse $4(x - \frac{1}{2})^2 + (\frac{y}{\varepsilon})^2 = 1$, the solution of the integral equation is

$$u(x, \varepsilon) = \frac{\varepsilon}{1 - 2\varepsilon} \frac{2 - 4x}{\sqrt{(\beta - x)(x - \alpha)}},$$

where $2\alpha = 1 - \sqrt{1 - 4\varepsilon^2}$ and $2\beta = 1 + \sqrt{1 - 4\varepsilon^2}$. For small ε, find the first two terms in the expansions for α, β, and u.

1.7 Uniformity

We have seen that other parameters or variables that a function depends on can interfere with the accuracy, or even existence, of an asymptotic expansion. To illustrate the situation, we consider the following equation for y:

$$(y - 1)(y - x) = \varepsilon y. \tag{1.63}$$

We are interested here in values of x in the interval $0 < x < 1$. For small ε the appropriate expansion of the solution is $y \sim y_0 + \varepsilon y_1 + \cdots$. Substituting this into the equation and carrying out the calculations, the following expansion is obtained for one of the solutions:

$$y \sim 1 + \frac{\varepsilon}{1 - x}. \tag{1.64}$$

There is no doubt that this is an asymptotic expansion of the solution; the only question is how accurate it is for values of x that are close to $x = 1$. Given the nature of the $O(\varepsilon)$ term in the expansion, it is expected that it will be necessary to have very small ε if such x are used. To say this in a slightly different way, no matter how small we take ε, there are x values in the interval $0 < x < 1$ where the second term is as large as, if not larger than, the first term. So the expansion does not remain well ordered as x approaches $x = 1$, and for this reason it is not uniformly valid for $0 < x < 1$. On the other hand, it is uniformly valid if, say, $0 < x < 3/4$.

The following definition is introduced to make precise what is meant by a uniform approximation. It is written for the specific case of small ε because that is the only limit we will be considering in this book when dealing with uniformity. Also, in what follows, I designates an interval on the x-axis.

Definition 1.4. Suppose $f(x, \varepsilon)$ and $\phi(x, \varepsilon)$ are continuous functions for $x \in I$ and $0 < \varepsilon < \varepsilon_1$. In this case, $\phi(x, \varepsilon)$ is a *uniformly valid* asymptotic approximation of $f(x, \varepsilon)$ for $x \in I$ as $\varepsilon \downarrow 0$ if, given any positive constant δ, there is an ε_2 (independent of x and ε) such that

$$|f - \phi| \le \delta|\phi| \quad \text{for } x \in I \quad \text{and} \quad 0 < \varepsilon < \varepsilon_2.$$

The critical point in this definition is that it is possible to find an open interval near $\varepsilon = 0$ (specifically, $0 < \varepsilon < \varepsilon_2$) so the inequality holds for all values of x that are under consideration. This is essentially the idea used to define uniform convergence, and this enables us to make several useful observations.

The lack of uniformity is often due to what happens at one of the endpoints of the interval I. This is true in (1.64), for x near one, and in Fig. 1.2, for x near zero. It is possible to use this observation to develop a simple test for uniformity. To explain, suppose that $f(x, \varepsilon) \sim \phi(x, \varepsilon)$ for $a < x < b$, where both f and ϕ are continuous functions for $a \leq x \leq b$. Also, letting x_0 be one of the endpoints, assume that $f(x_0, \varepsilon) \sim \phi_0(\varepsilon)$. If the approximation is uniform, then it is natural to expect that $\phi(x_0, \varepsilon)$ and $\phi_0(\varepsilon)$ are somehow related to one another. For those who might guess that if the approximation is uniform then it must be that $\phi_0(\varepsilon) = \phi(x_0, \varepsilon)$, the answer is no. These functions only need to be *equal asymptotically*, that is,

$$\lim_{\varepsilon \downarrow 0} \frac{\phi(x_0, \varepsilon)}{\phi_0(\varepsilon)} = 1. \tag{1.65}$$

It is assumed here that ϕ_0 is nonzero, so the quotient makes sense. The proof of (1.65) incorporates many of the same arguments used to establish the properties coming from uniform convergence and is left as an exercise (see also Exercise 1.55).

The usefulness of (1.65) is that it gives us a negative result. In particular, if

$$\lim_{\varepsilon \downarrow 0} \frac{\phi(x_0, \varepsilon)}{\phi_0(\varepsilon)} \neq 1, \tag{1.66}$$

then $f(x, \varepsilon) \sim \phi(x, \varepsilon)$ is not uniform for $a < x < b$. An example of how this is used is given in the next example.

Example

In Sect. 1.4 it was found that $x + e^{-x/\varepsilon} \sim x$ for $0 < x < 1$. It was also stated that this approximation is not uniform and the problem is at $x = 0$ (Fig. 1.2). To verify this, note that $f(x, \varepsilon) = x + e^{-x/\varepsilon}$ and $\phi(x, \varepsilon) = x$. Taking $x_0 = 0$, then $f(x_0, \varepsilon) = \phi_0(\varepsilon) = 1$ and $\phi(x_0, \varepsilon) = 0$. With this we have that

$$\lim_{\varepsilon \downarrow 0} \frac{\phi(x_0, \varepsilon)}{\phi_0(\varepsilon)} = 0.$$

The conclusion is, therefore, that the approximation is not uniformly valid on the interval $0 < x < 1$. ∎

As illustrated in the previous example, the result in (1.66) is useful for showing that an expansion is nonuniform. There are also simple tests for proving uniformity. We will only state one of these; to do this, assume the functions are continuous. It is a common mistake to think that if an expansion holds on the closed interval $a \leq x \leq b$, then it must be uniformly valid (this is often stated when constructing what are known as composite expansions using the method of matched asymptotic expansions). An example that illustrates this is given in Exercise 1.51. What is true is that if $|\phi|$ is a monotonically decreasing function of ε for $a < x < b$, and if the asymptotic approximation holds for $a \leq x \leq b$, then it is uniformly valid. Again the proof of this is very similar to the analogous result for uniform convergence and is left to the reader.

To summarize the preceding results, we have the following theorem.

Theorem 1.4. *Assume $f(x, \varepsilon)$, $\phi(x, \varepsilon)$, and $\phi_0(\varepsilon)$ are continuous for $a \leq x \leq b$ and $0 < \varepsilon < \varepsilon_1$.*

(a) If $f \sim \phi$ for $a \leq x \leq b$, and if $|\phi(x, \varepsilon)|$ is monotonically decreasing in ε, then this asymptotic approximation is uniformly valid for $a \leq x \leq b$.

(b) Suppose $f(x, \varepsilon) \sim \phi(x, \varepsilon)$ for $a < x < b$ and $f(x_0, \varepsilon) \sim \phi_0(\varepsilon)$ for $x_0 = a$ or for $x_0 = b$. If $\phi_0(\varepsilon)$ is nonzero and (1.66) holds, then $f \sim \phi$ is not uniformly valid for $a < x < b$.

The idea of uniformity can be extended to asymptotic expansions. To explain how, suppose we have a function $f(x, \varepsilon)$ and

$$f(x, \varepsilon) \sim a_1(x)\phi_1(x, \varepsilon) + a_2(x)\phi_2(x, \varepsilon) \text{ as } \varepsilon \downarrow 0. \qquad (1.67)$$

In other words, we have expanded f in terms of the basis functions ϕ_1 and ϕ_2, and they possibly depend on x (in addition to ε). To be able to say that this expansion is uniformly valid, we must have, at a minimum, that it remains well ordered for all values of x under consideration. The exact statement is as follows.

Definition 1.5.   The expansion (1.67) is uniformly valid for $x \in I$ if, given any positive constant δ, we can find an ε_2 (independent of x and ε) such that, for $0 < \varepsilon < \varepsilon_2$,

$$|f(x, \varepsilon) - a_1(x)\phi_1(x, \varepsilon)| \leq \delta |a_1(x)\phi_1(x, \varepsilon)|$$

and

$$|f(x, \varepsilon) - a_1(x)\phi_1(x, \varepsilon) - a_2(x)\phi_2(x, \varepsilon)| \leq \delta |a_2(x)\phi_2(x, \varepsilon)|$$

for all $x \in I$.

This definition extends in a straightforward manner to expansions involving an arbitrary number of scale functions.

As a comment in passing, in the preceding discussion the variable x was taken to be a scalar. Nevertheless, the definitions and theorem that have been given hold (without change) for the case where x is a vector.

As usual, given a function $f(x, \varepsilon)$, one of the main tools that is used to establish uniformity is Taylor's theorem, and this is illustrated in the following examples.

Examples

1. If $f = \sin(x + \varepsilon)$, then $f \sim \sin(x)$ is uniformly valid for $\frac{\pi}{4} < x < \frac{3\pi}{4}$. This can be established using Taylor's theorem (about $\varepsilon = 0$), which gives us that

$$f = \sin(x) + \varepsilon \cos(x + \xi).$$

Thus, $\lim_{\varepsilon \downarrow 0}[f - \sin(x)] = 0$, and from this we have that $f \sim \sin(x)$ for $\frac{\pi}{4} \leq x \leq \frac{3\pi}{4}$. Also, since $\phi \equiv \sin(x)$ is independent of ε, it follows from Theorem 1.3 that the approximation is uniformly valid over the stated interval. What may be surprising is that it is not uniform for $0 < x < \frac{\pi}{2}$, even though (1.66) holds. The easiest way to see this is to note $f \sim \varepsilon$ at $x = 0$, and so $\phi_0(\varepsilon) = \varepsilon$ and $\phi(0, \varepsilon) = 0$. Thus, the limit in (1.66) is not equal to 1, which means the approximation is nonuniform on this interval. ∎

2. A solution of the differential equation

$$\varepsilon y'' + 2y' + 2y = 0, \quad \text{for } 0 < x < 1, \tag{1.68}$$

is

$$y(x) = e^{\alpha x/\varepsilon} + e^{\beta x/\varepsilon}, \tag{1.69}$$

where $\alpha = -1 + \sqrt{1 - 2\varepsilon}$ and $\beta = -1 - \sqrt{1 - 2\varepsilon}$. To obtain an asymptotic approximation of this solution for small ε, note $\alpha \sim -\varepsilon - \varepsilon^2/2$ and $\beta \sim -2 + \varepsilon$. Thus, for $0 < x < 1$,

$$y(x) \sim e^{-(1+\varepsilon/2)x} + e^{(-2+\varepsilon)x/\varepsilon}$$

$$\sim e^{-x} + e^{-2x/\varepsilon} \tag{1.70}$$

$$\sim e^{-x}. \tag{1.71}$$

The approximation in (1.71) is not uniform for $0 < x < 1$. This is because the exponential that was dropped in going from (1.70) to (1.71) is of the same order as (1.71) when $x = 0$ [one can also use (1.66) to show nonuniformity]. If we were to retain this term, that is, if we used the approximation in (1.70), then we would have a uniformly valid approximation over the interval $0 < x < 1$ (Exercise 1.48). To illustrate the differences between these approximations, they are plotted in Fig. 1.14 in the case of $\varepsilon = 10^{-2}$. Both give a very good description of the solution away from $x = 0$, but only the uniform approximation works over the entire interval. The nonuniform approximation breaks down in the region near $x = 0$, in particular, in the region where $x = O(\varepsilon)$. This region is an example of what is known as a boundary layer, and these will be investigated extensively in the next chapter. ∎

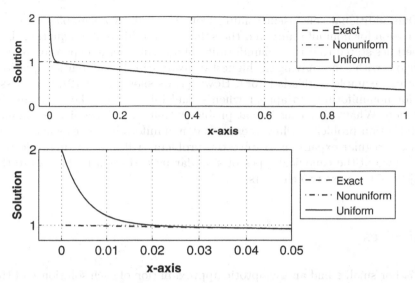

Figure 1.14 Comparison between the exact solution, (1.69), a nonuniform approximation, (1.71), and a uniform approximation, (1.70), of the solution. The *lower plot* shows the three functions in the region near $x = 0$. In both plots the exact and uniform curves are so close as to be indistinguishable from each other. In these calculations, $\varepsilon = 10^{-2}$

3. A solution of the differential equation

$$y'' + 2\varepsilon y' + y = 0, \quad \text{for } 0 < t, \tag{1.72}$$

is

$$y(t) = e^{-\varepsilon t} \sin\left(t\sqrt{1 - \varepsilon^2}\right). \tag{1.73}$$

Given that $\sqrt{1 - \varepsilon^2} \sim 1 - \frac{1}{2}\varepsilon^2$ and $e^{-\varepsilon t} \sim 1 - \varepsilon t$, it follows that

$$y(t) \sim e^{-\varepsilon t} \sin\left(t - \frac{1}{2}\varepsilon^2 t\right) \tag{1.74}$$

$$\sim (1 - \varepsilon t + \cdots)\left(\sin(t) - \frac{1}{2}\varepsilon^2 t \cos(t) + \cdots\right)$$

$$\sim \sin(t) - \varepsilon t \sin(t). \tag{1.75}$$

This example differs from the others because it concerns the approximation over an unbounded interval, where $0 < t < \infty$. In this sense Theorem 1.4 does not apply. However, the underlying ideas are still applicable. That is, in (1.75), as t increases, the second term eventually becomes as large as the first term, and this means the expansion is not well ordered. This is the core idea in nonuniformity and the reason that (1.75) is not a uniform asymptotic expansion for (1.73) for $0 \le t < \infty$. On the other hand, it is uniform over any bounded interval $0 \le t \le T$. ∎

One additional conclusion coming from the last two examples concerns the reason for nonuniformity and the solution of a differential equation. It is sometimes stated that the nonuniformity of a regular series expansion is synonymous with ε multiplying the highest derivative in the differential equation. A prime example of this is (1.68). However, as shown in (1.72), it is possible for nonuniformity to appear when ε multiplies some other term in the equation. What this means is that problems that we will classify as singular perturbation problems, which means that nonuniformity arises when trying to use a regular expansion to solve the problem, will arise for multiple reasons. Two of the canonical types of singular perturbation problems are the subject of the next two chapters.

Exercises

1.47. For small ε find an asymptotic approximation of each solution y of the following equations and determine if it is uniform (in x) over the indicated interval:

(a) $y^2 + (1 + \varepsilon - x)y - x = 0$, for $0 < x < 1$.

(b) $y^3 + (1 - 2x)y^2 + (x^2 - 2x - \varepsilon)y + x^2 = 0$, for $0 < x < 1$.

1.48. Show that (1.70) is a uniform approximation of (1.69) for $0 < x < 1$.

1.49. This problem investigates the time variable and its effect on the uniformity of an expansion.

(a) Assuming $0 \leq t \leq T$, where T is independent of ε, show that $\sin(t)$ is a uniform approximation of (1.73) and has error that is $O(\varepsilon)$.

(b) Assuming $0 \leq t \leq T$, where T is independent of ε, show that (1.74) is a uniform approximation of (1.73) and has error that is $O(\varepsilon^4)$. Also, explain why it is not uniform for $0 \leq t < \infty$.

(c) Explain why the expansion of the projectile problem given in (1.46) is uniform even though it contains a term like $\varepsilon\tau^3$.

1.50. If $f(\varepsilon) = 1$ and $\phi(\varepsilon) = 1 + \frac{x}{\varepsilon}e^{-x/\varepsilon}$, then $f \sim \phi$ for $0 \leq x \leq 1$ as $\varepsilon \downarrow 0$. Is this a uniform approximation? Does this violate Theorem 1.4?

1.51. For the function in (1.69) show that

$$\lim_{\varepsilon \downarrow 0} \lim_{x \downarrow 0} y(x) \neq \lim_{x \downarrow 0} \lim_{\varepsilon \downarrow 0} y(x).$$

This observation is an indication that (1.68) is a singular perturbation problem, and these types of problems are the subject of the next chapter.

1.52. For the following functions, assuming $\varepsilon \ll 1$, find an asymptotic expansion of the form $f(x, \varepsilon) \sim \phi(x)$ that holds for $-1 < x < 1$. Sketch $f(x, \varepsilon)$ and $\phi(x)$, and then explain why the approximation is not uniform for $-1 < x < 1$.

(a) $f(x, \varepsilon) = x + \exp((x^2 - 1)/\varepsilon)$.

(b) $f(x, \varepsilon) = x + \tanh(x/\varepsilon)$.

(c) $f(x, \varepsilon) = x + \operatorname{sech}(x/\varepsilon)$.

1.53. For the following functions, assuming $\varepsilon \ll 1$, find an asymptotic expansion of the form $f(t, \varepsilon) \sim \phi(t)$ that holds for $0 \le t < \infty$. Explain why the approximation is not uniform for $0 \le t < \infty$.

(a) $f(t, \varepsilon) = \exp(-\varepsilon t) \sin(t)$.

(b) $f(t, \varepsilon) = \sin(\omega t)$ where $\omega = 1 + \varepsilon$.

1.54. As an astute reader you probably noticed that the example used to introduce nonuniformity was never used in the subsequent examples. This exercise rectifies that oversight.

(a) Find the exact solution of (1.63) that corresponds to the expansion in (1.64).

(b) Find the first four terms in the expansion of the solution from part (a) in the case where $x = 1$. Comment on the differences between this expansion and that in (1.64).

(c) For both cases, the first-term approximation is $y \sim 1$. Is this uniform for $0 < x < 1$?

1.55. This problem explores different ways one can show that an approximation is not uniformly valid. Given a point x_0 satisfying $a \le x_0 \le b$, assume the following limits exist and are uniform:

$$\Phi(\varepsilon) = \lim_{x \to x_0} \phi(x, \varepsilon) \quad \text{and} \quad F(\varepsilon) = \lim_{x \to x_0} f(x, \varepsilon).$$

Also, assume that f and ϕ are continuous for $a < x < b$ and $0 < \varepsilon < \varepsilon_1$. In this case, if $f \sim \phi$ is uniformly valid for $a < x < b$, and if $\phi(x, \varepsilon)$ is bounded, then it must be that

$$\lim_{\varepsilon \downarrow 0} \Phi(\varepsilon) = \lim_{\varepsilon \downarrow 0} F(\varepsilon). \tag{1.76}$$

In this problem it is assumed that $x_0 = a$.

(a) The statement that $\Phi(\varepsilon) = \lim_{x \to a} \phi(x, \varepsilon)$ is uniform means that there is an interval $0 < \varepsilon < \varepsilon_1$ so that given any $\delta > 0$ it is possible to find η (independent of ε) so

$$|\phi(x, \varepsilon) - \Phi(\varepsilon)| < \delta \quad \text{if } a < x_0 < a + \eta.$$

Prove (1.76).

(b) Assuming the limit $\Phi(\varepsilon) = \lim_{x \to a} \phi(x, \varepsilon)$ exists, show that it is uniform if there is an interval $a < x < a + \eta$ such that $|\phi(x, \varepsilon) - \Phi(c)|$ is monotonically decreasing as a function of ε.

(c) Show that the condition in part (b) holds if $\phi(x, \varepsilon) \geq \Phi(\varepsilon)$ and

$$\frac{\mathrm{d}}{\mathrm{d}\varepsilon} (\phi(x, \varepsilon) - \Phi(\varepsilon)) \leq 0.$$

(d) Use (1.76) to show that $x + e^{-x/\varepsilon} \sim x$ is not uniform for $0 < x < 1$.

1.8 Symbolic Computing

Anyone who uses perturbation methods is struck almost immediately by the amount of algebra that is sometimes necessary to find an approximation. This raises the question of whether it might be possible to do this with a computer. Because one of the objectives of perturbation methods is to have an answer that contains the parameters in a problem, it is not convenient to use a computing language like FORTRAN or C to carry out these calculations since they require all variables (and parameters) to be evaluated. An alternative is to use a symbolic language like Maple or Mathematica. These systems are capable of manipulating mathematical expressions without having to evaluate parameters. This ability has a tremendous potential for constructing asymptotic expansions, and so we will illustrate some simple applications. More extensive presentations can be found in Boccara (2007) and Richards (2002).

The simplest example is in the use of Taylor's theorem. For example, suppose we want a two-term expansion of the function

$$f(\varepsilon) = \sinh\left(\frac{1}{10\alpha + \cos(\varepsilon)}\right),$$

where α is a constant. If we expect that the expansion of this function is a regular power series, that is,

$$f \sim f_0 + \varepsilon f_1 + \varepsilon^2 f_2 + \cdots,$$

then we can obtain the coefficients using a Taylor expansion. To do this, one must enter the formula for f, and in what follows ep designates ε and a represents α. In Maple, the command for the taylor series expansion is then

$$\mathrm{taylor}(f, \mathrm{ep}, 3). \tag{1.77}$$

In a fraction of a second the machine dutifully responds with

$$\frac{1}{2} \exp\left(\frac{1}{10a + 1}\right) - \frac{1}{2} \exp\left(-\frac{1}{10a + 1}\right)$$

$$+ \left(\frac{1}{4}\frac{\exp(\frac{1}{10a+1})}{(10a + 1)^2} + \frac{1}{4}\frac{\exp(-\frac{1}{10a+1})}{(10a + 1)^2}\right)\mathrm{ep}^2 + O(\mathrm{ep}^3).$$

Clearly, the effort involved here is minimal, and this makes the use of a symbolic manipulator quite attractive. However, at the same time certain cautions are necessary. It is natural to think that since this is so easy, why not go for lots of terms and change the 3 in (1.77) to, say, 1,000. There may actually be situations where something like this is necessary, but one would be running the risk of what is known as intermediate expression swell. This is the situation where the expressions that are being manipulated are so large that the machine is incapable of handling them with the allotted memory, even though the final result is relatively simple. In terms of asymptotic approximations, such Herculean feats as deriving 1,000 terms of an expansion is generally of little value. In fact, as we saw earlier, the results may actually get worse. Something else that should be pointed out is that even though these systems can be big time savers, there are certain things they do not do easily. One is simplify certain expressions. For example, the preceding expansion can be rewritten as

$$\sinh\left(\frac{1}{10a+1}\right) + \frac{\cosh(\frac{1}{10a+1})}{2(10a+1)^2}\,\mathrm{ep}^2 + O(\mathrm{ep}^3).$$

It can be very difficult to get the program to "notice" this fact. Also, the program does not warn the user that this expansion is singular when $10a+1 = 0$. The point here is that one must not stop thinking about the problem, or its solution, just because the work is being done by a machine.

   The expansions for the solutions of many of the algebraic and differential equations studied in this chapter can be solved quite easily using a symbolic manipulator. To illustrate, consider the projectile problem (1.41). The Maple commands that can be used to construct the expansion are given in Table 1.1. The CPU time needed to carry out these commands depends on the machine being used; however, it should not be more than about a couple of seconds. Also, even though the example presented here uses Maple, it is possible to carry out similar calculations using most other symbolic programs.

   It is strongly recommended that anyone using asymptotic expansions experiment with a symbolic computing system using a program like the one given previously.

Exercises

1.56. Given a function $f(\varepsilon)$, suppose its expansion has the form $f(\varepsilon) \sim a_1\varepsilon^{\alpha} + a_2\varepsilon^{\beta} + \cdots$.
(a) Determine a sequence of operations that will find (in order) α, a_1, β, and a_2.
(b) Using a symbolic manipulator employ your procedure to try to find the expansions for the functions of Exercise 1.7.

Command	Explanation
`y := y0(t) + ep * y1(t) + ep * ep * y2(t);`	As in (1.43), this defines the terms in the expansion.
`de := diff(y, t, t) + 1/(1 + ep * y) ^ 2;`	As in (1.41), this defines the differential equation.
`DE := taylor(de, ep, 3);`	As in (1.44), this expands the equation in powers of ep.
`de0 := coeff(DE, ep, 0);`	Find the $O(1)$ equation.
`de1 := coeff(DE, ep, 1);`	Find the $O(ep)$ equation.
`y0(t) := rhs(dsolve(de0 = 0, y0(0) = 0, D(y0)(0) = 1, y0(t)));`	Solve the $O(1)$ problem using the initial conditions given in (1.42).
`y1(t) := rhs(dsolve(de1 = 0, y1(0) = 0, D(y1)(0) = 0, y1(t)));`	Solve the $O(ep)$ problem.

Table 1.1 Maple commands that can be used to find a two-term expansion of the projectile problem (1.41), (1.42)

1.57. Consider the boundary value problem

$$y'' = f(\varepsilon, x, y, y') \text{ for } a < x < b,$$

where $y(a) = \alpha$ and $y(b) = \beta$. Assume f is a smooth function and the constants a, b, α, and β are independent of ε. It is also assumed that the problem has a unique solution.
(a) Construct an algorithm that will find the first two terms in the expansion of y for small ε.
(b) Use your procedure to find the expansions for the problems of Exercise 1.18(a)–(c).

Chapter 2
Matched Asymptotic Expansions

2.1 Introduction

The ideas underlying an asymptotic approximation appeared in the early 1800s when there was considerable interest in developing formulas to evaluate special functions. An example is the expansion of Bessel's function, given in (1.15), that was derived by Poisson in 1823. It was not until later in the century that the concept of an asymptotic solution of a differential equation took form, and the most significant efforts in this direction were connected with celestial mechanics. The subject of this chapter, what is traditionally known as matched asymptotic expansions, appeared somewhat later. Its early history is strongly associated with fluid mechanics and, specifically, aerodynamics. The initial development of the subject is credited to Prandtl (1905), who was concerned with the flow of a fluid past a solid body (such as an airplane wing). The partial differential equations for viscous fluid flow are quite complicated, but he argued that under certain conditions the effects of viscosity are concentrated in a narrow layer near the surface of the body. This happens, for example, with air flow across an airplane wing, and a picture of this situation is shown in Fig. 2.1. This observation allowed Prandtl to go through an order-of-magnitude argument and omit terms he felt to be negligible in the equations. The result was a problem that he was able to solve. This was a brilliant piece of work, but it relied strongly on his physical intuition. For this reason there were numerous questions about his reduction that went unresolved for decades. For example, it was unclear how to obtain the correction to his approximation, and it is now thought that Prandtl's derivation of the second term is incorrect (Lagerstrom, 1988). This predicament was resolved when Friedrichs (1941) was able to show how to systematically reduce a boundary-layer problem. In analyzing a model problem (Exercise 2.1) he used a stretching transformation to match inner and outer

M.H. Holmes, *Introduction to Perturbation Methods*, Texts in Applied
Mathematics 20, DOI 10.1007/978-1-4614-5477-9_2,
© Springer Science+Business Media New York 2013

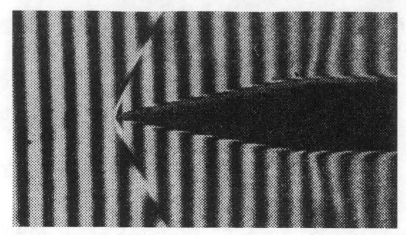

Figure 2.1 Supersonic air flow, at Mach 1.4, over a wedge. The high speed flow results in a shock layer in front of the wedge across which the pressure undergoes a rapid transition. Because of its position in the flow, the shock is an example of an interior layer (Sect. 2.5). There are also boundary layers present (Sect. 2.2). These can be seen near the surface of the wedge; they are thin regions where the flow drops rapidly to zero (which is the speed of the wedge). From Bleakney et al. (1949)

solutions, which is the basis of the method that is discussed in this chapter. This procedure was not new, however, as demonstrated by the way in which Gans (1915) used some of these ideas to solve problems in optics.

The golden age for matched asymptotic expansions was in the 1950s, and it was during this period that the method was refined and applied to a wide variety of physical problems. A short historical development of the method is presented in O'Malley (2010). The popularity of matched asymptotic expansions was also greatly enhanced with the appearance of two very good books, one by Cole (1968), the other by Van Dyke (1975). The method is now one of the cornerstones of applied mathematics. At the same time it is still being extended, both in the type of problems it is used to resolve as well as in the theory.

2.2 Introductory Example

The best way to explain the method of matched asymptotic expansions is to use it to solve a problem. The example that follows takes several pages to complete because it is used to introduce the ideas and terminology. As the procedure becomes more routine, the derivations will become much shorter.

The problem we will study is

$$\varepsilon y'' + 2y' + 2y = 0, \quad \text{for } 0 < x < 1, \tag{2.1}$$

where

$$y(0) = 0 \qquad (2.2)$$

and

$$y(1) = 1. \qquad (2.3)$$

This equation is similar to that used in Sect. 1.7 to discuss uniform and nonuniform approximations. The difference is that we will now derive the approximation directly from the problem rather than from a formula for the solution (Exercise 2.7).

An indication that this problem is not going to be as straightforward as the differential equations solved in Sect. 1.6 is that if $\varepsilon = 0$, then the problem is no longer second order. This leads to what is generally known as a singular perturbation problem, although singularity can occur for other reasons. In any case, to construct a first-term approximation of the solution for small ε, we will proceed in four steps. The fifth step will be concerned with the derivation of the second term in the expansion.

2.2.1 Step 1: Outer Solution

To begin, we will assume that the solution can be expanded in powers of ε. In other words,

$$y(x) \sim y_0(x) + \varepsilon y_1(x) + \cdots . \qquad (2.4)$$

Substituting this into (2.1) we obtain

$$\varepsilon(y_0'' + \varepsilon y_1'' + \cdots) + 2(y_0' + \varepsilon y_1' + \cdots) + 2(y_0 + \varepsilon y_1 + \cdots) = 0.$$

The $O(1)$ equation is therefore

$$y_0' + y_0 = 0, \qquad (2.5)$$

and the general solution of this is

$$y_0(x) = ae^{-x}, \qquad (2.6)$$

where a is an arbitrary constant. Looking at the solution in (2.6) we have a dilemma because there is only one arbitrary constant but two boundary conditions – (2.2), (2.3). What this means is that the solution in (2.6) and the expansion in (2.4) are incapable of describing the solution over the entire interval $0 \le x \le 1$. At the moment we have no idea which boundary condition, if any, we should require $y_0(x)$ to satisfy, and the determination of this will have to come later. This leads to the question of what to do next. Well, (2.6) is similar to using (1.71) to approximate the solution of (1.68). In looking at the comparison in Fig. 1.14, it is a reasonable working hypothesis to assume that (2.6) describes the solution over most of the interval, but there is a boundary layer at either $x = 0$ or $x = 1$, where a different approximation must be used. Assuming for the moment that it is at $x = 0$, and in looking at the reduction from (1.70) to (1.71), then we are probably missing a term like $e^{-\beta x/\varepsilon}$. The derivation of this term is the objective of the next step. Because we are going to end up with approximations of the solution over different regions, we will refer to (2.6) as the first term in the expansion of the *outer solution*.

2.2.2 Step 2: Boundary Layer

Based on the assumption that there is a boundary layer at $x = 0$, we introduce a *boundary-layer coordinate* given as

$$\bar{x} = \frac{x}{\varepsilon^\alpha}, \tag{2.7}$$

where $\alpha > 0$. From our earlier discussion it is expected that $\alpha = 1$, and this will be demonstrated conclusively subsequently. After changing variables from x to $\bar{x}$ we will take $\bar{x}$ to be fixed when expanding the solution in terms of ε. This has the effect of stretching the region near $x = 0$ as ε becomes small. Because of this, (2.7) is sometimes referred to as a *stretching transformation*.

From the change of variables in (2.7), and from the chain rule, we have that

$$\frac{\mathrm{d}}{\mathrm{d}x} = \frac{\mathrm{d}\bar{x}}{\mathrm{d}x}\frac{\mathrm{d}}{\mathrm{d}\bar{x}} = \frac{1}{\varepsilon^\alpha}\frac{\mathrm{d}}{\mathrm{d}\bar{x}}.$$

Letting $Y(\bar{x})$ denote the solution of the problem when using this boundary-layer coordinate, (2.1) transforms to

$$\varepsilon^{1-2\alpha}\frac{\mathrm{d}^2 Y}{\mathrm{d}\bar{x}^2} + 2\varepsilon^{-\alpha}\frac{\mathrm{d}Y}{\mathrm{d}\bar{x}} + 2Y = 0, \tag{2.8}$$

where, from (2.2),

$$Y(0) = 0. \tag{2.9}$$

The boundary condition at $x = 0$ has been included here because the boundary layer is at the left end of the interval.

The appropriate expansion for the *boundary-layer solution* is now

$$Y(\bar{x}) \sim Y_0(\bar{x}) + \varepsilon^\gamma Y_1(\bar{x}) + \cdots, \tag{2.10}$$

where $\gamma > 0$. As stated previously, in this expansion $\bar{x}$ is held fixed as ε goes to zero (in the same way that x is held fixed in the outer expansion). Substituting the expansion in (2.10) into (2.8) we get that

$$\varepsilon^{1-2\alpha}\frac{d^2}{d\bar{x}^2}(Y_0 + \cdots) + 2\varepsilon^{-\alpha}\frac{d}{d\bar{x}}(Y_0 + \cdots) + 2(Y_0 + \cdots) = 0. \qquad (2.11)$$

$$\underset{①}{} \qquad\qquad\qquad \underset{②}{} \qquad\qquad\qquad \underset{③}{}$$

Just as with the algebraic equations studied in Sect. 1.5, it is now necessary to determine the correct balancing in (2.11). The balance between terms ② and ③ was considered in Step 1, and so the following possibilities remain (also see Exercise 2.6):

(i) ① $\sim$ ③ and ② is higher order.
   The condition ① $\sim$ ③ requires that $1 - 2\alpha = 0$, and so $\alpha = \frac{1}{2}$. With this we have that ①,③ $= O(1)$ and ② $= O(\varepsilon^{-1/2})$. This violates our assumption that ② is higher order (i.e., ② $\ll$ ①), so this case is not possible.

(ii) ① $\sim$ ② and ③ is higher order.
   The condition ① $\sim$ ② requires that $1 - 2\alpha = -\alpha$, and so $\alpha = 1$. With this we have that ①,② $= O(\varepsilon^{-1})$ and ③ $= O(1)$. In this case, the conclusion is consistent with the original assumptions, and so this is the balancing we are looking for.

With this we have the following problem to solve:

$$O(\tfrac{1}{\varepsilon}) \quad Y_0'' + 2Y_0' = 0 \quad \text{for} \ 0 < \bar{x} < \infty,$$
$$Y_0(0) = 0.$$

The general solution of this problem is

$$Y_0(\bar{x}) = A(1 - e^{-2\bar{x}}), \qquad (2.12)$$

where A is an arbitrary constant. It should be observed that the differential equation for Y_0 contains at least one term of the outer-layer Eq. (2.5). This is important for the successful completion of Step 3.

The boundary-layer expansion in (2.10) is supposed to describe the solution in the immediate vicinity of the endpoint $x = 0$. It is therefore not unreasonable to expect that the outer solution (2.6) applies over the remainder of the interval (this is assuming there are no other layers). This means that the outer solution should satisfy the boundary condition at $x = 1$. From (2.6) and (2.3) one finds that

$$y_0(x) = e^{1-x}. \qquad (2.13)$$

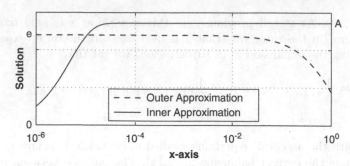

Figure 2.2 Sketch of inner solution, (2.12), and outer solution, (2.13). Note the overlap region along the x-axis where both solutions are essentially constant. Since these approximations are supposed to be describing the same continuous function, it must be that these constants are the same. Hence, $A = e^1$

2.2.3 Step 3: Matching

It remains to determine the constant A in the first-term approximation of the boundary-layer solution (2.12). To do this, the approximations we have constructed so far are summarized in Fig. 2.2. The important point here is that both the inner and outer expansions are approximations of the same function. Therefore, in the transition region between the inner and outer layers we should expect that the two expansions will give the same result. This is accomplished by requiring that the value of Y_0 as one comes out of the boundary layer (i.e., as $\bar{x} \to \infty$) is equal to the value of y_0 as one comes into the boundary layer (i.e., as $x \to 0$). In other words, we require that

$$\lim_{\bar{x} \to \infty} Y_0 = \lim_{x \to 0} y_0. \tag{2.14}$$

In this text, the preceding equation will usually be written in the more compact form of $Y_0(\infty) = y_0(0^+)$. However it is expressed, this is an example of a matching condition, and from it we find that $A = e^1$. With this (2.12) becomes

$$Y_0(\bar{x}) = e^1 - e^{1-2\bar{x}}. \tag{2.15}$$

This completes the derivation of the inner and outer approximations of the solution of (2.1). The last step is to combine them into a single expression.

Before moving on to the construction of the composite expansion, a word of caution is needed about the matching condition given in (2.14). Although we will often use this condition, it is limited in its applicability, and for more complex problems a more sophisticated matching procedure is often required. This will be discussed in more detail once the composite expansion is calculated.

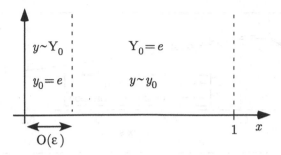

Figure 2.3 Sketch of inner and outer regions and values of approximations in those regions

2.2.4 Step 4: Composite Expansion

Our description of the solution consists of two pieces, one that applies near $x = 0$, the other that works everywhere else. Because neither can be used over the entire interval, they are not uniformly valid for $0 \le x \le 1$. The question we consider now is whether it is possible to combine them to produce a uniform approximation, that is, one that works over the entire interval. The position we are in is summarized in Fig. 2.3. The inner and outer solutions are constant outside their intervals of applicability, and the constant is the same for both solutions. The value of the constant can be written as either $y_0(0)$ or $Y_0(\infty)$, and the fact that they are equal is a consequence of the matching condition (2.14). This observation can be used to construct a uniform approximation, namely, we just add the approximations together and then subtract the part that is common to both. The result is

$$y \sim y_0(x) + Y_0\left(\frac{x}{\varepsilon}\right) - y_0(0)$$
$$\sim e^{1-x} - e^{1-2x/\varepsilon}. \tag{2.16}$$

The fact that the composite expansion gives a very good approximation of the solution over the entire interval is shown in Fig. 2.4. Note, however, that it satisfies the boundary condition at $x = 0$ exactly, but the one at $x = 1$ is only satisfied asymptotically. This is not of particular concern since the expansion also satisfies the differential equation in an asymptotic sense. However, an alternative expansion that satisfies both boundary conditions is developed in Exercise 2.14.

2.2.5 Matching Revisited

Because of the importance of matching, the procedure needs to be examined in more detail. The fact is that, even though the matching condition in

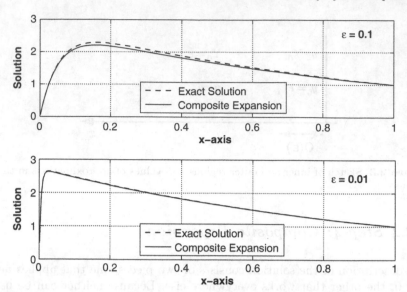

Figure 2.4 Graph of exact solution of (2.1) and composite expansion given in (2.16) in the case where $\varepsilon = 10^{-1}$ and where $\varepsilon = 10^{-2}$. Note the appearance of the boundary layer, as well as the convergence of the composite expansion to the solution of the problem, as ε decreases

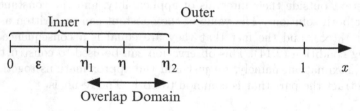

Figure 2.5 Schematic of domains of validity of inner and outer expansions as assumed in matching procedure. The intermediate variable is to be located within the overlap region

(2.14) is often used when calculating the first term in the expansion, there are situations where it is inapplicable. One of the more common examples of this occurs when either the inner or outer expansions are unbounded functions of their respective variable, so its limit does not exist. Examples of this will appear later in the chapter (e.g., Sect. 2.6), as well in succeeding chapters (e.g., Sect. 4.3). Another complication arises when constructing the second term in an expansion. What this all means is that we need a more sophisticated matching procedure, and finding one is the objective of this section.

The fundamental idea underlying matching concerns the overlap, or transition, region shown in Fig. 2.2. To connect the expansions on either side of this region, we introduce an intermediate variable $x_\eta = x/\eta(\varepsilon)$ that is positioned within this region. In particular, it is between the $O(1)$ coordinate of the outer layer and the $O(\varepsilon)$ coordinate of the inner layer (Fig. 2.5). This means that $\eta(\varepsilon)$ satisfies $\varepsilon \ll \eta \ll 1$. To match the expansions, the inner and outer approximations must give the same result when they are expressed in this transition layer coordinate.

The precise conditions imposed on $\eta(\varepsilon)$ and on the expansions are stated explicitly in the following matching procedure:

(i) Change variables in the outer expansion (from x to x_η) to obtain y_{outer}. It is assumed that there is an $\eta_1(\varepsilon)$ such that y_{outer} still provides a first-term expansion of the solution for any $\eta(\varepsilon)$ that satisfies $\eta_1(\varepsilon) \ll \eta(\varepsilon) \le 1$.

(ii) Change variables in the inner expansion (from $\bar{x}$ to x_η) to obtain y_{inner}. It is assumed that there is an $\eta_2(\varepsilon)$ such that y_{inner} still provides a first-term expansion of the solution for any $\eta(\varepsilon)$ that satisfies $\varepsilon \le \eta(\varepsilon) \ll \eta_2$.

(iii) It is assumed that the domains of validity of the expansions for y_{outer} and y_{inner} overlap, that is, $\eta_1 \ll \eta_2$. In this overlap region, for the expansions to match, it is required that the first terms from y_{outer} and y_{inner} be equal.

The assumptions contained in (i) and (ii) can be proved under fairly mild conditions; they are the essence of Kaplun's extension theorem (Lagerstrom, 1988). Actually, one is seldom interested in determining η_1 or η_2 but only that there is an interval for $\eta(\varepsilon)$ such that y_{inner} and y_{outer} match. It is important, however, that the matching not depend on the specific choice of $\eta(\varepsilon)$. For example, if one finds that matching can only occur if $\eta(\varepsilon) = \varepsilon^{1/2}$, then there is no overlap domain, and the procedure is not applicable. In comparison to the situation for (i) and (ii), the assumption on the existence of an overlap domain in (iii) is a different matter, and a satisfactory proof has never been given. For this reason it has become known as Kaplun's hypothesis on the domain of validity (Lagerstrom, 1988).

Examples

1. To use the foregoing matching procedure on our example problem, we introduce the intermediate variable x_η, defined as

$$x_\eta = \frac{x}{\varepsilon^\beta}, \tag{2.17}$$

where $0 < \beta < 1$. This interval for β comes from the requirement that the scaling for the intermediate variable must lie between the outer scale, $O(1)$, and the inner scale, $O(\varepsilon)$. Actually, it may be that in carrying out the matching of y_{inner} and y_{outer} we must reduce this interval for β. To see if the expansions match, note that the inner solution, from (2.10) and (2.12), becomes

$$y_{\text{inner}} \sim A(1 - e^{-2x_\eta/\varepsilon^{1-\beta}}) + \cdots$$

$$\sim A + \cdots, \tag{2.18}$$

and the outer solution, from (2.4) and (2.13), becomes

$$y_{\text{outer}} \sim e^{1-x_\eta \varepsilon^\beta} + \cdots$$

$$\sim e^1 + \cdots. \tag{2.19}$$

The expansions in (2.18) and (2.19) are supposed to agree to the first term in the overlap domain, and therefore $A = e^1$. ∎

2. Suppose that

$$y = \sqrt{1 + x + \frac{\varepsilon}{\varepsilon + x}}, \tag{2.20}$$

where $0 \leq x \leq 1$. The outer expansion is found by fixing x, with $0 < x \leq 1$, and expanding for small ε to obtain

$$y \sim \sqrt{1 + x} + \frac{\varepsilon}{2x\sqrt{1 + x}} + \cdots. \tag{2.21}$$

The boundary-layer expansion is found by setting $\bar{x} = x/\varepsilon$ and expanding to obtain

$$Y \sim \sqrt{\frac{2 + \bar{x}}{1 + \bar{x}}} + \frac{1}{2}\varepsilon\bar{x}\sqrt{\frac{1 + \bar{x}}{2 + \bar{x}}} + \cdots. \tag{2.22}$$

Given that (2.21) and (2.22) are expansions for a known function, we do not use matching to determine an unknown constant as in the previous example. Rather, the objectives here are to demonstrate how to use an intermediate variable to match the first two terms in an expansion and to provide an example that shows that condition (2.14), although very useful, has limited applicability. Its limitations are evident in this example by looking at what happens to the second term in the expansions when letting $x \to 0$ in (2.21) and letting $\bar{x} \to \infty$ in (2.22). Because both terms become unbounded, it is necessary to use a more refined matching method than a simple limit condition. To verify the matching principle, we substitute (2.17) into (2.21) to obtain

$$y_{\text{outer}} \sim \sqrt{1 + \varepsilon^\beta x_\eta} + \frac{\varepsilon^{1-\beta}}{2x_\eta\sqrt{1 + \varepsilon^\beta x_\eta}} + \cdots$$

$$\sim 1 + \frac{1}{2}\varepsilon^\beta x_\eta + \cdots + \varepsilon^{1-\beta}\frac{1}{2x_\eta} + \cdots.$$

Similarly, from (2.22) we get

$$y_{\text{inner}} \sim 1 + \varepsilon^{1-\beta} \frac{1}{2x_\eta} + \cdots + \frac{1}{2}\varepsilon^\beta x_\eta + \cdots .$$

Comparing these two expansions it is evident that they match. Another observation is that there is a strong coupling between the various terms in the expansions. For example, the first term from y_1 is what matches with the second term coming from Y_0. This coupling can cause difficulties in determining if expansions match, particularly when computing several terms in an expansion. A hint of this will be seen when calculating the second term in the next example. ∎

The interested reader may consult Lagerstrom (1988) for a more extensive discussion of the subtleties of matching using an intermediate variable. Also, there are other ways to match, and a quite popular one, due to Van Dyke (1975), is discussed in Exercise 2.12. His procedure is relatively simple to use but can occasionally lead to incorrect results (Fraenkel, 1969).

As the final comment about matching, it is a common mistake to think that it is equivalent to the requirement that y_0 and Y_0 must intersect. To show that this is incorrect, note that $Y_0 = 1 - e^{-x/\varepsilon}$ and $y_0 = 1 + x$ are inner and outer approximations, respectively, of $y = 1 + x - e^{-x/\varepsilon}$. However, even though $Y_0(\infty) = y_0(0^+)$, the two function never intersect.

2.2.6 Second Term

Generally, to illustrate a method, we will only derive the first term in an expansion. However, the second term is important as it gives a measure of the error. The procedure to find the second term is very similar to finding the first, so only the highlights will be given here.

Substituting the outer expansion (2.4) into the problem and collecting the $O(\varepsilon)$ terms one finds that $y_1' + y_1 = -\frac{1}{2}y_0''$, with $y_1(1) = 0$. The solution of this problem is

$$y_1 = \frac{1}{2}(1 - x)e^{1-x}.$$

Similarly, from the boundary-layer Eq. (2.11) we get that $\gamma = 1$ and $Y_1'' + 2Y_1' = -2Y_0$, with $Y_1(0) = 0$. The general solution of this is

$$Y_1 = B(1 - e^{-2\bar{x}}) - \bar{x}e^1(1 + e^{-2\bar{x}}),$$

where B is an arbitrary constant. To match the expansions, we use the intermediate variable given in (2.17). The outer expansion in this case takes the form

$$y_{\text{outer}} \sim e^{1-x_\eta \varepsilon^\beta} + \frac{\varepsilon}{2}(1 - x_\eta \varepsilon^\beta)e^{1-x_\eta \varepsilon^\beta} + \cdots$$

$$\sim e^1 - \varepsilon^\beta x_\eta e^1 + \frac{1}{2}\varepsilon e^1 + \frac{1}{2}\varepsilon^{2\beta}e^1 x_\eta^2 + \cdots, \tag{2.23}$$

and, setting $\xi = -2x_\eta/\varepsilon^{1-\beta}$, the boundary-layer expansion becomes

$$y_{\text{inner}} \sim e^1(1 - e^\xi) + \varepsilon\left[B(1 - e^\xi) - \frac{x_\eta e^1}{\varepsilon^{1-\beta}}(1 + e^\xi)\right] + \cdots$$

$$\sim e^1 - \varepsilon^\beta x_\eta e^1 + B\varepsilon + \cdots. \tag{2.24}$$

Matching these we get that $B = \frac{1}{2}e^1$. Note, however, that these expansions do not appear to agree since (2.23) contains a $O(\varepsilon^{2\beta})$ term that (2.24) does not have. To understand why this occurs, note that both expansions produce a $O(\varepsilon^\beta)$ term that does not contain an arbitrary constant. If this term is not identical for both expansions, then there is no way the expansions will match. In the outer expansion this term comes from the $O(1)$ problem, and in the boundary layer it comes from the $O(\varepsilon)$ solution. In a similar manner, one finds that the x_η^2 term in (2.23) also comes from the first term. However, for the boundary layer it comes from the $O(\varepsilon^2)$ problem (the verification of this is left as an exercise). Therefore, the expansions match.

It is now possible to construct a two-term composite expansion. The basic idea is to add expansions and then subtract the common part. This yields the following result:

$$y \sim y_0 + \varepsilon y_1 + Y_0 + \varepsilon Y_1 - \left(e^1 - x_\eta e^1\sqrt{\varepsilon} + \frac{\varepsilon}{2}e^1\right)$$

$$\sim e^{1-x} - (1 + x)e^{1-2x/\varepsilon} + \frac{\varepsilon}{2}\left[(1 - x)e^{1-x} - e^{1-2x/\varepsilon}\right].$$

Note that the common part in this case contains the terms in (2.23) and (2.24) except for the x_η^2 term in (2.23).

Occasionally it happens that an expansion (inner or outer) produces a term of an order, or form, that the other does not have. A typical example of this occurs when trying to expand in powers of ε. It can happen that to be able to match one expansion with another expansion from an adjacent layer, it is necessary to include other terms such as those involving $\ln(\varepsilon)$. This process of having to insert scales into an expansion because of what is happening in another layer is called *switchbacking*. Some of the more famous examples of this involve logarithmic scales; these are discussed in Lagerstrom (1988). We will come across switchbacking in Sects. 2.4 and 6.9 when including transcendentally small terms in the expansions.

2.2.7 Discussion

The importance of matching cannot be overemphasized. Numerous assumptions went into the derivation of the inner and outer approximations, and matching is one of the essential steps that supports these assumptions. If they had not matched, it would have been necessary to go back and determine where the error had occurred. The possibilities when this happens are almost endless, but it would be helpful to start looking in the following places.

1. The boundary layer is at $x = 1$, not at $x = 0$. In this case the boundary-layer coordinate is

$$\bar{x} = \frac{x - 1}{\varepsilon^\alpha}. \tag{2.25}$$

   This will be considered in Sect. 2.3. For certain problems it may be necessary to replace the denominator with a function $\mu(\varepsilon)$, where $\mu(\varepsilon)$ is determined from balancing or matching. A heuristic argument to help determine the location of the boundary layer is given in Sect. 2.5.
2. There are boundary layers at both ends of an interval. See Sect. 2.3.
3. There is an interior layer. In this case the stretching transformation is

$$\bar{x} = \frac{x - x_0}{\varepsilon^\alpha}, \tag{2.26}$$

   where x_0 is the location of the layer (this may depend on ε) (Sects. 2.5 and 2.6).
4. The form of the expansion is incorrect. For example, the outer expansion may have the form $y \sim \xi(\varepsilon)y_0(x) + \cdots$, where ξ is a function determined from the balancing or matching in the problem; see Exercise 2.2(b) and Sect. 2.6.
5. The solution simply does not have a layer structure and other methods need to be used (Exercise 2.4 and Chap. 3).

Occasionally it happens that it is so unclear how to proceed that one may want to try to solve the problem numerically to get an insight into the structure of the solution. The difficulty with this is that the presence of boundary or interior layers can make it hard, if not nearly impossible, to obtain an accurate numerical solution. Nice illustrations of this can be found in Exercises 2.29 and 2.40, and by solving (2.105) with $k = 1$. Another way to help guide the analysis occurs when the problem originates from an application and one is able to use physical intuition to determine the locations of the layers. As an example, in solving the problem associated with Fig. 2.1 one would expect a boundary layer along the surface of the wedge and an interior layer at the location of the shock.

Exercises

2.1. The Friedrichs model problem for a boundary layer in a viscous fluid is
(Friedrichs, 1941)
$$\varepsilon y'' = a - y' \quad \text{for } 0 < x < 1,$$
where $y(0) = 0$, $y(1) = 1$, and a is a given positive constant with $a \neq 1$.
(a) After finding the first term of the inner and outer expansions, derive a
    composite expansion of the solution of this problem.
(b) Derive a two-term composite expansion of the solution of this problem.

2.2. Find a composite expansion of the solution of the following problems:
(a) $\varepsilon y'' + 2y' + y^3 = 0$ for $0 < x < 1$, where $y(0) = 0$ and $y(1) = 1/2$.
(b) $\varepsilon y'' + e^x y' + \varepsilon y = 1$ for $0 < x < 1$, where $y(0) = 0$ and $y(1) = 1$.
(c) $\varepsilon y'' + y(y' + 3) = 0$ for $0 < x < 1$, where $y(0) = 1$ and $y(1) = 1$.
(d) $\varepsilon y'' = f(x) - y'$ for $0 < x < 1$, where $y(0) = 0$ and $y(1) = 1$. Also, $f(x)$
    is continuous.
(e) $\varepsilon y'' + (1 + 2x)y' - 2y = 0$ for $0 < x < 1$, where $y(0) = \varepsilon$ and $y(1) = \sin(\varepsilon)$.
(f) $\varepsilon y'' + y' + y = \displaystyle\int_0^1 K(\varepsilon x, s)y(s)ds$ for $0 < x < 1$, where $y(0) = 1$ and
    $y(1) = -1$. Also, $K(x, s) = e^{-s(1+x)}$.
(g) $\varepsilon y'' = e^{\varepsilon y'} + y$ for $0 < x < 1$, where $y(0) = 1$ and $y(1) = -1$.
(h) $\varepsilon y'' - y^3 = -1 - 7x^2$ for $0 < x < 1$, where $y(0) = 0$ and $y(1) = 2$.

2.3. Consider the problem of solving
$$\varepsilon^2 y'' + ay' = x^2 \quad \text{for } 0 < x < 1,$$
where $y'(0) = \lambda$, $y(1) = 2$, and a and λ are positive constants.
(a) Find a first-term composite expansion for the solution. Explain why the
    approximation does not depend on λ.
(b) Find the second terms in the boundary layer and outer expansions and
    match them. Be sure to explain your reasoning in matching the two ex-
    pansions.

2.4. A small parameter multiplying the highest derivative does not guarantee that the solution will have a boundary layer for small values of ε. As
demonstrated in this problem, this can be due to the form of the differential
equation or the particular boundary conditions used in the problem.
(a) After solving each of the following problems, explain why the solution
    does not have a boundary layer.

   (i) $\varepsilon y'' + 2y' + 2y = 2(1 + x)$ for $0 < x < 1$, where $y(0) = 0$ and $y(1) = 1$.
   (ii) $\varepsilon^2 y'' + \omega^2 y = 0$ for $0 < x < 1$ and $\omega > 0$.

(b) Consider the equation $\varepsilon^2 y'' - xy' = 0$ for $0 < x < 1$. From the exact solution, show that there is no boundary layer if the boundary conditions are $y(0) = y(1) = 2$, while there is a boundary layer if the boundary conditions are $y(0) = 1$ and $y(1) = 2$.

2.5. It is possible for a solution to have boundary-layer-like properties, but the form of the expansions is by no means obvious. The following examples illustrate such situations.

(a) $\varepsilon^2 y'' = y'$ for $0 < x < 1$, where $y'(0) = -1$ and $y(1) = 0$. Solve this problem and explain why there is a boundary layer at $x = 1$ but the expansion for the outer region is not given by (2.4).

(b) $\varepsilon y' = (x - 1)y$ for $0 < x$, where $y(0) = 1$. There is a boundary layer at $x = 0$. Use the methods of this section to derive a composite expansion of the solution. Find the exact solution and explain why the approximation you derived does not work.

2.6.(a) For (2.11) consider the balance of ① $\gg$ ②, ③. This case is not a distinguished limit because the order (α) is not unique. Explain why the solutions from this region are contained in (2.12).

(b) Discuss the case of ② $\gg$ ①, ③ in conjunction with the outer solution (this also is not a distinguished limit).

2.7. The exact solution of (2.1)–(2.3) is

$$y(x) = \frac{e^{r_+ x} - e^{r_- x}}{e^{r_+} - e^{r_-}},$$

where $\varepsilon r_\pm = -1 \pm \sqrt{1 - 2\varepsilon}$. Obtain the inner, outer, and composite expansions directly from this formula.

2.8. Consider the problem

$$\varepsilon y'' + y' + xy = 0 \quad \text{for } \alpha(\varepsilon) < x < \beta(\varepsilon),$$

where $y(\alpha) = 1$ and $y(\beta) = 0$. One way to deal with this ε-dependent interval is to change coordinates and let $s = (x - \alpha)/(\beta - \alpha)$. This fixes the domain and puts the problem into a Lagrange-like viewpoint.

(a) Find the transformed problem.

(b) Assuming $\alpha \sim \varepsilon \alpha_1 + \cdots$ and $\beta \sim 1 + \varepsilon \beta_1$, find a first-term composite expansion of the solution of the transformed problem. Transform back to the variable x and explain why the first-term composite expansion is unaffected by the perturbed domain.

(c) Find the second term in the composite expansion of the solution of the transformed problem. Transform back to the variable x and explain how the two-term composite expansion is affected by the perturbed domain.

2.9. Consider the problem

$$\varepsilon y'' + p(x)y' + q(x)y = f(x) \quad \text{for } 0 < x < 1,$$

where $y(0) = \alpha$ and $y(1) = \beta$. Assume $p(x)$, $q(x)$, and $f(x)$ are continuous and $p(x) > 0$ for $0 \le x \le 1$.
(a) In the case where $f = 0$, show that

$$y \sim \beta \exp\left(\int_x^1 \frac{q(s)}{p(s)}\, ds \right) + \left[\alpha - \beta \exp\left(\int_0^1 \frac{q(s)}{p(s)}\, ds \right) \right] h(x),$$

where $h(x) = e^{-p(0)x/\varepsilon}$.
(b) In the case where $f = 0$, but using the WKB method [see Exercise 4.3(b)], one obtains the result in part (a) except that

$$h(x) = \frac{p(0)}{p(x)} \exp\left(\int_0^x \frac{q(s)}{p(s)}\, ds - \frac{1}{\varepsilon} \int_0^x p(s)\, ds \right).$$

In Ou and Wong (2003) it is stated that this, and not the expression in part (a), is the "correct asymptotic approximation." Comment on this statement. (Hint: If they are correct, then the material covered in this section can be ignored.)
(c) Find a composite expansion of the solution in the case where $f(x)$ is not zero.
(d) Suppose $p(x) < 0$ for $0 \le x \le 1$. Show that the transformation $\hat{x} = 1 - x$ and the result from part (a) can be used to obtain a composite expansion of the solution.

2.10. Consider the problem

$$\varepsilon y'' + 6\sqrt{x}\, y' - 3y = -3 \quad \text{for } 0 < x < 1,$$

where $y(0) = 0$ and $y(1) = 3$.
(a) Find a composite expansion of the problem.
(b) Find a two-term composite expansion.

2.11. This problem is concerned with the integral equation

$$\varepsilon y(x) = -q(x) \int_0^x [y(s) - f(s)] s\, ds \quad \text{for } 0 \le x \le 1,$$

where $f(x)$ is smooth and positive.
(a) Taking $q(x) = 1$ find a composite expansion of the solution $y(x)$.
(b) Find a composite expansion of the solution in the case where $q(x)$ is positive and continuous but not necessarily differentiable.

2.12. Another way to match inner and outer expansions comes from Van Dyke (1975). To understand the procedure, suppose two terms have been calculated in both regions and the boundary-layer coordinate is $\bar{x} = x/\varepsilon^\alpha$, then do the following:

(i) Substitute x/ε^α for $\bar{x}$ into the inner expansion and expand the result to two terms (with x fixed).
(ii) Substitute $\varepsilon^\alpha \bar{x}$ for x into the outer expansion and expand the result to two terms (with $\bar{x}$ fixed).

After the results from (i) and (ii) are rewritten in terms of x, the matching condition states that the two expansions should agree exactly (to two terms).

(a) Using this matching procedure find a two-term composite expansion of the solution of (2.1).
(b) Using this matching procedure find a two-term composite expansion of the solution of

$$\varepsilon y'' = f(x) - y' \quad \text{for } 0 < x < 1,$$

where $y(0) = 0$, $y(1) = 1$, and $f(x)$ is a given smooth function.

2.13. As seen in Fig. 2.4, in the boundary layer the solution of (2.1) is concave down (i.e., $y'' < 0$). This observation is useful for locating layers; this is discussed further in Sect. 2.5. However, not all boundary layers have strict concavity properties, and this problem considers such an example. The interested reader is referred to Howes (1978) for an extended discussion of this situation.
(a) Find a composite expansion for the solution of

$$\varepsilon^2 y'' = (x - y)(y - 2) \quad \text{for } 0 < x < 1,$$

where $y(0) = 3$ and $y(1) = 1$.
(b) Explain why the solution of this problem does not have a boundary layer that is strictly concave up or concave down but has one that might be identified as concave–convex.

2.14. Some consider it bothersome that a composite expansion generally does not satisfy boundary conditions exactly. One procedure that has been used to correct this situation is to note that the composite expansion for (2.1), before imposing boundary condition (2.3), is $y \sim a(e^{-x} - e^{-2x/\varepsilon})$. Substituting this into (2.3) we then find that $a = e^1/(1 - e^{-3/\varepsilon})$.
(a) This violates our assumption, as expressed in (2.4), that $y_0(x)$ is independent of ε. However, is the result still an asymptotic approximation of the solution for $0 \leq x \leq 1$?
(b) Use this idea to find a first-term composite expansion (that satisfies the boundary conditions exactly) for the solution of the problem

$$\varepsilon y'' = f(x) - y' \quad \text{for } 0 < x < 1,$$

where $y(0) = 0$, $y(1) = 1$ and $f(x)$ is a given smooth function.

2.3 Examples Involving Boundary Layers

Almost all of the principal ideas underlying the method of matched asymptotic expansions were introduced in the example of Sect. 2.2. In the remainder of this chapter these ideas are applied, and extended, to more complicated problems. The extensions considered in this section are what happens when there are multiple boundary layers, and what can happen when the problem is nonlinear.

Example 1

To investigate an example where there is a boundary layer at each end of an interval, consider the problem of solving

$$\varepsilon^2 y'' + \varepsilon x y' - y = -e^x \quad \text{for } 0 < x < 1, \tag{2.27}$$

where

$$y(0) = 2, \text{ and } y(1) = 1. \tag{2.28}$$

Unlike the example in Sect. 2.2, one of the coefficients of the preceding equation depends on x. This is not responsible for the multiple boundary layers, but, as we will see, it does result in different equations for each layer.

2.3.1 Step 1: Outer Expansion

The expansion of the solution in this region is the same as the last example, namely, $y \sim y_0 + \cdots$. From this and (2.27) one obtains the first-term approximation

$$y_0 = e^x. \tag{2.29}$$

Clearly this function is incapable of satisfying either boundary condition, an indication that there are boundary layers at each end. An illustration of our current situation is given in Fig. 2.6. The solid curve is the preceding approximation, and the boundary conditions are also shown. The dashed curves are tentative sketches of what the boundary-layer solutions look like.

2.3.2 Steps 2 and 3: Boundary Layers and Matching

For the left endpoint we introduce the boundary-layer coordinate $\bar{x} = x/\varepsilon^\alpha$, in which case (2.27) becomes

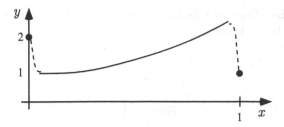

Figure 2.6 The *solid curve* is the outer approximation (2.29), and the *dashed curves* are guesses on how the boundary-layer solutions connect y_0 with the boundary conditions

$$\varepsilon^{2-2\alpha}\frac{d^2Y}{d\bar{x}^2} + \varepsilon\bar{x}\frac{dY}{d\bar{x}} - Y = -e^{\varepsilon^\alpha\bar{x}}. \qquad (2.30)$$

$$\underset{①}{} \qquad \underset{②}{} \quad \underset{③}{} \qquad \underset{④}{}$$

In preparation for balancing, note that

$$e^{\varepsilon^\alpha\bar{x}} \sim 1 + \varepsilon^\alpha\bar{x} + \cdots .$$

Also, as before, $Y(\bar{x})$ is used to designate the solution in this boundary-layer region. The balance in this layer is between terms ①, ③, ④, and so $\alpha = 1$. The appropriate expansion for Y is $Y \sim Y_0(\bar{x}) + \cdots$, and from (2.30) and the boundary condition at $\bar{x} = 0$ we have that

$$Y_0'' - Y_0 = -1 \quad \text{for } 0 < \bar{x} < \infty, \qquad (2.31)$$

where

$$Y_0(0) = 2. \qquad (2.32)$$

Note that (2.31) has at least one term in common with the equation for the outer region (which we should expect if there is to be any hope of matching the inner and outer expansions). The general solution is

$$Y_0(\bar{x}) = 1 + Ae^{-\bar{x}} + (1 - A)e^{\bar{x}}. \qquad (2.33)$$

This must match with the outer solution given in (2.29). The matching condition is $Y_0(\infty) = y_0(0)$, and so from (2.29) we have that $A = 1$.

To determine the solution in the boundary layer at the other end, we introduce the boundary-layer coordinate

$$\widetilde{x} = \frac{x - 1}{\varepsilon^\beta}. \qquad (2.34)$$

In this region we will designate the solution as $\widetilde{Y}(\widetilde{x})$. Introducing (2.34) into (2.27) one obtains the equation

$$\varepsilon^{2-2\beta}\frac{d^2\widetilde{Y}}{d\widetilde{x}^2} + (1 + \varepsilon\widetilde{x})\varepsilon^{1-\beta}\frac{d\widetilde{Y}}{d\widetilde{x}} - \widetilde{Y} = -e^{1+\varepsilon^\beta\widetilde{x}}. \qquad (2.35)$$

The distinguished limit in this case occurs when $\beta = 1$. So the expansion $\widetilde{Y} \sim \widetilde{Y}_0(\widetilde{x})$ yields the problem

$$\widetilde{Y}_0'' + \widetilde{Y}_0' - \widetilde{Y}_0 = -e \quad \text{for} \ -\infty < \widetilde{x} < 0, \tag{2.36}$$

where

$$\widetilde{Y}_0(0) = 1. \tag{2.37}$$

It is important to notice that this boundary-layer equation has at least one term in common with the equation for the outer region. In solving this problem the general solution is found to be

$$\widetilde{Y}_0(\widetilde{x}) = e + Be^{r_+\widetilde{x}} + (1 - e - B)e^{r_-\widetilde{x}}, \tag{2.38}$$

where $2r_\pm = -1 \pm \sqrt{5}$.

The matching requirement is the same as before, which is that when approaching the boundary layer from the outer region you get the same value as when you leave the boundary layer and approach the outer region. In other words, it is required that $\widetilde{Y}_0(-\infty) = y_0(1)$. Hence, from (2.38), $B = 1 - e$. As a final comment, note that boundary-layer Eq. (2.36) differs from (2.31), and this is due to the x dependence of the terms in the original problem.

2.3.3 Step 4: Composite Expansion

The last step is to combine the three expansions into a single expression. This is done in the usual way of adding the expansions together and then subtracting the common parts. From (2.29), (2.33), and (2.38), a first-term expansion of the solution over the entire interval is

$$y \sim y_0(x) + Y_0(\bar{x}) - Y_0(\infty) + \widetilde{Y}_0(\widetilde{x}) - \widetilde{Y}_0(-\infty)$$
$$\sim e^x + e^{-x/\varepsilon} + (1 - e)e^{-r(1-x)/\varepsilon}, \tag{2.39}$$

where $2r = -1 + \sqrt{5}$. This approximation is shown in Fig. 2.7 along with the numerical solution. One can clearly see the boundary layers at the endpoints of the interval as well as the accuracy of the asymptotic approximation as ε decreases. ∎

Example 2

Some interesting complications arise when using matched asymptotic expansions with nonlinear equations. For example, it is not unusual for the solution

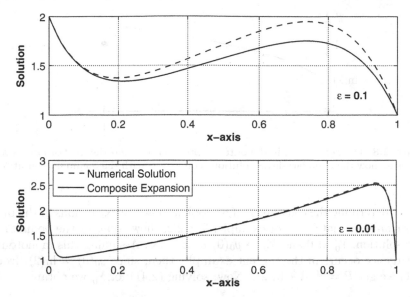

Figure 2.7 Graph of numerical solution of (2.27), (2.28) and the composite expansion given in (2.39) in the case where $\varepsilon = 10^{-1}$ and where $\varepsilon = 10^{-2}$. Note the appearance of the boundary layers, as well as the convergence of the composite expansion to the solution of the problem, as ε decreases

of a nonlinear equation to be defined implicitly. To understand this situation, consider the problem

$$\varepsilon y'' + \varepsilon y' - e^y = -2 - x \quad \text{for } 0 < x < 1, \tag{2.40}$$

where $y(0) = 0$ and $y(1) = 1$.

The first term in the outer expansion can be obtained by simply setting $\varepsilon = 0$ in (2.40). This yields $e^{y_0} = 2 + x$, and so $y \sim \ln(x + 2)$. Given that this does not satisfy either boundary condition, the outer expansion is assumed to hold for $0 < x < 1$. An illustration of our current situation is given in Fig. 2.8. The solid curve is the outer approximation, and the boundary conditions are also shown. The dashed curves are tentative sketches of what the boundary-layer solutions look like.

For the boundary layer at $x = 0$ the coordinate is $\bar{x} = x/\sqrt{\varepsilon}$, and one finds that $Y \sim Y_0$, where

$$Y_0'' - e^{Y_0} = -2$$

for $Y_0(0) = 0$. Multiplying the differential equation by Y_0' and integrating yields

$$\frac{1}{2}\left(\frac{d}{d\bar{x}}Y_0\right)^2 = B - 2Y_0 + e^{Y_0}. \tag{2.41}$$

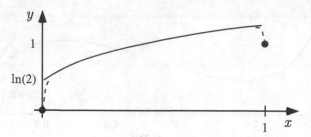

Figure 2.8 The *solid curve* is the outer approximation, and the *dashed curves* are guesses on how the boundary-layer solutions connect y_0 with the boundary conditions

We can determine the constant B if we look at how the expansions in the previous examples matched. For example, Fig. 2.2, shows that for the boundary-layer solution, $Y_0' \to 0$ and $Y_0 \to y_0(0)$ as $\bar{x} \to \infty$. Assuming this monotonic convergence occurs in the present example, then, since $y_0(0) = \ln(2)$, from (2.41) we get $B = 2[-1 + \ln(2)]$. Now, solving (2.41) for Y_0' we obtain

$$Y_0' = \pm\sqrt{2(B - 2Y_0 + e^{Y_0})}\,.$$

To determine which sign, it is evident from Fig. 2.8 that the boundary-layer solution increases from $Y_0(0) = 0$ to $Y_0(\infty) = y_0(0) = \ln(2)$, and so we take the $+$ sign. Doing this, separating variables, and then integrating yields the following result:

$$\int_0^{Y_0} \frac{ds}{\sqrt{2(B - 2s + e^s)}} = \bar{x}. \tag{2.42}$$

This is the solution of the $O(1)$ problem in the boundary layer and it defines Y_0 implicitly in terms of $\bar{x}$. It is important to note that the assumptions that were made to derive this result, such as $Y_0' \to 0$ and $Y_0 \to y_0(0)$ as $\bar{x} \to \infty$, hold for this solution.

The derivation of the expansion of the solution for the boundary layer at $x = 1$ is very similar (Exercise 2.26). The boundary-layer coordinate is $\tilde{x} = (x - 1)/\sqrt{\varepsilon}$. One finds that $\tilde{Y} \sim \tilde{Y}_0$, where

$$\int_1^{\tilde{Y}_0} \frac{ds}{\sqrt{2(A - 3s + e^s)}} = -\tilde{x} \tag{2.43}$$

for $A = 3[-1 + \ln(3)]$.

Even though the boundary-layer solutions are defined implicitly, it is still possible to write down a composite approximation. Adding the expansions together and then subtracting the common parts we obtain

$$y \sim y_0(x) + Y_0(\bar{x}) - y_0(0) + \tilde{Y}_0(\tilde{x}) - y_0(1)$$

$$\sim \ln\left(\frac{1}{6}(x+2)\right) + Y_0(\bar{x}) + \tilde{Y}_0(\tilde{x}).$$

A few additional comments and results for this example can be found in Exercise 2.26. ∎

Example 3

The ideas underlying matched asymptotic expansions are not limited to boundary-value problems. As an example, in studying the dynamics of auto-catalytic reactions one comes across the problem of solving (Gray and Scott, 1994)

$$\varepsilon \frac{du}{dt} = e^{-t} - uv^2 - u, \tag{2.44}$$

$$\frac{dv}{dt} = uv^2 + u - v, \tag{2.45}$$

where $u(0) = v(0) = 1$. In this case, there is an initial layer (near $t = 0$), as well as an outer solution that applies away from $t = 0$.

For the outer solution, assuming $u \sim u_0 + \varepsilon u_1 + \cdots$ and $v \sim v_0 + \varepsilon v_1 + \cdots$ one finds that the first-order problem is

$$0 = e^{-t} - u_0 v_0^2 - u_0,$$

$$\frac{d}{dt} v_0 = u_0 v_0^2 + u_0 - v_0.$$

The solution of this system is $v_0 = (t + a)e^{-t}$ and $u_0 = e^{-t}/(v_0^2 + 1)$, where a is a constant determined from matching.

For the initial layer, letting $\tau = t/\varepsilon$ and assuming that $U \sim U_0(\tau) + \cdots$ and $V \sim V_0(\tau) + \cdots$ one obtains the problem of solving

$$U_0' + \cdots = e^{-\varepsilon\tau} - (U_0 + \cdots)(V_0 + \cdots)^2 - (U_0 + \cdots),$$

$$V_0' + \cdots = \varepsilon\left[(U_0 + \cdots)(V_0 + \cdots)^2 + (U_0 + \cdots) - (V_0 + \cdots)\right].$$

From this it follows that

$$U_0' = 1 - U_0 V_0^2 - U_0,$$

$$V_0' = 0.$$

Solving these equations and using the given initial conditions that $U_0(0) = V_0(0) = 1$ one finds that $U_0 = \frac{1}{2}(1 + e^{-2\tau})$ and $V_0 = 1$.

The matching conditions are simply that $u_0(0) = U_0(\infty)$ and $v_0(0) = V_0(\infty)$, from which one finds that $a = 1$. Therefore, composite expansions for the first terms are

$$u \sim \frac{1}{2}e^{-2t/\varepsilon} + \frac{e^{-t}}{1 + (1+t)^2 e^{-2t}},$$
$$v \sim (1+t)e^{-t}.$$

The preceding derivation was relatively straightforward, but the time variable can cause some interesting complications. For example, the time interval is usually unbounded, and this can interfere with the well-ordering of an expansion, which is discussed extensively in Chap. 3. There are also questions related to the stability, or instability, of a steady-state solution, and this is considered in Chap. 6 . ■

Example 4

It is possible to have multiple layers in a problem and have them occur on the same side of an interval. An example of this arises when solving

$$\varepsilon^3 y'' + x^3 y' - \varepsilon y = x^3, \quad \text{for } 0 < x < 1, \tag{2.46}$$

where

$$y(0) = 1, \text{ and } y(1) = 3. \tag{2.47}$$

To help explain what happens in this problem, the numerical solution is shown in Fig. 2.9. It is the hook region, near $x = 0$, that is the center of attention in this example.

The outer expansion is $y \sim x + 2$, and this holds for $0 < x \le 1$. This linear function is clearly seen in Fig. 2.9. To investigate what is happening near $x = 0$, set $\bar{x} = x/\varepsilon^\alpha$. The differential equation in this case becomes

$$\varepsilon^{3-2\alpha}\frac{\mathrm{d}^2}{\mathrm{d}\bar{x}^2}Y + \varepsilon^{2\alpha}\bar{x}^3\frac{\mathrm{d}}{\mathrm{d}\bar{x}}Y - \varepsilon Y = \varepsilon^{3\alpha}\bar{x}^3. \tag{2.48}$$
$$\quad ① \qquad\qquad ② \qquad ③ \qquad ④$$

There are two distinguished limits (other than the one for the outer region). One comes when ① $\sim$ ③, in which case $\alpha = 1$. This is an inner–inner layer, and in Fig. 2.9 this corresponds to the monotonically decreasing portion of the curve in the immediate vicinity of $x = 0$. The other balance occurs when ② $\sim$ ③, in which case $\alpha = \frac{1}{2}$. This is an inner layer, and in Fig. 2.9 this corresponds to the region where the solution goes through its minimum value. Carrying out the necessary calculations, one obtains the following first-term composite expansion (Exercise 2.16):

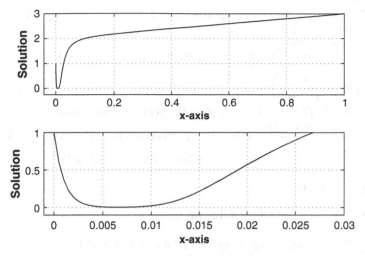

Figure 2.9 Numerical solution of (2.46) and (2.47), in the case where $\varepsilon = 10^{-3}$. The *lower plot* shows the solution in the hook region near $x = 0$

$$y \sim \begin{cases} x + e^{-x/\varepsilon} + 2e^{-\varepsilon/(2x^2)} & \text{for } 0 < x \le 1, \\ 1 & \text{for } \quad x = 0. \end{cases} \tag{2.49}$$

To reiterate a point made in the earlier examples, note that the equation for the inner–inner layer has no terms in common with the equation for the outer region, but the inner layer equation shares terms with both. ∎

Exercises

2.15. Find a composite expansion of the solution of the following problems and sketch the solution:

(a) $\varepsilon y'' + \varepsilon(x+1)^2 y' - y = x - 1$ for $0 < x < 1$, where $y(0) = 0$ and $y(1) = -1$.

(b) $\varepsilon y'' - y' + y^2 = 1$ for $0 < x < 1$, where $y(0) = 1/3$ and $y(1) = 1$.

(c) $\varepsilon y'' - e^x y = f(x)$ for $0 < x < 1$, where $y(0) = 1$ and $y(1) = -1$.

(d) $\varepsilon y'' + \varepsilon y' - y^2 = -1 - x^2$ for $0 < x < 1$, where $y(0) = 2$ and $y(1) = 2$.

(e) $\varepsilon y'' - y(y' + 1) = 0$ for $0 < x < 1$, where $y(0) = 3$ and $y(1) = 3$.

(f) $\varepsilon y'' + y(y' + 3) = 0$ for $0 < x < 1$, where $y(0) = 4$ and $y(1) = 4$.

(g) $\varepsilon y'' + y(1 - y)y' - y = 0$ for $0 < x < 1$, where $y(0) = -2$ and $y(1) = -2$.

(h) $\varepsilon^2 y'' + \varepsilon(3 - x^2)y' - 4y = 4x$ for $-1 < x < 1$, where $y(-1) = 2$ and $y(1) = 3$.

(i) $\varepsilon y'' - y(1 + y)y' - 3y = 0$ for $0 < x < 1$, where $y(0) = 2$ and $y(1) = 2$.

(j) $\varepsilon y'''' - (1+x)y' = 1 - x^2$ for $0 < x < 1$, where $y(0) = y'(0) = y(1) = y'(1) = 0$.

(k) $\varepsilon^3 y'' = \dfrac{(y - e^x)^3}{(1 + \varepsilon y')^2}$ for $0 < x < 1$, where $y(0) = 0$ and $y(1) = 4$.

(l) $\varepsilon \dfrac{d}{dx}\left(E(x)\dfrac{u'}{1 - \varepsilon u'}\right) - u = f(x)$ for $0 < x < 1$, where $u(0) = u(1) = 0$.
Also, $E(x)$, $f(x)$ are known, smooth, positive functions.

(m) $\varepsilon(x^2 y')' = \dfrac{x^2(y - 1)}{y}$ for $0 < x < 1$, where $y'(0) = 0$ and $y(1) = 2$.

(n) $(y^\varepsilon)'' - y = -e^x$ for $0 < x < 1$, where $y(0) = 2$ and $y(1) = 3$.

2.16. Derive (2.49).

2.17. Consider the boundary value problem

$$\varepsilon y'' - xy' - \kappa y = -1 \quad \text{for } -1 < x < 1,$$

where $y(-1) = y(1) = 0$. Assuming

$$\kappa = \int_{-1}^{1} y^2 dx,$$

find the first term in an expansion of κ for small ε.

2.18. The Reynolds equation from the gas lubrication theory for slider bearings is (DiPrima, 1968; Shepherd, 1978)

$$\varepsilon \frac{d}{dx}(H^3 yy') = \frac{d}{dx}(Hy) \quad \text{for } 0 < x < 1,$$

where $y(0) = y(1) = 1$. Here $H(x)$ is a known, smooth, positive function with $H(0) \neq H(1)$.
(a) Find a composite expansion of the solution for small ε. Note the boundary-layer solution will be defined implicitly, but it is still possible to match the expansions.
(b) Show that if the boundary layer is assumed to be at the opposite end from what you found in part (a), then the inner and outer expansions do not match.

2.19. This exercise considers the problem of a beam with a small bending stiffness. This consists in solving (Denoel and Detournay, 2010)

$$\varepsilon y'' = (1 - x)\sin y - \cos y \quad \text{for } 0 < x < 1,$$

where $y(0) = y(1) = \pi/2$. In this problem, y is an angular variable that is assumed to satisfy $0 \leq y \leq \pi$, and $\sin y + (1 - x)\cos y \geq 0$. Find the first term in the expansions in (i) the outer layer, (ii) the boundary layer at $x = 0$, and (iii) a composite expansion.

2.20. The Michaelis–Menten reaction scheme for an enzyme catalyzed reaction is (Holmes, 2009)

$$\frac{ds}{dt} = -s + (\mu + s)c,$$

$$\varepsilon\frac{dc}{dt} = s - (\kappa + s)c,$$

where $s(0) = 1$ and $c(0) = 0$. Here $s(t)$ is the concentration of substrate, $c(t)$ is the concentration of the chemical produced by the catalyzed reaction, and μ, κ are positive constants with $\mu < \kappa$. Find the first term in the expansions in (i) the outer layer, (ii) the initial layer, and (iii) a composite expansion.

2.21. The Poisson–Nernst–Planck model for flow of ions through a membrane consists of the following equations (Singer et al., 2008): for $0 < x < 1$,

$$\frac{dp}{dx} + p\frac{d\phi}{dx} = -\alpha,$$

$$\frac{dn}{dx} - n\frac{d\phi}{dx} = -\beta,$$

$$\varepsilon^2\frac{d^2\phi}{dx^2} = -p + n.$$

The boundary conditions are $\phi(0) = 1$, $\phi(1) = 0$, $p(0) = 4$, and $n(0) = 1$. In these equations, p and n are the concentrations of the ions with valency 1 and -1, respectively, and ϕ is the potential. Assume that α and β are positive constants that satisfy $\kappa < 1$, where

$$\kappa = \frac{\alpha + \beta}{2\sqrt{p(0)n(0)}}.$$

Also, you can assume that $\alpha \neq \beta$.
(a) Assuming there is a boundary layer at $x = 0$, derive the outer and boundary-layer approximations. Explain why, if the outer approximation for ϕ is required to satisfy $\phi(1) = 0$, the approximations you derived do not match.
(b) There is also a boundary layer at $x = 1$. Derive the resulting approximations and complete the matching you started in part (a). From this show that

$$p(1) \sim p(0)e^{\phi(0)}(1 - \kappa)^{2\beta/(\alpha+\beta)}$$

and

$$n(1) \sim n(0)e^{-\phi(0)}(1 - \kappa)^{2\alpha/(\alpha+\beta)}.$$

2.22. A modified version of the Grodsky model for insulin release is to find $y = y(t, \lambda)$, which satisfies (Carson et al., 1983)

$$\varepsilon \frac{dy}{dt} = -y + f(t) + \int_0^\infty y(t,s)e^{-\gamma s}ds \quad \text{for } 0 < t < \infty,$$

where $y = g(\lambda)$ when $t = 0$. Also, $\gamma > 1$.

(a) Find a composite expansion of the solution for small ε.

(b) Derive the composite expansion you obtained in part (a) from the exact solution, which is

$$y = \left[g(\lambda) - \gamma g_0 (1 - e^{t/\kappa}) \right] e^{-t/\varepsilon} + \frac{1}{\varepsilon} \int_0^t f(\tau) e^{-(\gamma-1)(t-\tau)/\kappa} d\tau,$$

where $\kappa = \varepsilon \gamma$ and $g_0 = \int_0^\infty g(s) e^{-\gamma s} ds$. Also, what is the composite expansion when $\gamma = 1$?

2.23. The eigenvalue problem for the vertical displacement, $u(x)$, of an elastic beam that is under tension is

$$\varepsilon^2 u'''' - u'' = \lambda u \quad \text{for } 0 < x < 1,$$

where $u = u' = 0$ at $x = 0, 1$. The question is, what values of λ produce a nonzero solution of the problem? In this context, λ is the eigenvalue, and it depends on ε.

(a) Find the first term in the expansions for $u(x)$ and λ.

(b) Find the second term in the expansions for $u(x)$ and λ.

2.24. Find a composite expansion of the solution of

$$\varepsilon^2 y'' + 2\varepsilon p(x)y' - q(x)y = f(x) \quad \text{for } 0 < x < 1,$$

where $y(0) = \alpha$ and $y(1) = \beta$. The functions $p(x)$, $q(x)$, and $f(x)$ are continuous and $q(x)$ is positive for $0 \le x \le 1$.

2.25. In the study of an ionized gas confined to a bounded domain Ω, the potential $\phi(\mathbf{x})$ satisfies

$$-\nabla^2 \phi + h\left(\frac{\phi}{\varepsilon}\right) = \alpha \quad \text{for } \mathbf{x} \in \Omega,$$

where conservation of charge requires

$$\int_\Omega h\left(\frac{\phi}{\varepsilon}\right) dV = \beta$$

and, assuming the exterior of the region is a conductor, $\partial_n \phi = \gamma$ on $\partial\Omega$. The function $h(s)$ is smooth and strictly increasing with $h(0) = 0$. The positive constants α and β are known (and independent of ε), and the constant γ is determined from the conservation of charge equation.

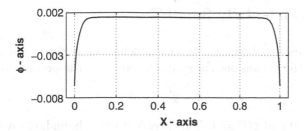

Figure 2.10 Solution of problem in Exercise 2.25 in the case where $\varepsilon = 10^{-3}$

(a) For the case of one dimension, suppose Ω is the interval $0 < x < 1$. What does the problem reduce to? Find γ in terms of α and β.

(b) Find the exact solution of the problem in part (a) when $h(s) = s$. Sketch the solution for $\gamma < 0$, and describe the boundary layers that are present.

(c) For the one-dimensional problem in part (a), find the first term in the inner and outer expansions of the solution. In doing this, take $h(s) = s^{2k+1}$, where k is a positive integer, and assume $\beta < \alpha$. For comparison the numerical solution is shown in Fig. 2.10 in the case where $k = 1$, $\alpha = 3$, $\beta = 1$, and $\varepsilon = 10^{-3}$.

(d) Discuss the steps needed to find a composite expansion involving the terms derived in part (c).

(e) For $\Omega \subset \mathbb{R}^n$, where $n > 1$, find γ.

2.26. This exercise examines Example 2 in more depth.

(a) Letting $f(s) = 2(B - 2s + e^s)$, where $B = 2[-1 + \ln(2)]$, sketch f for $-\infty < s < \infty$.

(b) Writing $f(s) = (\ln(2) - s)^2 g(s)$, show that $g(s)$ is positive and $g(\ln(2)) = 2$. It is also possible to show that $g(s)$ is monotone increasing for $-\infty < s < \infty$.

(c) With part (b), (2.42) can be written as

$$\int_0^{Y_0} \frac{ds}{(\ln(2) - s)\sqrt{g(s)}} = \bar{x}.$$

Use this to prove the monotonicity assumed in the derivation of (2.42).

(d) Derive (2.43).

(e) Use the ideas developed in parts (a) and (b) to show that (2.43) can be written as

$$\int_1^{\widetilde{Y}_0} \frac{ds}{(\ln(3) - s)\sqrt{h(s)}} = -\widetilde{x},$$

where h is positive with $h(\ln(3)) = 3$.

2.4 Transcendentally Small Terms

Even in the simplest boundary-layer problems, a question arises about what is missing in the expansion. For example, consider the problem of solving

$$\varepsilon y'' = 2 - y' \quad \text{for } 0 < x < 1, \tag{2.50}$$

where $y(0) = 0$ and $y(1) = 1$. This problem has a boundary layer at $x = 0$. For the outer approximation, assuming the usual power series expansion

$$y \sim y_0(x) + \varepsilon y_1(x) + \varepsilon^2 y_2(x) + \cdots, \tag{2.51}$$

one finds that $y_0 = 2x - 1$ and $y_1 = y_2 = \cdots = 0$. In other words, the entire outer expansion is just $y \sim 2x - 1$. Given that $y = 2x - 1$ is not the exact solution, the expansion (2.51) is missing something. Whatever this something is, it is transcendentally small compared to the power functions. Because of this, we have ignored this part of the approximation in all of the examples and exercises in the earlier sections of this chapter. There are, however, occasional situations where transcendentally small terms in the outer region must be accounted for. The objective here is to explore how this is done. The first example demonstrates the basic ideas, even though it is not necessary to include transcendentally small terms to obtain an accurate approximation. The second example is more challenging and involves a problem where such terms are used to complete the approximation.

Example 1

The first example is (2.50). As stated earlier, when assuming an outer expansion of the form (2.51), one finds that $y_0 = 2x - 1$ and $y_1 = y_2 = \cdots = 0$. For the boundary layer at $x = 0$, the expansion is found to be

$$Y(\bar{x}) \sim A_0\left(1 - e^{-\bar{x}}\right) + \varepsilon\left[2\bar{x} + A_1\left(1 - e^{-\bar{x}}\right)\right] + \varepsilon^2 A_2\left(1 - e^{-\bar{x}}\right) + \cdots, \tag{2.52}$$

where $\bar{x} = x/\varepsilon$. Note that the preceding expansion satisfies the boundary condition at $x = 0$ but has not yet been matched with the outer expansion.

To determine what the outer expansion is missing, we replace (2.51) with the assumption that

$$y \sim y_0 + \varepsilon y_1 + \varepsilon^2 y_2 + \cdots + z_0(x, \varepsilon) + z_1(x, \varepsilon) + \cdots, \tag{2.53}$$

where the z_i are well-ordered and transcendentally small compared to the power functions. Specifically, $z_j \ll z_i$, $\forall i < j$, and $z_i \ll \varepsilon^n$, $\forall i, n$. At this point we have no idea how z_i depends on ε, but we will determine this in the analysis to follow. It is important to note, however, that because (2.53) is the outer expansion, it holds for $\varepsilon \to 0$ with x held fixed with $0 < x \leq 1$.

Substituting (2.53) into (2.50) yields

$$\varepsilon(y_0'' + \varepsilon y_1'' + \cdots + z_0'' + z_1'' + \cdots) = 2 - (y_0' + \varepsilon y_1' + \cdots + z_0' + z_1' + \cdots).$$

The problems for the y_i are unaffected, and so as before we obtain $y_0 = 2x - 1$ and $y_1 = y_2 = \cdots = 0$. To determine the equation for z_0, remember that we are not certain of how this function depends on ε. Because of this we will retain everything involving z_0, and this means that the equation is $\varepsilon z_0'' = -z_0'$. The general solution is $z_0 = a(\varepsilon) + b(\varepsilon) \exp(-x/\varepsilon)$. Imposing the boundary condition $z_0 = 0$ at $x = 1$ yields

$$z_0 = b(\varepsilon)\left(e^{-x/\varepsilon} - e^{-1/\varepsilon}\right).$$

It remains to match the expansions. Introducing the intermediate layer variable $x_\eta = x/\varepsilon^\beta$, where $0 < \beta < 1$, (2.53) then becomes

$$y_{\text{outer}} \sim -1 + 2x_\eta \varepsilon^\beta + \cdots + b(\varepsilon)\left(e^{-x_\eta/\varepsilon^{1-\beta}} - e^{-1/\varepsilon}\right) + \cdots,$$

and the boundary-layer expansion (2.52) takes the form

$$y_{\text{inner}} \sim A_0\left(1 - e^{-x_\eta/\varepsilon^{1-\beta}}\right) + \varepsilon\left[2x_\eta/\varepsilon^{1-\beta} + A_1\left(1 - e^{-x_\eta/\varepsilon^{1-\beta}}\right)\right] + \cdots.$$

(2.54)

Matching these expressions, it follows that $A_0 = -1$ and $b = -A_0 = 1$. With this, we have that

$$z_0 = e^{-x/\varepsilon} - e^{-1/\varepsilon}.$$

(2.55)

It is not hard to show that this term is transcendentally small compared to the power functions for any given value of x satisfying $0 < x \leq 1$. The resulting outer expansion is therefore

$$y \sim 2x - 1 + \cdots + e^{-x/\varepsilon} - e^{-1/\varepsilon} + \cdots.$$

(2.56)

As you would expect, the transcendentally small terms contribute very little to the numerical value of the outer solution. To illustrate, if $\varepsilon = 0.01$ and $x = 3/4$, then $y_0 = 1/2$ while $z_0 \approx 2.6 \times 10^{-33}$. This is why we have ignored this portion of the outer expansion in the earlier sections of this chapter. The fact that you cannot always do this is demonstrated in the next example. ∎

The matching in the preceding example helps explain the reason for the transcendentally small terms in the outer expansion. In particular, the first transcendentally small term in the boundary-layer expansion (2.54) is what generates the need for z_0. In other words, the boundary layer causes our having to include z_0 in the outer expansion. Actually, information flows in both directions. If you calculate the higher terms in the expansions, you will find that it is also necessary to include transcendentally small terms in the boundary-layer expansion. This is explored in more depth in Exercise 2.27.

The expansion (2.53) is not very specific about how the z_i depend on ε, and we determined this dependence in the reduction of the problem. Some prefer a more explicit form of the expansion and make specific assumptions about the z_i. For example, an assumption often used to construct a composite expansion is that $z_0 = A(x)e^{-g(x)/\varepsilon}$ (Latta, 1951). This will not produce the result in (2.55), and so the assumption must be modified to account for transcendentally small terms. Examples can be found in MacGillivray (1997) and Howls (2010). A somewhat different approach is explored in Exercise 2.30.

Example 2

Consider the problem of solving

$$\varepsilon y'' - xy' + \varepsilon xy = 0 \quad \text{for } -1 < x < 1, \tag{2.57}$$

where $y(-1) = y_L$ and $y(1) = y_R$ are assumed to be specified. There are boundary layers at both ends, and the analysis is very similar to Example 1 in Sect. 2.3. For the outer expansion one assumes $y \sim y_0 + \cdots$, and from the differential equation it follows that $y_0 = c$ (i.e., the first term is just a constant). For the boundary layer at the left end, one lets $\bar{x} = (x+1)/\varepsilon$ and from this finds that $Y_0'' + Y_0' = 0$. Solving this, imposing the boundary condition $Y_0(0) = y_L$, and then matching one finds that $Y_0 = c + (y_L - c)e^{-\bar{x}}$. For the boundary layer at the right end, where $\tilde{x} = (x-1)/\varepsilon$, one finds that $\tilde{Y}_0 = c + (y_R - c)e^{\tilde{x}}$. Putting the results together we have that

$$y \sim \begin{cases} c + (y_L - c)e^{-\bar{x}} & \text{boundary layer at } x = -1, \\ c & \text{outer region}, \\ c + (y_R - c)e^{\tilde{x}} & \text{boundary layer at } x = 1. \end{cases} \tag{2.58}$$

What is unusual about this problem is that we have carried out the boundary-layer analysis but still have an unknown constant (i.e., c). How to deal with this depends on the problem. For some problems it is enough to look at the second term in the expansion (e.g., Exercise 2.57), for other linear problems a WKB type argument can be used (see Chap. 4), and for still others one can use a symmetry property (e.g., Sect. 2.5). What is going to be shown here is that the transcendentally small terms in the outer region can be used to determine c. This approach has the disadvantage of being somewhat more difficult mathematically but has the distinct advantage of being able to work on a wide range of linear and nonlinear problems (e.g., we will use this approach when studying metastability in Chap. 6).

The remedy is to use (2.53) instead of the regular power series expansion. In this case the differential equation becomes

$$\varepsilon(y_0'' + \cdots + z_0'' + \cdots) - x(y_0' + \cdots + z_0' + \cdots) + \varepsilon x(y_0 + \cdots + z_0 + \cdots) = 0.$$

The problem for y_0 is unaffected, and we only need to concentrate on z_0. In the previous example we used the dictum that everything involving z_0 is retained, which in this case means that $\varepsilon z_0'' - x z_0' + \varepsilon x z_0 = 0$. However, in this problem it is wise to think about what terms are actually needed in this equation. In fact, our construction will mimic the procedure used for (1.36), where the form of the expansion was determined in the derivation. Retaining $\varepsilon z_0''$ and $x z_0'$ is reasonable because it is very possible that a transcendentally small term of the form $e^{-x/\varepsilon}$ will occur, just as it did in the previous example. On the other hand, it is expected that $\varepsilon x z_0$ is higher order than the other two terms and for this reason contributes to the higher-order equations (e.g., the problem for z_1). This ad hoc reasoning is necessary because we have not yet determined the order for z_0, and the resulting reduction helps to simplify the analysis. What is going to be necessary, once z_0 is determined, is to check that these assumptions are correct.

Solving the equation $\varepsilon z_0'' - x z_0' = 0$, one obtains the general solution

$$z_0 = b(\varepsilon) + a(\varepsilon) \int_{-1}^{x} e^{s^2/(2\varepsilon)} ds.$$

The matching proceeds in the usual manner using intermediate variables.

Matching at Left End

The intermediate variable is $x_\eta = (x + 1)/\varepsilon^\beta$, where $0 < \beta < 1$. The boundary-layer approximation becomes

$$y_{\text{inner}} \sim c + (y_L - c) e^{-x_\eta/\varepsilon^{1-\beta}} + \cdots. \tag{2.59}$$

Before writing down the outer expansion, note that the integral in z_0 is going to take the form

$$\int_{-1}^{-1+\varepsilon^\beta x_\eta} e^{s^2/(2\varepsilon)} ds.$$

The expansion of this integral can be obtained using integration by parts by writing

$$e^{s^2/(2\varepsilon)} = \frac{\varepsilon}{s} \frac{d}{ds} e^{s^2/(2\varepsilon)}.$$

In this case, setting $q = \varepsilon^\beta x_\eta$, we have that

$$\int_{-1}^{-1+q} e^{s^2/(2\varepsilon)} ds = \frac{\varepsilon}{s} e^{s^2/(2\varepsilon)} \Big|_{s=-1}^{-1+q} + \int_{-1}^{-1+q} \frac{\varepsilon}{s^2} e^{s^2/(2\varepsilon)} ds$$

$$= \frac{\varepsilon}{s} \left(1 + \frac{\varepsilon}{s^2} \right) e^{s^2/(2\varepsilon)} \Big|_{s=-1}^{-1+q} + \int_{-1}^{-1+q} \frac{3\varepsilon^2}{s^4} e^{s^2/(2\varepsilon)} ds$$

$$= \frac{\varepsilon}{s}\left(1 + \frac{\varepsilon}{s^2} + \frac{3\varepsilon^2}{s^4} + \cdots\right)e^{s^2/(2\varepsilon)}\Bigg|_{s=-1}^{-1+q}$$

$$\sim \varepsilon e^{1/(2\varepsilon)}\left(1 - e^{-q/\varepsilon}\right).$$

In the last step it is assumed that the β interval is reduced to $1/2 < \beta < 1$. With this the outer expansion becomes

$$y_{\text{outer}} \sim c + \cdots + b(\varepsilon) + \varepsilon a(\varepsilon)e^{1/(2\varepsilon)}\left(1 - e^{-q/\varepsilon}\right) + \cdots. \qquad (2.60)$$

Matching (2.59) and (2.60) it follows that $\varepsilon a(\varepsilon)e^{1/(2\varepsilon)} = c - y_{\text{L}}$ and $b(\varepsilon) = -\varepsilon a(\varepsilon)e^{1/(2\varepsilon)}$.

Matching at Right End

The intermediate variable is $x_\eta = (x-1)/\varepsilon^\beta$, where $0 < \beta < 1$. The boundary-layer approximation becomes

$$y_{\text{outer}} \sim c + (y_{\text{R}} - c)e^{x_\eta/\varepsilon^{1-\beta}} + \cdots. \qquad (2.61)$$

Using an integration by parts argument similar to what was done for the layer at the left end, one finds that

$$y_{\text{inner}} \sim c + \cdots + b(\varepsilon) + \varepsilon a(\varepsilon)e^{1/(2\varepsilon)}\left(1 + e^{q/\varepsilon}\right) + \cdots, \qquad (2.62)$$

where $q = \varepsilon^\beta x_\eta$. Matching (2.61) and (2.62) it follows that $\varepsilon a(\varepsilon)e^{1/(2\varepsilon)} = y_{\text{R}} - c$ and $b(\varepsilon) = -\varepsilon a(\varepsilon)e^{1/(2\varepsilon)}$.

The condition we are seeking comes from the matching conditions, which require that $\varepsilon a(\varepsilon)e^{1/(2\varepsilon)} = c - y_{\text{L}}$ and $\varepsilon a(\varepsilon)e^{1/(2\varepsilon)} = y_{\text{R}} - c$. Equating these conditions it follows that

$$c = \frac{1}{2}\left(y_{\text{L}} + y_{\text{R}}\right). \qquad (2.63)$$

As an example, if $y(-1) = 2$ and $y(1) = -2$, then $c = 0$ and the outer expansion (2.51) becomes simply $y \sim 0$. We will obtain such an approximation in some of the other examples and exercises in this chapter. This does not mean the solution is zero. Rather, it means that the solution is transcendentally small compared to the power functions. ∎

The last example is interesting because it shows that what appear to be inconsequential terms in an expansion can affect the value of the first-order approximation. Although this situation is not rare, it is not common. Therefore, in the remainder of this text we will mostly use power functions for the scale functions. If something remains underdetermined, as it did in Example 2, we will then entertain the idea that transcendentally small terms are needed.

Exercises

2.27. This problem concerns the higher terms in the expansions from Example 1.

(a) Verify (2.52), and, by matching with (2.51), show that $A_0 = 1$ and $A_1 = A_2 = 0$.

(b) Explain why your result from part (a) does not match with (2.56).

(c) The result from part (b) shows that the outer expansion is the reason transcendentally small terms must also be included in the boundary-layer expansion. Assume that

$$Y \sim Y_0(\bar{x}) + \varepsilon Y_1(\bar{x}) + \cdots + Z_0(\bar{x}, \varepsilon) + Z_1(\bar{x}, \varepsilon) + \cdots,$$

where $Z_j \ll Z_i$, $\forall i < j$, and $z_i \ll \varepsilon^n$, $\forall i, n$ (with $\bar{x}$ fixed). Show that $Z_0 = B(\varepsilon)(1 - e^{-\bar{x}})$. From matching show that $B = -e^{-1/\varepsilon}$.

(d) Explain why Z_0 is the reason z_1 is needed in the outer expansion.

2.28. This problem completes some of the details in the derivation of Example 2.

(a) Use Laplace's approximation (Appendix C) to show that

$$\int_{-1}^{1} e^{s^2/(2\varepsilon)} \, ds \sim 2\varepsilon e^{1/(2\varepsilon)}.$$

(b) Use part (a) to help derive (2.62).

(c) The matching shows that $b = \frac{1}{2}(y_L - y_R)$, which appears to contradict the assumption that z_0 is transcendentally small compared to the power functions. Explain why there is, in fact, no contradiction in this result.

(d) Show that the assumption that $\varepsilon x z_0$ can be ignored compared to $\varepsilon z_0''$ and $x z_0'$ holds for the function z_0.

2.29. Consider the problem of solving

$$\varepsilon y'' - xy' = 0 \quad \text{for } a < x < b,$$

where $y(a) = y_L$ and $y(b) = y_R$. Also, assume that $a < 0$ and $b > 0$.

(a) Using the usual boundary-layer arguments show that

$$y \sim \begin{cases} c + (y_L - c)e^{-\bar{x}} & \text{boundary layer at } x = a, \\ c & \text{outer region}, \\ c + (y_R - c)e^{\tilde{x}} & \text{boundary layer at } x = b, \end{cases}$$

where c is an arbitrary constant, $\bar{x} = (x - a)/\varepsilon$, and $\tilde{x} = (x - b)/\varepsilon$.

(b) Find the exact solution of the problem.

(c) Using the result from part (b) and Laplace's approximation (Appendix C) show that

$$c = \begin{cases} y_L & \text{if } |a| < b, \\ \frac{1}{2}(y_L + y_R) & \text{if } |a| = b, \\ y_R & \text{if } b < |a|, \end{cases}$$

Comment on what happens to the assumed boundary layers in the problem.

(d) Sketch (by hand) the solution when $a = -2$, $b = 1$, $y_L = 1$, and $y_R = 3$. Comment on whether you used part (a) or part (b), and why.

2.30. There is a question whether it is possible to account for the transcendentally small terms without having to use the z_i in (2.53). One possibility is to modify the boundary layer method by first constructing a composite expansion, as described in Sect. 2.2, and then imposing the boundary conditions (Exercise 2.14). This is what is done when using the WKB method, and one of the reasons WKB has had some success in handling transcendentally small terms.

(a) Show that reversing the order for (2.50), with $y(0) = 0$ and $y(1) = 1$, results in the approximation

$$y \sim 2x + \frac{1}{1 - e^{-1/\varepsilon}}\left(-1 + e^{-\bar{x}}\right).$$

Explain why this reduces to (2.56).

(b) Explain why this modified boundary-layer method does not solve the unknown-constant problem that appears in (2.58).

2.31. Consider the problem

$$\varepsilon^2 y'' + 2\varepsilon y' + 2(y - xg)^2 = \varepsilon h(x) \quad \text{for } 0 < x < 1,$$

where $y(0) = \text{sech}^2(1/(2\varepsilon))$ and $y(1) = 1 + \text{sech}^2(1/(2\varepsilon))$. Also, $g(x) = e^{\varepsilon(x-1)}$ and $h(x) = [\varepsilon^2 + (2 + \varepsilon^2)(1 + \varepsilon x)]g(x)$.

(a) Suppose one were to argue that the exponentially small terms in the boundary conditions can be ignored and the usual power series expansion of the solution can be used. Based on this assumption, find the first two terms of a composite expansion of the solution.

(b) The exact solution of the problem is

$$y(x) = xe^{\varepsilon(x-1)} + \text{sech}^2\left(\frac{2x - 1}{2\varepsilon}\right).$$

Discuss this solution in connection with your expansion from part (a).

2.5 Interior Layers

The rapid transitions in the solution that are characteristic of a boundary layer do not have to occur only at the boundary. When this happens, the problems tend to be somewhat harder to solve simply because the location of the layer is usually not known until after the expansions are matched. However, the expansion procedure is essentially the same as in the previous examples. To understand how the method works, consider the problem

$$\varepsilon y'' = yy' - y, \quad \text{for } 0 < x < 1, \tag{2.64}$$

where

$$y(0) = 1 \tag{2.65}$$

and

$$y(1) = -1. \tag{2.66}$$

2.5.1 Step 1: Outer Expansion

The appropriate expansion in this region is the same as it usually is, in other words,

$$y(x) \sim y_0(x) + \cdots. \tag{2.67}$$

From (2.64) one finds that

$$y_0 y_0' - y_0 = 0, \tag{2.68}$$

and so either $y_0 = 0$ or else

$$y_0 = x + a, \tag{2.69}$$

where a is an arbitrary constant. The fact that we have two possible solutions means that the matching might take somewhat longer than previously because we will have to determine which of these solutions matches to the inner expansion.

2.5.2 Step 1.5: Locating the Layer

Generally, when one first begins trying to solve a problem, it is not known where the layer is, or whether there are multiple layers. If we began this problem like the others and assumed there is a boundary layer at either one of the endpoints, we would find that the expansions do not match. This is a lot of effort for no results, but fortunately there is a simpler way to come to

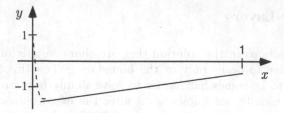

Figure 2.11 Schematic of solution if there is a (convex) boundary layer at $x = 0$ and the linear function in (2.69) is the outer solution

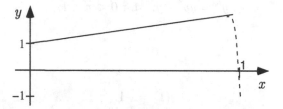

Figure 2.12 Schematic of solution if there is a concave boundary layer at $x = 1$ and the linear function in (2.69) is the outer solution

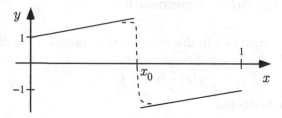

Figure 2.13 Schematic of linear functions that make up outer expansion and interior layer solution connecting them

the same conclusion. To illustrate how, suppose it is assumed that there is a boundary layer at $x = 0$ and (2.69) is the outer solution. This situation is shown in Fig. 2.11. If the solution behaves like the other example problems, then in the boundary layer it is expected that y'' is positive (i.e., y is concave up), y' is negative, and y is both positive and negative. In other words, in such a boundary layer, the right-hand side of (2.64) is positive while the left-hand side can be negative. This is impossible, and so there is not a boundary layer as indicated in Fig. 2.11.

It is possible to rule out a boundary layer at $x = 1$ in the same way. In particular, as illustrated in Fig. 2.12, in the boundary layer y'' and $y' - 1$ are negative, while y is both positive and negative. Using a similar argument one can rule out having boundary layers at both ends.

Another possibility is that the layer is interior to the interval, and this is illustrated in Fig. 2.13. To check, in the layer region to the left of x_0, $y' - 1$ is negative, y is positive, and y'' is negative. This is consistent with (2.64).

Similarly, to the right of x_0, both $y' - 1$ and y are negative, and y'' is positive. This too is consistent with (2.64). It also indicates that $y(x_0) = 0$, a result we will need later to complete the derivation of the interior layer solution.

It should be pointed out that these are only plausibility arguments and do not prove anything. What they do is guide the analysis and, hopefully, reduce the work necessary to obtain the solution. A more expanded version of this analysis is explored in Exercise 2.34.

2.5.3 Steps 2 and 3: Interior Layer and Matching

Based on the preceding observations, we investigate the possibility of an interior layer. This is done by introducing the interior-layer coordinate

$$\bar{x} = \frac{x - x_0}{\varepsilon^\alpha}, \tag{2.70}$$

where $0 < x_0 < 1$. The location of the layer, $x = x_0$, is not known and will be determined subsequently. Actually, the possibilities of either $x_0 = 0$ or $x_0 = 1$ could be included here, but we will not do so. Also, note that since $0 < x_0 < 1$, there are two outer regions, one for $0 \le x < x_0$, the other for $x_0 < x \le 1$ (Fig. 2.13). Now, substituting (2.70) into (2.64) yields

$$\varepsilon^{1-2\alpha} Y'' = \varepsilon^{-\alpha} Y Y' - Y. \tag{2.71}$$

The distinguished limit here occurs when $\alpha = 1$ (i.e., the first and second terms balance). Also, as in the previous examples, we are using Y to designate the solution in the layer. Expanding the interior-layer solution as

$$Y(\bar{x}) \sim Y_0(\bar{x}) + \cdots \tag{2.72}$$

it follows from (2.71) that

$$Y_0'' = Y_0 Y_0'. \tag{2.73}$$

Integrating this one obtains

$$Y_0' = \frac{1}{2} Y_0^2 + A.$$

There are three solutions of this first-order equation, corresponding to A's being positive, negative, or zero. The respective solutions are

$$Y_0 = B \frac{1 - D e^{B\bar{x}}}{1 + D e^{B\bar{x}}}, \tag{2.74}$$

$$Y_0 = B \tan(C - B\bar{x}/2),$$

and

$$Y_0 = \frac{2}{C - \bar{x}},$$

where B, C, and D are arbitrary constants. The existence of multiple so-lutions makes the problem interesting, but it also means that the matching procedure is not as straightforward as it was for the linear equations studied earlier. This is because for the linear problems we relied on being able to find the general solution in each region and then determining the constants by matching. For nonlinear problems the concept of a general solution has little meaning, and because of this it can sometimes be difficult to obtain a solution that is general enough to be able to match to the outer expansion(s).

Of the solutions to (2.73), the one given in (2.74) is capable of matching to the outer expansions as $\bar{x} \to \pm\infty$. Again it should be remembered that the working hypothesis here is that $0 < x_0 < 1$. Thus, the outer expansion for $0 \leq x < x_0$ should satisfy $y(0) = 1$. From this it follows that

$$y_0 = x + 1, \quad \text{for } 0 \leq x < x_0. \tag{2.75}$$

Similarly, the outer region on the other side of the layer should satisfy the boundary condition at $x = 1$, and this yields

$$y_0 = x - 2, \quad \text{for } x_0 < x \leq 1. \tag{2.76}$$

Now, for (2.74) to be able to match to either (2.75) or (2.76) it is going to be necessary that both B and D in (2.74) be nonzero (in fact, without loss of generality, we will take B to be positive). The requirements imposed in the matching are very similar to those obtained for boundary layers. In particular, we must have that $Y_0(\infty) = y_0(x_0^+)$ and $Y_0(-\infty) = y_0(x_0^-)$. From (2.74) and (2.75) we get that

$$B = x_0 + 1,$$

and from (2.74) and (2.76) we have that

$$-B = x_0 - 2.$$

Solving these equations one finds that $B = \frac{3}{2}$ and $x_0 = \frac{1}{2}$.

2.5.4 Step 3.5: Missing Equation

From matching we have determined the location of the layer and one of the constants in the layer solution. However, the matching procedure did not determine D in (2.74). Fortunately, from the discussion in Step 1.5, we are able to determine its value. In particular, we found that $y(x_0) = 0$, and for this to happen it must be that $D = 1$. Therefore,

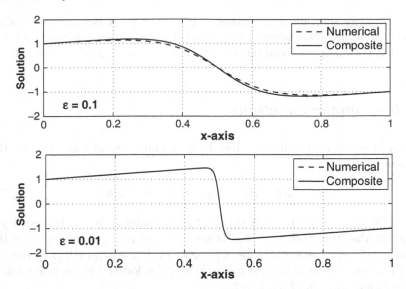

Figure 2.14 Graph of numerical solution of (2.64)–(2.66) and composite expansion given in (2.78) in the case where $\varepsilon = 10^{-1}$ and where $\varepsilon = 10^{-2}$

$$Y(\bar{x}) \sim \frac{3}{2} \frac{1 - e^{3\bar{x}/2}}{1 + e^{3\bar{x}/2}} . \tag{2.77}$$

Having an undetermined constant left after matching also occurred in the previous section (Example 2). Unlike that earlier example, we were able to determine D using the properties of the differential equation and boundary conditions. Our argument was heuristic, and those who prefer a more mathematical proof that $D = 1$ should consult Exercise 2.42. Another approach to evaluating the undetermined constant is given in Exercise 2.37.

2.5.5 Step 4: Composite Expansion

It can be more difficult to construct a composite expansion when the outer solutions are discontinuous across the interior layer like they are in this problem. What is done is to find one for the interval $0 \le x \le x_0$ and then another for $x_0 \le x \le 1$. As it turns out, for this example the expansions on either side are the same, and the result is

$$y \sim x + 1 - \frac{3}{1 + e^{-3(2x-1)/4\varepsilon}} \quad \text{for } 0 \le x \le 1 \tag{2.78}$$

This composite expansion is graphed in Fig. 2.14 to illustrate the nature of the interior layer and how it appears as ε decreases. The rapid transition from one outer solution to the other is typical of what is sometimes called a shock solution. Also shown in Fig. 2.14 is the numerical solution, and it is clear that the composite and numerical solutions are in good agreement.

2.5.6 Kummer Functions

Interior layers can arise in linear problems. To understand this, consider the problem of solving

$$\varepsilon y'' + (3x - 1)y' + xy = 0 \quad \text{for } 0 < x < 1, \tag{2.79}$$

where $y(0) = 1$ and $y(1) = 2$. A tip-off that this might have an interior layer is the fact that the coefficient of y' is zero at a point in the interval. For the moment, this observation is more of a curiosity, but it is worth pointing out that the interior layer of the last example was also located at the point where the coefficient of y' in (2.64) is zero.

There is nothing particularly unusual about this problem, so we will assume a regular expansion for the outer solution. In particular, assuming $y \sim y_0(x) + \varepsilon y_1(x) + \cdots$, the $O(1)$ equation is $(3x - 1)y_0' + xy_0 = 0$. Solving this one finds that

$$y_0 = \frac{a}{(3x - 1)^{1/9}} e^{-x/3}. \tag{2.80}$$

Given that the denominator is zero at $x = 1/3$, it should not come as a surprise that there is an interior layer located at $x = 1/3$. This means there are two outer solutions, and we have that

$$y_0(x) = \begin{cases} \dfrac{a_\mathrm{l}}{(1 - 3x)^{1/9}} e^{-x/3} & \text{for } 0 \leq x < \frac{1}{3}, \\[2mm] \dfrac{a_\mathrm{r}}{(3x - 1)^{1/9}} e^{-x/3} & \text{for } \frac{1}{3} < x \leq 1. \end{cases} \tag{2.81}$$

Satisfying the boundary conditions, one finds that $a_\mathrm{l} = 1$ and $a_\mathrm{r} = 2^{10/9} e^{1/3}$.

We will use the interior-layer coordinate given in (2.70), with $x_0 = 1/3$ and $\alpha = 1/2$. The problem in this case becomes

$$Y'' + \bar{x}Y' + \left(\frac{1}{3} + \varepsilon^{1/2}\right)Y = 0, \quad \text{for } -\infty < \bar{x} < \infty. \tag{2.82}$$

The $O(1)$ equation coming from this is

$$Y_0'' + 3\bar{x}Y_0' + \frac{1}{3}Y_0 = 0, \quad \text{for } -\infty < \bar{x} < \infty. \tag{2.83}$$

This equation is why this subsection is titled "Kummer Functions"; it also provides the motivation for the next paragraph.

What we have shown is that to determine the solution in the interior layer, we must be able to solve an equation of the form

$$y'' + \alpha x y' + \beta y = 0, \quad \text{for} \quad -\infty < x < \infty,$$

where α and β are nonzero constants. This equation can be solved using a power series expansion or the Laplace transform. Doing this shows that the general solution can be written as

$$y(x) = A_0 \, M\left(\frac{\beta}{2\alpha}, \frac{1}{2}, -\frac{1}{2}\alpha x^2\right) + B_0 \, x M\left(\frac{\alpha+\beta}{2\alpha}, \frac{3}{2}, -\frac{1}{2}\alpha x^2\right), \qquad (2.84)$$

where $M(a, b, z)$ is *Kummer's function* and its definition and basic properties are given in Appendix B. As an example, if $\alpha = \beta$, then the solution is

$$y(x) = A_0 e^{-\alpha x^2/2} + B_0 \frac{1}{x} \int_0^x e^{\alpha(s^2 - x^2)/2} \, ds.$$

For us to be able to match with the outer solutions, we need to know what happens to (2.84) when $x \to \pm\infty$. This depends on whether α is positive or negative. Using the formulas in Appendix B, for $x^2 \to \infty$,

$$y(x) \sim \sqrt{\pi} \left[\frac{A_0}{\Gamma\left(\frac{1}{2} - \kappa\right)} \pm \frac{B_0}{\sqrt{2\alpha}\,\Gamma(1 - \kappa)}\right] \eta^{-\kappa} \quad \text{if } \alpha > 0 \qquad (2.85)$$

and

$$y(x) \sim \sqrt{\pi} \left[\frac{A_0}{\Gamma(\kappa)} \pm \frac{B_0}{\sqrt{-2\alpha}\,\Gamma\left(\frac{1}{2} + \kappa\right)}\right] (-\eta)^{\kappa - \frac{1}{2}} e^{-\eta} \quad \text{if } \alpha < 0, \qquad (2.86)$$

where $\kappa = \beta/(2\alpha)$ and $\eta = \frac{1}{2}\alpha x^2$. In the preceding expressions, the $+$ is taken when $x > 0$ and the $-$ when $x < 0$. Also, the arguments of the Gamma functions are assumed not to be nonpositive integers. What happens in those cases is interesting and discussed toward the end of Sect. 2.6.

Based on the preceding discussion, the general solution of the interior-layer Eq. (2.83) can be written as

$$Y_0 = A_0 \, M\left(\frac{1}{18}, \frac{1}{2}, -\frac{3}{2}\bar{x}^2\right) + B_0 \, \bar{x} M\left(\frac{5}{9}, \frac{3}{2}, -\frac{3}{2}\bar{x}^2\right). \qquad (2.87)$$

The constants in this expression are determined by matching to the outer solution. To do this we use the intermediate variable $x_\eta = (x - x_0)/\varepsilon^\gamma$, where $0 < \gamma < 1/2$. Introducing this into the outer solution (2.81) yields

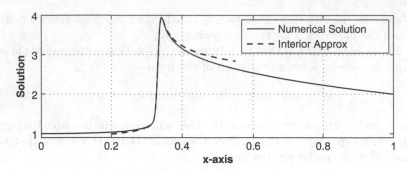

Figure 2.15 Graph of numerical solution of (2.79) and interior-layer solution given in (2.87) in the case where $\varepsilon = 10^{-4}$

$$y_{\text{outer}} \sim \begin{cases} \dfrac{a_l}{(3e)^{1/9}} (\varepsilon^\gamma |x_\eta|)^{-1/9} & \text{for} \quad x_\eta < 0, \\[2mm] \dfrac{a_r}{(3e)^{1/9}} (\varepsilon^\gamma |x_\eta|)^{-1/9} & \text{for} \quad 0 < x_\eta. \end{cases} \qquad (2.88)$$

For the interior-layer solution (2.87) we use (2.85) to obtain

$$y(x) \sim \sqrt{\pi} \left[\frac{A_0}{\Gamma(\frac{4}{9})} \pm \frac{B_0}{\sqrt{6}\,\Gamma(\frac{17}{18})} \right] \left(\frac{2\varepsilon}{3} \right)^{1/18} (\varepsilon^\gamma |x_\eta|)^{-1/9}, \qquad (2.89)$$

where the $+$ is taken if $x_\eta > 0$ and the $-$ if $x_\eta < 0$. Matching the inner and outer approximations, it follows that

$$A_0 = \frac{1}{2\sqrt{\pi}} (a_r + a_l) \Gamma\left(\frac{4}{9} \right) (6\varepsilon e^2)^{-1/18}$$

and

$$B_0 = \sqrt{\frac{3}{2\pi}} (a_r - a_l) \Gamma\left(\frac{17}{18} \right) (6\varepsilon e^2)^{-1/18}.$$

The coefficients have ended up depending on ε. What this means is that the original assumption that $Y \sim Y_0 + \cdots$ should have been $Y \sim \varepsilon^{-1/18}(Y_0 + \cdots)$. Because this problem is linear, the need for the multiplicative factor $\varepsilon^{-1/18}$ does not affect the validity of the asymptotic approximation.

To give a sense of how well this approximation does, the numerical solution and the interior layer approximation are plotted in Fig. 2.15. It is evident that the latter is asymptotic to the exponential functions that make up the outer solution (2.81), and it gives a very accurate approximation of the solution in the layer. What is also interesting is that, unlike the other layer examples considered so far, the interior-layer solution is not monotone. An analysis of more complex nonmonotone interior layers can be found in DeSanti (1987).

Exercises

2.32. Find a first-term expansion of the solution of each of the following problems. It should not be unexpected that for the nonlinear problems the solutions are defined implicitly or that the transition layer contains an undetermined constant.

(a) $\varepsilon y'' = -(x^2 - \frac{1}{4})y'$ for $0 < x < 1$, where $y(0) = 1$ and $y(1) = -1$.

(b) $\varepsilon y'' + 2xy' + (1 + \varepsilon x^2)y = 0$ for $-1 < x < 1$, where $y(-1) = 2$ and $y(1) = -2$.

(c) $\varepsilon y'' = yy' - y^3$ for $0 < x < 1$, where $y(0) = \frac{3}{5}$ and $y(1) = -\frac{2}{3}$.

(d) $3\varepsilon y'' + 3xy' + (1 + x)y = 0$ for $-1 < x < 1$, where $y(-1) = 1$ and $y(1) = -1$.

(e) $\varepsilon y'' + y(1 + y^2)y' - \frac{1}{2}y = 0$ for $0 < x < 1$, where $y(0) = -1$ and $y(1) = 1$.

(f) $\varepsilon y'' + y(y' + 3) = 0$ for $0 < x < 1$, where $y(0) = -1$ and $y(1) = 2$.

2.33. Consider the problem

$$\varepsilon y'' = yy' \quad \text{for } 0 < x < 1,$$

where $y(0) = a$ and $y(1) = -a$. Also, a is positive.

(a) Prove that $y(\frac{1}{2}) = 0$. (Hint: Use the method described in Exercise 2.42.)

(b) Find a composite expansion of the solution.

(c) Show that the exact solution has the form

$$y = A\frac{1 - Be^{Ax/\varepsilon}}{1 + Be^{Ax/\varepsilon}},$$

where, for small ε, $A \sim a(1 + 2e^{-a/(2\varepsilon)})$ and $B \sim e^{-a/(2\varepsilon)}$. Comment on how this compares with your result from part (b).

2.34. This problem explores various possibilities for the layer solutions of (2.64). You do not need to derive the expansions to answer these questions, but you do need to know the general forms for the outer solutions as well as the general form of the layer solution (2.74).

(a) The plausibility argument used to rule out boundary layers (e.g., Fig. 2.11) did not consider the other possible outer solution, namely, $y_0 = 0$. If this is the outer solution, then there must be a boundary layer at both ends. Explain why this cannot happen.

(b) To examine how the position of the interior layer depends on the boundary conditions, suppose that $y(0) = a$ and $y(1) = b$, where $-1 < a + b < 1$ and $b < 1 + a$. What is the value of x_0 in this case? Also, state how you use the stated inequalities on a and b.

(c) What happens to the layer(s) if the boundary conditions are $y(0) = y(1) = a$?

(d) Suppose $y(0) = -1/2$ and $y(1) = 1/4$. Use the plausibility argument to show that there are multiple possible solutions.

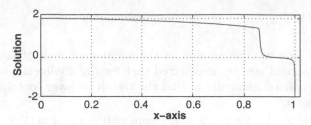

Figure 2.16 Solution of problem in Exercise 2.35

2.35. Consider the problem

$$\varepsilon y'' + y(1-y)y' - xy = 0 \quad \text{for } 0 < x < 1,$$

where $y(0) = 2$ and $y(1) = -2$. The numerical solution is shown in Fig. 2.16 in the case where $\varepsilon = 10^{-3}$. This will prove useful when deriving a first-term approximation of the solution.

(a) Find the first term in the expansion of the outer solution. Assume that this function satisfies the boundary condition at $x = 0$.

(b) Assume there is a boundary layer at $x = 1$. After finding a first-term approximation in the boundary layer show that it does not match with the outer solution you found in part (a).

(c) Assuming there is an interior layer across which the solution jumps from one outer solution to another, find a first-term approximation in the layer. From the matching show that the layer is located at $x_0 = \sqrt{3}/2$. Note that your layer solution will contain an undetermined constant.

(d) Correct the boundary-layer analysis in part (b) based on your result from part (c).

2.36. This problem examines the solution of the boundary-value problem

$$\varepsilon y'' = -f(x)y' \quad \text{for } 0 < x < 1,$$

where $y(0) = a$ and $y(1) = -b$. Assume that a and b are positive constants. Also, assume that $f(x)$ is smooth with $f'(x) > 0$ and $f(x_0) = 0$ for $0 < x_0 < 1$.

(a) Explain why there must be at least one point in the interval $0 < x < 1$ where $y(x) = 0$.

(b) Find the exact solution of the problem and then write down the equation that must be solved to determine where $y(x) = 0$. From this explain why there is exactly one solution of $y(x) = 0$.

(c) Using your result from part (b), find a two-term expansion of the solution of $y(x) = 0$. The second term will be defined implicitly and involves solving an equation of the form

$$\text{erf}(\mu_0\sqrt{f'(x_0)/2}) = \frac{a-b}{a+b}$$

for μ_0. Note that Laplace's approximation (Appendix C) will be useful here.

(d) Find a two-term expansion of the solution of $y(x) = 0$ by first constructing an asymptotic expansion of the solution of the boundary-value problem.

2.37. One way to resolve the problem of having an undetermined constant after matching is to use a variational principle. To understand this approach, consider the problem

$$\varepsilon y'' + p(x,\varepsilon)y' + q(x,\varepsilon)y = 0 \quad \text{for } 0 < x < 1,$$

where $y(0) = a$ and $y(1) = b$. Associated with this is the functional

$$I(v) = \int_0^1 L(v,v')dx, \quad \text{where } L = \frac{1}{2}[\varepsilon(v')^2 - qv^2]e^{\frac{1}{\varepsilon}\int_0^x p(s,\varepsilon)ds}.$$

In this variational formulation L is a Lagrangian for the equation.

(a) Show that if $\frac{d}{dr}I(y + ru) = 0$, at $r = 0$, for all smooth functions $u(x)$ satisfying $u(0) = u(1) = 0$, then $y(x)$ is a solution of the preceding differential equation. In other words, an extremal of the functional is a solution of the differential equation.

(b) Consider the problem

$$\varepsilon y'' - (2x - 1)y' + 2y = 0 \quad \text{for } 0 < x < 1,$$

where $y(0) = 1$ and $y(1) = -3$. The solution of this problem has a boundary layer at each end of the interval. Find a composite expansion of the solution for $0 \le x \le 1$. Your solution will contain an arbitrary constant that will be designated as k in what follows.

(c) From your result in part (b) derive an expansion for the Lagrangian L.

(d) Explain why the constant k should be such that $\frac{d}{dk}I(y) = 0$. From this determine k.

2.38. In the Langmuir–Hinshelwood model for the kinetics of a catalyzed reaction the following problem appears:

$$\varepsilon\frac{dy}{dx} = 1 - \frac{1}{x}F(y) \quad \text{for } 0 < x < 1,$$

where $F(y) = 2(1 - y)(\alpha + y)/y$ and $y(1) = 0$. Also, $0 < \alpha < 1$. In this problem ε is the deactivation rate parameter and $y(x)$ is the concentration of the reactant (Kapila, 1983).

(a) For small ε, find a first-term expansion of the solution in the outer region and in the boundary layer.

(b) Find a composite expansion of the solution for $0 < x \le 1$.

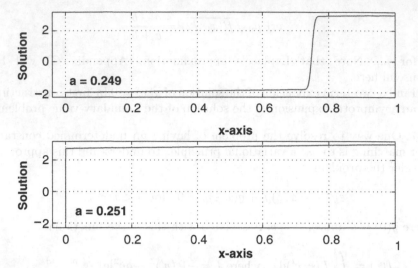

Figure 2.17 Graph of solution for Exercise 2.40 for two slightly different values of a. In this calculation, $b = 0.75$ and $\varepsilon = 10^{-4}$

2.39. This problem examines a differential-difference equation. The specific problem is (Lange and Miura, 1991)

$$\varepsilon^2 y''(x) - y(x) + q(x)y(x-1) = f(x) \quad \text{for } 0 < x < 3/2,$$

where $y(x) = 0$ for $-1 \le x \le 0$ and $y(3/2) = 1$. The functions $q(x)$ and $f(x)$ are assumed smooth. What is significant here is that the solution is evaluated at $x - 1$ in one of the terms of the equation. To answer the questions below, keep in mind that $y(x)$ and $y'(x)$ are continuous for $0 \le x \le 3/2$.
(a) There is a layer at $x = 0^+$, at $x = 1^{\pm}$, and at $x = 3/2$. Use this information to find a first-term approximation of the solution. To do this, you should consider $x < 1$ and $1 < x$ separately and then require smoothness at $x = 1$. Also, you will have to find the first two terms in the layer at $x = 1$ to get the expansions to match.
(b) Find the exact solution of the problem in the case where $f(x) = 0$, and compare the result with the expansion from part (a).

2.40. For some problems, locating the layer(s) can be difficult. To understand this, consider the following problem:

$$\varepsilon y'' + (x - a)(x - b)(4y' - 1) = 0 \quad \text{for } 0 < x < 1,$$

where $y(0) = -2$ and $y(1) = 3$. Also, $0 < a < b < 1$. The graph of the solution of this problem is shown in Fig. 2.17.

(a) Using the plausibility argument given in the discussion for Fig. 2.11, explain why there is no boundary layer at $x = 1$ but there might be one at $x = 0$.

(b) Interior layers can appear at points where the coefficient of y' is zero. Using the plausibility argument explain why there is no layer at $x = a$ but there might be one at $x = b$.

(c) Assuming the layer is at $x = 0$, calculate the first term in the expansions. Also, assuming the layer is at $x = b$, calculate the first term in the expansions. Explain why it is not possible to determine the position for the layer from these expansions.

(d) Find the exact solution. Use this to show that the layer is at $x = 0$ if $\frac{b}{3} \le a < b$ and at $x = b$ if $0 < a < \frac{b}{3}$. Note that Appendix C will be helpful here.

2.41. In the study of explosions of gaseous mixtures one finds a model where the (nondimensional) temperature $T(t)$ of the gas satisfies (Kassoy, 1976; Kapila, 1983)

$$T' = \varepsilon(T_\infty - T)^n \exp\left(\frac{T-1}{\varepsilon T}\right)$$

for $T(0) = 1$. Here $T_\infty > 1$ is a constant known as the adiabatic explosion temperature. Also, n is a positive integer (it is the overall reaction order). Assuming a high activation energy, the parameter ε is small.

(a) What is the steady-state temperature?

(b) Find the first two terms in a regular expansion of the temperature. This expansion satisfies the initial condition and describes the solution in what is known as the ignition period. Explain why the expansion is not uniform in time. Also, toward the end of the ignition period the solution is known to undergo a rapid transition to the steady state. Use your expansion to estimate when this occurs.

(c) To understand how the solution makes the transition from the rapid rise in the transition layer to the steady state, let

$$\tau = \frac{t - t_0}{\mu(\varepsilon)},$$

where t_0 is the time where the transition takes place and $\mu(\varepsilon)$ is determined from balancing in the layer. Assuming that $T \sim T_\infty - \varepsilon T_1(\tau) + \cdots$, find μ and T_1. Although T_1 is defined implicitly, use its direction field to determine what happens when $\tau \to \infty$ and $\tau \to -\infty$.

It is worth pointing out that there is a second internal layer in this problem, and it is located between the one for the ignition region and the layer you found in part (c). The matching of these various layers is fairly involved; the details can be found in Kapila (1983) for the case where $n = 1$. Also, it is actually possible to solve the original problem in closed form, although the solution is not simple (Parang and Jischke, 1975).

2.42. This problem outlines a proof, using a symmetry argument, that $D = 1$ in (2.74). Basically, the proof is based on the observation that in Fig. 2.13, if the solution were flipped around $y = 0$ and then flipped around $x = 1/2$, one would get the solution back again. In this problem, instead of (2.64), (2.65), suppose the boundary conditions are $y(0) = a$ and $y(1) = b$. In this case the solution can be written as $y = f(x, a, b)$.

(a) Change variables and let $s = 1 - x$ (which produces a flip around $x = 1/2$) and $z = -y$ (which produces a flip around $y = 0$) to obtain

$$\varepsilon z'' = zz' - z \quad \text{for } 0 < s < 1,$$

where $z(0) = -b$ and $z(1) = -a$.

(b) Explain why the solution of the problem in part (a) is $z = f(s, -b, -a)$.

(c) Use part (b) to show that $y = -f(1 - x, -b, -a)$, and from this explain why $f(x, a, b) = -f(1 - x, -b, -a)$.

(d) Use part (c) to show that in the case where $a = 1$ and $b = -1$, $y(\frac{1}{2}) = 0$. It follows from this that $D = 1$.

2.6 Corner Layers

One of the distinguishing features of the problems we have studied in this chapter is the rapid changes in the solution in the layer regions. The problems we will now investigate are slightly different because the rapid changes will be in the slope, or derivatives of the solution, and not in the value of the solution itself. To illustrate this, we consider the following problem:

$$\varepsilon y'' + \left(x - \frac{1}{2}\right)p(x)y' - p(x)y = 0 \quad \text{for } 0 < x < 1, \tag{2.90}$$

where

$$y(0) = 2 \tag{2.91}$$

and

$$y(1) = 3. \tag{2.92}$$

The function $p(x)$ is assumed to be smooth and positive, with $p(1/2) = 1$. For example, one could take $p(x) = x + 1/2$ or $p(x) = e^{2x-1}$. It should also be noted that the coefficient of y' is zero at $x = 1/2$. Because of this, given the observations of the previous section, it should not come as a surprise that the layer in this example is located at $x = 1/2$.

2.6.1 Step 1: Outer Expansion

The solution in this region is expanded in the usual power series as follows:

$$y(x) \sim y_0(x) + \varepsilon y_1(x) + \cdots . \tag{2.93}$$

From (2.90) one then finds that

$$y_0 = a\left(x - \frac{1}{2}\right), \tag{2.94}$$

where a is an arbitrary constant. As usual, we are faced with having to satisfy two boundary conditions with only one integration constant.

2.6.2 Step 2: Corner Layer

We begin by determining whether there are boundary layers. These can be ruled out fairly quickly using the plausibility argument presented in the previous section. For example, if there is a boundary layer at $x = 0$ and (2.94) is the outer solution, then we have a situation similar to that shown in Fig. 2.11. In the boundary layer $y'' > 0$, $y' < 0$, and there is a portion of the curve where $y < 0$. This means that $\varepsilon y'' > 0 > -(x - \frac{1}{2})p(x)y' + p(x)y$. It is therefore impossible to satisfy the differential Eq. (2.90) and have a boundary layer as indicated in Fig. 2.11. Using a similar argument, and the fact that the coefficient of the y' term changes sign in the interval, one can also argue that there is not a boundary layer at the other end.

It therefore appears that there is an interior layer. To investigate this, we introduce the stretched variable

$$\bar{x} = \frac{x - x_0}{\varepsilon^\alpha}. \tag{2.95}$$

With an interior layer there is an outer solution for $0 \leq x < x_0$ and one for $x_0 < x \leq 1$. Using boundary conditions (2.91), (2.92) and the general solution given in (2.94), we have that the outer solution is

$$y \sim \begin{cases} -4(x - \frac{1}{2}) & \text{if } 0 \leq x < x_0, \\ 6(x - \frac{1}{2}) & \text{if } x_0 < x \leq 1. \end{cases} \tag{2.96}$$

It will make things easier if we can determine x_0 before undertaking the layer analysis. If $0 < x_0 < \frac{1}{2}$ or if $\frac{1}{2} < x_0 < 1$, then the outer solution (2.96) is discontinuous at x_0. Using a plausibility argument similar to that presented in Sect. 2.5, one can show that neither case is possible (Exercise 2.46). In other words, $x_0 = \frac{1}{2}$. With this we get that the outer solution is continuous,

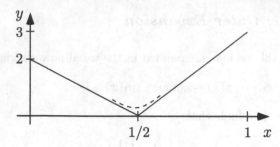

Figure 2.18 Schematic of corner layer formed by outer solution in (2.96). The solution from the corner-layer region should provide a smooth transition between these linear functions

but it is not differentiable at $x = \frac{1}{2}$. This situation is shown in Fig. 2.18. It is for this reason that we have what is called a *corner region*, or derivative layer, at $x_0 = \frac{1}{2}$.

Now, substituting (2.95) into (2.90), and letting Y designate the solution in this region, we get

$$\varepsilon^{1-2\alpha} Y'' + \bar{x}\, p\left(\frac{1}{2} + \varepsilon^\alpha \bar{x}\right) Y' - p\left(\frac{1}{2} + \varepsilon^\alpha \bar{x}\right) Y = 0. \tag{2.97}$$

To determine the distinguished limit note that

$$p\left(\frac{1}{2} + \varepsilon^\alpha \bar{x}\right) = p\left(\frac{1}{2}\right) + \varepsilon^\alpha \bar{x} p'\left(\frac{1}{2}\right) + \cdots$$

$$= 1 + O(\varepsilon^\alpha).$$

With this, from balancing the terms in (2.97) we obtain $\alpha = \frac{1}{2}$. Unlike what we assumed in previous examples, we now take

$$Y \sim y_0(x_0) + \varepsilon^\gamma Y_0 + \cdots. \tag{2.98}$$

For this example $y_0(x_0) = 0$. Also, the multiplicative factor ε^γ is needed to be able to match to the outer solution (the constant γ will be determined from the matching). Thus, substituting this into (2.97) yields

$$Y_0'' + \bar{x}\, Y_0' - Y_0 = 0, \quad \text{for } -\infty < \bar{x} < \infty. \tag{2.99}$$

It is possible to solve this equation using power series methods. However, a simpler way is to notice that $Y_0 = \bar{x}$ is a solution, and so using the method of reduction of order, one finds that the general solution is

$$Y_0 = A\bar{x} + B\left[e^{\bar{x}^2/2} + \bar{x} \int_0^{\bar{x}} e^{-s^2/2} ds\right]. \tag{2.100}$$

2.6.3 Step 3: Matching

In the examples from the previous two sections, the layer analysis and matching were carried out in a single step. This is not done here because the matching in this problem is slightly different and is worth considering in more detail. To do the matching, we introduce the intermediate variable

$$x_\eta = \frac{x - 1/2}{\varepsilon^\kappa}, \tag{2.101}$$

where $0 < \kappa < \frac{1}{2}$. Rewriting the outer solution (2.96) in this variable yields

$$y \sim \begin{cases} -4\varepsilon^\kappa x_\eta & \text{if} \quad x_\eta < 0, \\ 6\varepsilon^\kappa x_\eta & \text{if} \quad 0 < x_\eta. \end{cases} \tag{2.102}$$

Also, using the fact that

$$\int_0^\infty e^{-s^2/2}ds = \sqrt{\frac{\pi}{2}},$$

it follows from (2.100) that

$$Y \sim \begin{cases} \varepsilon^{\gamma+\kappa-1/2}x_\eta(A - B\sqrt{\frac{\pi}{2}}) & \text{if} \quad x_\eta < 0, \\ \varepsilon^{\gamma+\kappa-1/2}x_\eta(A + B\sqrt{\frac{\pi}{2}}) & \text{if} \quad 0 < x_\eta. \end{cases} \tag{2.103}$$

To be able to match (2.102) and (2.103) we must have $\gamma = \frac{1}{2}$. In this case,

$$A - B\sqrt{\frac{\pi}{2}} = -4$$

and

$$A + B\sqrt{\frac{\pi}{2}} = 6,$$

from which it follows that $A = 1$ and $B = 5\sqrt{2/\pi}$.

2.6.4 Step 4: Composite Expansion

Even though the situation is slightly more complicated than before, the construction of a composite expansion follows the same rules as in the earlier examples. For example, for $0 \le x \le \frac{1}{2}$,

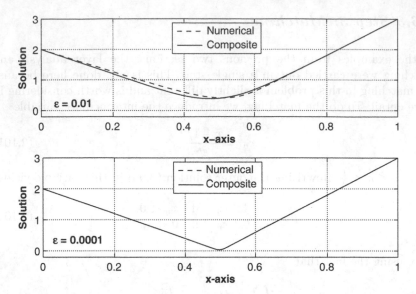

Figure 2.19 Composite expansion (2.104) and the numerical solution of (2.90)–(2.92) when $\varepsilon = 10^{-2}$ and when $\varepsilon = 10^{-4}$. Also, $p(x) = e^{5(2x-1)}$

$$y \sim -4\left(x - \frac{1}{2}\right) + \varepsilon^{1/2}\left[\bar{x} + 5\sqrt{\frac{2}{\pi}}\left(e^{-\bar{x}^2/2} + \bar{x}\int_0^{\bar{x}} e^{-s^2/2}ds\right)\right] + 4\varepsilon^\kappa x_\eta$$

$$= \left(x - \frac{1}{2}\right)\left[1 + 5\,\mathrm{erf}\left(\frac{x - \frac{1}{2}}{\sqrt{2\varepsilon}}\right)\right] + 5\sqrt{\frac{2\varepsilon}{\pi}}\,e^{-(2x-1)^2/(8\varepsilon)}, \qquad (2.104)$$

where $\mathrm{erf}(\cdot)$ is the error function. One finds that this is also the composite expansion for $\frac{1}{2} \leq x \leq 1$. Thus, (2.104) is a composite expansion over the entire interval. This function is shown in Fig. 2.19, and it is clear that it is in very good agreement with the numerical solution. In fact, when $\varepsilon = 10^{-3}$, the two curves are indistinguishable.

The equation studied in this section is related to one that has received a great deal of attention because the solution has been found to have some rather interesting properties. To understand the situation, consider

$$\varepsilon y'' - \left(x - \frac{1}{2}\right)y' + ky = 0 \quad \text{for } 0 < x < 1, \qquad (2.105)$$

where k is a constant. Assume that $y(0) = y(1) = 1$; the solution is then

$$y(x) = \frac{M\left(-\frac{k}{2}, \frac{1}{2}, \frac{1}{8\varepsilon}(2x-1)^2\right)}{M\left(-\frac{k}{2}, \frac{1}{2}, \frac{1}{8\varepsilon}\right)}, \qquad (2.106)$$

where M is Kummer's function (Appendix B). It is assumed here that the value of ε is such that the denominator in (2.106) is nonzero. Using the known asymptotic properties of M one finds that for small ε and $x \neq 1/2$,

$$y(x) \sim \begin{cases} e^{-x(1-x)/2\varepsilon} & \text{for} \quad k \neq 0, 2, 4, \ldots, \\ (2x-1)^k & \text{for} \quad k = 0, 2, 4, \ldots. \end{cases}$$

This shows that there is a boundary layer at each endpoint for all but a discrete set of values for the constant k. When there are boundary layers, the solution in the outer region is transcendentally small and goes to zero as $\varepsilon \downarrow 0$. What is significant is that for $k = 0, 2, 4, \ldots$ this does not happen. This behavior at a discrete set of points for the parameter k is reminiscent of resonance, and this has become known as Ackerberg–O'Malley resonance. Those interested in pursuing this topic further are referred to the original paper by Ackerberg and O'Malley (1970) and a later study by De Groen (1980).

Exercises

2.43. Find a composite expansion of the solutions of the following problems:
(a) $\varepsilon y'' + (y')^2 - 1 = 0$ for $0 < x < 1$, where $y(0) = 1$, and $y(1) = 1$.
(b) $\varepsilon y'' + (y')^2 - 1 = 0$ for $0 < x < 1$, where $y(0) = 1$, and $y(1) = 1/2$.
(c) $\varepsilon y'' = 9 - (y')^2$ for $0 < x < 1$, where $y(0) = 0$, and $y(1) = 1$.
(d) $\varepsilon y'' + 2xy' - (2 + \varepsilon x^2)y = 0$ for $-1 < x < 1$, where $y(-1) = 2$ and $y(1) = -2$.

2.44. Consider the problem

$$\varepsilon y'' = x^2[1 - (y')^2] \quad \text{for } 0 < x < 1,$$

where $y(0) = y(1) = 1$.
(a) Assuming there is a corner-layer solution, explain why there are two possible outer solutions. Each one is piecewise linear, much like the outer solution in (2.96). Use the plausibility argument to rule out one of them.
(b) After finding the corner-layer solution, construct a composite expansion.

2.45. Consider the problem

$$\varepsilon y'' + xp(x)y' - q(x)y = 0 \quad \text{for } -1 < x < 1,$$

where $y(-1) = a$ and $y(1) = b$. The functions $p(x)$ and $q(x)$ are continuous, $p(x) \neq 0$, and $q(x) > 0$ for $-1 \leq x \leq 1$.
(a) If $p(0) < 0$, then there is a boundary layer at each end. Find a composite expansion of the solution.
(b) If $p(0) > 0$, then there is an interior layer. Find the approximations in the layer and outer regions.

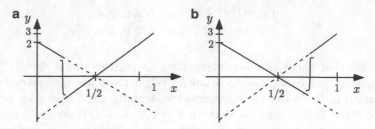

Figure 2.20 Schematic of situations considered in Exercise 2.46. The outer solution (*solid lines*) is determined from (2.94)

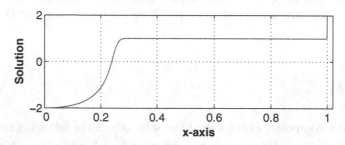

Figure 2.21 Solution of problem in Exercise 2.47

2.46. This problem demonstrates that $x_0 = \frac{1}{2}$ in (2.96).
(a) Use the plausibility argument given in Sect. 2.5 to show that it is not possible that $0 < x_0 < \frac{1}{2}$. This situation is shown in Fig. 2.20a.
(b) Use the plausibility argument given in Sect. 2.5 to show that it is not possible that $\frac{1}{2} < x_0 < 1$. This situation is shown in Fig. 2.20b.

2.47. Consider the problem

$$\varepsilon y'' - (x - a)(x - b)y' - x(y - 1) = 0 \quad \text{for } 0 < x < 1,$$

where $y(0) = -2$ and $y(1) = 2$. The numerical solution is shown in Fig. 2.21 in the case where $a = 1/4$, $b = 3/4$, and $\varepsilon = 10^{-4}$. Based on this information derive a first-term approximation of the solution for arbitrary $0 < a < b < 1$.

2.48. Corner layers can occur within a boundary layer. As an example of this, consider the problem

$$\varepsilon y'' + \tanh(y') - y = -1 \quad \text{for } 0 < x < 1,$$

where $y(0) = 3$ and $y(1) = 5$. The numerical solution of the problem is shown in Fig. 2.22 when $\varepsilon = 10^{-4}$.
(a) Find a first-term expansion of the solution. You do not need to solve the problem for the corner layer but you do need to explain why the solution matches with the neighboring layer solutions.

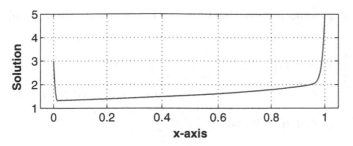

Figure 2.22 Solution of problem in Exercise 2.48

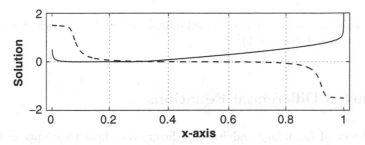

Figure 2.23 Solution of problem in Exercise 2.49

(b) As a test of the effectiveness of the numerical solvers currently available, find the numerical solution and compare it to your result in part (a) in the case where $\varepsilon = 10^{-4}$.

2.49. Consider the problem (Lorentz, 1982)

$$\varepsilon y'' + y(1 - y^2)y' - y = 0 \quad \text{for } 0 < x < 1,$$

where $y(0) = \alpha$ and $y(1) = \beta$. Find a first-term approximation of the solution in the following cases.

(a) $\alpha = 1/2$ and $\beta = 2$. The numerical solution of the problem is shown in Fig. 2.23 (solid curve), when $\varepsilon = 0.0008$. You do not need to solve the problem for the corner layer (which is located at $x_0 = 1/3$).

(b) $\alpha = 3/2$ and $\beta = -3/2$. The numerical solution of the problem is in Fig. 2.23 (dashed curve), when $\varepsilon = 0.08$.

(c) $\alpha = 3/2$ and $\beta = 2$.

2.50. This exercise concerns the shock wave produced by a cylinder that is expanding with a constant radial velocity of εa_0, where a_0 is the velocity of sound in still air (Lighthill, 1949). The velocity potential in this case has the form $\phi(r, t) = a_0^2 t f(\eta)$, where r is the radial distance from the center of the expanding cylinder, $\eta = r/(a_0 t)$, and $f(\eta)$ is determined below. Also, the radius of the cylinder is $r = \varepsilon a_0 t$, and the radius of the shock wave is $r = \alpha a_0 t$, where α is a constant determined below. The problem that determines $f(\eta)$ and α is

$$\left[1 - (\gamma - 1)\left(f - \eta f' + \frac{1}{2}(f')^2\right)\right](f' + \eta f'')$$
$$= \eta(\eta - f')^2 f'' \quad \text{for } \varepsilon < \eta < \alpha,$$

where $f'(\varepsilon) = \varepsilon$, $f(\alpha) = 0$ and

$$f'(\alpha) = \frac{2(\alpha - 1/\alpha)}{\gamma + 1}.$$

In this problem γ is a positive constant called the adiabatic index. Show that, for small ε, $\alpha \sim 1 + \frac{3}{8}(\gamma + 1)^2 \varepsilon^4$.

2.7 Partial Differential Equations

The subject of boundary and interior layers and how they appear in the solutions of partial differential equations is enormous. We will examine a couple of examples that lend themselves to matched asymptotic expansions.

2.7.1 Elliptic Problem

The first example concerns finding the function $u(x, y)$ that is the solution of the following boundary-value problem:

$$\varepsilon \nabla^2 u + \alpha \partial_x u + \beta \partial_y u + u = f(x, y) \quad \text{for } (x, y) \in \Omega, \tag{2.107}$$

where

$$u = g(x, y), \quad \text{for } (x, y) \in \partial\Omega. \tag{2.108}$$

The domain Ω is assumed to be bounded, simply connected, and have a smooth boundary $\partial\Omega$. The coefficients α and β are constant with at least one of them nonzero. The functions $f(x, y)$ and $g(x, y)$ are assumed to be continuous.

To illustrate what a solution of (2.107) looks like, consider the special case where $f(x, y) = a$ and $g(x, y) = b$, where a and b are constants. Taking Ω to be the unit disk and using polar coordinates $x = \rho \cos\varphi$, $y = \rho \sin\varphi$, the problem can be solved using separation of variables. The resulting solution is

$$u(\rho, \varphi) = a + \frac{1}{F(\rho, \varphi)} \sum_{n=-\infty}^{\infty} I_n(\chi\rho)[a_n \sin(n\varphi) + b_n \cos(n\varphi)]/I_n(\chi), \tag{2.109}$$

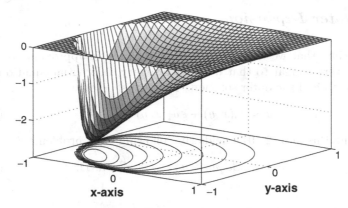

Figure 2.24 Plot of solution (2.109) and its associated contour map when $\alpha = \beta = a = 1$, $b = 0$, and $\varepsilon = 0.05$. The domain for the problem is $x^2 + y^2 < 1$, but to generate the plot, the solution was extended to the unit square $0 \leq x, y \leq 1$ by setting $u = 0$ for $x^2 + y^2 \geq 1$. It is apparent there is a boundary layer, and its presence is particularly pronounced in the region of the boundary near $x = y = -1/\sqrt{2}$.

where $F(\rho, \varphi) = \exp(\rho(\alpha \cos \varphi + \beta \sin \varphi)/(2\varepsilon))$, $\chi = \sqrt{\alpha^2 + \beta^2 - 4\varepsilon}/(2\varepsilon)$, $I_n(z)$ is a modified Bessel function,

$$a_n = \frac{b-a}{2\pi} \int_0^{2\pi} F(1, \varphi) \sin(n\varphi) d\varphi,$$

and

$$b_n = \frac{b-a}{2\pi} \int_0^{2\pi} F(1, \varphi) \cos(n\varphi) d\varphi.$$

This solution is shown in Fig. 2.24. A few observations can be made from this figure that will make the derivation of the asymptotic approximation easier to follow. First, a boundary layer is clearly evident in the solution. For example, if one starts at $x = y = -1/\sqrt{2}$ and then moves into the domain, then the solution undergoes a rapid transition to what appears to be an outer solution. It is also seen that the layer is not present around the entire boundary $\partial \Omega$ but only over a portion of $\partial \Omega$. Moreover, if one follows the solution around the edge of the boundary, then it is not easy to identify exactly where the solution switches between the boundary layer and the outer region. However, wherever it is, the transition is relatively smooth. This latter observation will be useful later when deciding on the importance of what are called tangency points.

Before jumping into the derivation of the asymptotic expansion, it is worth noting that when $\varepsilon = 0$, the differential equation in (2.107) reduces to a first-order hyperbolic equation. This change in type, from an elliptic to a hyperbolic equation, has important consequences for the analysis, and it helps explain some of the steps that are taken below.

2.7.2 Outer Expansion

The procedure that we will use to find an asymptotic approximation of the solution is very similar to that used for ordinary differential equations. The first step is to find the outer solution. To do this, assume

$$u \sim u_0(x, y) + \varepsilon u_1(x, y) + \cdots .$$ (2.110)

Substituting this into (2.107) yields the following $O(1)$ problem:

$$\alpha \partial_x u_0 + \beta \partial_y u_0 + u_0 = f(x, y) \quad \text{in } \Omega,$$ (2.111)

where

$$u_0 = g(x, y) \text{ on } \partial \Omega_o.$$ (2.112)

When constructing an asymptotic approximation it is not always immediately clear which boundary condition, if any, the outer solution should satisfy. In (2.112) the portion of the boundary where u_0 satisfies the original boundary condition has been identified as $\partial \Omega_o$. This is presently unknown and, in fact, could turn out to be empty (e.g., Exercise 2.52).

To solve (2.111), we change coordinates from (x, y) to (r, s) , where s is the characteristic direction for (2.111). Specifically, we let $x = \alpha s + \xi(r)$ and $y = \beta s + \eta(r)$. The functions ξ and η can be chosen in a number of different ways, and our choice is based on the desire to have a simple coordinate system. In particular, we take

$$x = \alpha s + \beta r \quad \text{and} \quad y = \beta s - \alpha r.$$ (2.113)

With this $\partial_s = \alpha \partial_x + \beta \partial_y$, and so (2.111) becomes

$$\partial_s u_0 + u_0 = f(\alpha s + \beta r, \beta s - \alpha r).$$ (2.114)

This is easy to solve, and the general solution is

$$u_0 = a_0(r) e^{-s} + \int^s e^{\tau - s} f(\alpha \tau + \beta r, \beta \tau - \alpha r) d\tau,$$ (2.115)

where $a_0(r)$ is arbitrary. Since we have not yet determined exactly what portion of the boundary condition, if any, the outer solution should satisfy, we are not yet in a position to convert back to (x, y) coordinates.

Before investigating the boundary layer, it is instructive to consider the change of coordinates given in (2.113). Fixing r and letting s increase, one obtains the directed lines shown in Figs. 2.25 and 2.26. So, starting at boundary point P, as the variable s increases, one crosses the domain Ω and arrives at boundary point P_*. In terms of our first-term approximation, we need to know what value u_0 starts with at P and what value it has when the point P_* is reached. However, we only have one integration constant in (2.115), so there is going to be a boundary layer at either P or P_*. As we will see below,

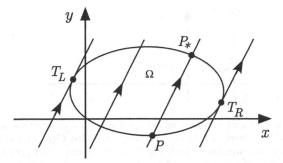

Figure 2.25 Schematic drawing of characteristic curves obtained from the problem for the outer solution. These curves are determined from (2.113) and are directed straight lines with an orientation determined by the direction of increasing s. As drawn, the coefficients α, β are assumed positive. Also note the tangency points T_r, T_L

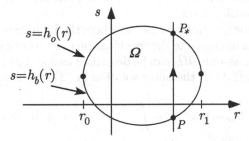

Figure 2.26 Schematic drawing of transformed region coming from the domain in Fig. 2.25

for our problem, the outer solution will satisfy the boundary condition at P_* and there will be a boundary layer of width ε at P. Based on these comments, we will assume that the points of the boundary can be separated into three disjoint sets:

- $\partial\Omega_o$. These are the boundary points where the characteristics leave Ω (like P_* in Figs. 2.25 and 2.26). In Fig. 2.26 these points make up the curve $s = h_o(r)$ for $r_0 < r < r_1$.
- $\partial\Omega_b$. There are the boundary points where the characteristics enter Ω (like P in Figs. 2.25 and 2.26). In Fig. 2.26 these points make up the curve $s = h_b(r)$ for $r_0 < r < r_1$.
- $\partial\Omega_t$. There are the tangency points. In Fig. 2.25 there are two such points, T_L and T_R, and in Fig. 2.26 they occur when $r = r_0$ and $r = r_1$. These can cause real headaches when constructing an asymptotic approximation and will be left until the end.

To keep the presentation simple, we will assume that the situation is as pictured in Fig. 2.25, i.e., the domain is convex. This means that if a

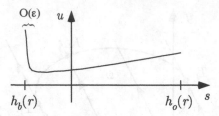

Figure 2.27 Schematic of solution as a function of the characteristic coordinate s as defined in (2.113). There is a boundary layer of width $O(\varepsilon)$ at the left end of the interval. Note that, in connection with Fig. 2.26, the point $s = h_b(r)$ can be thought of as corresponding to the point P, and $s = h_o(r)$ can be thought of as corresponding to the point P_*

characteristic curve enters the domain, then the only other time it intersects the boundary is when it leaves the domain. Moreover, $\partial\Omega_o$ and $\partial\Omega_b$ are assumed to be smooth curves.

Now that we know (or think we know) where the boundary layer is located, we are in a position to complete the specification of the outer solution given in (2.115). To do this, assume $\partial\Omega_o$ can be described as $s = h_o(r)$ for $r_0 < r < r_1$. In this case, using (2.112), the outer solution in (2.115) becomes

$$u_0 = g(\alpha h_o + \beta r, \beta h_o - \alpha r)e^{h_o - s} - \int_s^{h_o(r)} e^{\tau - s} f(\alpha\tau + \beta r, \beta\tau - \alpha r)d\tau. \quad (2.116)$$

This solution does not apply along $\partial\Omega_b$ or on $\partial\Omega_t$. To be more specific, suppose the curve $s = h_b(r)$ for $r_0 < r < r_1$ describes $\partial\Omega_b$. In this case (2.116) holds for $r_0 < r < r_1$ and $h_b(r) < s \leq h_o(r)$.

2.7.3 Boundary-Layer Expansion

To find out what goes on in a boundary layer, we introduce the boundary coordinate (Fig. 2.27)

$$S = \frac{s - h_b(r)}{\varepsilon}. \quad (2.117)$$

This coordinate, by necessity, depends on r since the boundary depends on r. This complicates the calculations in making the change of variables from (r, s) to (r, S). One finds using the chain rule that the derivatives transform as follows:

$$\partial_s \to \frac{1}{\varepsilon}\partial_S, \qquad \partial_r \to -\frac{h_b'}{\varepsilon}\partial_S + \partial_r,$$

$$\partial_s^2 \to \frac{1}{\varepsilon^2}\partial_S^2, \qquad \partial_r^2 \to \frac{(h_b')^2}{\varepsilon^2}\partial_S^2 - \frac{h_b''}{\varepsilon}\partial_S - \frac{2h_b'}{\varepsilon}\partial_S\partial_r + \partial_r^2.$$

To convert the original problem into boundary-layer coordinates, we must first change from (x, y) to (r, s) coordinates in (2.107). This is relatively easy since from (2.113) one finds that

$$\partial_x = \frac{1}{\gamma}(\alpha\partial_s + \beta\partial_r) \quad \text{and} \quad \partial_y = \frac{1}{\gamma}(\beta\partial_s - \alpha\partial_r),$$

where $\gamma = \alpha^2 + \beta^2$. With this, (2.107) becomes

$$\varepsilon(\partial_s^2 + \partial_r^2)u + \gamma\partial_s u + \gamma u = \gamma f. \tag{2.118}$$

Now, letting $U(r, S)$ denote the solution in the boundary layer, substituting the boundary-layer coordinates into (2.118) yields

$$[\mu\partial_S^2 + \gamma\partial_S + O(\varepsilon)]U = \varepsilon\gamma f, \tag{2.119}$$

where

$$\mu(r) = 1 + (h_b')^2. \tag{2.120}$$

In keeping with our usual assumptions, $h_b(r)$ is taken to be a smooth function. However, note that at the tangency points shown in Fig. 2.25, $h_b'(r) = \infty$. For this reason, these points will have to be dealt with separately after we finish with the boundary layer.

We are now in a position to expand the solution in the usual power series expansion, and so let

$$U(r, S) \sim U_0(r, S) + \cdots . \tag{2.121}$$

Substituting this into (2.119) yields the equation

$$\mu\partial_S^2 U_0 + \gamma\partial_S U_0 = 0.$$

The general solution of this is

$$U_0(r, S) = A(r) + B(r)e^{-\gamma S/\mu}, \tag{2.122}$$

where $A(r)$ and $B(r)$ are arbitrary. Now, from the boundary and matching conditions we must have $U_0(r, 0) = g$ and $U_0(r, \infty) = u_0(r, h_b)$. Imposing these on our solution in (2.122) yields

$$U_0(r, S) = u_0(r, h_b) + [g(\alpha h_b + \beta r, \beta h_b - \alpha r) - u_0(r, h_b)]e^{-\gamma S/\mu}, \tag{2.123}$$

where

$$u_0(r, h_b) = g(\alpha h_o + \beta r, \beta h_o - \alpha r)e^{h_o - h_b}$$
$$- \int_{h_b}^{h_o} e^{\tau - h_b(r)} f(\alpha\tau + \beta r, \beta\tau - \alpha r)d\tau.$$

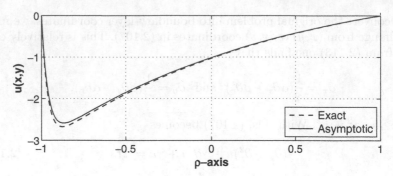

Figure 2.28 Comparison between exact solution (2.109) and composite expansion given in (2.125) in the case where $\varepsilon = 0.05$. The values of each function are given along the line $x = \rho\cos(\pi/4), y = \rho\sin(\pi/4)$ for $-1 \le \rho \le 1$

2.7.4 Composite Expansion

It is not difficult to put together a composite expansion that will give us a first-term approximation of the solution in the outer and boundary-layer regions. Adding (2.116) and (2.123) together and then subtracting their common part yields

$$u \sim g(\alpha h_o + \beta r, \beta h_o - \alpha r)e^{h_o - s} - \int_s^{h_o} e^{\tau - s} f(\alpha\tau + \beta r, \beta\tau - \alpha r)\mathrm{d}\tau$$

$$+ [g(\alpha h_b + \beta r, \beta h_b - \alpha r) - u_0(r, h_b)]\, e^{-\gamma S/\mu}. \qquad (2.124)$$

This approximation holds for $r_0 < r < r_1$ and $h_b(r) \le s \le h_o(r)$. This result may not be pretty, but it does give us a first-term approximation of the solution in the boundary layer and outer domain. It is also considerably simpler than the formula for the exact solution, an example of which is given in (2.109). What is interesting is that we have been able to patch together the solution of an elliptic problem using solutions to hyperbolic and elliptic problems. Readers interested in the theoretical foundation of the approximations constructed here are referred to Levinson (1950), Eckhaus and Jager (1966), and Il'in (1992).

Example

To apply the result to the domain used in Fig. 2.24, note that the change of variables in (2.113) transforms $\partial\Omega$, which is the unit circle $x^2 + y^2 = 1$, to the circle $r^2 + s^2 = 1/(\alpha^2 + \beta^2)$. In this case, $h_o(r)$ is the upper half of the circle, where $s > 0$, and $h_b(r)$ is the lower half, where $s < 0$. Taking $f = a$

and $g = b$, then (2.124) reduces to

$$u \sim a + (b - a)\left[e^{h_o - s} + \left(1 - e^{h_o - h_b}\right)e^{-\gamma S/\mu}\right], \qquad (2.125)$$

where $h_o = \sqrt{1/\gamma - r^2}$, $h_b = -h_o$, $\mu = 1 + r^2/h_o$, and $\gamma = \alpha^2 + \beta^2$. This can be expressed in terms of (x, y) using the formulas $r = (\beta x - \alpha y)/\gamma$ and $s = (\alpha x + \beta y)/\gamma$. With (2.125) we have a relatively simple expression that is a composite approximation of the exact solution given in (2.109). To compare them, let $f = 1$, $g = 0$, and $\alpha = \beta = 1$. These are the same values used in Fig. 2.24. The resulting approximation obtained from (2.125), along with the exact solution given in (2.109), is shown in Fig. 2.28 for a slice through the surface. Based on this graph, it seems that we have done reasonably well with our approximation. However, we are not finished as (2.124) does not hold in the immediate vicinity of the tangency points T_R and T_L, which are shown in Fig. 2.25. ∎

2.7.5 Parabolic Boundary Layer

To complete the construction of a first-term approximation of the solution of (2.107), it remains to find out what happens near the tangency points shown in Figs. 2.25 and 2.26. We will concentrate on T_L. To do this, let (r_0, s_0) be its coordinates in the (r, s) system. Also, suppose the smooth curve $r = q(s)$ describes the boundary $\partial \Omega$ in this region. In this case $r_0 = q(s_0)$ and $q'(s_0) = 0$. It will be assumed here that $q''(s_0) \neq 0$. The boundary-layer coordinates are now

$$\tilde{r} = \frac{r - q(s)}{\varepsilon^\alpha} \quad \text{and} \quad \tilde{s} = \frac{s - s_0}{\varepsilon^\beta}. \qquad (2.126)$$

The transformation formulas for the derivatives are similar to those derived earlier for the boundary layer along $\partial \Omega_b$, so they will not be given. The result is that (2.118) takes the form

$$\varepsilon(\varepsilon^{-2\alpha}\partial_{\tilde{r}}^2 + \varepsilon^{-2\beta}\partial_{\tilde{s}}^2 + \cdots)\tilde{U}$$
$$+ \gamma(\varepsilon^{-\beta}\partial_{\tilde{s}} - \varepsilon^{\beta-\alpha}\tilde{s}q_0''\partial_{\tilde{r}} + \cdots)\tilde{U} + \gamma\tilde{U} = \gamma f.$$

Here $\tilde{U}(\tilde{r}, \tilde{s})$ is the solution in this region, and we have used the Taylor series expansion $q'(s_0 + \varepsilon^\beta \tilde{s}) \sim \varepsilon^\beta \tilde{s}q_0''$ to obtain the preceding result. There are at least two balances that need to be considered. One is $\alpha = \beta = 1$, the other is $2\beta = \alpha = 2/3$. The latter is the one of interest, and assuming

$$\tilde{U}(\tilde{r}, \tilde{s}) \sim \tilde{U}_0(\tilde{r}, \tilde{s}) + \cdots \qquad (2.127)$$

one obtains the equation

$$(\partial_{\tilde{r}}^2 - \gamma \tilde{s} q_0'' \partial_{\tilde{r}} + \gamma \partial_{\tilde{s}}) \widetilde{U}_0 = 0 \tag{2.128}$$

for $0 < \tilde{r} < \infty$ and $-\infty < \tilde{s} < \infty$. This is a parabolic equation, and for this reason this region is referred to as a *parabolic boundary layer*. The solution is required to match to the solutions in the adjacent regions and should satisfy the boundary condition at $\tilde{r} = 0$. The details of this calculation will not be given here but can be found in van Harten (1976). The papers by Cook and Ludford (1971, 1973) should also be consulted as they have an extensive analysis of such parabolic layers and how they appear in problems where the domain has a corner. The theory necessary to establish the uniform validity of the expansions when corners are present in the boundary, and a historical survey of this problem, can be found in Shih and Kellogg (1987).

2.7.6 Parabolic Problem

To illustrate the application of boundary-layer methods to parabolic equations, we consider the problem of solving

$$u_t + u u_x = \varepsilon u_{xx} \quad \text{for} \quad -\infty < x < \infty \text{ and } 0 < t, \tag{2.129}$$

where $u(x, 0) = \phi(x)$. It is assumed here that $\phi(x)$ is smooth and bounded except for a jump discontinuity at $x = 0$. Moreover, $\phi'(x) \leq 0$ for $x \neq 0$ and $\phi(0^-) > \phi(0^+)$.

As an example, suppose that

$$u(x, 0) = \begin{cases} 1 & \text{for } x < 0, \\ 0 & \text{for } 0 < x. \end{cases}$$

This type of initial condition generates what is known as a Riemann problem, and the resulting solution is shown in Fig. 2.29. It is a traveling wave, and the smaller the value of ε, the sharper the transition from $u = 0$ to $u = 1$. In the limit of $\varepsilon \to 0$, the transition becomes a jump, producing a solution containing a shock wave.

The nonlinear diffusion equation in (2.129) is known as *Burger's equation*. It has become the prototype problem for studying shock waves and for the use of what are called viscosity methods for finding smooth solutions to such problems. In what follows we will concentrate on the constructive aspects of finding asymptotic approximations to the solution. The theoretical underpinnings of the procedure can be found in Il'in (1992) and its extension to systems in Goodman and Xin (1992).

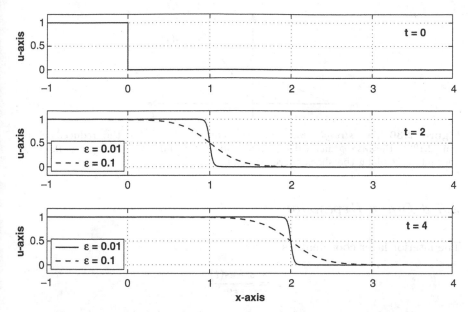

Figure 2.29 Solution of Burger's equation (2.129) for two values of ϵ, showing the traveling-wave nature of the solution as well as the sharpening of the wave as ϵ approaches zero

2.7.7 Outer Expansion

The first step is to find the outer solution. To do this, assume $u \sim u_0(x,t) + \cdots$. Substituting this into (2.129) produces the first-order hyperbolic equation

$$\partial_t u_0 + u_0 \partial_x u_0 = 0. \tag{2.130}$$

The characteristics for this equation are the straight lines $x = x_0 + \phi(x_0)t$, and the solution $u_0(x,t)$ is constant along each of these lines (Fig. 2.30). Therefore, given a point (x,t), then $u_0(x,t) = \phi(x_0)$, where x_0 is determined from the equation $x = x_0 + \phi(x_0)t$. This construction succeeds if the characteristics do not intersect and they cover the upper half-plane. For us the problem is with intersections. As illustrated in Fig. 2.30, the characteristics coming from the negative x-axis intersect with those from the positive x-axis. From the theory for nonlinear hyperbolic equations it is known that this generates a curve $x = s(t)$, known as a shock wave, across which the solution has a jump discontinuity (Holmes, 2009). The solution in this case is determined by the characteristics up to when they intersect the shock. The complication here is that the position of the interior layer is moving and centered at $x = s(t)$. As it turns out, the formula for $s(t)$ can be determined from (2.130), but we will derive it when examining the solution in the transition layer.

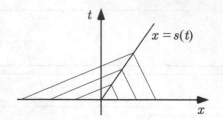

Figure 2.30 The *straight lines* are the characteristics for the reduced problem (2.130). Because ϕ has a discontinuity at $x = 0$, and $\phi' \geq 0$ for $x \neq 0$, these lines intersect along the shock curve $x = s(t)$

2.7.8 Inner Expansion

The interior layer coordinate is

$$\bar{x} = \frac{x - s(t)}{\varepsilon^\alpha} .$$

Letting $U(\bar{x}, t)$ denote the solution of the problem in this layer, (2.129) takes the form

$$\partial_t U - \varepsilon^{-\alpha} s'(t) \partial_{\bar{x}} U + \varepsilon^{-\alpha} U \partial_{\bar{x}} U = \varepsilon^{1-2\alpha} \partial_{\bar{x}}^2 U. \qquad (2.131)$$

Balancing the terms in this equation, one finds $\alpha = 1$. Thus, the appropriate expansion of the solution is $U \sim U_0(\bar{x}, t) + \cdots$, and substituting this into (2.131) yields

$$- s'(t) \partial_{\bar{x}} U_0 + U_0 \partial_{\bar{x}} U_0 = \partial_{\bar{x}}^2 U_0.$$

Integrating this, one finds that

$$\partial_{\bar{x}} U_0 = \frac{1}{2} U_0^2 - s'(t) U_0 + A(t). \qquad (2.132)$$

The boundary conditions to be used come from matching with the outer solution on either side of the layer. They are

$$\lim_{\bar{x} \to -\infty} U_0 = u_0^- \quad \text{and} \quad \lim_{\bar{x} \to \infty} U_0 = u_0^+, \qquad (2.133)$$

where

$$u_0^\pm = \lim_{x \to s(t)^\pm} u_0(x, t). \qquad (2.134)$$

From (2.132) it follows that $A(t) = -\frac{1}{2}(u_0^-)^2 + s'(t) u_0^-$ and

$$s'(t) = \frac{1}{2}(u_0^+ + u_0^-). \qquad (2.135)$$

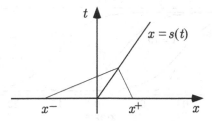

Figure 2.31 Schematic showing a shock and two characteristics that intersect on the shock

The differential equation in (2.135) determines the position of the shock and is known as the Rankine–Hugoniot condition. Its solution requires an initial condition, and, because of the assumed location of the discontinuity in $\phi(x)$, we take $s(0) = 0$.

To complete the analysis of the first-order problem in the shock layer, we separate variables in (2.132) and then integrate to obtain, for the case where $u_0^+ < u_0^-$,

$$U_0(\bar{x}, t) = \frac{u_0^+ + b(\bar{x}, t)u_0^-}{1 + b(\bar{x}, t)}, \tag{2.136}$$

where

$$b(\bar{x}, t) = B(t)e^{\bar{x}(u_0^+ - u_0^-)/2} \tag{2.137}$$

and $B(t)$ is an arbitrary nonzero function.

The indeterminacy in $U_0(\bar{x}, t)$, as given by the unspecified function in (2.137), is the same difficulty we ran into when investigating interior layers in Sect. 2.5. However, for this nonlinear diffusion problem it is possible to determine $B(t)$, up to a multiplicative factor, by examining the next order problem. To state the result, note that given a point $(x, t) = (s(t), t)$ on the shock, $u_0^+ = \phi(x^+)$ and $u_0^- = \phi(x^-)$. Here $x^\pm$ are the initial points on the x-axis for the two characteristics that intersect at $(s(t), t)$ (Fig. 2.21). In this case, one finds from the $O(\varepsilon)$ problem that (Exercise 2.57)

$$B(t) = B_0\sqrt{\frac{1 + t\phi'(x^+)}{1 + t\phi'(x^-)}}, \tag{2.138}$$

where the constant B_0 is found from the initial condition.

To determine B_0 in (2.138), one must match the shock layer solution with the solution from the initial layer located near $x = x_0$. The appropriate coordinate transformation for this layer is $\bar{x} = (x - s(t))/\varepsilon$ and $\tau = t/\varepsilon$. The steps involved in this procedure are not difficult, and one finds that $B_0 = 1$ (Exercise 2.57). With this the first-term approximation of the solution in the shock layer is determined.

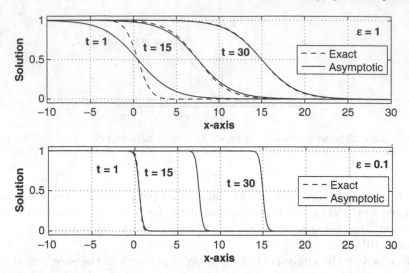

Figure 2.32 Graph of exact solution of Burger's equation and asymptotic approximation in the case where $\phi(x) = 1$ if $x < 0$ and $\phi(x) = 0$ if $x > 0$. The solutions are compared at $t = 1, 15, 30$, for both $\varepsilon = 1$ and $\varepsilon = 10^{-1}$

To demonstrate the accuracy of the asymptotic approximation, in Fig. 2.32 the approximation is shown along with the exact solution at two nonzero values of t and ε. The exact solution in this case is given in Exercise 2.56. For $\varepsilon = 0.1$ the asymptotic and exact solutions are so close that they are essentially indistinguishable in the graph. They are even in reasonable agreement when $\varepsilon = 1$, although the differences are more apparent for the smaller value of t.

In this section we have considered elliptic and parabolic problems. Matched asymptotic expansions can also be applied to hyperbolic problems, and several examples are worked out in Kevorkian and Cole (1981). However, there are better methods for wave problems, particularly when one is interested in the long time behavior of the wave. This topic will be taken up in Chaps. 3 and 4.

Exercises

2.51. A special case of (2.107) is the problem of solving

$$\varepsilon \nabla^2 u + \partial_y u = 2 \text{ in } \Omega,$$

where $u = x + y$ on $\partial\Omega$. Let Ω be the unit disk $x^2 + y^2 < 1$.

(a) Sketch the domain and characteristic curves for the outer solution (Fig. 2.25). Identify the points T_L and T_R.
(b) What is the composite expansion, as in (2.124), for this problem?
(c) What is the parabolic layer equation, as in (2.128), in this case?

2.52. Find a first-term composite expansion, for the outer region and boundary layers, of the solution of

$$\varepsilon \nabla^2 u + u = 1 \text{ in } \Omega,$$

where $u = g(x, y)$ on $\partial \Omega$. Let Ω be the unit disk $x^2 + y^2 < 1$.

2.53. In this exercise, variations of the elliptic boundary-value problem (2.107) are considered.
(a) If the coefficients α and β in (2.107) are negative, how does the composite expansion in (2.124) change?
(b) If one of the coefficients α and β in (2.107) is negative and the other is positive, how does the composite expansion in (2.124) change?

2.54. Consider the problem of solving

$$u_t = \varepsilon u_{xx} - c u_x \quad \text{for } 0 < x \text{ and } 0 < t,$$

where $u(0, t) = u_\ell$ and $u(x, 0) = u_r$. Assume c, u_ℓ, and u_r are constants with c positive and $u_r \neq u_\ell$.
(a) Find the first term in the outer expansion. Explain why this shows that there is an interior layer located at $x = ct$.
(b) Find the first term in the inner expansion. From this find a first-term composite expansion of the solution.
(c) Where is the assumption that $c > 0$ used in parts (a) or (b)? What about the assumption that $u_r \neq u_\ell$? Note that the case where $c < 0$ is considered in Exercise 2.55.
(d) The exact solution is

$$u(x, t) = u_r + \frac{1}{2}(u_\ell - u_r) \left[\text{erfc}\left(\frac{x - ct}{2\sqrt{\varepsilon t}} \right) + e^{cx/\varepsilon} \text{erfc}\left(\frac{x + ct}{2\sqrt{\varepsilon t}} \right) \right].$$

Verify that this satisfies the differential equation as well as the boundary and initial conditions.
(e) Explain how your composite expansion in part (b) can be obtained from the solution in part (d).

2.55. The equation of one-dimensional heat conduction in a material with a low conductivity is (Plaschko, 1990)

$$u_t = \varepsilon u_{xx} + v(t) u_x \quad \text{for } 0 < x \text{ and } 0 < t,$$

where $u(0,t) = g(t)$, $u(\infty,t) = 0$, and $u(x,0) = h(x)$. Assume that the functions $v(t)$, $g(t)$, and $h(x)$ are smooth with $0 < v(t)$ for $0 \leq t < \infty$, $g(0) = h(0)$, and $h(\infty) = 0$.

(a) Find a first-term composite expansion of the solution.

(b) Find the second term in the composite expansion. Is the expansion uniformly valid over the interval $0 \leq t < \infty$? What conditions need to be placed on the functions $h(x)$ and $v(t)$? A method for constructing uniformly valid approximations in a case like this is the subject of Chap. 3.

2.56. Using the Cole–Hopf transformation it is possible to solve Burger's equation (2.129) (Whitham, 1974). In the case where

$$u(x,0) = \begin{cases} u_1 & \text{if } x < 0, \\ u_2 & \text{if } 0 < x, \end{cases}$$

where $u_1 > u_2$ are constants, one finds $u(x,t) = \frac{u_2 + K(x,t)u_1}{1 + K(x,t)}$, where

$$K(x,t) = \frac{\text{erfc}\left(\frac{x - u_1 t}{2\sqrt{\varepsilon t}}\right)}{\text{erfc}\left(-\frac{x - u_2 t}{2\sqrt{\varepsilon t}}\right)} e^{(x - v_0 t)(u_2 - u_1)/2\varepsilon}$$

and $v_0 = \frac{1}{2}(u_1 + u_2)$. Compare this with the first-term approximation derived for (2.129). Make sure to comment on the possible differences for small and for large values of t.

2.57. The function $B(t)$ in (2.137) can be found by matching the second term in the inner and outer expansions. This exercise outlines the necessary steps.

(a) Show that the second term in the outer expansion is

$$u_1(x,t) = \frac{t\phi''(\xi)}{(1 + t\phi'(\xi))^2},$$

where the value of ξ is determined from the equation $x = \xi + t\phi(\xi)$.

(b) By changing variables from $\bar{x}$ to $z = (1 - b)/(1 + b)$, where b is given in (2.137), show that the equation for U_1 becomes

$$(1 - z^2)\partial_z^2 U_1 + 2U_1 = \frac{4(s'' + r'z)}{r^2(1 - z^2)} - \frac{2}{r^2}\left(\frac{rB'}{B} + r'\ln\left(\frac{1 - z}{B(1 + z)}\right)\right),$$

where $r = \frac{1}{2}(u_0^+ - u_0^-)$.

(c) Solve the equation in part (b), and from this obtain the first two terms in the expansion of U_1 for $z \to 1$ and for $z \to -1$. (Hint: Because of the nature of this calculation, the use of a symbolic computer program is recommended.)

(d) To match the inner and outer expansions, introduce the intermediate variable $\bar{x}_\eta = (x - s(t))/\varepsilon^\eta$. With this, show that the outer expansion expands as follows:

$$u \sim \phi(\xi) + \varepsilon^\eta \frac{\bar{x}_\eta \phi'(\xi)}{1 + t\phi'(\xi)} + \varepsilon \frac{t\phi''(\xi)}{(1 + t\phi'(\xi))^2} + \cdots,$$

where ξ is determined from the equation $s(t) = \xi + t\phi(\xi)$. Do the same for the inner expansion, and by matching the two derive the result in (2.138).

(e) To find the constant B_0, introduce the initial-layer coordinates $\bar{x} = (x - s(t))/\varepsilon$ and $\tau = t/\varepsilon$. Find the first term in this layer, and then match the result with (2.136) to show $B_0 = 1$.

2.58. This exercise involves modifications of the expansions for Burger's equation.

(a) Discuss the possibility of obtaining a composite expansion for the solution of (2.129).

(b) The center of the shock wave is where u is half-way between u_0^+ and u_0^-. If this is located at $x = X(t)$, then find the first two terms in the expansion of $X(t)$. A discussion of this, and other aspects of the problem, can be found in Lighthill (1956).

(c) Because the position of the shock is determined by the solution, then it should, presumably, depend on ε. How do things change if one allows for the possibility that $s(t) \sim s_0(t) + \varepsilon s_1(t) + \cdots$?

2.59. Consider the linear diffusion problem

$$u_t + \alpha u_x + \beta u = \varepsilon u_{xx} \quad \text{for } -\infty < x < \infty \text{ and } 0 < t,$$

where $u(x, 0) = \phi(x)$. Assume that $\phi(x)$ has the same properties as the initial condition for (2.129) and that α and β are positive constants.

(a) Find the first terms in the inner and outer expansions of the solution.

(b) Comment on the differences between the characteristics of the shock layer for Burger's equation and the one you found in part (a).

2.60. Find the first term in the inner and outer expansions of the solution of

$$u_t + f(u)u_x = \varepsilon u_{xx} \quad \text{for } -\infty < x < \infty \text{ and } 0 < t,$$

where $u(x, 0) = \phi(x)$. Assume $\phi(x)$ has the same properties as the initial condition for (2.129) and $f(r)$ is smooth with $f'(r) > 0$.

2.61. Consider the problem of solving

$$\frac{\varepsilon}{r} \partial_r \left[\frac{1}{r} \partial_r (r^2 w) \right] = \partial_t w + \mu w + \kappa e^{-2t} \quad \text{for } 0 \leq r < 1, 0 < t,$$

where $w(r, 0) = 0$, $\partial_r w(0, t) = 0$, and $w(1, t) = 0$. Also, μ and κ are positive constants with $\mu \neq 2$. This problem arises in the study of the rotation of a

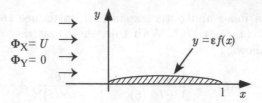

Figure 2.33 Schematic of air flow over an airplane wing as assumed in Exercise 2.62

cylindrical container filled with a dilute suspension (Ungarish, 1993). In this context, $w(r, t)$ is the angular velocity of the suspension and the boundary layer is known as a Stewartson shear layer. For small ε, find a first-term composite expansion of the solution.

2.62. The equation for the velocity potential $\phi(x, y)$ for steady air flow over an airplane wing is (Cole and Cook, 1986)

$$(a^2 - \Phi_x^2)\Phi_{xx} - 2\Phi_x\Phi_y\Phi_{xy} + (a^2 - \Phi_y^2)\Phi_{yy} = 0,$$

where

$$a^2 = a_\infty^2 + \frac{1}{2}(\gamma - 1)(U^2 - \Phi_x^2 - \Phi_y^2).$$

Here $a_\infty > 0$, $U > 0$, and $\gamma > 1$ are constants. The wing is assumed to be described by the curve $y = \varepsilon f(x)$ for $0 < x < 1$ (Fig. 2.33). In this case, the boundary conditions are that $\phi = Ux$ as $x \to -\infty$ and

$$\Phi_y = \begin{cases} \varepsilon f'(x)\Phi_x & \text{when} \quad y = \varepsilon f(x) \text{ and } 0 < x < 1, \\ 0 & \text{when} \quad y = 0 \text{ and } -\infty < x < 0 \text{ or } 1 < x. \end{cases}$$

(a) The thickness ε of the wing is small, and this is the basis of what is known as small disturbance theory. The appropriate expansion for the potential in this case has the form

$$\Phi \sim Ux + \varepsilon^\alpha \phi_1 + \varepsilon^\beta \phi_2 + \cdots.$$

Find α, and then determine what problem ϕ_1 satisfies.
(b) Find ϕ_1 in the case where $M_\infty > 1$, where $M_\infty = U/a_\infty$.
(c) For the case where $M_\infty > 1$, find ϕ_2 and explain why for the expansion to be valid it must be that $\varepsilon \ll (M_\infty^2 - 1)^{3/2}$. (Hint: Use characteristic coordinates $\xi = x - y\sqrt{M_\infty^2 - 1}$, $\eta = x + y\sqrt{M_\infty^2 - 1}$.)

2.63. For a semiconductor to function properly, one must be concerned with the level of impurities that diffuse in from the outer surface and occupy vacant locations in the crystalline structure of the semiconductor. A problem for the concentration of impurities $c(x, t)$ and vacancies $v(x, t)$ is (King et al., 1992)

$$\left. \begin{array}{l} \partial_t c = \partial_x(v\partial_x c - c\partial_x v) \\ \partial_t v + r\partial_t c = \varepsilon^2 \partial_x^2 v \end{array} \right\} \quad \text{for } 0 < x < \infty \text{ and } 0 < t,$$

where $c = 0$ and $v = 1$ when $t = 0$, $c = 1$, and $v = \mu$ when $x = 0$, and $c \to 0$ and $v \to 1$ as $x \to \infty$. Also, r and μ are positive constants. For small ε derive a composite expansion of the solution of this problem.

2.8 Difference Equations

Up until now, when discussing boundary-layer problems, we have dealt almost exclusively with differential equations. We will now expand our horizons and investigate what happens with singularly perturbed difference equations. As will be seen, many of the ideas developed in the first part of this chapter will reappear when analyzing difference equations, but there are subtle and interesting differences.

Our starting point is the boundary-value problem

$$\varepsilon y_{n+1} + \alpha_n y_n + \beta_n y_{n-1} = 0 \quad \text{for } n = 1, 2, \ldots, N - 1, \qquad (2.139)$$

where

$$y_0 = a, \quad y_N = b. \qquad (2.140)$$

What we have here is a second-order linear difference equation with prescribed values at the ends (where $n = 0, N$). In what follows it is assumed that N is fixed, and we will investigate how the solution behaves for small ε. It should also be pointed out that we will be assuming that the α_n's and β_n's are nonzero.

There are a couple of observations about the problem that should be made before starting the derivation of the asymptotic approximation of the solution. First, it is clear that the problem is singular for small ε since the reduced equation $\alpha_n y_n + \beta_n y_{n-1} = 0$ is first order and cannot be expected to satisfy both boundary conditions. The second observation can be made by considering an example. If $\alpha_n = 2$ and $\beta_n = a = b = 1$, then the solution of (2.139), (2.140) is

$$y_n = \left(\frac{1 - m_1^N}{m_2^N - m_1^N} \right) m_2^n - \left(\frac{1 - m_2^N}{m_2^N - m_1^N} \right) m_1^n,$$

where $m_1 = -(1 + \sqrt{1 - \varepsilon})/\varepsilon$ and $m_2 = -(1 - \sqrt{1 - \varepsilon})/\varepsilon$. For small ε this reduces to

$$y_n \sim \left(-\frac{1}{2} \right)^n + \left[1 - \left(-\frac{1}{2} \right)^N \right] \left(-\frac{\varepsilon}{2} \right)^{N-n}, \quad \text{for } n = 1, 2, \ldots, N. \qquad (2.141)$$

This shows boundary-layer type of behavior near the end $n = N$ in the sense that if one starts at $n = N$ and then considers the values at $n = N-1, N-2$, ..., then the $O(\varepsilon^{N-n})$ term in (2.141) rapidly decays. Moreover, away from the immediate vicinity of the right end, this term is small in comparison to the other term in the expansion.

2.8.1 Outer Expansion

We now derive an asymptotic approximation of the solution of (2.139), (2.140). The easiest component to obtain is the outer expansion, and this is determined by simply assuming an expansion of the form

$$y_n \sim \bar{y}_n + \varepsilon \bar{z}_n + \cdots . \qquad (2.142)$$

Substituting this into (2.139) and then equating like powers of ε, one finds that

$$\alpha_n \bar{y}_n + \beta_n \bar{y}_{n-1} = 0. \qquad (2.143)$$

Based on the observations made earlier, we expect the boundary layer to be at $n = N$. Thus, we require $\bar{y}_0 = a$. Solving (2.143) and using this boundary condition, one finds that

$$\bar{y}_n = \kappa_n a \quad \text{for } n = 0, 1, 2, 3, \ldots, \qquad (2.144)$$

where $\kappa_0 = 1$, and for $n \neq 0$

$$\kappa_n = \prod_{j=1}^{n} \left(-\frac{\beta_j}{\alpha_j} \right). \qquad (2.145)$$

Except for special values of α_n and β_n, the solution in (2.144) does not satisfy the boundary condition at $n = N$. How to complete the construction of the approximate solution when this happens is the objective of what follows.

2.8.2 Boundary-Layer Approximation

Now the question is, how do we deal with the boundary layer at the right end? To answer this, one needs to remember that (2.144) is a first-term approximation of the solution in the outer region. As given in (2.142), the correction to this approximation in the outer region is $O(\varepsilon)$. The correction at the right end, however, is $O(1)$. This is because the exact solution satisfies $y_N = b$, and (2.144) does not do this.

Our approach to finding the boundary-layer approximation is to first rescale the problem by letting

$$y_n = \varepsilon^{\gamma(n)} Y_n. \tag{2.146}$$

Once we find $\gamma(n)$ and Y_n, the general solution of the original problem will consist of the addition of the two approximations. Specifically, the composite approximation will have the form

$$y_n \sim \bar{y}_n + \varepsilon^{\gamma(n)} \overline{Y}_n, \tag{2.147}$$

where the tilde over the variables indicates the first-term approximation from the respective region. Given that $y_N = b$, and $\bar{y}_n$ does not satisfy this boundary condition, then from (2.147) we will require that

$$\gamma(N) = 0 \tag{2.148}$$

and

$$\overline{Y}_N = b - y_N. \tag{2.149}$$

The exponent $\gamma(n)$ is determined by balancing, and to do this we substitute (2.146) into (2.139) to obtain

$$\underset{①}{\varepsilon^{1+\gamma(n+1)} Y_{n+1}} + \underset{②}{\alpha_n \varepsilon^{\gamma(n)} Y_n} + \underset{③}{\beta_n \varepsilon^{\gamma(n-1)} Y_{n-1}} = 0. \tag{2.150}$$

In the outer region, the balancing takes place between terms ② and ③. For the boundary layer we have two possibilities to investigate:

(i) ① $\sim$ ③ and ② is higher order.
   The condition ① $\sim$ ③ requires that $\gamma(n+1) = \gamma(n-1)-1$. Thus, if $n = 2k$, then $\gamma(2k) = \gamma(0) - k$, and if $n = 2k+1$, then $\gamma(2k+1) = \gamma(1) - k$. In the case where $n = 2k$, we have ①, ③ $= O(\varepsilon^{1+\gamma(1)-k})$, ② $= O(\varepsilon^{\gamma(0)-k})$, and if $n = 2k + 1$, then we have ①, ③ $= O(\varepsilon^{\gamma(0)-k})$, ② $= O(\varepsilon^{\gamma(1)-k})$. To be consistent with the assumed balancing, we require that $1 + \gamma(1) \leq \gamma(0)$ and $\gamma(0) \leq \gamma(1)$. From this it follows that it is not possible to pick values for $\gamma(0)$ and $\gamma(1)$ that are consistent with our original assumption that ② is higher order. Thus, this balance is not possible.

(ii) ① $\sim$ ② and ③ is higher order.
   The condition ① $\sim$ ② requires that $\gamma(n + 1) = \gamma(n) - 1$, and so $\gamma(n+1) = \gamma(0) - n$. From this it follows that ①, ② $= O(\varepsilon^{1+\gamma(0)-n})$ and ③ $= O(\varepsilon^{2+\gamma(0)-n})$. This is consistent with the original assumption, and so this is the balancing we are looking for.

The balancing argument has shown that $\gamma(n) = 1 + \gamma(0) - n$. From this and (2.148) it follows that $\gamma(0) = N - 1$, and so

$$\gamma(n) = N - n.$$

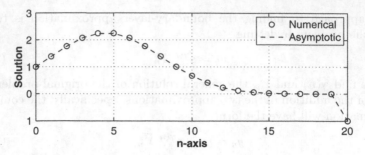

Figure 2.34 Comparison between asymptotic expansion in (2.155) and the numerical solution of (2.139), (2.140) in the case where $\alpha_n = -(1 + \frac{n}{N})$, $\beta_n = \frac{3}{2} - \frac{n}{N}$, $a = 1$, $b = -1$, $N = 20$, and $\varepsilon = 10^{-2}$

In this case, (2.150) takes the form

$$Y_{n+1} + \alpha_n Y_n + \varepsilon \beta_n Y_{n-1} = 0 \quad \text{for } n = N - 1, N - 2, \ldots. \tag{2.151}$$

The appropriate expansion of the boundary-layer solution is

$$Y_n \sim \overline{Y}_n + \varepsilon \overline{Z}_n + \cdots. \tag{2.152}$$

Introducing this into (2.151) yields the equation

$$\overline{Y}_{n+1} + \alpha_n \overline{Y}_n = 0.$$

The solution of this that also satisfies the boundary condition (2.149) is

$$\overline{Y}_n = \lambda_{N-n}(b - \kappa_N a), \tag{2.153}$$

where $\lambda_0 = 1$, and for $k \neq 0$

$$\lambda_k = \prod_{j=1}^{k} \left(-\frac{1}{\alpha_{N-j+1}} \right). \tag{2.154}$$

Using (2.147) we have found that a composite expansion of the solution is

$$y_n \sim \kappa_n a + \varepsilon^{N-n} \lambda_{N-n}(b - \kappa_N a). \tag{2.155}$$

It is possible to prove that this does indeed give us an asymptotic approximation of the solution (Comstock and Hsiao 1976). A demonstration of the accuracy of the approximation is given in Fig. 2.34. It is seen that the numerical solution of the difference equation and the asymptotic expansion are in very close agreement.

2.8.3 Numerical Solution of Differential Equations

There is a very interesting connection between the difference equation in (2.139) and the numerical solution of an associated differential equation. To understand this, consider the boundary value problem

$$\varepsilon y'' + p(x)y' + q(x)y = 0 \quad \text{for } 0 < x < 1, \tag{2.156}$$

where $y(0) = a$ and $y(1) = b$. It is assumed that the functions $p(x)$ and $q(x)$ are continuous. Now, if $p(x) < 0$ for $0 \le x \le 1$, then there is a boundary layer at $x = 1$ with width $O(\varepsilon)$ (Exercise 2.10). Given this fact, suppose we want to solve the problem numerically using finite differences. The standard centered-difference approximation will be used for the second derivative, but for the first derivative we will consider using either the forward difference approximation (Holmes, 2007)

$$y'(x_n) \approx \frac{y_{n+1} - y_n}{h} \tag{2.157}$$

or the backward difference approximation

$$y'(x_n) \approx \frac{y_n - y_{n-1}}{h}. \tag{2.158}$$

Using the backward difference we get from (2.156) that

$$\varepsilon y_{n+1} + (\alpha_n - 2\varepsilon)y_n + (\beta_n + \varepsilon)y_{n-1} = 0, \tag{2.159}$$

where $\alpha_n = hp_n + h^2 q_n$ and $\beta_n = -hp_n$. Because this equation differs from (2.139) only in the addition of higher-order terms in the coefficients, a composite expansion of the solution is still given in (2.155). Thus, the difference equation has a boundary layer in the same location as the associated differential equation. This is good if one expects the numerical solution to have any resemblance to the solution of the original problem. What is interesting is that the forward difference (2.157) results in a difference equation with a boundary layer at $x = 0$ and not at $x = 1$ (Exercise 2.64). This observation is strong evidence that one should use (2.158) rather than (2.157) to solve this problem.

Another way to approximate the first derivative is to use a centered difference. This would seem to be a better choice because it is a $O(h^2)$ approximation while the approximations in (2.157) and (2.158) are $O(h)$. However, its major limitation is that it cannot delineate a boundary layer at either end of an interval.

Before completing the discussion of the numerical solution of (2.156), it is worth making a comment about the order of the stepsize h. Presumably h should be relatively small for the finite difference equation to be an accurate approximation of the differential equation. This introduces a second small parameter into the problem, and one must be careful about its size in comparison to ε. For example, since $\alpha_n = O(h)$ and $\beta_n = O(h)$, then it should not be unexpected that one must require $\varepsilon \ll h$ to guarantee the expansion is well ordered. We will not pursue this topic, but interested readers should consult the articles by Brown and Lorenz (1987), Farrell (1987), and Linss et al. (2000).

Exercises

2.64. (a) Find a composite expansion of the difference equation

$$\omega_n y_{n+1} + \alpha_n y_n + \varepsilon y_{n-1} = 0 \quad \text{for } n = 1, 2, \ldots, N-1,$$

where $y_0 = a$ and $y_N = b$. Also, the ω_n are nonzero.
(b) Suppose one uses the forward difference approximation given in (2.157) to solve (2.156). Show that you get a difference equation like the one in part (a), and write down the resulting composite expansion of the solution.
(c) At which end does the difference equation you found in part (b) have a boundary layer? What condition should be placed on $p(x)$ so this numerical approximation can be expected to give an accurate approximation of the solution?

2.65. Find a composite expansion of the difference equation

$$\varepsilon y_{n+1} + \alpha_n y_n + \varepsilon \beta_n y_{n-1} = 0 \quad \text{for } n = 1, 2, \ldots, N-1,$$

where $y_0 = a$ and $y_N = b$. It is suggested that you first solve the problem in the case where α_n and β_n are constants. With this you should be able to find the expansion of the solution of the full problem.

2.66. This problem investigates the forward and backward stability of certain difference equations. In what follows, the coefficients α_n, β_n, and ω_n are assumed to be nonzero and bounded.
(a) Consider the initial value problem

$$\varepsilon y_{n+1} + \alpha_n y_n + \beta_n y_{n-1} = 0 \quad \text{for } n = 1, 2, \ldots,$$

where $y_0 = a$ and $y_1 = b$. Explain why the solution of this problem becomes unbounded as n increases. You can do this, if you wish, by making specific choices for the coefficients α_n and β_n.
(b) Consider the initial value problem

$$\omega_n y_{n+1} + \alpha_n y_n + \varepsilon y_{n-1} = 0 \quad \text{for } n = 1, 2, \ldots,$$

where $y_0 = a$ and $y_1 = b$. Explain why the solution of this problem is bounded as n increases. You can do this, if you wish, by making specific choices for the coefficients w_n and α_n.

(c) Now suppose the problems in parts (a) and (b) are to be solved for $n = 0, -1, -2, -3, \ldots$. Explain why the solution of the problem in (a) is bounded as $n \to -\infty$, but the solution of the problem from (b) is unbounded as $n \to -\infty$. These observations give rise to the statement that the equation in (a) is backwardly stable, while the equation in (b) is forwardly stable. These properties are reminiscent of what is found for the heat equation.

2.67. This problem examines the use of centered-difference approximations to solve the singularly perturbed boundary value problem in (2.156).

(a) Find a first-term composite expansion of the solution of the difference equation

$$(\alpha + \varepsilon)y_{n+1} + 2(\beta - \varepsilon)y_n - (\alpha - \varepsilon)y_{n-1} = 0 \text{ for } n = 1, 2, \ldots, N - 1,$$

where $y_0 = a$, $y_N = b$, and $\alpha \neq 0$. Are there any boundary layers for this problem?

(b) Suppose one uses the centered-difference approximation of the first derivative to solve (2.156). Letting $p(x)$ and $q(x)$ be constants, show that you get a difference equation like the one in part (a), and write down the resulting composite expansion of the solution. Which terms in the differential equation do not contribute to the composite expansion?

(c) The solution of (2.156) has a boundary layer at $x = 0$ if $p(x) > 0$ for $0 \leq x \leq 1$ and one at $x = 1$ if $p(x) < 0$ for $0 \leq x \leq 1$. Comment on this and the results from parts (a) and (b).

Chapter 3
Multiple Scales

3.1 Introduction

When one uses matched asymptotic expansions, the solution is constructed in different regions that are then patched together to form a composite expansion. The method of multiple scales differs from this approach in that it essentially starts with a generalized version of a composite expansion. In doing this, one introduces coordinates for each region (or layer), and these new variables are considered to be independent of one another. A consequence of this is that what may start out as an ordinary differential equation is transformed into a partial differential equation. Exactly why this helps to solve the problem, rather than make it harder, will be discussed as the method is developed in this chapter.

The history of multiple scales is more difficult to delineate than, say, boundary-layer theory. This is because the method is so general that many apparently unrelated approximation procedures are special cases of it. One might argue that the procedure got its start in the first half of the nineteenth century. For example, Stokes (1843) used a type of coordinate expansion in his calculations of fluid flow around an elliptic cylinder. Most of these early efforts were limited, and it was not until the latter half of the nineteenth century that Poincaré (1886), based on the work of Lindstedt (1882), made more extensive use of the ideas underlying multiple scales in his investigations into the periodic motion of planets. He found that the approximation obtained from a regular expansion accurately described the motion for only a few revolutions of a planet, after which the approximation becomes progressively worse. The error was due, in part, to the contributions of the second term of the expansion. He referred to this difficulty as the presence of a secular term. To remedy the situation, he expanded the independent variable with the intention of making the approximation uniformly valid by removing the secular term. This idea is also at the heart of the modern version of the method. What Poincaré was missing was the introduction of multiple

M.H. Holmes, *Introduction to Perturbation Methods*, Texts in Applied
Mathematics 20, DOI 10.1007/978-1-4614-5477-9_3,
© Springer Science+Business Media New York 2013

independent variables based on the expansion parameter. This step came much later; the most influential work in this regard is by Kuzmak (1959) and Cole and Kevorkian (1963).

3.2 Introductory Example

As in the last chapter, we will introduce the ideas underlying the method by going through a relatively simple example. The problem to be considered is to find the function y(t) that satisfies

$$y'' + \varepsilon y' + y = 0 \quad \text{for} \ \ 0 < t, \tag{3.1}$$

where

$$y(0) = 0 \ \ \text{and} \ \ y'(0) = 1. \tag{3.2}$$

This equation corresponds to an oscillator (i.e., a mass–spring–dashpot) with weak damping, where the time variable has been scaled by the period of the undamped system (Exercise 1.42). This is the classic example used to illustrate multiple scales. Our approach to constructing the approximation comes from the work of Reiss (1971).

3.2.1 Regular Expansion

It appears that this is not a layer type of problem since ε is not multiplying the highest derivative in (3.1). It seems, therefore, that the solution has a regular power series expansion. Thus, assuming

$$y(t) \sim y_0(t) + \varepsilon y_1(t) + \cdots , \tag{3.3}$$

one finds from (3.1) and (3.2) that

$$y(t) \sim \sin(t) - \frac{1}{2}\varepsilon t \sin(t). \tag{3.4}$$

In comparison, the exact solution of the problem is

$$y(t) = \frac{1}{\sqrt{1 - \varepsilon^2/4}} \, e^{-\varepsilon t/2} \sin\left(t\sqrt{1 - \varepsilon^2/4}\right). \tag{3.5}$$

These functions are plotted in Fig. 3.1. It is apparent that we have not been particularly successful in finding a very accurate approximation of the solution for $0 \le t < \infty$. This is because the second term in the expansion is as large as the first term once $\varepsilon t \approx 1$. For this reason, the second term is said

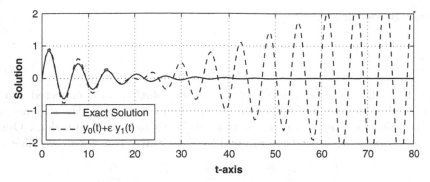

Figure 3.1 Comparison between the regular perturbation expansion in (3.4) and the exact solution given in (3.5) when $\varepsilon = 10^{-1}$. The two curves are reasonably close at the beginning (i.e., up to $t \approx 10$), but they differ significantly for larger values of t

to be a *secular term*, and for the expansion to be well ordered we must have $\varepsilon t << 1$. The problem with our approximation is that the solution (3.5) decays but the first term in (3.4) does not. The second term tries to compensate for this, and in the process it eventually becomes as large as the first term.

Another way to look at what is happening in the problem is to consider the energy. Multiplying (3.1) by the velocity y', integrating, and then using the initial conditions (3.2), one obtains the equation

$$H(y, y') = H(0, 1) - \varepsilon \int_0^t [y'(\tau)]^2 d\tau, \qquad (3.6)$$

where

$$H(y, y') \equiv \frac{1}{2}(y')^2 + \frac{1}{2}y^2. \qquad (3.7)$$

The function H is the Hamiltonian and consists of the sum of the kinetic and potential energies for the system. Thus, (3.6) shows that the energy equals the amount included at the start minus whatever is lost due to damping. To relate this to the multiple-scale expansion, recall that a system is said to be conservative, or energy is conserved, if H is independent of t. On the other hand, if $\frac{d}{dt}H < 0$, so that energy is lost, then the system is dissipative. Using these ideas, we have that (3.1) is dissipative but the problem for $y_0(t)$ is conservative. Therefore, the first term in the approximation in (3.4) is a conservative approximation of a weakly dissipative problem. This is one of many situations where the multiple-scale method is generally needed.

3.2.2 Multiple-Scale Expansion

The solution of the problem has an oscillatory component that occurs on a time scale that is $O(1)$, and it also has a slow variation that takes place on a time scale $O(\frac{1}{\varepsilon})$. To incorporate two time scales into the problem, we introduce the variables

$$t_1 = t \tag{3.8}$$

and

$$t_2 = \varepsilon^\alpha t. \tag{3.9}$$

Based on the earlier discussion of the exact solution, it is expected that $\alpha = 1$. However, we will derive this later when constructing the expansion of the solution. These two time scales will be treated as independent. One consequence of this is that the original time derivative transforms as follows:

$$\frac{d}{dt} \rightarrow \frac{dt_1}{dt} \frac{\partial}{\partial t_1} + \frac{dt_2}{dt} \frac{\partial}{\partial t_2} = \frac{\partial}{\partial t_1} + \varepsilon^\alpha \frac{\partial}{\partial t_2}.$$

Substituting this into (3.1) and (3.2) yields

$$\left(\partial_{t_1}^2 + 2\varepsilon^\alpha \partial_{t_1} \partial_{t_2} + \varepsilon^{2\alpha} \partial_{t_2}^2\right)y + \varepsilon\left(\partial_{t_1} + \varepsilon^\alpha \partial_{t_2}\right)y + y = 0, \tag{3.10}$$

where

$$y = 0 \quad \text{and} \quad \left(\partial_{t_1} + \varepsilon^\alpha \partial_{t_2}\right)y = 1 \quad \text{for } t_1 = t_2 = 0. \tag{3.11}$$

To simplify the notation, we use the symbol ∂_{t_1} in place of $\frac{\partial}{\partial t_1}$ (and similarly for ∂_{t_2}). The benefits of introducing these two time variables are not yet apparent. In fact, one might argue that we have made the problem harder since the original ordinary differential equation has been turned into a partial differential equation. The reasons for doing this will become evident in the discussion that follows. It should be pointed out that the solution of (3.10), (3.11) is not unique and that we need to impose more conditions for uniqueness on the solution. This freedom will enable us to prevent secular terms from appearing in the expansion (at least over the time scales we are using).

We will now use a power series expansion of the form

$$y \sim y_0(t_1, t_2) + \varepsilon y_1(t_1, t_2) + \cdots. \tag{3.12}$$

Substituting this into (3.10) yields

$$\left(\partial_{t_1}^2 + 2\varepsilon^\alpha \partial_{t_1} \partial_{t_2} + \varepsilon^{2\alpha} \partial_{t_2}^2\right)(y_0 + \varepsilon y_1 + \cdots)$$
$$+ \varepsilon(\partial_{t_1} + \varepsilon^\alpha \partial_{t_2})(y_0 + \cdots) + y_0 + \varepsilon y_1 + \cdots = 0. \tag{3.13}$$

From this and (3.11) we obtain the first problem to solve:

$O(1)$ $\quad (\partial_{t_1}^2 + 1)y_0 = 0$
$\qquad y_0 = 0, \ \partial_{t_1} y_0 = 1 \quad \text{when } t_1 = t_2 = 0.$

The general solution of this problem is

$$y_0 = a_0(t_2) \sin(t_1) + b_0(t_2) \cos(t_1), \tag{3.14}$$

where

$$a_0(0) = 1 \quad \text{and} \quad b_0(0) = 0. \tag{3.15}$$

For the moment, the coefficients $a_0(t_2)$ and $b_0(t_2)$ are arbitrary functions of t_2 except that they are required to satisfy the initial conditions given in (3.15).

To go any further, we need to determine α in the slow time scale (3.9). To do this, note that in the $O(\varepsilon)$ equation obtained from (3.13) we will have the term $\partial_{t_1} y_0$. It is not hard to check that this will cause a secular term to appear in the expansion. However, this can be prevented if we make use of the dependence of y_0 on the time scale t_2. In (3.13) the only term available to us that can be used to prevent a secular term is $2\varepsilon^\alpha \partial_{t_1} \partial_{t_2} y_0$. Balancing this with $\varepsilon \partial_{t_1} y_0$ gives us $\alpha = 1$. With this and the initial conditions in (3.11), we get the following problem:

$$O(\varepsilon) \quad (\partial_{t_1}^2 + 1)y_1 = -2\partial_{t_1}\partial_{t_2} y_0 - \partial_{t_1} y_0$$
$$y_1 = 0 \quad \text{and} \quad \partial_{t_1} y_1 = -\partial_{t_2} y_0 \quad \text{when} \quad t_1 = t_2 = 0.$$

From (3.14), the differential equation for y_1 is

$$\left(\partial_{t_1}^2 + 1\right) y_1 = (2b_0' + b_0) \sin(t_1) - (2a_0' + a_0) \cos(t_1). \tag{3.16}$$

The general solution of this problem is found to be

$$y_1 = a_1(t_2) \sin(t_1) + b_1(t_2) \cos(t_1)$$
$$- \frac{1}{2}(2b_0' + b_0) t_1 \cos(t_1) - \frac{1}{2}(2a_0' + a_0) t_1 \sin(t_1), \tag{3.17}$$

where

$$a_1(0) = a_0'(0) \quad \text{and} \quad b_1(0) = 0. \tag{3.18}$$

In (3.17) we have the possibility of secular terms in the expansion. However, the functions a_0 and b_0 can be chosen to prevent this. From (3.17) and (3.15) we find that

$$\begin{array}{ccc} 2b_0' + b_0 = 0 \\ 2a_0' + a_0 = 0 \end{array} \Rightarrow \begin{array}{c} b_0(t_2) = \beta_0 e^{-t_2/2} \\ a_0(t_2) = \alpha_0 e^{-t_2/2} \end{array} \Rightarrow \begin{array}{c} b_0(t_2) = 0 \\ a_0(t_2) = e^{-t_2/2}. \end{array} \tag{3.19}$$

In the last step, we impose the initial conditions given in (3.15).

Putting our results together, we have found that

$$y \sim e^{-\varepsilon t/2} \sin(t). \tag{3.20}$$

This is a first-term approximation that is valid up to at least $\varepsilon t = O(1)$. This approximation and the exact solution are shown in Fig. 3.2, from which it is clear that the multiple-scale result is quite good. In fact, one can prove that this is a uniformly valid asymptotic approximation of the solution for

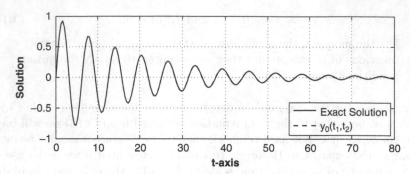

Figure 3.2 Comparison between the multiple-scale approximation in (3.20) and the exact solution given in (3.5). The two curves are so close that they are essentially indistinguishable from one another. In these calculations $\varepsilon = 10^{-1}$

$0 \le t \le O\left(\frac{1}{\varepsilon}\right)$ (Exercise 3.10). Also, the procedure is easily extendable to more general oscillators, and an example can be found in Exercise 3.9.

3.2.3 Labor-Saving Observations

There are two observations that need to be made to help reduce the work required to construct a multiple-scale expansion. The first concerns the equation

$$y'' + \omega^2 y = 0, \tag{3.21}$$

where ω is a positive constant. The general solution of this equation can be written as

$$y(t) = a \sin(\omega t) + b \cos(\omega t) \tag{3.22}$$

or as

$$y(t) = A \cos(\omega t + \theta) \tag{3.23}$$

or as

$$y(t) = A e^{i\omega t} + \bar{A} e^{-i\omega t}. \tag{3.24}$$

In the last equation, A is understood to be a complex-valued constant and the overbar designates complex conjugate. In many of the examples and exercises in this chapter, Eq. (3.21) must be solved to obtain the first term in the approximation. Indeed, this is exactly what happened in the earlier example, and it resulted in the solution given in (3.14). In this example, where the original problem, as given in (3.1), is linear, all of the preceding three expressions can be used without difficulty. However, this is not true if the original problem is nonlinear. It is recommended that whenever using multiple scales on a nonlinear problem and (3.21) appears as the $O(1)$ problem, then the solution be written as in (3.24) or as in (3.23). The reason is that these two representations make finding the secular terms much easier.

The second labor-saving observation concerns removing secular terms without actually having to solve the $O(\varepsilon)$ equation. In the preceding example the solution was written out in (3.17), and from this the secular-removing conditions in (3.19) were obtained. However, it is known that the solution of the differential equation

$$y'' + \omega^2 y = f(t),\tag{3.25}$$

where ω is a positive constant, has a secular term if $f(t)$ contains an individual term involving a solution of the associated homogeneous equation, in other words, if $f(t)$ contains any of the functions in (3.22)–(3.24). For example, a secular term will be obtained if $f(t) = 2\sin(\omega t)$ or if $f(t) = -5\cos(\omega t) + \cos(3\omega t)$, but it will not contain a secular term if $f(t) = 7\cos(3\omega t)$. Based on this observation, a secular term can be prevented by simply requiring that the coefficients of $\sin(t_1)$ and $\cos(t_1)$ in (3.16) be zero. Our ability to use this to avoid having to solve the $O(\varepsilon)$ equation depends on finding the $\sin(\omega t)$ and $\cos(\omega t)$ contributions to $f(t)$. For example, this is easy if $f(t) = -5\cos(\omega t) + \cos(3\omega t)$ but not so if $f(t) = \sin(\omega t)\cos(\omega t)$ or $f(t) = \cos^3(\omega t)$. For the latter two, one needs to use trig identities to realize that

$$f(t) = \sin(\omega t)\cos(\omega t) = \frac{1}{2}(1 + \sin(2\omega t))$$

will not produce a secular term, whereas

$$f(t) = \cos^3(\omega t) = \frac{1}{4}(3\cos(\omega t) + \cos(3\omega t))$$

will cause a secular term to appear. This is why it is recommended that you use (3.24) or (3.23), because they reduce the need for the more esoteric trig identities. There are other ways to eliminate secular terms, and these are examined in Exercises 3.13 and 3.14.

3.2.4 Discussion

Now that we have gone through the analysis of a problem using multiple scales, it is worth reflecting on what we did. The approach used to construct the expansion in (3.20) was to introduce the two time scales $t_1 = t$ and $t_2 = \varepsilon t$. For this reason, the method is sometimes called *two-timing*, and t_1 is said to be the fast time scale and t_2 the slow scale. This terminology implies that the method is used only with the time variable, which is definitely not the case. As we will see in Sect. 3.7, it is possible to solve some of the boundary-layer problems in Chap. 2 by using these same ideas.

The time scales that should be used depend on the problem and the interval over which one wants an approximation of the solution. The list of possible scales is endless, but some possibilities include the following.

1. Several time scales: for example, one may need to use the four time scales $t_1 = t/\varepsilon, t_2 = t, t_3 = \varepsilon t, t_4 = \varepsilon^2 t$. Some of the issues involved with such expansions are explored in Sect. 3.3.

2. More complex dependence on ε: for example, $t_1 = (1 + w_1\varepsilon + w_2\varepsilon^2 + \cdots)t$ and $t_2 = \varepsilon t$, where the w_n are determined while solving the problem (this is called the *method of strained coordinates* or *Lindstedt's method*). In the previous problem, one would find that $t_1 = (1 - \varepsilon^2/8 + \cdots)t$.

3. The correct scaling may not be immediately apparent, and one starts off with something like $t_1 = \varepsilon^\alpha t$ and $t_2 = \varepsilon^\beta t$, where $\alpha < \beta$.

4. Nonlinear time dependence: for example, one may have to assume $t_1 = f(t, \varepsilon)$ and $t_2 = \varepsilon t_1$, where the function $f(t, \varepsilon)$ is determined from the problem (this type of scaling is used in Sect. 3.6).

Multiple scales are introduced to keep the expansion well ordered. This should not be interpreted to mean that they are always to remove unbounded terms, although many times this is what happens (as in the preceding example). In fact, whatever freedom there is in the dependence on t_2 is used to minimize the error in the approximation. The role of such a minimization principle in the construction of the multiple-scale expansion is explained in Sect. 3.3.

Exercises

In the following exercises, when asked to find a first-term expansion of the solution that is valid for large t, you should introduce a time-scale that removes the first secular term that would appear in a regular expansion.

3.1. Find a first-term expansion of the solutions of the following problem that is valid for large t.
(a) $y'' + \varepsilon(y')^3 + y = 0$ for $0 < t$, where $y(0) = 0$ and $y'(0) = 1$.
(b) $\varepsilon y'' + y' + y = 0$ for $0 < t$, where $y(0) = 0$ and $y'(0) = 1$.
(c) $\varepsilon y'' + \varepsilon y' + y = \cos(t)$ for $0 < t$, where $y(0) = 0$ and $y'(0) = 0$.
(d) $\varepsilon y'' + \varepsilon \alpha y' + y + \varepsilon y^3 = 0$ for $0 < t$, where $y(0) = 0$, $y'(0) = 1$, and α is a positive constant.

3.2. The equation for an oscillator can be written as

$$y'' + V'(y) = 0 \quad \text{for } 0 < t,$$

where $y(0) = \varepsilon$ and $y'(0) = 0$. Also, $V(y)$ is the potential function. In what follows, you are to find a first-term expansion of the solution that is valid for large t for particular potential functions.

(a) For a pendulum, $V(y) = -\cos(y)$. In this case, $y(t)$ is the angular displacement of the pendulum.

(b) For a Morse oscillator, $V(y) = (1 - e^{-\alpha y})^2$ and α is a positive constant.

(c) For a Toda oscillator, $V(y) = e^y - y$.

(d) For a Lennard–Jones oscillator, $V(y) = (1 + y)^{-12} - (1 + y)^{-6}$.

3.3. Consider the weakly nonlinear oscillator problem of solving

$$y'' + y + \varepsilon y^n = 0 \quad \text{for } 0 < t,$$

where $y(0) = a$ and $y'(0) = 0$. One of the objectives of this exercise is to show that even-order nonlinearities can take longer to produce secular terms than those of odd-order. Note that there are problems where this does not hold, and an example can be found in Exercise 3.18.

(a) If $n = 3$, then a secular term appears in the second term of a regular expansion (you do not need to show this). Find a first-term expansion of the solution that is valid for large t.

(b) Taking $n = 2$ and using a regular expansion, show that a secular term does not appear in the second term but does in the third term.

(c) Taking $n = 2$, find a first-term expansion of the solution that is valid for large t.

(d) Suppose n is a positive integer. Using your results from parts (b) and (c), as well as those from Exercise 3.11, explain why the time-scale over which a secular term appears does not depend on the magnitude of n but, rather, on whether n is odd or even.

3.4. A generalization of (3.1), (3.2) is

$$y'' + \omega^2 y = \varepsilon f(y, y') \quad \text{for } 0 < t,$$

where $y(0) = a$ and $y'(0) = b$. Also, ω is a positive constant. Assume here that (3.8), (3.9), and (3.12) apply with $\alpha = 1$. Also, assume f is a smooth function. The objective is to find a first-term expansion of the solution that is valid for large t.

(a) Show that the solution of the $O(1)$ problem has the form $y_0 = A\cos(t_1 + \theta)$, where A and θ are functions of t_2. What initial conditions do they satisfy?

(b) Determine the $O(\varepsilon)$ problem.

(c) The function $F(\phi) = f(A\cos\phi, -\omega A \sin\phi)$ is 2π periodic. Expanding it in a Fourier series and using your result from part (b), show that to prevent secular terms in the expansion it is required that

$$A' = -\frac{1}{2\pi\omega} \int_0^{2\pi} f(A\cos\phi, -\omega A \sin\phi)\sin\phi\,d\phi,$$

$$A\theta' = -\frac{1}{2\pi\omega} \int_0^{2\pi} f(A\cos\phi, -\omega A \sin\phi)\cos\phi\,d\phi.$$

(d) Find A and θ from part (c) when $f(y, y') = -y^2$ and when $f(y, y') = -y^3$. Explain why the time scale does not necessarily remove the first secular term that appears in a regular expansion. What assumption should be made about the integrals in part (c) so it does remove the first secular term?

3.5. Consider the problem of solving

$$y' = y - y^3, \quad \text{for } 0 < t,$$

where $y(0) = \varepsilon$.

(a) Sketch the direction field for this problem and from this determine what happens to the solution as $t \to \infty$.

(b) Suppose one assumes a regular expansion of the form $y \sim \varepsilon y_0(t) + \varepsilon^\alpha y_1(t) + \cdots$. After finding y_0 and y_1, explain why y_0 is not expected to be an accurate approximation as $t \to \infty$. Also, explain why a multiple-scale expansion should be used but (3.9) will not work for this problem.

(c) Suppose $t_2 = \varepsilon^\alpha f(t)$, where $f(t)$ is determined from the secular-removing condition and the requirement that $f(0) = 0$. Show that

$$y \sim \frac{\varepsilon e^t}{\sqrt{1 + \varepsilon^2 (e^{2t} - 1)}}.$$

3.6. In the study of Josephson junctions, the following problem appears (Sanders, 1983):

$$\phi'' + \varepsilon(1 + \gamma \cos \phi)\phi' + \sin \phi = \varepsilon\alpha \quad \text{for } 0 < t,$$

where $\phi(0) = \phi'(0) = 0$ and γ is a positive constant. Find a first-term approximation of $\phi(t)$ that is valid for large t.

3.7. The equation for a relativistic damped oscillator is (Struble and Harris, 1964)

$$\frac{d}{dt}\left(\frac{u_t}{\sqrt{1 + \varepsilon\mu (u_t)^2}}\right) + 2\alpha\varepsilon u_t + u = 0,$$

where α and μ are positive constants. The initial conditions are $u(0) = 1$ and $u_t(0) = 0$. Find a first-term approximation of $u(t)$ that is valid for large t.

3.8. In the study of quantum jumps, single-ion trapping experiments are carried out using something called a Paul trap (Cook, 1990). In nondimensional form, the equation governing the ion motion in such a device is

$$\varepsilon (i\phi_t + \phi_{xx}) = v(x)\phi \cos\left(\frac{t}{\varepsilon}\right) \quad \text{for } -\infty < x < \infty \text{ and } 0 < t,$$

where $v(x)$ is a specified function. Note that, for small ε, the coefficient of the right-hand side of this equation is rapidly varying. In the theory of quantum jumps, this observation serves as the starting point for something called a rotating wave approximation. The result is a reduced equation for $\phi(x,t)$ where the coefficients of the equation are independent of t. You are asked in this problem to derive the result using multiple scales.

(a) Find a first-term approximation of $\phi(x,t)$ that is valid for $t = O(1)$. Note that you need to derive only the differential equation that determines the first-term approximation.

(b) The multivariable version of the quantum jump equation is

$$\varepsilon \left(i\phi_t + \nabla^2 \phi \right) = v(\mathbf{x})\phi \cos\left(\frac{t}{\varepsilon}\right).$$

This problem can be reduced in exactly the same way as was done in part (a). Instead, let $\phi(\mathbf{x},t) = \psi(\mathbf{x},t)\exp[-iv(\mathbf{x})\sin(t/\varepsilon)]$. Then find a first-term expansion of $\psi(\mathbf{x},t)$ that is valid for $t = O(1)$. Compare your result with the one found in (a).

3.9. Consider the weakly nonlinear oscillator problem of solving

$$y'' + y + \varepsilon f(y) = 0 \quad \text{for } 0 < t,$$

where $y(0) = a$ and $y'(0) = 0$. The results from Exercise 3.11 will be useful here, and the conclusions of Exercise 3.3 should be reviewed.

(a) Taking $f(y) = \sinh(y)$, show that a first-term expansion of the solution that is valid for large t has the form $y \sim a\cos(\omega t)$, where $\omega = 1 + \varepsilon\lambda$ and λ is an infinite series that depends on a.

(b) Suppose $f(0) = 0$, and assume that the Maclaurin series of $f(y)$ converges for all y. What additional condition on $f(y)$ is needed so the conclusion from part (a) holds? Also, find a formula for λ using information determined from the Maclaurin series.

3.10. In this problem, you are to prove that (3.20) is a uniformly valid approximation of the solution for $0 \le t \le T/\varepsilon$, where $T > 0$ is fixed. Do this by showing that $|y(t) - y_0(t, \varepsilon t)| = O(\varepsilon)$ for $0 \le t \le T/\varepsilon$.

(a) Find the exact solution, and then use Taylor's theorem to prove the result.

(b) Let $z(t) = y(t) - y_0(t, \varepsilon t)$. After finding the problem satisfied by $z(t)$, prove the result.

3.11. In what follows, n and m are positive integers.

(a) Using the representation $\cos\varphi = \frac{1}{2}(e^{i\varphi} + e^{-i\varphi})$ and the binomial expansion, show that

$$\cos^n \varphi = \frac{1}{2^n} \sum_{j=0}^{n} \binom{n}{j} \cos((2j - n)\varphi).$$

(b) Using part (a), show that

$$\cos^{2m+1}\varphi = \alpha_0 \cos\varphi + \sum_{j=1}^{m} \alpha_j \cos((2j+1)\varphi),$$

where

$$\alpha_0 = \frac{1}{2^{2m}} \frac{(2m+1)!}{m!(m+1)!}.$$

(c) Using part (a), show that

$$\cos^{2m}\varphi = \sum_{j=0}^{m} \gamma_j \cos(2j\varphi).$$

(d) Show that

$$\sin^{2m+1}\varphi = \alpha_0 \sin\varphi + \sum_{j=1}^{m} (-1)^j \alpha_j \sin((2j+1)\varphi)$$

and

$$\sin^{2m}\varphi = \sum_{j=0}^{m} (-1)^j \gamma_j \cos(2j\varphi),$$

where the α_j and γ are the same as those used in parts (b) and (c).

3.12. This problem examines when a secularlike term appears in the solution of

$$y'' + \omega^2 y = f(t),$$

where $\omega > 0$. In this problem it is useful to be aware of the results from the previous exercise as well as to remember that a continuous periodic function is bounded.

(a) Suppose $f(t) = \cos^n(\omega t)$, where n is a positive integer. Show that if n is even, then the solution is bounded, while if n is odd, then $y = \frac{\alpha_1}{2\omega} t \sin(\omega t) + q(t)$, where $q(t)$ is a bounded function and α_1 is defined in Exercise 3.11.

(b) Suppose $f(t) = \sin^n(\omega t)$, where n is a positive integer. Show that if n is even, then the solution is bounded, while if n is odd, then $y = -\frac{\alpha_1}{2\omega} t \cos(\omega t) + q(t)$, where $q(t)$ is a bounded function.

(c) Suppose $f(t) = \sin^n(\omega t) \cos^m(\omega t)$. Show that if n and m are both even, or if they are both odd, then the solution is bounded.

3.13. This problem examines the weakly nonlinear oscillator equation

$$y'' + y = \varepsilon f(y, y') \quad \text{for } 0 < t,$$

where $y(0) = a$ and $y'(0) = b$. Also, assume f is a smooth function.

(a) Consider the equation $u'' + \omega^2 u = f(t)$, where $f(t)$ is $2\pi/\omega$ periodic. If $u(t)$ is $2\pi/\omega$ periodic, show that $\int_0^{2\pi/\omega} f(t) \sin(\omega t) dt = \int_0^{2\pi/\omega} f(t) \cos(\omega t) dt = 0$.

(b) Show that the solution of the $O(1)$ problem has the form $y_0 = A\cos(t_1 + \theta)$. A common assumption about nonlinear oscillators is that the solution is periodic in the fast variable. Use this and the result from part (a) to show that

$$A' = p\cos\theta + q\sin\theta,$$
$$A\theta' = -p\sin\theta + q\cos\theta,$$

where p and q are integrals that depend on A and θ.

(c) Show that the result in part (b) is the same as that given in Exercise 3.9(c).

3.14. This problem considers how to apply the method from the previous exercise to the first-order system

$$\mathbf{y}' = \mathbf{A}\mathbf{y} + \mathbf{f}(\mathbf{y}, \varepsilon),$$

where $\mathbf{y}(0) = \mathbf{a}$. It is assumed $\mathbf{f}$ is smooth with $\mathbf{f}(\mathbf{y}, 0) = \mathbf{0}$. Also, letting $\mathbf{y}^T = (u, v)$, it is assumed that

$$\mathbf{A} = \begin{bmatrix} \alpha & \beta \\ -\gamma & -\alpha \end{bmatrix},$$

where $\alpha^2 < \beta\gamma$. The matrix $\mathbf{A}$ and vector $\mathbf{a}$ do not depend on ε.

(a) Show that the weakly damped oscillator equation given in (3.1) can be written in this form.

(b) Introduce the time scales $t_1 = t$ and $t_2 = \varepsilon t$, and then show that the first term in the expansion of the solution has the form

$$\mathbf{y}_0(t_1, t_2) = \mathbf{q}(t_2)e^{i\omega t_1} + \overline{\mathbf{q}}(t_2)e^{-i\omega t_1},$$

where $\omega = \sqrt{\beta\gamma - \alpha^2}$ and the overbar indicates complex conjugate. Here $\mathbf{q}$ is an eigenvector associated with a first-order problem. Also, what is $\mathbf{q}(0)$?

(c) Suppose the solution is known to be $2\pi/\omega$ periodic in t_1. This information can be used to remove secular terms. To show how, let $\mathbf{p}(t_1)$ be a solution of the adjoint equation $\mathbf{p}' = -\mathbf{A}^T\mathbf{p}$. From the equation for the $O(\varepsilon)$ term, show that

$$\int_0^{2\pi/\omega} (\partial_{t_2}\mathbf{y}_0 - \partial_\varepsilon\mathbf{f}(\mathbf{y}_0, 0)) \cdot \mathbf{p}\, dt_1 = 0.$$

Explain why this is a first-order differential equation that completes the determination of the vector $\mathbf{q}$.

(d) Show that (b) and (c) produce (3.20) when applied to the weakly damped oscillator equation from part (a).

(e) Consider the system of equations

$$u' = u + v - \varepsilon u(2 + \varepsilon uv),$$
$$v' = -4u - v + \varepsilon u(2 + \varepsilon uv).$$

After identifying the matrix $\mathbf{A}$ and the function $\mathbf{f}$, use the results of parts (b) and (c) to find a first-term approximation of $\mathbf{y}(t)$ that is valid for large t [assume the solution satisfies the periodicity condition mentioned in (c)].

(f) In the study of chemical oscillators, the following system of equations is found (Schnakenberg, 1979):

$$u' = u + v + \varepsilon u(2u + 2v + \varepsilon uv),$$
$$v' = -4u - v - \varepsilon u(2u + 2v + \varepsilon uv).$$

In this case, show that the result in part (c) indicates that the wrong time scale has been introduced. Extend the results of parts (b) and (c) to include this situation, and in the process find a first-term approximation of $\mathbf{y}(t)$ that is valid for large t [assume the solution satisfies the periodicity condition mentioned in part (c)].

3.3 Introductory Example (continued)

Most of those who use multiple scales are interested in finding a first-term approximation involving two time scales. The basic idea used to construct the approximation, namely, preventing a secular term from appearing in the expansion, is easy to understand. It also serves as the basis for most of the examples and exercises in this chapter. This section, however, is an exception. The objective here is to examine some of the underlying assumptions used in multiple scales and, in the process, develop a deeper understanding of how and why the method works.

In the previous section two time scales, $t_1 = t$ and $t_2 = \varepsilon t$, were used. With them, it was shown that a first-term approximation of the equation

$$y'' + \varepsilon y' + y = 0, \tag{3.26}$$

where

$$y(0) = 0 \text{ and } y'(0) = 1, \tag{3.27}$$

is given as

$$y \sim e^{-t_2/2} \sin(t_1). \tag{3.28}$$

We now ask an apparently simple question: can we do better than this? To answer this we need to investigate some of the possible extensions of what was done earlier. This includes using an additional scale and looking at including more than the first term in the expansion. We begin with the additional time scale.

3.3.1 Three Time Scales

The approximation (3.28) holds up to $\varepsilon t = O(1)$, that is, it holds for $0 \le \varepsilon t \le T$, where T is fixed. What if one wants an approximation that holds over a longer time interval, say up to $\varepsilon^2 t = O(1)$? The appropriate time scales in this case are

$$t_1 = t,$$
$$t_2 = \varepsilon t,$$
$$t_3 = \varepsilon^2 t.$$

The time derivative now transforms as

$$\frac{d}{dt} \to \frac{\partial}{\partial t_1} + \varepsilon \frac{\partial}{\partial t_2} + \varepsilon^2 \frac{\partial}{\partial t_3}.$$

With this, (3.26) becomes

$$[\partial_{t_1}^2 + 2\varepsilon\partial_{t_1}\partial_{t_2} + \varepsilon^2 \left(\partial_{t_2}^2 + 2\partial_{t_1}\partial_{t_3}\right) + O(\varepsilon^3)]y$$
$$+ \varepsilon \left(\partial_{t_1} + \varepsilon\partial_{t_2} + \varepsilon^2\partial_{t_3}\right)y + y = 0, \quad (3.29)$$

and the initial conditions are

$$y = 0 \quad \text{and} \quad \left(\partial_{t_1} + \varepsilon\partial_{t_2} + \varepsilon^2\partial_{t_3}\right)y = 1 \quad \text{for } t_1 = t_2 = t_3 = 0. \quad (3.30)$$

Similar to what was done earlier, we use a power series expansion of the form

$$y \sim y_0(t_1, t_2, t_3) + \varepsilon y_1(t_1, t_2, t_3) + \varepsilon^2 y_2(t_1, t_2, t_3) + \cdots. \quad (3.31)$$

The $O(1)$ and $O(\varepsilon)$ problems are basically the same as before. Consequently, from (3.14) and (3.15) we have that

$$y_0 = a_0(t_2, t_3) \sin(t_1) + b_0(t_2, t_3) \cos(t_1), \quad (3.32)$$

where $a_0(0,0) = 1$ and $b_0(0,0) = 0$. Also, from (3.17) and (3.18)

$$y_1 = a_1(t_2, t_3) \sin(t_1) + b_1(t_2, t_3) \cos(t_1)$$
$$- \frac{1}{2} \left(2\partial_{t_2}b_0 + b_0\right) t_1 \cos(t_1) - \frac{1}{2} \left(2\partial_{t_2}a_0 + a_0\right) t_1 \sin(t_1), \quad (3.33)$$

where $a_1(0,0) = \partial_{t_1} a_0(0,0)$, $b_1(0,0) = 0$. Removing the secular terms in the preceding expression yields

$$a_0(t_2, t_3) = \alpha_0(t_3)\, e^{-t_2/2}, \tag{3.34}$$

$$b_0(t_2, t_3) = \beta_0(t_3)\, e^{-t_2/2}, \tag{3.35}$$

where $\alpha_1(0) = 1$ and $\beta_1(0) = 0$. The preceding expressions are expected generalizations of (3.19).

To find $\alpha_0(t_3)$ and $\beta_0(t_3)$, it is necessary to consider the $O(\varepsilon^2)$ equation:

$$O(\varepsilon^2)\ \ (\partial_{t_1}^2 + 1)y_2 = -2\partial_{t_1}\partial_{t_2} y_1 - (\partial_{t_2}^2 + 2\partial_{t_1}\partial_{t_3})y_0 - \partial_{t_1} y_1 - \partial_{t_2} y_0.$$

With the solutions for y_0 and y_1 given in (3.32) and (3.33), to prevent secular terms in the expansion, we must require that

$$a_1 = \alpha_1(t_3)\, e^{-t_2/2} + \frac{1}{8}\,(\beta_0 - 8\alpha_0')\, t_2 e^{-t_2/2} \tag{3.36}$$

and

$$b_1 = \beta_1(t_3)\, e^{-t_2/2} - \frac{1}{8}\,(\alpha_0 + 8\beta_0')\, t_2 e^{-t_2/2}. \tag{3.37}$$

To summarize, our expansion is

$$y \sim \left[\alpha_0(t_3) + \varepsilon\left(\alpha_1(t_3) + \frac{1}{8}\,(\beta_0 - 8\alpha_0')\, t_2\right)\right] e^{-t_2/2} \sin(t_1)$$

$$+ \left[\beta_0(t_3) + \varepsilon\left(\beta_1(t_3) - \frac{1}{8}\,(\alpha_0 + 8\beta_0')\, t_2\right)\right] e^{-t_2/2} \cos(t_1). \tag{3.38}$$

This is a different situation than we had previously because we have removed the secular terms yet still have not determined α_0 and β_0. These coefficients can be found if we remember that our objective is to obtain an accurate asymptotic approximation, which means minimizing the error. One way to look at this is that to improve the accuracy of y_0 we should reduce, as much as possible, the contribution of y_1. In (3.33), we have already removed the $t_1 \cos t_1$ and $t_1 \sin t_1$ terms. We can reduce the contributions of a_1 and b_1, as given in (3.36) and (3.37), by taking $\beta_0 - 8\alpha_0' = 0$ and $\alpha_0 + 8\beta_0' = 0$. From this and the initial conditions (3.15) it follows that $\alpha_0 = \cos(t_3/8)$ and $\beta_0 = -\sin(t_3/8)$. Therefore, a first-term approximation valid up to $\varepsilon^2 t = O(1)$ is

$$y \sim e^{-\varepsilon t/2} \sin(\omega t), \tag{3.39}$$

where $\omega = 1 - \frac{1}{8}\varepsilon^2$.

3.3.2 Two-Term Expansion

Instead of using three time scales, one can try improving the approximation in (3.28) by using two time scales but including the second term y_1. The expansion is the same as in (3.31), except that the terms do not depend on t_3. For this reason, (3.38) applies except that the coefficients α_0, β_0, α_1, β_1 are independent of t_3. Working out the details it is found that the two-term expansion is

$$y \sim e^{-\varepsilon t/2}\left(\sin(t) - \frac{1}{2}\varepsilon^2 t \cos(t)\right). \qquad (3.40)$$

3.3.3 Some Comparisons

We now have three different approximations for the solution:

(i) $y \sim y_0(t_1, t_2)$, as given in (3.28);

(ii) $y \sim y_0(t_1, t_2) + \varepsilon y_1(t_1, t_2)$, as given in (3.40);

(iii) $y \sim y_0(t_1, t_2, t_3)$, as given in (3.39).

With three different approximations of the same function, it is natural to ask how they compare.

(i) vs. (iii): (i) is guaranteed to hold up to $\varepsilon t = O(1)$ while (iii) holds up to $\varepsilon^2 t = O(1)$. Consequently, (iii) holds over a longer time interval. However, there is no reason to expect that (iii) is more accurate than (i) over the time interval where they both apply, that is, for $0 \le \varepsilon t \le O(1)$.

(i) vs. (ii): (ii) should give a more accurate approximation than (i) as ε becomes small and $0 \le \varepsilon t \le O(1)$.

(ii) vs. (iii): (ii) should give a more accurate approximation than (iii) as ε becomes small and $0 \le \varepsilon t \le O(1)$. However, (iii) holds for $\varepsilon^2 t = O(1)$, while (ii) only holds for $\varepsilon t = O(1)$. Note that the effort to find (ii) and (iii) is about the same since the $O(\varepsilon^2)$ problem must be examined to be able to find y_1 as well as remove the secular terms associated with the t_3 time scale.

3.3.4 Uniqueness and Minimum Error

The method of multiple scales has the flexibility to work on a wide variety of problems. This capability is also the source of some confusion about the method. To illustrate, suppose that the exact solution is found to be

$$y(t) = 1 + \varepsilon^2 t e^{-t} + \varepsilon^4 \sin(t).$$

Also, suppose the time scales used in the construction of the multiple-scale expansion are $t_1 = t$, $t_2 = \varepsilon t$, and $t_3 = \varepsilon^2 t$. If the objective is to find an expansion up through the $O(\varepsilon^2)$ terms, then the following are possible:

(i) $y \sim y_0 + \varepsilon^2 y_2 + \cdots$, where $y_0 = 1$ and $y_2 = t_1 e^{-t_1}$;

(ii) $y \sim y_0 + \varepsilon y_1 + \cdots$, where $y_0 = 1$ and $y_1 = t_2 e^{-t_1}$;

(iii) $y \sim y_0 + \cdots$, where $y_0 = 1 + t_3 e^{-t_1}$;

(iv) $y \sim y_0 + \varepsilon y_1 + \varepsilon^2 y_2 + \cdots$, where $y_0 = 1 + t_3 e^{-t_1}$, $y_1 = -t_2 e^{-t_1}$, and $y_2 = t_1 e^{-t_1}$.

As the preceding list demonstrates, if you identify the time scales and then write down the form of the expansion, the result is not unique. It is for this reason that the method of multiple scales includes a principle that is used to determine the preferred expansion. In the damped oscillator example of Sect. 3.2, the choice was clear because of the appearance of a secular term. This principle does not apply directly to the preceding expansions because, as a function of t, none of them have secular terms. One might argue that the removal of secular terms can still be used if you make use of the assumption that the three time scales are independent. Based on this, the expansions in (ii) and (iv) are not uniform in t_2. For this reason both can be eliminated from consideration. This leaves (i) and (iii), and this means that secularity is not enough.

A principle that can help filter out unwanted multiple-scale expansions involves minimizing the error. To explain, in determining y_0 we want to minimize the value of the error

$$E_0 = \max \left| y(t) - y_0 \left(t, \varepsilon t, \varepsilon^2 t \right) \right|,$$

where the max is taken over the interval $0 \le t \le O\left(1/\varepsilon^2\right)$ (i.e., over the time interval $0 \le t \le T/\varepsilon^2$, where T is fixed). In the preceding list of possible expansions, $E_0 = O(\varepsilon^2)$ for (i) and $E_0 = O(\varepsilon^4)$ for (iii). Therefore, according to the minimum error requirement, we should take the y_0 given in (iii). Once this result is established, then in determining y_1 we want the error

$$E_1 = \max \left| y(t) - y_0 \left(t, \varepsilon t, \varepsilon^2 t \right) - \varepsilon y_1 \left(t, \varepsilon t, \varepsilon^2 t \right) \right|$$

to be minimized. One finds that $y_1 = 0$, and the other terms can be obtained using a similar procedure. A demonstration of how to use arguments like this to prove the validity of an expansion can be found in Smith (1985).

Although there is merit to using the error, it has the distinct disadvantage of requiring more information about the exact solution than is usually known. What is more typical is the result in (3.36) and (3.37), where the coefficients α_0 and β_0 are undetermined. The idea of minimizing the error translates into the condition of minimizing the contribution of y_1. Given that α_1 and β_1 do not depend on t_2, the requirement of minimizing y_1 leads to the conclusion that $\beta_0 - 8\alpha_0' = 0$ and $\alpha_0 + 8\beta_0' = 0$.

Exercises

3.15. This problem considers the case where the logistic equation has a slowly varying carrying capacity $c(\varepsilon t)$. The equation is (Shepherd and Stojkov, 2007)

$$y' = ry\left(1 - \frac{y}{c(\varepsilon t)}\right) \quad \text{for } 0 < t,$$

where $y(0) = \alpha$ and r and α are positive constants. It is assumed that $c(\tau)$ is a smooth positive function. Show that a first-term approximation that is valid for large t is

$$y \sim \frac{c(\varepsilon t)}{1 + A_0 c(\varepsilon t) e^{-rt}},$$

where $A_0 = \frac{1}{\alpha} - \frac{1}{c(0)}$. Make sure to explain your reasoning in how you determine the dependence of y_0 on the slow time variable $t_2 = \varepsilon t$.

3.16. Consider Eq. (3.26), subject to the initial conditions $y(0) = a$ and $y'(0) = b$.
(a) Using the time scales $t_1 = 1$ and $t_2 = \varepsilon t$, and assuming that $y \sim y_0(t_1, t_2) + \varepsilon y_1(t_1, t_2) + \cdots$, show that

$$y \sim [B\sin(t) + A\cos(t)]\, e^{-\varepsilon t/2},$$

where $B = b + \frac{1}{8}\varepsilon(4 + t_2)a$ and $A = a - \frac{1}{8}\varepsilon t_2 b$.
(b) Using the time scales $t_1 = 1$, $t_2 = \varepsilon t$, and $t_3 = \varepsilon^2 t$, and assuming that $y \sim y_0(t_1, t_2, t_3) + \cdots$, show that

$$y \sim [b\sin(\omega t) + a\cos(\omega t)]\, e^{-\varepsilon t/2},$$

where $\omega = 1 - \frac{1}{8}\varepsilon^2$.
(c) Extend the result in part (b) by finding y_1 and show that

$$y \sim [c\sin(\omega t) + a\cos(\omega t)]\, e^{-\varepsilon t/2},$$

where $c = b + \frac{1}{2}a\varepsilon$ and $\omega = 1 - \frac{1}{8}\varepsilon^2$.

3.17. Consider the differential equation $y'' + y + \varepsilon y^3 = 0$, subject to the initial conditions $y(0) = a$ and $y'(0) = b$. This nonlinear equation is known as Duffing's equation.

(a) Using the time scales $t_1 = 1$ and $t_2 = \varepsilon t$, and assuming that $y \sim y_0(t_1, t_2) + \varepsilon y_1(t_1, t_2) + \cdots$, show that

$$y_0 = b \sin(\omega t) + a \cos(\omega t),$$

where $\omega = 1 + \frac{3}{8}(a^2 + b^2)\varepsilon$.

(b) Using the time scales $t_1 = 1$, $t_2 = \varepsilon t$, and $t_3 = \varepsilon^2 t$, and assuming that $y \sim y_0(t_1, t_2, t_3) + \cdots$, show that

$$y_0 = b \sin(\omega t) + a \cos(\omega t),$$

where $\omega = 1 + \frac{3}{8}(a^2 + b^2)\varepsilon - \kappa\varepsilon^2$ and $\kappa = \frac{3}{256}(7a^4 + 46a^2b^2 + 23b^4)$.

(c) The exact solution is $y = A\,\mathrm{cn}(\alpha t + \beta, k)$, where $\mathrm{cn}(x, k)$ is a Jacobian elliptic function (Exercise 1.26), $\alpha = \sqrt{1 + \varepsilon A^2}$, and $\alpha k = A\sqrt{\varepsilon/2}$. To satisfy the initial conditions it is required that A and β satisfy

$$A\,\mathrm{cn}(\beta, k) = a,$$
$$-\alpha A\,\mathrm{sn}(\beta, k)\sqrt{1 - k^2\mathrm{sn}^2(\beta, k)} = b.$$

The resulting solution is periodic with period $T = 4K/\alpha$, where

$$K = \int_0^{\pi/2} \frac{dt}{\sqrt{1 - k^2 \sin^2 t}}.$$

Using the approximations in Exercise 1.26, show that a two-term approximation of T for small ε is consistent with the value of ω given in part (a). With a bit more effort, it can also be shown that a three-term expansion is consistent with the result in part (b).

3.4 Forced Motion Near Resonance

The problem to be considered corresponds to a damped, nonlinear oscillator that is forced at a frequency near resonance. In particular, the problem that is investigated is

$$y'' + \varepsilon\lambda y' + y + \varepsilon\kappa y^3 = \varepsilon \cos(1 + \varepsilon\omega)t \quad \text{for } 0 < t, \tag{3.41}$$

where

$$y(0) = 0 \quad \text{and} \quad y'(0) = 0. \tag{3.42}$$

In the differential equation, the damping ($\varepsilon\lambda y'$), nonlinearity ($\varepsilon\kappa y^3$), and forcing ($\varepsilon\cos(1 + \varepsilon\omega)t$) are small. Also, ω, λ, and κ are constants, with λ and

κ nonnegative. This oscillator has been studied extensively and is an example of a forced Duffing equation (Exercise 1.42).

Because of the small forcing and the zero initial conditions, it is natural to expect the solution to be small. For example, if we consider the simpler equation

$$y'' + y = \varepsilon \cos(\Omega t), \qquad (3.43)$$

then the solution is, assuming $\Omega \neq \pm 1$,

$$y(t) = \frac{\varepsilon}{1 - \Omega^2}[\cos(\Omega t) - \cos(t)]. \qquad (3.44)$$

Therefore, the amplitude of the solution is proportional to the amplitude of the forcing. This observation is not true when the driving frequency $\Omega \approx 1$. To investigate this situation, suppose $\Omega = 1 + \varepsilon w$. A particular solution of (3.43) in this case is, for $w \neq 0, -2/\varepsilon$,

$$y = \frac{-1}{w(2 + \varepsilon w)} \cos(1 + \varepsilon w)t, \qquad (3.45)$$

and for $w = 0$ or $w = -2/\varepsilon$,

$$y = \frac{1}{2}\varepsilon t \sin(t). \qquad (3.46)$$

In both cases a relatively small, order $O(\varepsilon)$, forcing results in at least an $O(1)$ solution. Also, the behavior of the solution is significantly different for $w = 0$ and $w = -2/\varepsilon$ than for other values of w. This is typical of a system that is driven at one of its characteristic frequencies, and it is associated with the phenomenon of resonance. Such behavior is also present in the nonlinear oscillator equation in (3.41). To demonstrate this, the values of $y(t)$ obtained from the numerical solution of (3.41), (3.42) are shown in Fig. 3.3. It is seen that the amplitude of the solution is an order of magnitude larger than the forcing.

Another observation that can be made from Fig. 3.3 is that there appear to be at least two time scales present in the problem. One is associated with the rapid oscillation in the solution, and the other is connected with the relatively slow increase in the amplitude. Thus, the multiple-scale method is a natural choice for finding an asymptotic approximation of the solution. We will try to give ourselves some flexibility and take $t_1 = t$ and $t_2 = \varepsilon^\alpha t$, where $\alpha > 0$. In this case (3.41) becomes

$$\left(\partial_{t_1}^2 + 2\varepsilon^\alpha \partial_{t_1} \partial_{t_2} + \varepsilon^{2\alpha} \partial_{t_2}^2\right)y + \varepsilon\lambda(\partial_{t_1} + \varepsilon^\alpha \partial_{t_2})y + y + \varepsilon\kappa y^3 = \varepsilon \cos(t_1 + \varepsilon w t_1). \qquad (3.47)$$

As stated earlier, because of the zero initial conditions (3.42) and the small forcing, one would expect the first term in the expansion of the solution to be $O(\varepsilon)$. However, near a resonant frequency the solution can be much larger. It

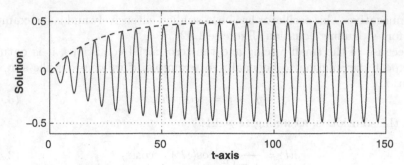

Figure 3.3 Numerical solution, shown with the *solid curve*, of the nonlinear oscillator in (3.41) and (3.42) when $\varepsilon = 0.05$, $\lambda = 2$, $\kappa = 4$, and $\omega = 3/8$. The *dashed curve* is the numerical solution of (3.52) and (3.53) for the amplitude function $A(\varepsilon t)$ obtained from the multiple-scale approximation (3.55)

is not clear what amplitude the solution actually reaches, and for this reason we will take

$$y \sim \varepsilon^\beta y_0(t_1, t_2) + \varepsilon^\gamma y_1(t_1, t_2) + \cdots, \tag{3.48}$$

where, to guarantee that the expansion is well ordered, $\beta < \gamma$. Also, because of the resonance effect, it is assumed that $\beta < 1$.

Substituting (3.48) into (3.47) yields the equation

$$(\partial_{t_1}^2 + 2\varepsilon^\alpha \partial_{t_1}\partial_{t_2} + \cdots)(\varepsilon^\beta y_0 + \varepsilon^\gamma y_1 + \cdots) + \varepsilon\lambda(\partial_{t_1} + \cdots)(\varepsilon^\beta y_0 + \cdots)$$
$$+ \varepsilon^\beta y_0 + \varepsilon^\gamma y_1 + \cdots + \varepsilon\kappa(\varepsilon^{3\beta} y_0^3 + \cdots) = \varepsilon \cos(t_1 + \varepsilon\omega t_1).$$

In writing out the preceding equation, an effort was made to include only the terms that might contribute to the first two terms of the expansion. Multiplying out the terms, we have that

$$\varepsilon^\beta \partial_{t_1}^2 y_0 + 2\varepsilon^{\alpha+\beta}\partial_{t_1}\partial_{t_2} y_0 + \varepsilon^\gamma \partial_{t_1}^2 y_1 + \cdots + \varepsilon^{1+\beta}\lambda\partial_{t_1} y_0 + \cdots$$
$$\overset{\textbf{❷}}{} \qquad\qquad \overset{\textbf{❶}}{} \qquad\qquad \overset{①}{}$$
$$+ \varepsilon^\beta y_0 + \varepsilon^\gamma y_1 + \cdots + \varepsilon^{1+3\beta}\kappa y_0^3 + \cdots = \varepsilon \cos(t_1 + \varepsilon\omega t_1). \tag{3.49}$$
$$\overset{\textbf{❶}}{} \qquad \overset{①}{} \qquad\qquad\qquad \overset{②}{}$$

With this and (3.42), the following problems result:

$$O(\varepsilon^\beta) \; (\partial_{t_1}^2 + 1)y_0 = 0,$$
$$y_0(0,0) = \partial_{t_1}y_0(0,0) = 0.$$

The general solution of this problem is

$$y_0 = A(t_2)\cos(t_1 + \theta(t_2)), \tag{3.50}$$

where

$$A(0) = 0. \tag{3.51}$$

The compact form of the solution is used in (3.50), rather than the form in (3.14), since it is easier to deal with the nonlinear terms that appear in the next problem. In particular, we will make use of the identity $\cos^3 \phi = \frac{1}{4}[3\cos(\phi) + \cos(3\phi)]$.

We need to determine α, β, and γ before going any further. With the preceding solution for y_0, we have two terms to deal with in (3.49), and they are labeled with a ①. There is also the forcing term ②. The problem for the second term in the expansion, which comes from the terms labeled with a ❶, must deal with one or more of these terms. It is possible to include all of them, which produces the most complete approximation, if we take $\gamma = 1$ and $\beta = 0$. The remaining observation to make is that all three terms, labeled with a ① or ②, will produce a secular term. Therefore, we must choose our slow time scale t_2 to deal with this. This means the term ❷ must balance with ① and ②, which gives us that $\alpha = 1$. With this, the right-hand side of (3.49) is rewritten as $\varepsilon \cos(t_1 + \omega t_2)$. Thus, the next order problem to solve is the following:

$$O(\varepsilon) \quad (\partial_{t_1}^2 + 1)y_1 = -2\partial_{t_1}\partial_{t_2}y_0 - \lambda\partial_{t_1}y_0 - \kappa y_0^3 + \cos(t_1 + \omega t_2)$$
$$= (2A' + \lambda A)\sin(t_1 + \theta) + 2\theta' A \cos(t_1 + \theta)$$
$$-\tfrac{\kappa}{4}A^3\left[3\cos(t_1 + \theta) + \cos 3(t_1 + \theta)\right] + \cos(t_1 + \omega t_2).$$

We want to determine which terms on the right-hand side of this equation produce secular terms – in other words, which terms are solutions of the associated homogeneous equation. This can be done by noting

$$\cos(t_1 + \omega t_2) = \cos(t_1 + \theta - \theta + \omega t_2)$$
$$= \cos(t_1 + \theta)\cos(\theta - \omega t_2) + \sin(t_1 + \theta)\sin(\theta - \omega t_2).$$

Thus, to remove the $\sin(t_1 + \theta)$ and $\cos(t_1 + \theta)$ terms, it is required that

$$2A' + \lambda A = -\sin(\theta - \omega t_2) \tag{3.52}$$

and

$$2\theta' A - \frac{3\kappa}{4}A^3 = -\cos(\theta - \omega t_2). \tag{3.53}$$

From (3.52) and (3.53), the initial condition in (3.51), and assuming $A'(0) > 0$, it follows that

$$\theta(0) = -\frac{\pi}{2}. \tag{3.54}$$

To summarize our findings, the first-term expansion of the solution of (3.41), (3.42) has the form

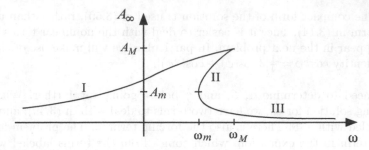

Figure 3.4 Schematic of positive solutions of (3.56) when λ and κ are positive (the exact shape of the *curve* depends on λ and κ). The *curve* as shown has three branches: (I) one for $-\infty < \omega \leq \omega_M$, (II) one for $\omega_m < \omega < \omega_M$, and (III) one for $\omega_m \leq \omega < \infty$. Note that $A_M = 1/\lambda$ and $\omega_M = 3\kappa/(8\lambda^2)$

$$y \sim A(t_2)\cos(t_1 + \theta(t_2)). \tag{3.55}$$

To find the amplitude A and phase function θ, it remains to solve (3.52) and (3.53) along with initial conditions (3.51) and (3.54). Unfortunately, the solution of this nonlinear system of equations is not apparent. In lieu of the exact solution, we can investigate what sort of solution may come from this system. In particular, we will investigate what value the amplitude A approaches if it goes to a steady state as $t \to \infty$. So, assuming $A' = 0$, let $A = A_\infty$, where A_∞ is a positive constant. From (3.52) it follows that $\theta = \omega t_2 + \theta_\infty$, where θ_∞ is a constant that satisfies $\sin(\theta_\infty) = -\lambda A_\infty$. From this and (3.53) the constant A_∞ is determined from the equation

$$A_\infty^2 \left[\lambda^2 + \left(2\omega - \frac{3\kappa}{4}A_\infty^2 \right)^2 \right] = 1. \tag{3.56}$$

An example of the possible solutions obtained from (3.56) is shown in Fig. 3.6 as a function of the frequency parameter ω. The interesting parts of the curve occur for $\omega_m < \omega < \omega_M$, where there are three possible steady states for A. The middle one, branch II, is unstable and cannot be reached unless the initial conditions are chosen so that the solution starts exactly at one of these points (the justification of this statement is given in Chap. 6, and Exercise 3.27 should also be consulted). The other two branches, I and III, are stable, and the solution will approach one of them depending on the initial conditions. This situation is known as *bistability*. Note that if one were to start on branch I and then increase the driving frequency to the point where $\omega > \omega_M$, then the amplitude would jump down to branch III. In a similar manner, once it is on the lower part of this solution curve and the frequency is decreased below ω_m, the amplitude then jumps up to branch I.

A comparison between the amplitude function $A(t_2)$ obtained from (3.52), (3.53) and the numerical solution of (3.41) is shown in Fig. 3.3. This type of comparison is made, rather than plotting all the values for the solution that are obtained from (3.55), because of the oscillatory nature of the solution. In any case, it is clearly seen that the agreement is quite good. However, one might argue that if the reduced problem (3.52), (3.53) must be solved numerically, then why not just solve the original problem numerically? It is certainly true that the numerical solution is not limited to small values of the parameter ε. In this regard it is superior to the asymptotic approximation. The asymptotic approximation, on the other hand, is capable of providing quick insight into how the solution depends on the various parameters of the problem. This sort of pro-and-con argument applies to almost all numerical/asymptotic solutions. What is special about the present example is the oscillatory nature of the response. In this case, the asymptotic approximation has the rapid oscillations built into the representation. Thus, the functions that need to be computed numerically are relatively smooth. In this sense, it is easier to solve (3.52), (3.53) numerically than the original Eq. (3.41).

The analysis here has been for what is sometimes called harmonic, or primary, resonance. There are also subharmonic and superharmonic resonances. To see where these originate, note that the solution in (3.44) is periodic if $\Omega = n$ or if $\Omega = 1/n$, where n is a positive integer with $n \neq 1$. For the first case, the solution has period 2π while the forcing has period $2\pi/n$. This corresponds to what is known as subharmonic oscillation. If $\Omega = 1/n$, then the solution and forcing have period $2\pi n$. This is known as superharmonic, or ultraharmonic, oscillation. With the nonlinear Duffing equation (3.41), it is possible for there to be a resonant response if Ω is equal to a subharmonic or superharmonic frequency. However, the amplitudes are generally smaller than what is found for harmonic resonance. This is demonstrated in Exercise 3.20, which examines the use of multiple scales to analyze the subharmonic response.

Exercises

3.18. In the relativistic mechanics of planetary motion around the Sun, one comes across the problem of solving

$$\frac{d^2u}{d\theta^2} + u = \alpha(1 + \varepsilon u^2) \quad \text{for } 0 < \theta < \infty,$$

where $u(0) = 1$ and $u'(0) = 0$. Here θ is the angular coordinate in the orbital plane; $u(\theta) = 1/r$, where r is the normalized radial distance of the planet from the Sun; and α is a positive constant. Note that if $\varepsilon = 0$, then one obtains the Newtonian description.
(a) Find a first-term approximation of the solution that is valid for large θ.
(b) Using the result from part (a), find a two-term expansion of the angle $\Delta\theta$ between successive perihelions, that is, the angle between successive maxima in $u(\theta)$.

(c) The parameters in the equation are $\varepsilon = 3(h/cr_c)^2$ and $\alpha = r_c/[a(1 - e^2)]$, where h is the angular momentum of the planet per unit mass, r_c is a characteristic orbital distance, c is the speed of light, a is the semimajor axis of the elliptical orbit, and e is the eccentricity of the orbit. For the planet Mercury, $h/c = 9.05 \times 10^3$ km, $r_c = a = 57.91 \times 10^6$ km, and $e = 0.20563$ (Nobili and Will, 1986). It has been observed that the precession of Mercury's perihelion, defined as $\Delta\phi \equiv \Delta\theta - 2\pi$, after a terrestrial century is $43.11'' \pm 0.45''$ (note that Mercury orbits the Sun in 0.24085 terrestrial years). How does this compare with your theoretical result in (b)?

(d) In the 11th edition of *Merriam-Webster's Collegiate Dictionary*, the second definition given for the word "perturbation" is "a disturbance of the regular and usually elliptical course of motion of a celestial body that is produced by some force additional to that which causes its regular motion." Comment on this in conjunction with your result from part (a). Also, how would you change the definition so that it encompasses the ideas in Chaps. 1 and 2?

3.19. What follows is the equation for van der Pol's oscillator (with small damping):
$$y'' - \varepsilon(1 - y^2)y' + y = 0 \quad \text{for } 0 < t,$$
where $y(0) = \alpha_0$ and $y'(0) = \beta_0$.

(a) Find a first-term approximation of the solution that is valid for large t.

(b) Sketch the first-term approximation in the phase plane as a function of t when $\alpha_0 = 1$ and $\beta_0 = 0$.

3.20. This problem investigates the subharmonic resonances of Duffing's equation,
$$y'' + \varepsilon\lambda y' + y + \varepsilon\kappa y^3 = \varepsilon\cos(\Omega t) \quad \text{for } 0 < t,$$
where $y(0) = y'(0) = 0$.

(a) Suppose one attempts to use the regular expansion
$$y \sim \varepsilon\left(y_0(t) + \varepsilon y_1(t) + \cdots\right).$$

Explain why this is nonuniform if $\Omega = \pm 1$ (primary resonance), or if $\Omega = \pm 3$ (subharmonic resonance). Also explain how your calculation shows that subharmonic resonance is due to the nonlinear interaction of the forcing with the primary resonance mechanism of the system.

(b) Setting $\Omega = 3 + \varepsilon\omega$, find a first-term approximation of the solution that is valid for large t. If you are not able to solve the problem that determines the t_2 dependence, then find the possible steady states (assume $\lambda > 0$).

(c) Compare the results from part (b) with those in Fig. 3.4.

3.21. The equation for a pendulum with a weak forcing is
$$\theta'' + \frac{1}{\varepsilon}\sin(\varepsilon\theta) = \varepsilon^2\sin(\omega t) \quad \text{for } 0 < t,$$

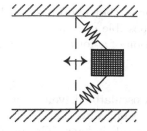

Figure 3.5 Oscillator studied in Exercise 3.22

where $\theta(0) = \theta'(0) = 0$.
(a) For fixed $\omega^2 \neq 1, 1/9, \ldots$ and $\omega^2 \neq 1, 9, \ldots$, find a two-term approximation of $\theta(t)$.
(b) For $\omega = 1 + \omega_0 \varepsilon^2$, find a first-term approximation of $\theta(t)$ that is valid for large t. If you are not able to solve the problem that determines the t_2 dependence, then find the possible steady states (if any).

3.22. Suppose a mass is situated between two parallel walls and is connected to the walls by springs (as shown in Fig. 3.5). For a small periodic forcing the equation for the transverse oscillations of the mass is (Stoker, 1950; Arnold and Case, 1982; Forbes, 1991)

$$y'' + y\left[1 - \frac{\lambda}{\sqrt{1 + y^2}}\right] = \varepsilon \cos(\kappa t + \varepsilon^{2/3}\omega t) \quad \text{for } 0 < t,$$

where λ and κ are positive constants with $0 < \lambda < 1$. Also, $y(0) = y'(0) = 0$. Set $\kappa^2 = 1 - \lambda$, and for small ε find a first-term approximation of the solution that is valid for large t. If you are not able to solve the problem that determines the t_2 dependence, then find, and sketch, the possible steady states (if any) for the amplitude.

3.23. In the study of Raman scattering, one comes across the equation for a forced Morse oscillator with small damping, given as

$$y'' + \varepsilon^2 \alpha y' + (1 - e^{-y})e^{-y} = \varepsilon^3 \cos(1 + \varepsilon^2 \omega)t,$$

where $y(0) = 0$ and $y'(0) = 0$. Also, $\alpha > 0$.
(a) Find a first-term approximation of $y(t)$ that is valid for large t. If you are not able to solve the problem that determines the t_2 dependence, then find the possible steady states (if any) for the amplitude.
(b) Lie and Yuan (1986) used numerical methods to solve this problem. They were interested in how important the value of the damping parameter α is for there to be multiple steady states for the amplitude. They were unable to answer this question because of the excessive computing time it took to solve the problem using the equipment available to them. However,

based on their calculations, they hypothesized that multiple steady states for the amplitude are possible even for small values of α. Sketching the graph of A_∞ as a function of ω for $\alpha > 0$ determine whether or not their hypothesis is correct.

3.24. Figure 3.6 shows an oscillator driven by a belt moving with velocity v. The block moves in the same direction as the belt until the spring has sufficient restoring force to pull it back. The block then retracts until the force in the spring reduces to the point where the frictional force between the block and belt causes the block to start moving again in the same direction as the belt. This back-and-forth motion is an example of what is known as a stick-slip problem. Among other things, this mechanism has been used to model squeaky doors and violin strings (Popp and Stelter, 1990; Gao and Kuhlmann-Wilsdorf, 1990). The equation of motion in this case is

$$y'' + \varepsilon \beta y' + y = f(v - \varepsilon y') \quad \text{for } 0 < t,$$

where β and v are positive constants. The function $f(v - \varepsilon y')$ accounts for the friction between the mass and the belt. It is an odd function of the relative velocity $V = v - \varepsilon y'$. In this problem, for $V > 0$, we will take $f(V) = \frac{1}{3}aV(V^2 - 3) + b$, where a and b are constants with $0 < 2a < 3b$. We will also take $y(0) = y'(0) = 0$.
(a) For $v^2 > 1 - \beta/a$, find a first-term approximation of the solution that is valid for large t.
(b) Find a first-term approximation of the solution that is valid for large t when $\frac{4}{5}(1 - \beta/a) < v^2 < 1 - \beta/a$. Make sure to point out where the stated conditions on v are used in the derivation.

3.25. The equation of a nonlinear beam with a small forcing is

$$\partial_x^4 u - \kappa \partial_x^2 u + \partial_t^2 u = \varepsilon f(t) \sin(\pi x) \quad \text{for } 0 < x < 1 \text{ and } 0 < t,$$

where $u = u'' = 0$ at $x = 0, 1$. Also,

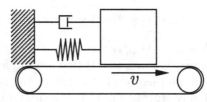

Figure 3.6 Forced oscillator used to model stick-slip problem in Exercise 3.24

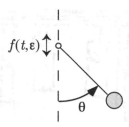

Figure 3.7 Forced pendulum studied in Exercise 3.26

$$\kappa = \frac{1}{4} \int_0^1 (u_x)^2 \mathrm{d}x.$$

(a) Find a first-term expansion of the solution that is valid for large t when $f(t) = \sin(t)$.
(b) If $f(t) = \sin[(\pi + \omega_0 \varepsilon)t]$, find a first-term expansion of the solution that is valid for large t.

3.26. The equation for a weakly damped pendulum whose support is subjected to a vertical excitation is (Fig. 3.7)

$$\theta'' + \varepsilon\mu\theta' + [1 - f(t, \varepsilon)] \sin\theta = 0,$$

where $\mu > 0$ is constant and $f(t, \varepsilon)$ is the prescribed excitation of the support. We will assume $\theta(0) = \varepsilon a_0$ and $\theta'(0) = \varepsilon b_0$, where a_0 and b_0 are constants independent of ε.
(a) For $f(t, \varepsilon) = \varepsilon\alpha \sin^2(t)$, where α is a positive constant, find a first-term expansion of the solution that is valid for large t.
(b) For $f(t, \varepsilon) = \alpha \sin(\varepsilon\omega t)$, where α and ω are positive constants, find a first-term expansion of the solution that is valid for large t.

3.27. This problem concerns the forced Duffing problem (3.41) and its asymptotic solution (3.55).
(a) Sketch the positive solutions of (3.56) when $\lambda = 0$. Identify which branch is in phase with the forcing and which branch is out of phase.
(b) One way to obtain a qualitative understanding of the solution of (3.52), (3.53) is to use the phase plane. Set $\phi = \theta - \omega l_2$, and sketch the direction fields in the A, ϕ-plane for $\lambda \neq 0$ and for $\lambda = 0$.
(c) Based on your results from part (b), comment on the possibility that A approaches a constant as $t_2 \to \infty$.

Figure 3.8 Sketch by Huygens, circa 1665, of two pendulum clocks attached to a beam supported by chairs

3.5 Weakly Coupled Oscillators

In this section we will use multiple scales to study systems involving multiple oscillators that are weakly coupled. This is a classic problem in applied mathematics, going back to Huygens' attempt to solve one of the more important unsolved technological problems in the middle 1600s. An accurate method was needed to determine longitude on a sailing ship, and Huygens thought he could use two of his pendulum clocks to produce such a device. A sketch of what he had in mind is shown in Fig. 3.8. It did not work as expected because the pendulums would become synchronous after running for about 30 min. In particular, he found that irrespective of how the pendulums started out, their motion would somehow change so they would eventually be 180° out of phase. After some experimentation Huygens discovered that this was due to the beam the pendulums were attached to, which allowed for a weak coupling between the oscillators. It is also worth commenting on how easy it is to obtain this type of response; the simplicity of the experiment shown in Fig. 3.8 is evidence of this (another example is shown in Fig. 6.22). In fact, it is so easy that Huygens claimed, in a letter to his father, that he did the experiment while lying in bed. It is therefore not surprising that synchronous behavior is found throughout nature. One example arises in neuroscience. The brain contains millions of neurons, and to certain stimuli groups of these neurons can respond in an oscillatory manner. It is thought that the resulting synchronized neural activity is important for cognitive function, although exactly how is still an open question (Poulet and Petersen, 2008; Klimesch et al., 2010). Another example, one most people are familiar with, is the synchronized clapping that can occur in a theater (Neda et al., 2000).

To investigate some of the consequences that weak coupling might have on the overall behavior of oscillators, we will consider a simple problem, one involving masses, springs, and dashpots. An example is shown in Fig. 3.9. Each oscillator, which consists of a single mass and spring, is coupled to every other oscillator through a spring and dashpot. The resulting equations of motion are

$$m_1 y_1'' + k_1 y_1 = \varepsilon \left(f_{12} + f_{13} \right), \tag{3.57}$$
$$m_2 y_2'' + k_2 y_2 = \varepsilon \left(-f_{12} + f_{23} \right), \tag{3.58}$$
$$m_3 y_3'' + k_3 y_3 = \varepsilon \left(-f_{13} - f_{23} \right), \tag{3.59}$$

where

$$f_{12} = k_{12}(y_2 - y_1) + c_{12}(y_2' - y_1'), \tag{3.60}$$
$$f_{13} = k_{13}(y_3 - y_1) + c_{13}(y_3' - y_1'), \tag{3.61}$$
$$f_{23} = k_{23}(y_3 - y_2) + c_{23}(y_3' - y_2'). \tag{3.62}$$

In the preceding equations, f_{ij} is the force of the coupling between the ith and jth oscillators and the minus signs appearing in (3.58) and (3.59) are due to Newton's third law. Also, these equations are nondimensionalized; an explanation of how this is done in given in Exercise 3.32. In this context, m_i and k_i are the scaled mass and spring constants, respectively, for the ith oscillator. Similarly, k_{ij} is the spring constant for the spring connecting the ith and jth oscillators, and c_{ij} is the constant for the dashpot connecting the ith and jth oscillators.

The equations of motion (3.57)–(3.62) can be written in matrix form as

$$\mathbf{M}y'' + \mathbf{K}y = -\varepsilon \left(\mathbf{A}y + \mathbf{B}y' \right), \tag{3.63}$$

where

$$\mathbf{M} = \begin{pmatrix} m_1 & 0 & 0 \\ 0 & m_2 & 0 \\ 0 & 0 & m_3 \end{pmatrix}, \qquad \mathbf{K} = \begin{pmatrix} k_1 & 0 & 0 \\ 0 & k_2 & 0 \\ 0 & 0 & k_3 \end{pmatrix},$$

$$\mathbf{A} - \begin{pmatrix} k_{12} + k_{13} & -k_{12} & -k_{13} \\ -k_{12} & k_{12} + k_{23} & -k_{23} \\ -k_{13} & -k_{23} & k_{13} + k_{23} \end{pmatrix},$$

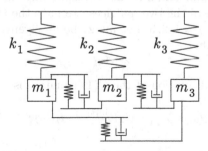

Figure 3.9 Three oscillators weakly coupled by springs and dashpots

$$\mathbf{B} = \begin{pmatrix} c_{12} + c_{13} & -c_{12} & -c_{13} \\ -c_{12} & c_{12} + c_{23} & -c_{23} \\ -c_{13} & -c_{23} & c_{13} + c_{23} \end{pmatrix}.$$

The left-hand side of (3.63) comes from the masses and springs of the oscillators, while the right-hand side contains the coupling terms. It is important to note that all four of the preceding matrices is symmetric. Moreover, assuming that the parameters m_i, k_i, k_{ij}, and c_{ij} are positive, the matrices $\mathbf{M}$ and $\mathbf{K}$ are positive definite, and $\mathbf{A}$ and $\mathbf{B}$ are positive semidefinite (Exercise 3.31).

We are assuming that the coupling is weak, and so ε is small. Our objective is to derive an approximation for the solution that holds for large t. Exactly what happens depends on the differences in the oscillators. In the calculations to follow we will assume they are all different in the sense that $m_i/k_i \neq m_j/k_j$ if $i \neq j$. In Exercise 3.28 the other extreme is considered, where the oscillators are all the same. Later, in Sect. 6.7, we will consider what happens when the oscillators are nonlinear, where each gives rise to what is known as a limit cycle solution.

The equation in (3.63) is similar enough to (3.1) that it should not be unexpected that we need to use a multiple-scale approximation to derive an approximation that holds for large t. With this in mind, we introduce the two time scales $t_1 = t$ and $t_2 = \varepsilon t$. In this case, (3.63) becomes

$$\mathbf{M}(\partial_{t_1}^2 + 2\varepsilon\partial_{t_1}\partial_{t_2} + \varepsilon^2\partial_{t_2}^2)\mathbf{y} + \mathbf{K}\mathbf{y} = -\varepsilon\left[\mathbf{A}\mathbf{y} + (\partial_{t_1} + \varepsilon\partial_{t_2})\mathbf{B}\mathbf{y}\right].$$

Assuming that

$$\mathbf{y} \sim \mathbf{y}_0(t_1, t_2) + \varepsilon\mathbf{y}_1(t_1, t_2) + \cdots$$

we obtain the following problem:

$O(1)$ $\mathbf{M}(\partial_{t_1}^2 \mathbf{y}_0) + \mathbf{K}\mathbf{y}_0 = \mathbf{0}$.

To find the solution, assume that $\mathbf{y}_0 = \mathbf{z}e^{irt_1}$. Substituting this into the equation yields $\mathbf{K}\mathbf{z} = r^2\mathbf{M}\mathbf{z}$. This is an eigenvalue equation, where r^2 is the eigenvalue and $\mathbf{z}$ is the corresponding eigenvector. Given the diagonal nature of the two matrices, we have the three eigenvalues $r_j^2 = k_j/m_j$ for $j = 1, 2, 3$. Letting $\mathbf{z}_j$ be an eigenvector for r_j^2, normalized so that

$$\mathbf{z}_j\mathbf{M}\mathbf{z}_k = \delta_{jk}, \tag{3.64}$$

then the general solution of the $O(1)$ equation is

$$\mathbf{y}_0 = \sum_j \left[\alpha_j(t_2)e^{ir_jt_1} + \beta_j(t_2)e^{-ir_jt_1}\right]\mathbf{z}_j, \tag{3.65}$$

where the dependence of the α_j and β_j on t_2 is determined from $O(\varepsilon)$ problem.

The solution in (3.65) was written so it applies to systems containing more that three oscillators. That is, it applies to the second-order differential equation given in (3.63), where $\mathbf{M}$ and $\mathbf{K}$ are symmetric and positive-definite matrices. The symmetry in this case guarantees that the eigenvectors $\mathbf{z}_j$ form a basis, and the positive-definiteness of these two matrices means that the eigenvalues r_j^2 are positive. What we had to assume explicitly was that the eigenvalues are distinct. This does not affect the form of the solution in (3.65), but it does have consequences for how the coupling is responsible for generating secular terms.

$$O(\varepsilon) \quad \mathbf{M}(\partial_{t_1}^2 \mathbf{y}_1) + \mathbf{K}\mathbf{y}_1 = -2\mathbf{M}(\partial_{t_1}\partial_{t_2}\mathbf{y}_0) - \mathbf{A}\mathbf{y}_0 - \mathbf{B}(\partial_{t_1}\mathbf{y}_0).$$

Given that the $\mathbf{z}_j$ form a basis, we can write

$$\mathbf{y}_1 = \sum_j \gamma_j(t_1, t_2)\mathbf{z}_j.$$

Substituting this into the $O(\varepsilon)$ equation and using the fact that $\mathbf{K}\mathbf{z}_j = r_j^2 \mathbf{M}\mathbf{z}_j$ yields

$$\sum_j (\partial_{t_1}^2 \gamma_j + r_j^2 \gamma_j)\mathbf{M}\mathbf{z}_j = \sum_j (2ir_j\beta_j'\mathbf{M}\mathbf{z}_j + ir_j\beta_j\mathbf{B}\mathbf{z}_j - \beta_j\mathbf{A}\mathbf{z}_j)e^{-ir_j t_1}$$
$$- \sum_j (2ir_j\alpha_j'\mathbf{M}\mathbf{z}_j + ir_j\alpha_j\mathbf{B}\mathbf{z}_j + \alpha_j\mathbf{A}\mathbf{z}_j)e^{ir_j t_1}.$$

Given the orthogonality condition (3.64) it follows that

$$\partial_{t_1}^2 \gamma_k + r_k^2 \gamma_k = -2ir_k\alpha_k'e^{ir_k t_1} + 2ir_k\beta_k'e^{-ir_k t_1}$$
$$- \sum_j \alpha_j(ir_j\mathbf{z}_k\mathbf{B}\mathbf{z}_j + \mathbf{z}_k\mathbf{A}\mathbf{z}_j)e^{ir_j t_1}$$
$$- \sum_j \beta_j(-ir_j\mathbf{z}_k\mathbf{B}\mathbf{z}_j + \mathbf{z}_k\mathbf{A}\mathbf{z}_j)e^{-ir_j t_1}.$$

The $e^{\pm ir_k t_1}$ terms on the right-hand side will generate secular terms, and so we require that

$$- 2ir_k\alpha_k' - \alpha_k(ir_k\mathbf{z}_k\mathbf{B}\mathbf{z}_k + \mathbf{z}_k\mathbf{A}\mathbf{z}_k) = 0 \qquad (3.66)$$

and

$$2ir_k\beta_k' - \beta_k(-ir_k\mathbf{z}_k\mathbf{B}\mathbf{z}_k + \mathbf{z}_k\mathbf{A}\mathbf{z}_k) = 0. \qquad (3.67)$$

Solving these equations yields

$$\alpha_k = a_k \exp\left(\frac{1}{2}\left(-\mathbf{z}_k\mathbf{B}\mathbf{z}_k + \frac{i}{r_k}\mathbf{z}_k\mathbf{A}\mathbf{z}_k\right)t_2\right) \quad \text{for } k = 1, 2, 3,$$

and

$$\beta_k = b_k \exp\left(\frac{1}{2}\left(-\mathbf{z}_k \mathbf{B} \mathbf{z}_k - \frac{i}{r_k} \mathbf{z}_k \mathbf{A} \mathbf{z}_k\right) t_2\right) \text{ for } k = 1, 2, 3,$$

where the constants a_k and b_k are determined from the initial conditions.

Based on the preceding calculations and the result from Exercise 3.31(c), the first-term approximation of the weakly coupled oscillator problem is

$$\mathbf{y} \sim \sum_j \left(a_j e^{i\theta_j t} + b_j e^{-i\theta_j t}\right) e^{-\varepsilon \kappa_j t} \mathbf{z}_j, \tag{3.68}$$

where

$$\kappa_j = \frac{1}{2} \mathbf{z}_j \mathbf{B} \mathbf{z}_j$$

$$= \frac{1}{2m_j} B_{jj}$$

and

$$\theta_j = r_j + \frac{\varepsilon}{2r_j} \mathbf{z}_j \mathbf{A} \mathbf{z}_j$$

$$= \sqrt{\frac{k_j}{m_j}} \left[1 + \frac{\varepsilon}{2} \frac{A_{jj}}{k_1}\right].$$

The preceding approximation holds for the general matrix Eq. (3.63), so long as the eigenvalues r_j^2 are distinct and have a geometric multiplicity of one. In other words, in the case of n oscillators, there are n eigenvalues with $r_i^2 \neq r_j^2$, $\forall i \neq j$.

The solution in (3.68) applies to the general form of the equation as given in (3.63). For the three oscillator systems first introduced in Fig. 3.9, the approximation gives us that the displacement for the first oscillator is

$$y_1 \sim A_1 e^{-\varepsilon \kappa_1 t} \cos(\theta_1 t + \varphi_1), \tag{3.69}$$

where

$$\kappa_1 = \frac{1}{2m_1} (c_{12} + c_{13})$$

and

$$\theta_1 = \sqrt{\frac{k_1}{m_1}} \left[1 + \frac{\varepsilon}{2} \frac{k_{12} + k_{13}}{k_1}\right].$$

The constants A_1 and φ_1 are determined from the initial conditions. There are similar expressions for the displacements of the other two oscillators.

In (3.69), given that κ_1 depends on c_{12} and c_{13}, the decay in the solution is due to the damping terms in the coupling. Similarly, the dependence of θ_1 on k_{12} and k_{13} shows that the spring coupling causes a shift in the frequency of the oscillation. For anyone familiar with a damped harmonic oscillator, this result is not particularly surprising. However, what might not be obvious is that this result hinges on the fact that the three oscillators have different resonant frequencies. As shown in Exercise 3.28, when they are the same, the time decay seen in (3.69) no longer occurs and the solution evolves into simple harmonic motion.

Exercises

3.28. This problem explores what happens when all of the oscillators are the same for the system shown in Fig. 3.9. Specifically, it is assumed that $m_i = k_i = 1$, $i = 1, 2, 3$ [Exercise 3.32(b)]. The coupling coefficients, however, are not assumed to be equal.

(a) Show that in constructing a first-term multiple-scale approximation one obtains (3.65), but instead of (3.66) and (3.67), it is necessary to solve first-order systems of the form $\boldsymbol{\alpha}' = \mathbf{Q}\boldsymbol{\alpha}$ and $\boldsymbol{\beta}' = \mathbf{P}\boldsymbol{\beta}$.

(b) Show that zero is an eigenvalue of $\mathbf{Q}$ and all other eigenvalues have negative real part. What about the eigenvalues of $\mathbf{P}$? With this information, write down the resulting first-term multiple-scale approximation of $\mathbf{y}$.

(c) A single mass–spring–dashpot system, as embodied in (3.1), is oscillatory but decays over time. The decay comes from the dissipative nature of the dashpot mechanism. One might think that the same is true with a multiple mass–spring–dashpot system. Explain why your answer in part (b) shows that this expectation is incorrect. Also, explain this situation by considering the contribution of the dashpots to the equations of motion, which is contained in the term $-\varepsilon \mathbf{B}\mathbf{y}'$.

(d) Show that as $t \to \infty$, the three oscillators are not only periodic but in-phase. In other words, they evolve into synchronous periodic motion.

3.29. This problem explores three pendulums that are coupled with springs, as shown in Fig. 3.10. It is assumed the pendulums are the same, so they have the same mass m and length L. It is also assumed that the two springs have the same spring constant k, and each is attached at a distance ℓ from the upper pivot (as measured along the rod of the pendulum).

(a) The equations of motion are

$$mL^2\theta_1'' = mgL\sin(\theta_1) + k\ell^2(\sin\theta_2 - \sin\theta_1)\cos\theta_1,$$
$$mL^2\theta_2'' = -mgL\sin(\theta_2) + \left(k\ell^2(\sin\theta_3 - \sin\theta_2) - k\ell^2(\sin\theta_2 - \sin\theta_1)\right)\cos\theta_2,$$
$$mL^2\theta_3'' = -mgL\sin(\theta_3) - k\ell^2(\sin\theta_3 - \sin\theta_2)\cos\theta_3.$$

Assuming a small angle, so $\sin\theta \approx \theta$ and $\cos\theta \approx 1$, nondimensionalize the resulting linear problem and show that it can be written as

$$\mathbf{y}'' + \mathbf{y} = -\varepsilon \mathbf{A}\mathbf{y},$$

where

$$\mathbf{A} = \begin{pmatrix} 1 & -1 & 0 \\ -1 & 2 & -1 \\ 0 & -1 & 1 \end{pmatrix}.$$

It is assumed that the coupling is weak, so ε is small.
(b) Derive a first-term multiple-scale approximation of $\mathbf{y}$.

3.30. The spring and dashpot coupling shown in Fig. 3.9 occurs in parallel. It is possible to put them in series. A two oscillator example is shown in Fig. 3.11. Assuming weak coupling, the equations of motion are (Holmes, 2009)

$$m_1 y_1'' + k_1 y_1 = \varepsilon f,$$
$$m_2 y_2'' + k_2 y_2 = -\varepsilon f,$$
$$\frac{1}{k_0} f' + \frac{1}{c_0} f = y_2' - y_1'.$$

In these equations, f is the force due to the coupling.
(a) The equations for two oscillators that have parallel coupling is obtained by setting $y_3 = k_{13} = c_{13} = k_{23} = c_{23} = 0$ in (3.57)–(3.62). Comment on the differences in the equations for series and parallel coupling. Also, show that they reduce to the same problem if the coupling has no dashpot, so $c_{12} = 0$ and $c_0 \to \infty$, or has no spring, so $k_{12} = 0$ and $k_0 \to \infty$.
(b) Derive a first-term multiple-scale approximation of the solution of the parallel problem in the case where $k_1/m_1 \neq k_2/m_2$.
(c) Show that the displacement for the first oscillator can be written as

$$y_1 \sim A_1 e^{-\varepsilon \bar{\kappa}_1 t} \cos(\bar{\theta}_1 t + \bar{\varphi}_1).$$

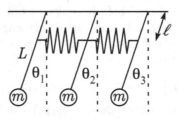

Figure 3.10 Three pendulums that are weakly coupled by springs, as described in Exercise 3.29

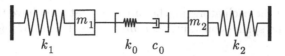

Figure 3.11 Two oscillators that are weakly coupled by a spring and dashpot in series, as described in Exercise 3.30

Comment on how the decay in the solution and the frequency shift differ from the parallel solution as given in (3.69). For example, does the solution for the series coupling decay faster or have a larger frequency shift than the solution for parallel coupling? Also, show that this solution agrees with (3.69) for the two special cases described in part (a).

(d) Derive a first-term multiple-scale approximation of the solution of the parallel problem in the case where $m_1 = m_2 = k_1 = k_2 = 1$.

(e) For your solution in part (d), explain what happens as $t \to \infty$ when the initial conditions satisfy $x_1(0) + x_2(0) = 0$ and $x_1'(0) + x_2'(0) = 0$. What happens when the initial conditions do not satisfy these conditions?

3.31. This problem develops some of the properties of the matrices and vectors used in this section. Assume here that the parameters m_i, k_i, k_{ij}, and c_{ij} are positive.

(a) Show that $\mathbf{M}$ and $\mathbf{K}$ are positive definite.

(b) Show that $\mathbf{A}$ and $\mathbf{B}$ are positive semidefinite. In particular, for each matrix, show that one eigenvalue is zero and the others are positive.

(c) Explain why, even with the normalization given in (3.64), $\mathbf{z}_j$ is not uniquely determined. Also, show that one choice is $(\mathbf{z}_j)_i = \delta_{ij}/\sqrt{m_j}$.

3.32. This problem derives the nondimensional oscillator problem (3.57)–(3.59). In dimensional variables the equation for the first oscillator is

$$M_1 Y_1'' + K_1 Y_1$$
$$= K_{12}(Y_2 - Y_1) + K_{13}(Y_3 - Y_1) + C_{12}(Y_2' - Y_1') + C_{13}(Y_3' - Y_1'),$$

where $Y_i(T)$ is the displacement of the ith oscillator, M_i its mass, and K_i its spring constant. Also, T is the time variable, and the K_{ij} and C_{ij} are the coupling constants. The capital letters designate dimensional quantities, and their lowercase counterparts will designate the nondimensional version of the respective variable. There are also equations for the other two oscillators, but this problem will concentrate only on the first one.

(a) Introducing the averages $M_a = \frac{1}{3}\sum_j M_j$ and $K_a = \frac{1}{3}\sum_j K_j$, let $m_i = M_i/M_a$ and $k_i = K_i/K_a$. Also, assume the nondimensionalization is $Y_i = Y_c y_i$ and $T = T_c t$, where Y_c and T_c are the same for the three oscillators. What does T_c need to obtain (3.57)? Explain why there is no condition imposed on Y_c (other than that it is independent of i).

(b) Show that if the oscillators are the same, so that they have the same mass and the same spring constant, then $m_i = k_i = 1$ in (3.57)–(3.59).

3.6 Slowly Varying Coefficients

To illustrate what happens when a more complicated time scale must be used, we present the following problem:

$$y'' + k^2(\varepsilon t)y = 0 \quad \text{for } 0 < t, \tag{3.70}$$

where

$$y(0) = a \text{ and } y'(0) = b. \tag{3.71}$$

This equation corresponds to an undamped oscillator with a spring that has a slowly varying restoring force. This is the sort of situation that could arise, for example, if the spring gradually fatigued or changed periodically in time. We will assume that $k(\tau)$ is a smooth positive function for $\tau \geq 0$. It is also assumed that it is not constant.

Before deriving a multiple-scale approximation of the solution of (3.70), (3.71), it is worth examining what a solution can look like. Therefore, in Fig. 3.12 the numerical solution is shown for a particular choice of the function $k(\varepsilon t)$. It is evident from this graph that there are (at least) two time scales, a fast time scale associated with the rapid oscillations in the solution and a slower time scale involved with the variation of the amplitude of the oscillation. Our objective is to determine exactly what these time scales are and how they depend on the coefficient function $k(\varepsilon t)$.

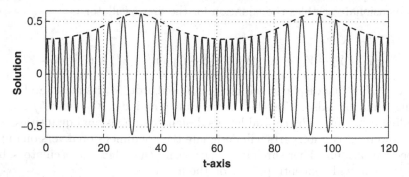

Figure 3.12 Numerical solution (*dashed curve*) of (3.70) and (3.71) when $k(\varepsilon t) = 2 + \cos(\varepsilon t)$, $\varepsilon = 10^{-1}$, $a = 0$, and $b = 1$. Also shown (*solid curve*) is the amplitude $A(t)$ of the oscillation determined from the multiple-scale approximation given in (3.81)

Based on our success with the weakly damped oscillator in Sect. 3.2, suppose we assume the two principal time scales for this problem are $t_1 = t$ and $t_2 = \varepsilon t$. Using the same expansion as in (3.12), one finds that

$$y_0 = a_0(t_2) \sin[k(t_2)\, t_1] + b_0(t_2) \cos[k(t_2)\, t_1].$$

To determine a_0 and b_0, we need to solve the $O(\varepsilon)$ equation. One finds that this equation is

$$(\partial_{t_1}^2 + k^2) y_1 = -2\left[(a_0 k)' - b_0 k k' t_1\right] \cos[k(t_2)\, t_1]$$
$$+ 2\left[(b_0 k)' + a_0 k k' t_1\right] \sin[k(t_2)\, t_1].$$

Looking at the right-hand side of this equation, one sees that terms like $t_1^2 \sin[k(t_2)\, t_1]$ and $t_1 \sin[k(t_2)\, t_1]$ are going to appear in the solution for y_1. Since a_0, b_0, and k are independent of t_1, the only way to prevent this is to take $a_0 = b_0 = 0$. In other words, the only solution we are able to find that does not contain secular terms is the zero solution. Clearly, we have not gotten very far with our simple expansion.

It is necessary to think a little harder on how to choose the time scales, and the first place to begin is with t_1. This is the faster time scale, so it should be a measure of the period of oscillation. In this problem, the period depends on $k(\varepsilon t)$. At the moment, it is unclear exactly how to define this time scale, and so we will let

$$t_1 = f(t, \varepsilon) \tag{3.72}$$

and

$$t_2 = \varepsilon t. \tag{3.73}$$

The function $f(t, \varepsilon)$ must satisfy the following conditions:

1. $f(t, \varepsilon)$ is nonnegative and increases with t,

2. $\varepsilon t \ll f$ as $\varepsilon \downarrow 0$ (i.e., t_1 is the fast time scale and t_2 is the slow scale), and

3. $f(t, \varepsilon)$ is smooth.

Now, with the multiple-scale transformation and the chain rule, we have

$$\frac{d}{dt} = \frac{\partial f}{\partial t} \frac{\partial}{\partial t_1} + \varepsilon \frac{\partial}{\partial t_2},$$
$$\frac{d^2}{dt^2} = f_t^2 \frac{\partial^2}{\partial t_1^2} + f_{tt} \frac{\partial}{\partial t_1} + 2\varepsilon f_t \frac{\partial^2}{\partial t_1 \partial t_2} + \varepsilon^2 \frac{\partial^2}{\partial t_2^2}.$$

Substituting these into (3.70) yields

$$(f_t^2 \partial_{t_1}^2 + f_{tt} \partial_{t_1} + 2\varepsilon f_t \partial_{t_1} \partial_{t_2} + \varepsilon^2 \partial_{t_2}^2) y + k^2(\varepsilon t) y = 0. \qquad (3.74)$$

①   ②

The oscillation takes place because of the balance between terms ① and ②. We therefore let $f_t = k(\varepsilon t)$. Integrating this yields

$$f(t, \varepsilon) = \int_0^t k(\varepsilon \tau) \mathrm{d}\tau. \qquad (3.75)$$

It is not hard to show that this choice satisfies the three requirements listed earlier. Also, with this, we have that $f_{tt} = \varepsilon k'(t_2)$. So with the expansion

$$y \sim y_0(t_1, t_2) + \varepsilon y_1(t_1, t_2) + \cdots, \qquad (3.76)$$

we obtain the following problems from (3.74) and (3.71):

$O(1)$ $\quad (\partial_{t_1}^2 + 1) y_0 = 0,$
$\qquad\quad y_0(0,0) = a, \quad k(0) \partial_{t_1} y_0(0,0) = b.$

The solution of this problem is

$$y_0 = a_0(t_2) \sin(t_1) + b_0(t_2) \cos(t_1), \qquad (3.77)$$

where $b_0(0) = a$ and $a_0(0) = b/k(0)$.

$O(\varepsilon)$ $\quad k^2(\partial_{t_1}^2 + 1) y_1 = -2k \partial_{t_1} \partial_{t_2} y_0 - k' \partial_{t_1} y_0$
$$\qquad\qquad\qquad\qquad = -(a_0 k' + 2k a_0') \cos(t_1) + (b_0 k' + 2k b_0') \sin(t_1).$$

To prevent secular terms, we let $a_0 k' + 2k a_0' = b_0 k' + 2k b_0' = 0$. Thus,

$$a_0 = \frac{\alpha_0}{\sqrt{k}}, \quad \text{and} \quad b_0 = \frac{\beta_0}{\sqrt{k}}, \qquad (3.78)$$

where α_0 and β_0 are constants determined from the initial conditions.

Our first-term approximation is therefore

$$y \sim \frac{1}{\sqrt{k(\varepsilon t)}} \left[\alpha_0 \sin\left(\int_0^t k(\varepsilon \tau) \mathrm{d}\tau \right) + \beta_0 \cos\left(\int_0^t k(\varepsilon \tau) \mathrm{d}\tau \right) \right]. \qquad (3.79)$$

The values of α_0 and β_0 are determined from the initial conditions, and one finds that $\beta_0 = y(0)\sqrt{k(0)}$ and $\alpha_0 = y'(0)/\sqrt{k(0)}$.

As a check on the accuracy of the multiple-scale approximation in (3.79), note that in the special case where $y(0) = 0$, (3.79) reduces to

$$y \sim A(t) \sin\left(\int_0^t k(\varepsilon \tau) \mathrm{d}\tau \right), \qquad (3.80)$$

where the amplitude function $A(t)$ is

$$A(t) = \frac{\alpha_0}{\sqrt{k(\varepsilon t)}}. \qquad (3.81)$$

To compare the multiple-scale approximation and the numerical solution of the problem, $A(t)$ is plotted in Fig. 3.12. The amplitude is used because the highly oscillatory nature of the solution makes a point-by-point comparison unwieldy. Based on how well the amplitude compares, it is is evident that the multiple-scale approximation gives a very accurate description of the solution.

Problems with slowly varying coefficients are not uncommon, and it is also not uncommon to see ad hoc approximations of the solution. Usually this involves solving the problem as if the coefficients were constant and then allowing the coefficients in the resulting solution to be slowly varying functions. This idea, and how badly it works, is explored in Exercise 3.34.

There are other ways to derive the approximation in (3.79). One that is relatively easy to apply is the WKB method (Exercise 4.2). The advantage of the multiple-scale approach is that it works on nonlinear problems. This is evident in the original paper on this problem by Kuzmak (1959), and it will be evident in the next section. Using the method, however, requires us to be able to recognize the appropriate scales and then determine what terms are responsible for secular terms.

Exercises

3.33. This problem concerns the slowly varying undamped oscillator (3.70).

(a) Suppose that one uses Taylor's theorem to expand $k(\varepsilon t)$ for small ε and then uses a standard multiple-scale transformation as in (3.8), (3.9). How does the first term in this case compare with (3.79)? Over what time interval is this expansion valid?

(b) Show that the fast time scale given in (3.75) satisfies the following conditions: (i) $f(t, \varepsilon)$ is positive and increases with t; (ii) $\varepsilon t \ll f$ as $\varepsilon \downarrow 0$; and (iii) $f(t, \varepsilon)$ is smooth. Is it necessary that $f(0, \varepsilon) = 0$?

(c) Suppose that instead of (3.73) one takes $t_2 = \varepsilon t_1$. Does this alter (3.79)?

3.34. This problem explores one of the more common, but flawed, methods for approximating the solution to a problem with slowly varying coefficients.

(a) Assuming k is constant, find the general solution of (3.70). The flawed approximation is obtained by taking this solution and letting k depend on εt. Use this expression to satisfy the initial conditions $y(0) = 0$ and $y'(0) = b$.

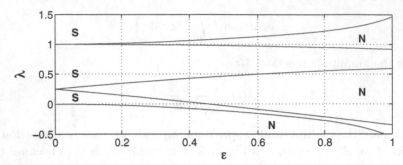

Figure 3.13 Stability regions in ε, λ-plane for Mathieu's equation. An S indicates a stable region and an N an unstable region. In Exercise 3.37 it is assumed that $\lambda > 0$

(b) How does the approximation in part (a) compare with the corresponding multiple-scale approximation in (3.80), (3.81)? Comment specifically on the differences in the amplitude and phase of the two expressions. Use the example in Fig. 3.12 to illustrate your conclusions.

3.35. Consider the problem

$$\frac{d}{dt}\left(D(\varepsilon t)\frac{dy}{dt}\right) + y = 0 \quad \text{for } 0 < t,$$

where $y(0) = \alpha$ and $y'(0) = \beta$. The coefficient $D(\tau)$ is a smooth positive function with $D' > 0$. Find a first-term approximation of the solution valid for large t.

3.36. The equation for the angular displacement $\theta(t)$ of a pendulum whose length is slowly varying is

$$\theta'' + \kappa(\varepsilon t)\sin(\theta) = 0 \quad \text{for } 0 < t,$$

where $\theta(0) = \varepsilon$ and $\theta'(0) = 0$. The function $\kappa(\tau)$ is smooth and positive. Find a first-term expansion of the solution that is valid for large t.

3.37. Mathieu's equation is

$$y'' + [\lambda + \varepsilon \cos(t)]y = 0 \quad \text{for } 0 < t,$$

where $y(0) = a$ and $y'(0) = b$. Also, λ is a positive constant. This equation describes the small-amplitude oscillations of a pendulum whose length varies periodically with time. If the pendulum's natural frequency is a particular multiple of the frequency of the length variation, then instability can occur. This is indicated in Fig. 3.13, which shows the regions in the ε, λ-plane where the motion is stable (S) and unstable (N). Equations for the boundaries of these regions are derived in this exercise.

(a) Assuming λ is independent of ε, use a regular expansion to show that a secular term appears in the second term of the expansion if $\lambda = 1/4$ and in the third term no matter what the value of λ.

(b) In the case where $\lambda = 1/4$, use multiple scales to remove the secular term in the second term of the expansion. Use this to explain why the solution can grow exponentially in time and is therefore unstable. By generalizing this analysis, it is possible to show that the solution may be unbounded, and hence unstable, if $\lambda = n^2/4$, where $n = 1, 2, 3, \ldots$ (you do not need to show this).

(c) Assuming $\lambda \neq n^2/4$, where $n = 1, 2, 3, \ldots$, use multiple scales to show that

$$y \sim a_0 \cos\left(\sqrt{\lambda}t + \frac{\varepsilon^2 t}{4\sqrt{\lambda}(1 - 4\lambda)} + \theta_0\right),$$

where a_0 and θ_0 are constants. This expression indicates what the solution looks like in the stable regions in Fig. 3.13.

(d) To investigate what happens for λ values near $1/4$, suppose $\lambda \sim \frac{1}{4}(1 + 2\varepsilon\lambda_1)$. Find a first-term approximation of the solution that is valid for large t. From this, show that the solution may be unbounded, depending on the initial conditions, if $|\lambda_1| < 1$. Moreover, irrespective of the initial conditions, the solution is bounded if $|\lambda_1| > 1$. Because of this, the curves $\lambda \sim \frac{1}{4}(1 \pm 2\varepsilon)$ form the stability boundaries in this region.

(e) To investigate what happens near $\lambda = 1$, suppose $\lambda \sim 1 + \varepsilon^2\lambda_1$. Find a first-term approximation of the solution that is valid for large t. From this, show that the solution may be unbounded, depending on the initial conditions, if $\frac{-1}{12} < \lambda_1 < \frac{5}{12}$.

3.38. This problem considers the case where the logistic equation has a slowly varying rate $r(\varepsilon t)$. The equation is

$$y' = r(\varepsilon t)y(1 - y) \quad \text{for } 0 < t,$$

where $y(0) = \alpha$ is a positive constant. It is assumed that $r(\tau)$ is a smooth positive function.

(a) Using the time scales $t_1 = t$ and $t_2 = \varepsilon t$, show that there is a problem when attempting to determine the dependence of y_0 on t_2.

(b) Using the time scales (3.72) and (3.73), derive a first-term approximation for the solution that is valid for large t.

3.39. Consider the following boundary-value problem for the function $u(x)$:

$$\varepsilon^2 u'' + [\omega^2 - k^2(x)]u = 0 \quad \text{for } 0 < x < \infty,$$

where $u(0) = 1$ and $u \to 0$ as $x \to \infty$. Also, $k(x)$ is smooth and positive, and ω is a positive constant.

(a) Find a first-term approximation of the solution if $k(x) > \omega$ for $0 \leq x < \infty$.

(b) Suppose $k'(x) > 0$ for $0 \leq x < \infty$ and $k(x_t) = \omega$, where $0 < x_t < \infty$. Find a first-term approximation of the solution for $0 \leq x < x_t$ and another approximation of the solution for $x_t < x < \infty$. The approximations from these two regions need to be matched; how to do this will be discussed in Sect. 4.3.

3.40. The equation for what is known as the Rayleigh oscillator is

$$y'' - \varepsilon\left(1 - \frac{1}{3}(y')^2\right)y' + y = 0 \quad \text{for } 0 < t,$$

where $y(0) = 0$ and $y'(0) = 1$. Find a first-term approximation of the solution valid for large t.

3.41. The approach used in this section can be adapted to other secular-producing approximations. As an example, consider the eigenvalue problem for the vertical displacement of an elastic string with variable density, given as

$$y'' + \lambda^2 \rho(x, \varepsilon)y = 0 \quad \text{for } 0 < x < 1, \quad \text{where } y(0) = y(1) = 0.$$

Both the eigenfunction $y(x)$ and associated eigenvalue λ are to be determined from this problem. The density ρ is assumed known, with $\rho \sim 1 + \varepsilon\mu(x)$, where $\mu(x)$ is smooth. The objective is to find an approximation of y and λ, for small ε, that is valid for both large and small values of λ.

(a) Assuming regular expansions of the form $y \sim y_0(x) + \varepsilon y_1(x) + \cdots$ and $\lambda \sim \lambda_0 + \varepsilon\lambda_1 + \cdots$, show that $y_0 = A_0 \sin(\lambda_0 x)$ and $\lambda_0 = n\pi$. Also, by examining the problem for the second terms, explain why y_1 contains a secular term for large n (you can assume $\mu = x$ to do this).

(b) In a similar manner as in (3.72), introduce a change of variables, $X = f(x, \varepsilon)$ and $Y(X) = y(x)$. Find f so the problem transforms into solving

$$Y'' + h(X, \varepsilon)Y' + \nu^2 Y = 0 \quad \text{for } 0 < X < 1,$$

where $Y = 0$ for $X = 0, 1$ and ν^2 is a positive constant. Also, for small ε, $h(X, \varepsilon) \sim \frac{1}{2}\varepsilon\mu'(X) + \cdots$.

(c) Assuming $\mu(x) = \alpha x$, find the first two terms in the expansion of Y and ν. Are secular terms present in the second terms? What are the corresponding expansions for y and λ?

(d) In the case where $\rho = 1/(1 + \varepsilon x)^2$, the exact eigenvalues are

$$\lambda_n^2 = \frac{\varepsilon^2(\pi n)^2}{\ln^2(1 + \varepsilon)} + \frac{\varepsilon}{4}.$$

Compare this with your result in part (c).

3.7 Boundary Layers

As stated earlier, the method of multiple scales is not limited to the time variable. To illustrate this, we consider one of the boundary-layer problems from the last chapter. Specifically, we take the following problem from Sect. 2.2:

$$\varepsilon y'' + 2y' + 2y = 0 \quad \text{for } 0 < x < 1, \tag{3.82}$$

where

$$y(0) = 0 \text{ and } y(1) = 1. \tag{3.83}$$

We set $x_1 = x/\varepsilon$ and $x_2 = x$, and (3.82) becomes

$$\left(\partial_{x_1}^2 + 2\varepsilon\partial_{x_1}\partial_{x_2} + \varepsilon^2\partial_{x_2}^2\right)y + 2\left(\partial_{x_1} + \varepsilon\partial_{x_2}\right)y + 2\varepsilon y = 0. \tag{3.84}$$

It is worth noting the intervals that are used for the space scales we have introduced. For the boundary-layer scale, we have that $0 \le x_1 < \infty$, and for the other coordinate we require $0 \le x_2 \le 1$. Now, assuming

$$y \sim y_0(x_1, x_2) + \varepsilon y_1(x_1, x_2) + \cdots , \tag{3.85}$$

one finds, from (3.84), that the general solution for the first term is

$$y_0 = a_0(x_2) + b_0(x_2)e^{-2x_1}. \tag{3.86}$$

To determine the coefficients, we need the $O(\varepsilon)$ equation from (3.84), and we find that

$$\left(\partial_{x_1}^2 + 2\partial_{x_1}\right)y_1 = -2\partial_{x_1}\partial_{x_2}y_0 - 2\partial_{x_2}y_0 - 2y_0. \tag{3.87}$$

The general solution of this is

$$y_1 = a_1(x_2) + b_1(x_2)e^{-2x_1} - (a_0' + a_0)x_1 + (b_0 - b_0')x_1 e^{-2x_1}. \tag{3.88}$$

Since $0 \le x_1 < \infty$, to remove the secular terms, we require $a_0' + a_0 = 0$, and to minimize the contribution of the second term, we take $b_0 - b_0' = 0$ (Exercise 3.43). Using the fact that $y_0(0,0) = 0$ and $y_0(\infty, 1) = 1$, it follows that the first-term approximation of the solution is

$$y \sim e^{1-x} - e^{1+x-2x/\varepsilon}. \tag{3.89}$$

In comparison, using matched asymptotics, we obtain, using the composite expansion (2.16),

$$y \sim e^{1-x} - e^{1-2x/\varepsilon}.$$

These expressions are unequal, and the difference is in the second exponential in (3.89). However, this term is exponentially small except in the boundary layer. Since the layer has thickness $O(\varepsilon)$, it follows that any contribution the x makes is of second order. In other words, the two expansions are asymptotically equivalent (to first order). As a final comment, note that when using multiple scales, the balancing needed to determine the correct space scale and the location of the boundary layer are hidden but are still a very important component of the solution procedure. Readers wishing to pursue the connections between matched asymptotic expansions and multiple scales should consult the papers by Wollkind (1977) and Bouthier (1984).

Exercises

3.42. Consider the problem

$$\varepsilon y'' + p(x)y' + q(x)y = f(x) \quad \text{for } 0 < x < 1,$$

where $y(0) = \alpha$ and $y(1) = \beta$. Here $p(x)$, $q(x)$, and $f(x)$ are smooth, and $p(x) > 0$ for $0 \leq x \leq 1$.
(a) Using matched asymptotic expansions, construct a composite expansion of the solution.
(b) Let

$$x_1 = \frac{1}{\varepsilon} \int_0^x p(s)ds$$

and $x_2 = x$. Find a first-term multiple-scale approximation.
(c) Discuss the differences between your results in parts (a) and (b). Comment on the specific case where $p(x)$ and $q(x)$ are constants.

3.43. Set $R(x, \varepsilon) = y(x) - y_0(x/\varepsilon, x)$, where y_0 is defined in (3.86) and $a_0(x)$ and $b_0(x)$ are arbitrary.
(a) Find the exact solution to (3.82), (3.83), and write down the resulting expression for R.
(b) Use Taylor's theorem, with remainder, to expand the terms in R for small ε. Use this to explain how a_0 and b_0 should be chosen to minimize the error for $0 \leq x \leq 1$.

3.8 Introduction to Partial Differential Equations

To begin the study of the application of multiple scales to partial differential equations, we consider the equation for an elastic string with weak damping. Specifically, we will study the problem

$$\partial_x^2 u = \partial_t^2 u + \varepsilon \partial_t u \quad \text{for } 0 < x < 1 \text{ and } 0 < t, \tag{3.90}$$

where

$$u = 0 \quad \text{for } x = 0, 1 \tag{3.91}$$

and

$$u(x, 0) = g(x) \quad \text{and} \quad \partial_t u(x, 0) = 0. \tag{3.92}$$

Even though a partial differential equation is involved, (3.90) is similar to the oscillator problem considered in Sect. 3.2. This is because the input is through the initial conditions (3.92), and there is weak damping. Consequently, it should not be surprising to find out that a regular expansion results in secular terms. (Another way to reach this conclusion is given in Exercise 3.45.)

To use multiple scales, we need to determine the appropriate time scales. Given the similarity with the weakly damped oscillator, it seems reasonable to assume that the same time scales can be used in this problem, and so we take $t_1 = t$ and $t_2 = \varepsilon t$. In this case, (3.90)–(3.92) becomes

$$\partial_x^2 u = \left(\partial_{t_1}^2 + 2\varepsilon \partial_{t_1} \partial_{t_2} + \varepsilon^2 \partial_{t_2}^2\right) u + \varepsilon \left(\partial_{t_1} + \varepsilon \partial_{t_2}\right) u, \tag{3.93}$$

where

$$u = 0 \quad \text{for } x = 0, 1 \tag{3.94}$$

and

$$u = g(x) \quad \text{and} \quad (\partial_{t_1} + \varepsilon \partial_{t_2}) u = 0 \quad \text{at } t_1 = t_2 = 0. \tag{3.95}$$

As before, the solution is not unique, and we will use this freedom to minimize the contribution of the second term in the expansion. This will prevent the appearance of secular terms.

The appropriate expansion of the solution is

$$u \sim u_0(x, t_1, t_2) + \varepsilon u_1(x, t_1, t_2) + \cdots. \tag{3.96}$$

Introducing this into (3.93)–(3.95) yields the $O(1)$ problem

$$\partial_x^2 u_0 = \partial_{t_1}^2 u_0, \tag{3.97}$$

where

$$u_0 = 0 \quad \text{for } x = 0, 1 \tag{3.98}$$

and

$$u_0(x, 0, 0) = g(x) \quad \text{and} \quad \partial_{t_1} u_0(x, 0, 0) = 0. \tag{3.99}$$

Using separation of variables, one finds that the general solution of (3.97), (3.98) is

$$u_0(x, t_1, t_2) = \sum_{n=1}^{\infty} [a_n(t_2) \sin(\lambda_n t_1) + b_n(t_2) \cos(\lambda_n t_1)] \sin \lambda_n x, \tag{3.100}$$

where $\lambda_n = n\pi$. The initial conditions in (3.99) will be imposed once we determine the coefficients $a_n(t_2)$ and $b_n(t_2)$.

The $O(\varepsilon)$ equation that comes from introducing (3.96) into (3.93) is

$$\partial_x^2 u_1 = \partial_{t_1}^2 u_1 + 2\partial_{t_1}\partial_{t_2} u_0 + \partial_{t_1} u_0$$

$$= \partial_{t_1}^2 u_1 + \sum_{n=1}^{\infty} A_n(t_1, t_2) \sin \lambda_n x, \qquad (3.101)$$

where
$$A_n = (2a_n' + a_n)\lambda_n \cos(\lambda_n t_1) - (2b_n' + b_n)\lambda_n \sin(\lambda_n t_1).$$

Given the boundary conditions (3.94), it is appropriate to expand the solution of (3.101) in a modal expansion like the one that appears in (3.100). Thus, the solution of (3.101) has the form

$$u_1 = \sum_{n=1}^{\infty} v_n(t_1, t_2) \sin \lambda_n x.$$

Substituting this into (3.101) and using (3.100), one finds that

$$\partial_{t_1}^2 v_n + \lambda_n^2 v_n = -(2a_n' + a_n)\lambda_n \cos(\lambda_n t_1) + (2b_n' + b_n)\lambda_n \sin(\lambda_n t_1).$$

Since the trigonometric functions on the right-hand side are solutions of the associated homogeneous equation, to prevent secular terms in the expansion we must take

$$2a_n' + a_n = 0 \text{ and } 2b_n' + b_n = 0.$$

Solving these equations and using the initial conditions (3.99), one finds that $a_n = 0$ and $b_n = \beta_n e^{-t_2/2}$, where

$$\beta_n = 2 \int_0^1 g(x) \sin(\lambda_n x) dx.$$

Therefore, our first-term approximation of the solution is

$$u(x, t) \sim \sum_{n=1}^{\infty} \beta_n e^{-\varepsilon t/2} \cos(\lambda_n t) \sin \lambda_n x. \qquad (3.102)$$

It appears from this example that the application of the method of multiple scales to partial differential equations is a relatively straightforward generalization of what was used for ordinary differential equations. It should be understood, however, that this was a very simple example and partial differential equations can easily produce different, and very interesting, uses of multiple scales.

Exercises

3.44. Consider the following problem:

$$\partial_x^2 u = \partial_t^2 u + \varepsilon u \quad \text{for } 0 < x < 1 \text{ and } 0 < t,$$

where $u(0,t) = u(1,t) = 0$, $u(x,0) = g(x)$, and $\partial_t u(x,0) = 0$. Using multiple scales, find a first-term approximation of the solution that is valid for large t.

3.45. This problem concerns the energy associated with the weakly damped string equation in (3.90).

(a) Multiplying (3.90) by the velocity u_t, show that the equation can be rewritten in the form

$$\partial_t H + \partial_x S = -\Phi.$$

In this case, H is the Hamiltonian, S the energy flux density, and Φ the dissipation function (the latter being the only one that depends explicitly on ε in this problem).

(b) Suppose a regular expansion of the solution is used. Explain why the first term is a conservative approximation to a dissipative problem.

(c) Expand the energy equation in part (a) using multiple scales. Can you interpret the approximation in (3.102) in terms of this result?

3.46. The problem for a weakly damped elastic string is

$$\partial_x^2 u = \partial_t^2 u + \varepsilon u_t + \varepsilon f(x) \sin(\omega t) \quad \text{for } 0 < x < 1 \text{ and } 0 < t,$$

where $u(0,t) = u(1,t) = 0$ and $u(x,0) = \partial_t u(x,0) = 0$. Assume $f(x)$ is smooth.

(a) When $\omega \neq n\pi$, where n is an integer, find a first-term approximation of the solution that is valid for large t.

(b) When $\omega = \pi$, find a first-term approximation of the solution that is valid for large t.

(c) Suppose one looks for time-periodic solutions of the form $u(x,t) = v(x)e^{i\omega t}$. Find $v(x)$ and then describe how this periodic solution compares to the large t behavior of the solutions found in parts (a) and (b). Assume that $f(x) = \sin(\pi x)$.

3.47. Consider the problem

$$\partial_x^2 u = \partial_t u + c(\varepsilon x)u \quad \text{for } 0 < x < 1 \text{ and } 0 < t,$$

where $u(0,t) = u(1,t) = 0$ and $u(x,0) = f(x)$. Assume $c(s)$ is smooth and positive, and $c'(0) \neq 0$. Find a first-term approximation of the solution that is valid for large t.

3.48. The equation for a weakly damped beam is

$$\partial_x^4 u + \varepsilon \partial_x^2 u + \partial_t^2 u + \varepsilon \partial_t u = 0 \quad \text{for } 0 < x < 1 \text{ and } 0 < t,$$

where $u = u_{xx} = 0$ at $x = 0, 1$. The initial conditions are $u(x,0) = f(x)$, and $u_t(x,0) = 0$.

(a) Find a first-term approximation of the solution that is valid for large t.
(b) Find the exact solution, and compare it with the result from part (a).

3.49. The Kirchhoff equation for the transverse vibrations of a weakly non-linear string is (Leissa and Saad, 1994)

$$T(t)\partial_x^2 u = \partial_t^2 u \quad \text{for } 0 < x < 1 \text{ and } 0 < t,$$

where

$$T(t) = 1 + \varepsilon \int_0^1 u_x^2 \mathrm{d}x.$$

Also, $u(0,t) = u(1,t) = 0$, $u(x,0) = g(x)$, and $\partial_t u(x,0) = h(x)$.
(a) Assuming that $g(x) = \sin(n\pi x)$, where n is a positive integer, and $h(x) = 0$, find a first-term approximation of the solution that is valid for large t.
(b) For the general case, show that the first-term approximation of the solution that is valid for large t has the form $u \sim \sum a_n \sin \lambda_n x$, where $a_n = \alpha_n \cos(\omega_n t) + \beta_n \sin(\omega_n t)$ for

$$\omega_n = \lambda_n \left(1 + \frac{3}{16}\varepsilon\lambda_n^2(\alpha_n^2 + \beta_n^2)\right).$$

(c) The result in part (b) appears to be nonuniform for large values of n. Show that this is not the case when $h(x) = 0$ and either $g(x) = |x - \frac{1}{2}|$ or $g(x) = 1 + H(x - \frac{1}{2})$, where H is the Heaviside step function. What about the general case?

3.50. This problem examines the piston-driven motion of a gas inside a cylinder. The Lagrangian equations of motion in this case are (Wang and Kassoy, 1990)

$$\left.\begin{array}{l} \partial_t \rho + \rho^2 \partial_x u = 0 \\ \partial_t u + \rho^{\gamma-1}\partial_x \rho = 0 \end{array}\right\} \quad \text{for } 0 < x < 1 \text{ and } 0 < t,$$

where $u(0,t) = \varepsilon U(\varepsilon t)$, $u(1,t) = 0$, and at $t = 0$ we have $\rho = 1$ and $u = 0$. Here $\rho(x,t)$ is the density of the gas, $u(x,t)$ is the velocity of the gas, and U is the prescribed velocity of the piston. The parameter ε is the ratio of the piston velocity to the speed of sound, and $\gamma > 1$ is constant. It is assumed that $U(0) = 0$, $U'(0) > 0$, and

$$\int_0^\infty U(\tau)\mathrm{d}\tau < 1.$$

(a) Using regular expansions for $u(x,t)$ and $\rho(x,t)$, show that secular terms are present in the second terms.
(b) Using multiple scales, find a first-term approximation of the solution that is valid for large t.

3.9 Linear Wave Propagation

To investigate how multiple scales can be used for traveling waves, we will again start with the equation for an elastic string with weak damping. The problem is

$$\partial_x^2 u = \partial_t^2 u + \varepsilon \partial_t u \quad \text{for} \quad -\infty < x < \infty \text{ and } 0 < t, \tag{3.103}$$

where

$$u(x, 0) = F(x) \quad \text{and} \quad \partial_t u(x, 0) = 0. \tag{3.104}$$

It is assumed that $F(x)$ is smooth and bounded.

To see what happens when the multiple-scale method is not used, we try the regular expansion

$$u \sim u_0(x, t) + \varepsilon u_1(x, t) + \cdots . \tag{3.105}$$

Substituting this into (3.103) and (3.104) yields the following problems:

$O(1)$ $\partial_x^2 u_0 = \partial_t^2 u_0$
      $u_0(x, 0) = F(x), \quad \partial_t u_0(x, 0) = 0.$

It is convenient at this point to switch to characteristic coordinates and let $\theta_1 = x - t$ and $\theta_2 = x + t$. In this case, $\partial_x = \partial_{\theta_1} + \partial_{\theta_2}$, $\partial_t = -\partial_{\theta_1} + \partial_{\theta_2}$, and $t = 0$ corresponds to $\theta_1 = \theta_2$. Also, the equation for u_0 reduces to $\partial_{\theta_1} \partial_{\theta_2} u_0 = 0$. Integrating this yields the general solution

$$u_0 = f_0(\theta_1) + g_0(\theta_2),$$

where $f_0(\theta_1)$ and $g_0(\theta_2)$ are determined from the initial conditions. From these conditions one finds that the solution of the $O(1)$ problem is

$$u_0 = \frac{1}{2} [F(\theta_1) + F(\theta_2)].$$

$O(\varepsilon)$ $\left(\partial_x^2 - \partial_t^2 \right) u_1 = \partial_t u_0.$

If we use characteristic coordinates and the $O(1)$ solution, the equation becomes

$$4 \partial_{\theta_1} \partial_{\theta_2} u_1 = -\frac{1}{2} F'(\theta_1) + \frac{1}{2} F'(\theta_2).$$

Integrating this, we obtain

$$u_1 = -\frac{1}{8} \theta_2 F(\theta_1) + \frac{1}{8} \theta_1 F(\theta_2) + f_1(\theta_1) + g_1(\theta_2),$$

where $f_1(\theta_1)$ and $g_1(\theta_2)$ are arbitrary functions. Since $-\infty < x < \infty$, this solution contains secular terms; from what we have done so far, there is no way to prevent this from occurring.

Because of the appearance of secular terms in the regular expansion, we will use multiple scales to find an asymptotic approximation of the solution. Unlike the problems examined earlier, this one appears to have secular terms in both space and time. Consequently, we can introduce several scales. For example, we can use one of the following:

(i) $t_1 = t$, $\quad t_2 = \varepsilon t$, $\quad x_1 = x$, $\quad x_2 = \varepsilon x$;

(ii) $\theta_1 = x - t$, $\quad \theta_2 = x + t$, $\quad \phi_1 = \varepsilon\theta_1$, $\quad \phi_2 = \varepsilon\theta_2$;

(iii) $\theta_1 = x - t$, $\quad \theta_2 = x + t$, $\quad t_2 = \varepsilon t$, $\quad x_2 = \varepsilon x$.

These are equivalent, but note that each list contains four variables. Unfortunately, this is usually necessary and means that there is going to be some serious symbol pushing ahead.

We will use the scales given in (iii), which means that

$$\partial_x \to \partial_{\theta_1} + \partial_{\theta_2} + \varepsilon\partial_{x_2} \text{ and } \partial_t \to -\partial_{\theta_1} + \partial_{\theta_2} + \varepsilon\partial_{t_2}. \qquad (3.106)$$

Substituting these into (3.103), we obtain the following equation:

$$\left\{ 4\partial_{\theta_1}\partial_{\theta_2} + 2\varepsilon \left[(\partial_{\theta_1} + \partial_{\theta_2})\partial_{x_2} - (-\partial_{\theta_1} + \partial_{\theta_2})\partial_{t_2} \right] + O(\varepsilon^2) \right\} u$$
$$= \varepsilon \left[-\partial_{\theta_1} + \partial_{\theta_2} + O(\varepsilon) \right] u.$$

Also note that the initial conditions at $t = 0$ now correspond to $\theta_1 = \theta_2 = x$ and $t_2 = 0$. In fact, the initial conditions in (3.104) become

$$u = F(\theta_1) \text{ and } (-\partial_{\theta_1} + \partial_{\theta_2} + \varepsilon\partial_{t_2})\, u = 0 \quad \text{when } \theta_1 = \theta_2 \text{ and } t_2 = 0.$$

The next step is to introduce the expansion for the solution. For this problem, it is simply

$$u \sim u_0(\theta_1, \theta_2, x_2, t_2) + \varepsilon u_1(\theta_1, \theta_2, x_2, t_2) + \cdots.$$

This leads to the following problems:

$O(1)$ $4\partial_{\theta_1}\partial_{\theta_2} u_0 = 0$,

$\qquad u_0 = F(\theta_1)$, $(-\partial_{\theta_1} + \partial_{\theta_2})u_0 = 0$ at $\theta_1 = \theta_2 = 0$ and $t_2 = 0$.

The general solution of the differential equation is

$$u_0 = A(\theta_1, x_2, t_2) + B(\theta_2, x_2, t_2).$$

This expression is required to satisfy the initial conditions, and this enables us to make a labor-saving observation. Since the initial conditions do not depend on x_2, it is not unreasonable to expect that u_0 does not depend on x_2. In other words, we can simplify the preceding solution to

$$u_0 = f_0(\theta_1, t_2) + g_0(\theta_2, t_2). \tag{3.107}$$

If this hypothesis is incorrect, then the problem will tell us when we attempt to remove secular terms.

$O(\varepsilon)$ $4\partial_{\theta_1}\partial_{\theta_2} u_1 = 2\left(-\partial_{\theta_1} + \partial_{\theta_2}\right)\partial_{t_2} u_0 + \left(-\partial_{\theta_1} + \partial_{\theta_2}\right)u_0.$

Substituting the $O(1)$ solution into this equation and solving, one finds that

$$u_1 = -\frac{1}{4}\theta_2\left(2\partial_{t_2} + 1\right)f_0 + \frac{1}{4}\theta_1\left(2\partial_{t_2} + 1\right)g_0 + f_1(\theta_1, t_2) + g_1(\theta_2, t_2).$$

To prevent secular terms, we take

$$(2\partial_{t_2} + 1)f_0 = 0 \quad \text{and} \quad (2\partial_{t_2} + 1)g_0 = 0.$$

Integrating these equations and using the $O(1)$ initial conditions yields

$$f_0 = \frac{1}{2}F(\theta_1)e^{-t_2/2} \quad \text{and} \quad g_0 = \frac{1}{2}F(\theta_2)e^{-t_2/2}. \tag{3.108}$$

Therefore, a first-term approximation of the solution that is valid for large t is

$$u(x, t) \sim \frac{1}{2}[F(x - t) + F(x + t)]e^{-\varepsilon t/2}. \tag{3.109}$$

The accuracy of this approximation is quite good and has to do with the fact that it is actually the exact solution of the problem. This does not happen when more general initial conditions are used, as demonstrated in Exercise 3.51.

Exercises

3.51. Instead of (3.104), suppose the initial conditions are $u(x, 0) = F(x)$ and $\partial_t u(x, 0) = G(x)$.

(a) Find a first-term approximation of the solution of (3.103), for these initial conditions, that is valid for large t.
(b) The exact solution is (Tychonov and Samarskii, 1970)

$$u = \frac{1}{2}[F(x+t) + F(x-t)]e^{-\varepsilon t/2}$$
$$+ \frac{1}{4}e^{-\varepsilon t/2} \int_{x-t}^{x+t} [K(x,t,z)F(z) + L(x,t,z)G(z)]dz,$$

where

$$K(x,t,z) = \varepsilon I_0\left(\frac{\varepsilon}{2}s\right) + \frac{\varepsilon t}{s} I_1\left(\frac{\varepsilon}{2}s\right),$$
$$L(x,t,z) = 2I_0\left(\frac{\varepsilon}{2}s\right),$$

$s = \sqrt{t^2 - (x-z)^2}$, and I_0 and I_1 are modified Bessel functions. Show that this reduces to your result in part (a).

3.52. This exercise explores what happens when the domain is semi-infinite. The problem is to solve

$$\partial_x^2 u = \partial_t^2 u + \varepsilon \partial_t u \quad \text{for } 0 < x < \infty \text{ and } 0 < t,$$

where $u(x,0) = a(x)$, $\partial_t u(x,0) = b(x)$, and $u(0,t) = c(t)$. It is assumed that $a(0) = c(0)$ and $b(0) = c'(0)$.
(a) Before finding a multiple-scale approximation, it is worth first considering the problem when $\varepsilon = 0$. In this case, information travels with speed $s = 1$. This means that at any given nonzero value of x, information from the boundary (where $x = 0$) does not arrive until $t = x$. Consequently, the boundary condition only affects the solution over the interval $0 \leq x \leq t$, and the solution for $t < x$ is determined entirely from the initial conditions. Based on these observations, the solution for $\varepsilon = 0$ can be written as

$$u(x,t) = \begin{cases} F(x-t) + G(x+t) & \text{if } 0 \leq t < x, \\ f(x-t) + g(x+t) & \text{if } 0 \leq x < t. \end{cases}$$

Explain why the preceding function satisfies the wave equation (you can assume that $x \neq t$). Also, determine F and G from the initial conditions, and then determine f and g from the boundary condition and the requirement that u be continuous at $x = t$.
(b) In the case of small ε, find a first-term approximation of the solution that is valid for large t. Note that the result from part (a) will be useful here, but it is necessary to assume that the functions also depend on the slow variables.
(c) The exact solution, when $0 \leq t \leq x$, is (Tychonov and Samarskii, 1970)

$$u = \frac{1}{2}[a(x+t) + a(x-t)]e^{-\varepsilon t/2}$$

$$+ \frac{1}{4}e^{-\varepsilon t/2} \int_{x-t}^{x+t} [K(x,t,z)a(z) + L(x,t,z)b(z)]dz,$$

where J and K are defined in Exercise 3.51(b). Show that this reduces to your result in part (b) when $0 \le t \le x$.

(d) The exact solution, when $0 \le x \le t$, is

$$u = u_c + c(t-x)e^{-\varepsilon x/2} + \frac{\varepsilon}{2}e^{-\varepsilon t/2} \int_0^{t-x} M(x,t,z)c(z)dz,$$

where u_c is the expression for the solution in part (c), and

$$M(x,t,z) = \frac{x}{s} J_1\left(\frac{\varepsilon}{2}s\right)e^{\varepsilon z/2}.$$

Show that this reduces to your result in part (b) when $0 \le x \le t$.

3.53. The wave equation with a slowly varying phase velocity is

$$c^2(\varepsilon t)\partial_x^2 u = \partial_t^2 u \quad \text{for} \quad -\infty < x < \infty \text{ and } 0 < t,$$

where $u(x,0) = f(x)$ and $u_t(x,0) = 0$. Find a first-term approximation of the solution that is valid for large t. It is assumed that $c(t)$ is a smooth positive function. (Hint: See Sect. 3.6.)

3.54. Consider the following advection problem with weak diffusion:

$$\varepsilon\partial_x^2 u = \partial_t u + \partial_x u \quad \text{for} \quad -\infty < x < \infty \text{ and } 0 < t,$$

where $u(x,0) = f(x)$. Using multiple scales, find a first-term approximation of the solution that is valid for large t.

3.55. In the study of acoustic wave propagation in a tube, one comes across the horn equation for the pressure $p(x,t)$. This equation can be written as (Rabbitt and Holmes, 1988)

$$p_{xx} + \varepsilon a(\varepsilon x)p_x = p_{tt} \quad \text{for} \quad -\infty < x < \infty \text{ and } 0 < t.$$

The coefficient $a(\varepsilon x)$ is included to account for a slowly varying cross section of the tube. Specifically, if the cross-sectional area is $A(\varepsilon x)$, then $a(\varepsilon x) = A'(\varepsilon x)/A(\varepsilon x)$.

(a) Using multiple scales, find a first-term approximation of the solution that is valid for large x.

(b) Find the second term in the expansion.

3.56. In the study of the quantum motion of an ion in a Paul trap, one is interested in finding what is called a Wigner function $u(x,y,t)$. In the case of a weakly harmonic potential, the equation for this function is (Stenholm, 1992)

$$u_t + u_x - xf(t)u_y = 0.$$

Here $f(t) = a - \varepsilon \cos(t)$, where a is a positive constant. Using multiple scales, find a first-term approximation of the solution that is valid for large t.

3.10 Nonlinear Waves

Traveling-wave solutions are found for a wide variety of nonlinear partial differential equations. This differs significantly from the situation for linear problems where wave propagation is limited to particular types of equations. In this section, examples are analyzed that illustrate some of the situations where nonlinear waves are possible and how to use multiple scales to find them.

3.10.1 Nonlinear Wave Equation

Consider the nonlinear wave equation

$$\partial_x^2 u = \partial_t^2 u + u + \varepsilon u^3 \quad \text{for} \quad -\infty < x < \infty \text{ and } 0 < t, \qquad (3.110)$$

where

$$u(x,0) = F(x) \quad \text{and} \quad \partial_t u(x,0) = G(x). \qquad (3.111)$$

This is known as the nonlinear Klein–Gordon equation, and it describes the motion of a string on an elastic foundation as well as the waves in a cold electron plasma. There is little hope of solving this problem in closed form, so our approach will be to investigate particular solutions that are in the form of traveling waves.

To understand the approach we will take, consider the linear equation that is obtained when $\varepsilon = 0$. Using the Fourier transform, one finds that the general solution has the form

$$u(x,t) = \int_{-\infty}^{\infty} A(k)e^{i(kx-\omega t)}\,dk + \int_{-\infty}^{\infty} B(k)e^{i(kx+\omega t)}\,dk, \qquad (3.112)$$

where $A(k)$ and $B(k)$ are determined from the initial conditions and $\omega = \sqrt{1+k^2}$. This shows that the solution of the linear equation can be written as the superposition of the plane wave solutions $u_\pm(x,t) = e^{i(kx\pm\omega t)}$. For this reason, it is worth investigating what the nonlinear equation does to one of these plane waves.

The analysis to follow investigates what happens to the right-running plane wave $u(x,t) = \cos(kx - \omega t)$, where $\omega = \sqrt{1 + k^2}$. To find an expansion of the solution, one might try something like

$$u(x,t) \sim u_0(kx - \omega t) + \varepsilon u_1(x,t) + \cdots .$$

However, it is not hard to show that this leads to secular terms. Consequently, we will use multiple scales, and this brings up choices for the scales similar to those we had for the linear wave problem (Sect. 3.9). Because we are investigating what effects the nonlinearity has on the right-running wave, for the scales we will take $\theta = kx - \omega t$, $x_2 = \varepsilon x$, and $t_2 = \varepsilon t$. In this development, we assume that $\omega = \sqrt{1 + k^2}$ and $k > 0$. Using the change of variable formulas given in (3.106), differential Eq. (3.110) becomes

$$\left[\partial_\theta^2 - 2\varepsilon(k\partial_{x_2} + \omega\partial_{t_2})\partial_\theta + O(\varepsilon^2)\right]u + u + \varepsilon u^3 = 0. \qquad (3.113)$$

The appropriate expansion of the solution is now

$$u(x,t) \sim u_0(\theta, x_2, t_2) + \varepsilon u_1(\theta, x_2, t_2) + \cdots . \qquad (3.114)$$

This leads to the following problems:

$O(1)$ $(\partial_\theta^2 + 1)u_0 = 0.$

The general solution of this problem is

$$u_0 = A(x_2, t_2)\cos[\theta + \phi(x_2, t_2)]. \qquad (3.115)$$

As usual, the functions A and ϕ are used to prevent secular terms from appearing in the higher-order terms in the expansion. Note that the initial conditions have not been listed in this problem. These will be discussed after the general form of the expansion has been determined.

$O(\varepsilon)$ $(\partial_\theta^2 + 1)u_1 = 2(k\partial_{x_2} + \omega\partial_{t_2})\partial_\theta u_0 - u_0^3$

Using (3.115), the equation takes the form

$$(\partial_\theta^2 + 1)u_1 = -2\left[(k\partial_{x_2} + \omega\partial_{t_2})A\right]\sin(\theta + \phi) - \frac{1}{4}A^3\cos 3(\theta + \phi)$$

$$- 2\left[(k\partial_{x_2} + \omega\partial_{t_2})\phi + \frac{3}{8}A^2\right]A\cos(\theta + \phi).$$

This is an ordinary differential equation much like what we saw earlier. To prevent secular terms, it is required that

$$(k\partial_{x_2} + \omega\partial_{t_2})A = 0 \qquad (3.116)$$

and

$$(k\partial_{x_2} + \omega\partial_{t_2})\phi + \frac{3}{8}A^2 = 0. \tag{3.117}$$

These first-order equations can be solved using characteristic coordinates; thus, let $r = \omega x_2 + kt_2$ and $s = \omega x_2 - kt_2$. With this (3.116) and (3.117) become

$$\partial_r A = 0 \quad \text{and} \quad \partial_r \phi = -\frac{3}{16\omega k}A^2.$$

From this we get $A = A(s)$ and $\phi = -\frac{3}{16\omega k}A^2 r + \phi_0(s)$.

We have found that a first-term approximation of the solution of the nonlinear Klein–Gordon equation is

$$u \sim A\cos\left[\theta - \frac{3}{16\omega k}(\omega x_2 + kt_2)A^2 + \phi_0\right], \tag{3.118}$$

where $A = A(\omega x_2 - kt_2)$ and $\phi_0 = \phi_0(\omega x_2 - kt_2)$ are arbitrary.

We began this example by asking what the nonlinearity does to the plane wave solution. To answer this, suppose the initial conditions correspond to the plane wave $u = \alpha\cos(kx - \omega t)$, so

$$u(x,0) = \alpha\cos(kx) \quad \text{and} \quad u_t(x,0) = \alpha\omega\sin(kx). \tag{3.119}$$

In terms of the multiple scales used to derive (3.118), these translate into the conditions that $u_0(\theta, x_2, 0) = \alpha\cos(\theta)$ and $\partial_\theta u_0(\theta, x_2, 0) = -\alpha\sin(\theta)$. In (3.118) this means $A(\omega x_2) = \alpha$ and $\phi_0(\omega x_2) = \frac{3}{16k}A^2 x_2$. Thus, a first-term approximation of the solution in this case is

$$u(x,t) \sim \alpha\cos(kx - \lambda\omega t), \tag{3.120}$$

where

$$\lambda = 1 + \frac{3\varepsilon\alpha^2}{8\omega^2}. \tag{3.121}$$

This shows that the phase velocity

$$v_{\text{ph}} = \left(1 + \frac{3\varepsilon\alpha^2}{8\omega^2}\right)\frac{\omega}{k}$$

is increased by the nonlinearity and that this increase is amplitude dependent. Specifically, the larger the amplitude, the faster it moves. This dependence of the velocity on amplitude is typical of nonlinear waves and does not occur for linear problems.

As it turns out, this example was one of the first wave problems to be analyzed using multiple scales (Luke, 1966). There are other ways to approach the problem; one of the more popular ways is to look for a solution of the form $u = f(kx - \omega t)$. This is known as a permanent-wave assumption, and it transforms (3.110) into an ordinary differential equation (Exercise 3.59). Although this can make the problem easier to analyze, it does not readily extend to more complicated wave problems.

3.10.2 Wave–Wave Interactions

Most initial conditions do not correspond to a single plane wave but consist of a superposition of many such waves. To help develop an understanding of the effects of the nonlinearity in such situations, we will consider what happens when there are two waves. From (3.112) we know that

$$u = \alpha_1 \cos(k_1 x - \omega_1 t) + \alpha_2 \cos(k_2 x - \omega_2 t),$$

where $\omega_1 = \sqrt{1 + k_1^2}$ and $\omega_2 = \sqrt{1 + k_2^2}$ is a solution of the linear $(\varepsilon = 0)$ Klein–Gordon Eq. (3.110). We will assume that the initial condition for the nonlinear problem corresponds to this solution. In other words, we will take

$$u(x, 0) = \alpha_1 \cos(k_1 x) + \alpha_2 \cos(k_2 x) \tag{3.122}$$

and

$$\partial_t u(x, 0) = \alpha_1 \omega_1 \sin(k_1 x) + \alpha_2 \omega_2 \sin(k_2 x). \tag{3.123}$$

It is assumed the waves are distinct, so $k_1 \neq k_2$.

   Given the multiple waves, the expansion in (3.114) will not work. Instead, we need to introduce the scales

$$x_1 = x, \quad x_2 = \varepsilon x,$$
$$t_1 = t, \quad t_2 = \varepsilon t.$$

With this, we have that

$$\partial_x \to \partial_{x_1} + \varepsilon \partial_{x_2},$$
$$\partial_t \to \partial_{t_1} + \varepsilon \partial_{t_2},$$

and (3.110) takes the form

$$\left(\partial_{x_1}^2 + 2\varepsilon \partial_{x_1} \partial_{x_2} + \varepsilon^2 \partial_{x_2}^2\right) u = \left(\partial_{t_1}^2 + 2\varepsilon \partial_{t_1} \partial_{t_2} + \varepsilon^2 \partial_{t_2}^2\right) u + u + \varepsilon u^3.$$

The next step is to introduce the expansion for the solution. For this problem it is simply

$$u \sim u_0(x_1, t_1, x_2, t_2) + \varepsilon u_1(x_1, t_1, x_2, t_2) + \cdots.$$

This leads to the following problems:

$O(1)$ $\partial_{x_1}^2 u_0 = \partial_{t_1}^2 u_0 + u_0,$
$u_0(x_1, 0, x_2, 0) = \alpha_1 \cos(k_1 x_1) + \alpha_2 \cos(k_2 x_1),$
$\partial_{t_1} u_0(x_1, 0, x_2, 0) = \alpha_1 \omega_1 \sin(k_1 x_1) + \alpha_2 \omega_2 \sin(k_2 x_1).$

Given the initial conditions and the known solution for the linear problem, we will assume that

$$u_0 = A_0(x_2, t_2) \cos(\phi_0) + A_1(x_2, t_2) \cos(\phi_1), \qquad (3.124)$$

where

$$\phi_0 = k_1 x_1 - \omega_1 t_1 + \theta_0(x_2, t_2)$$

and

$$\phi_1 = k_2 x_1 - \omega_2 t_1 + \theta_1(x_2, t_2).$$

Also, A_0, A_1, θ_0, and θ_1 are arbitrary functions of x_2 and t_2, except for the stipulation that they satisfy the initial conditions. This means that they satisfy

$$A_0(x_2, 0) = \alpha_0, \quad A_1(x_2, 0) = \alpha_1, \quad \theta_0(x_2, 0) = \theta_1(x_2, 0) = 0. \quad (3.125)$$

$O(\varepsilon)$ $\partial_{x_1}^2 u_1 = \partial_{t_1}^2 u_1 + u_1 - 2\partial_{x_1}\partial_{x_2} u_0 + 2\partial_{t_1}\partial_{t_2} u_0 - u_0^3 .$

From (3.124), and a little algebra, one finds that

$$u_0^3 = \frac{3}{4} \left(A_0^2 + 2A_1^2 \right) A_0 \cos(\phi_0) + \frac{3}{4} \left(A_1^2 + 2A_0^2 \right) A_1 \cos(\phi_1)$$
$$+ \frac{1}{4} A_0^3 \cos(3\phi_0) + \frac{1}{4} A_1^3 \cos(3\phi_1)$$
$$+ \frac{3}{4} A_0^2 A_1 [\cos(2\phi_0 - \phi_1) + \cos(2\phi_0 + \phi_1)]$$
$$+ \frac{3}{4} A_0 A_1^2 [\cos(2\phi_1 - \phi_0) + \cos(2\phi_1 + \phi_0)] .$$

In the preceding expression, any cosine term that satisfies $\omega = \sqrt{1 + k^2}$ will generate a secular term in the expansion. The only ones satisfying this requirement are the first two terms. These can be removed from the equation, in the usual manner, using the terms coming from $-2\partial_{x_1}\partial_{x_2} u_0 + 2\partial_{t_1}\partial_{t_2} u_0$. This leads to the following equations:

$$(k_i \partial_{x_2} + \omega_i \partial_{t_2}) A_i = 0 \qquad (3.126)$$

and

$$(k_i \partial_{x_2} + \omega_i \partial_{t_2}) \theta_i = -\frac{3}{8} \left(A_i^2 + 2A_j^2 \right) . \qquad (3.127)$$

In the last equation, if $i = 1$, then $j = 0$, while if $i = 0$, then $j = 1$. The general solution of the linear wave Eq. (3.126) is $A_i = A_i(\omega_i x_2 - k_i t_2)$. Given the initial conditions (3.125), it follows that A_i is constant and is given as $A_i = \alpha_i$. With this, the general solution of the linear wave Eq. (3.127) is

$$(k_i\omega_j - k_j\omega_i)\,\theta_i = -\frac{3}{8}\left(A_i^2 + 2A_j^2\right)(\omega_j x_2 - k_j t_2) + c_i(\omega_i x_2 - k_i t_2).$$

The function c_i is determined from (3.125). From this one finds that

$$\theta_i = -\frac{3}{8\omega_i}\left(A_i^2 + 2A_j^2\right)t_2. \tag{3.128}$$

We have found that a first-term approximation of the solution of the weakly nonlinear Klein–Gordon equation, subject to the initial conditions (3.122) and (3.123), is

$$u \sim \alpha_0 \cos(k_0 x - \lambda_0 \omega_0 t) + \alpha_1 \cos(k_1 x - \lambda_1 \omega_1 t), \tag{3.129}$$

where

$$\lambda_0 = 1 + \frac{3\varepsilon}{8\omega_0^2}\left(\alpha_0^2 + 2\alpha_1^2\right) \tag{3.130}$$

and

$$\lambda_1 = 1 + \frac{3\varepsilon}{8\omega_1^2}\left(\alpha_1^2 + 2\alpha_0^2\right). \tag{3.131}$$

The expression in (3.129) is the expected generalization of the single-wave solution, given in (3.120). However, what is new is the change in the frequency multiplier λ. What we have found is that for the wave–wave problem, the amplitude of the second wave affects the multiplier, as given in (3.130) and (3.131). For example, the presence of a second wave causes the first wave to travel faster than it would if it were by itself.

3.10.3 Nonlinear Diffusion

In the study of wave propagation in chemical and biological systems, one comes across the problem of having to solve

$$u_t = \varepsilon u_{xx} + f(u) \quad \text{for} \quad -\infty < x < \infty \text{ and } 0 < t, \tag{3.132}$$

where

$$u(x,0) = g(x). \tag{3.133}$$

There are restrictions on the functions $f(u)$ and $g(x)$, and these will be given later. What is interesting is that the solution of this nonlinear diffusion problem can evolve into a traveling wave of the form $u(x,t) = U(x - vt)$, where v is the velocity of the wave. Such a solution is impossible with linear diffusion, and this is one of the reasons why nonlinear diffusion problems have received a lot of attention over the last few years. We will use multiple scales to find an asymptotic approximation of the solution of (3.132), and our approach will be based on the work of Berman (1978). An introduction to the theory and some of the applications of this equation can be found in Edelstein-Keshet (2005) and Murray (2003).

We first need to list the restrictions made on the nonlinearity in (3.132). In doing so, we specify only what happens for $0 \leq u \leq 1$. This interval for $u(x, t)$ is the one of physical interest since this function is a nondimensional concentration or density. In what follows, we will assume that $f(u)$ is smooth for $0 \leq u \leq 1$ and positive for $0 < u < 1$. Moreover, $f(0) = f(1) = 0$, $f'(0) > 0$, and $f'(1) < 0$. Perhaps the simplest example, and the most well studied, is the quadratic function $f(u) = u(1 - u)$. This gives rise to what is known as Fisher's equation or the Kolmogorov–Petrovsky–Piskunov (KPP) equation. It has been used to model such phenomena as the spatial spread of certain genetic characteristics, the propagation of cosmic rays in the atmosphere, and the evolution of a neutron population in a reactor. It is also worth pointing out that this quadratic function is the same as appears in the logistic equation, which is studied in connection with population dynamics.

The initial condition in (3.133) must also be consistent with the physics. The requirements on this function are that $0 < g(x) < 1$, $g(x) \to 1$ as $x \to -\infty$, and $g(x) \to 0$ as $x \to \infty$. In what follows, we will simply take

$$g(x) = \frac{1}{1 + e^{\lambda x}}, \tag{3.134}$$

where $\lambda > 0$ is a given constant.

The origin of all these conditions may seem mysterious, and it may be unclear what connections, if any, there are between the choices we are making for $f(u)$ and $g(x)$. To explain what we have done, note that our requirements on $f(u)$ result in two steady-state solutions of the differential equation, namely, $u = 0$ and $u = 1$. The function $f(u)$ has also been set up so that $u = 1$ is stable and $u = 0$ is unstable. (The tools needed to show this are developed in Chap. 6.) Therefore, by requiring that $g \to 1$ as $x \to -\infty$, and $g \to 0$ as $x \to \infty$, the initial condition effectively connects these two steady states. The solution of the problem in this case is fairly simple since at any given point on the x-axis, $u(x, t)$ simply moves away from $u = 0$ and approaches $u = 1$. What is important is that the speed at which it approaches $u = 1$ depends nonlinearly on the solution. The resulting movement toward $u = 1$ is what gives rise to the traveling wave. This is a very simplistic explanation of what occurs, but hopefully it will provide some insight into the behavior of the solution.

Our assumption that ε is small in (3.132) means that there is weak diffusion, and this introduces a second time scale into the problem. To account for this, we introduce the two time scales $t_1 = t$ and $t_2 = \varepsilon t$. The equation in (3.132) then takes the form

$$(\partial_{t_1} + \varepsilon\partial_{t_2})u = \varepsilon u_{xx} + f(u). \tag{3.135}$$

There is nothing unusual about this equation, and so the appropriate expansion for the solution is a regular power series of the form

$$u \sim u_0(x, t_1, t_2) + \varepsilon u_1(x, t_1, t_2). \tag{3.136}$$

The $O(1)$ problem that comes from substituting (3.136) into (3.135) and (3.133) is

$$\partial_{t_1} u_0 = f(u_0), \tag{3.137}$$

where

$$u_0(x, 0, 0) = g(x). \tag{3.138}$$

Separating variables in (3.137) one finds that

$$\int_{1/2}^{u_0} \frac{dr}{f(r)} = t_1 + \theta(x, t_2), \tag{3.139}$$

where $\theta(x, t_2)$ is arbitrary. Imposing the initial condition (3.138) yields

$$\theta(x, 0) = \int_{1/2}^{g(x)} \frac{dr}{f(r)}. \tag{3.140}$$

Without knowing exactly what the function $f(u)$ is, we will have to leave $u_0(x, t_1, t_2)$ defined implicitly in (3.139). However, note that we can write the solution in the form $u_0 = U_0(t_1 + \theta(x, t_2))$, where $U_0(s)$ is a function that satisfies

$$\int_{1/2}^{U_0} \frac{dr}{f(r)} = s.$$

This form will prove handy when removing secular terms.

To determine the second term in the expansion, the $O(\varepsilon)$ equation obtained from (3.135) is

$$\partial_{t_1} u_1 = f'(u_0)u_1 + \partial_x^2 u_0 - \partial_{t_2} u_0$$
$$= f'(u_0)u_1 + f(u_0)\left[\theta_{xx} - \theta_{t_2} + f'(u_0)\theta_x^2\right].$$

Equations (3.139) and (3.137) were used in the last step. Fortunately, this equation for u_1 is linear and is not hard to solve. To do this, it is worth pointing out that, from (3.137), $f'(u_0) = \partial_{t_1} \ln(f(u_0))$. Using this result, one finds that the solution of the preceding equation for u_1 is

$$u_1 = \left[A(x, t_2) + (t_1 + \theta)(\theta_{xx} - \theta_{t_2}) + \theta_x^2 \ln(f(u_0))\right] f(u_0), \tag{3.141}$$

where $A(x, t_2)$ is arbitrary. From a quick inspection of this solution, it would appear that secular terms are going to arise unless we restrict $\theta(x, t_2)$. However, determining the condition on θ requires some care because the logarithm term depends on t_1 and θ.

We need to consider what happens with the logarithm term in (3.141) if $f(u_0)$ approaches zero. Given the assumptions made on the function $f(u)$, this only happens if u_0 approaches either 1 or 0. From (3.139) this means that $t_1 + \theta \to \pm\infty$. In the case where $t_1 + \theta \to \infty$, because of the multiplicative factor of $f(u_0)$ in (3.141), a secular term does not occur (Exercise 3.71). Just the opposite conclusion is made when $t_1 + \theta \to -\infty$. In this limit, $u_0 \to 0$ and $\ln(f(u_0)) \sim \kappa(t_1 + \theta)$, where $\kappa = f'(0)$ is a positive constant. Thus, $\ln(f(u_0))$ produces a secular term in (3.141) that is of the same order as the other $t_1 + \theta$ term in that expression.

Based on the observations in the previous paragraph, we require

$$\theta_{xx} - \theta_{t_2} + \kappa\theta_x^2 = 0. \tag{3.142}$$

This equation can be solved if we make the change of variables $w(x, t_2) = e^{\kappa\theta(x,t_2)}$. In this case (3.142) reduces to the linear diffusion equation

$$w_{xx} = w_{t_2} \quad \text{for} \quad -\infty < x < \infty \text{ and } 0 < t_2. \tag{3.143}$$

The solution of this is

$$w(x, t_2) = \frac{1}{\sqrt{\pi}} \int_{-\infty}^{\infty} h\left(x + 2r\sqrt{t_2}\right) e^{-r^2} dr, \tag{3.144}$$

where $h(x) = w(x, 0)$. From (3.140) we get

$$h(x) = \exp\left(\kappa \int_{1/2}^{g(x)} \frac{dr}{f(r)}\right). \tag{3.145}$$

In stating that the solution of (3.143) is given by the formula in (3.144), we are assuming that the function $h(x)$ is well enough behaved that the improper integral in (3.144) is defined for all x and t_2. As shown in Exercise 3.71, with the conditions we have imposed on $f(u)$ and the initial condition in (3.134), the solution in (3.144) is well defined.

From the preceding analysis we have found that a first-term approximation of the solution of (3.132) is $u \sim u_0(x, t_1, t_2)$, where u_0 is defined implicitly through the equation

$$\int_{1/2}^{u_0} \frac{dr}{f(r)} = t_1 + \frac{1}{\kappa} \ln\left[w(x, t_2)\right], \tag{3.146}$$

where $w(x, t_2)$ is given in (3.144) and $\kappa = f'(0)$.

3.10.3.1 Example: Fisher's Equation

As stated earlier, nonlinear diffusion problems can give rise to traveling-wave solutions. The asymptotic approximation of the solution given in (3.146) can be used to demonstrate this. To show how, we will take $f(u) = u(1 - u)$. In this case, we have from (3.139) that $u_0 = U_0(t_1 + \theta(x, t_2))$, where

$$U_0(s) = \frac{1}{1 + e^{-s}}.$$

This can be rewritten as

$$u_0(x, t_1, t_2) = \frac{w(x, t_2)}{w(x, t_2) + e^{-t_1}},$$

where, from (3.144),

$$w(x, t_2) = e^{-\lambda x} \sqrt{\frac{t_2}{\pi}} \int_{-\infty}^{\infty} e^{-t_2 s(s - 2\lambda)} ds$$

$$= e^{-\lambda x + \lambda^2 t_2}.$$

Therefore, the asymptotic approximation of the solution of Fisher's equation is

$$u \sim \frac{1}{1 + e^{\lambda x - (1 + \lambda^2 \varepsilon)t}}, \tag{3.147}$$

where the positive constant λ is specified in the initial condition (3.134). This is a traveling wave with velocity

$$v \sim \frac{1 + \lambda^2 \varepsilon}{\lambda}. \tag{3.148}$$

The fact that the velocity of the wave depends on the shape of the initial profile is not unusual for a nonlinear problem. What is interesting is that as a function of λ there is a minimum wave speed of $2\sqrt{\varepsilon}$, which occurs when $\lambda = 1/\sqrt{\varepsilon}$. This agrees with the theoretical limit for the wave speed, which has been proven to satisfy the inequality $2\sqrt{\varepsilon} \le v < \infty$ (Kolmogorov et al., 1937). To illustrate the character of the traveling wave that comes from the nonlinear diffusion equation, both the asymptotic and numerical solutions of Fisher's equation are shown in Fig. 3.14. To help compare these two graphs, the time labels are at the same coordinates in both graphs. It is seen that as ε decreases, the wave slows down, which is expected given (3.148). Also, the asymptotic and numerical solutions are in such agreement for $\varepsilon = 0.01$ that the two curves are indistinguishable in the graph. In fact, they are in rather good agreement even for $\varepsilon = 0.1$. ∎

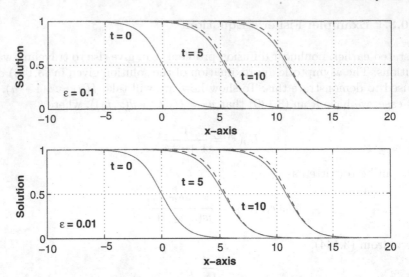

Figure 3.14 Traveling-wave solution of Fisher's equation as determined from the asymptotic solution (3.147) for two values of ε. The asymptotic solution is shown by a *dashed curve* at $t = 5$, and 10. For comparison, the numerical solution of the original problem (3.132)–(3.134) is shown by *solid curves*. In these calculations, $\lambda = 1$

Exercises

3.57. Consider the weakly nonlinear Klein–Gordon equation

$$\partial_x^2 u = \partial_t^2 u + u + \varepsilon u^2 \quad \text{for} \quad -\infty < x < \infty \text{ and } 0 < t,$$

where the initial conditions are given in (3.119) with $\omega = \sqrt{1 + k^2}$.
(a) Find a multiple-scale expansion for the right-running wave that removes the secular term coming from the nonlinearity.
(b) The time and space scales needed to remove the secular term in this problem are different than what were needed for the cubically nonlinear problem. Comment on what would be needed if the u^2 term is replaced with u^n, where n is a positive integer.

3.58. Consider the nonlinear wave equation

$$u_x + u_t = \varepsilon f(u) \quad \text{for} \quad -\infty < x < \infty \text{ and } 0 < t,$$

where $u(x, 0) = g(x)$.
(a) Find an asymptotic approximation of the solution that is valid for large t. Assume $f(u)$ and $g(x)$ are smooth and positive.
(b) What is the approximation when $f(u) = u(1 - u)$ and $g(x)$ is given in (3.134)? Sketch the approximate solution for a few values of t, and comment on how it compares with the solution of Fisher's equation shown in Fig. 3.11.

3.59. Suppose one looks for traveling-wave solutions of (3.110) of the form $u(x,t) = f(\theta)$, where $\theta = x - vt$.

(a) After finding the equation satisfied by $f(\theta)$, show that a regular expansion of $f(\theta)$ contains secular terms.

(b) Expanding both $f(\theta)$ and v, use multiple scales to find a first-term approximation valid for large θ. Explain why the expansion for v is not necessary to remove secular terms.

(c) Suppose the initial conditions are given in (3.119). Find the first-term approximation from part (b), and compare the result to that given in (3.120).

3.60. A model for collective motion is (Lee, 2011)

$$\partial_t u = -2c\partial_x v,$$
$$\partial_t v = -c\partial_x u + \varepsilon(\gamma uv - v),$$

where $u(x,0) = g(x)$ and $v(x,0) = 0$. Also, c and γ are positive constants, and the variables u and v are nonnegative. Assume that $\gamma g(x) \le 2$.

(a) Show that if $\epsilon = 0$, then the general solution of the problem has the form $u = F(\theta_1) + G(\theta_2)$, $v = [F(\theta_1) - G(\theta_2)]/\sqrt{2}$, where $\theta_1 = x - \sqrt{2}ct$ and $\theta_2 = x + \sqrt{2}ct$.

(b) For small ϵ, find an asymptotic approximation of the solution that is valid for large t.

(c) What happens to your approximation if the assumption $\gamma g(x) \le 2$ is not made?

3.61. The Korteweg–deVries (KdV) equation is

$$\partial_t u + \partial_x u + \alpha u \partial_x u + \beta \partial_x^3 u = 0 \quad \text{for} \quad -\infty < x < \infty \text{ and } 0 < t,$$

where α and β are positive constants and $u(x,0) = \varepsilon f(x)$ for $0 < \varepsilon \ll 1$.

(a) Setting $\theta = kx - \omega t$, suppose one looks for traveling-wave solutions using an expansion of the form

$$u(x,t) \sim \varepsilon[u_0(\theta) + \varepsilon u_1(\theta) + \cdots],$$

where $\omega = k - \beta k^3$ and k is a positive constant. Show that this can lead to secular terms.

(b) Multiple scales can be used to prevent the secular terms in the expansion in part (a). Do this, and find a first-term expansion that is valid for large t. What form must $f(x)$ have to be able to generate this traveling wave? The analysis in this problem is simplified greatly if you remember that $f(x)$ is independent of ε.

(c) Suppose $u(x,0) = \varepsilon[\mu + A(\varepsilon x)\cos[\varphi(x, \varepsilon x)]]$, where μ is a nonzero constant and A and φ are smooth bounded functions. Redo part (b), and find a first-term expansion of the traveling-wave solution that is valid for large t. What form must $\varphi(x, \varepsilon x)$ have to be able to generate this traveling wave?

3.62. In the study of waves in shallow water, one comes across the Boussinesq equation

$$u_{tt} = c^2 u_{xx} + \alpha u_{xxxx} + \varepsilon \partial_x^2 u^2 \quad \text{for} \quad -\infty < x < \infty \text{ and } 0 < t,$$

where c and α are positive constants. Proceeding as in the nonlinear Klein–Gordon example, find a first-term approximation, valid for large t, of the waves traveling to the right.

3.63. In the theory of nonlinear sound propagation in a fluid one comes across the Westervelt equation for acoustic pressure (Westervelt, 1963)

$$p_{xx} - \mu^2 p_{tt} = -\varepsilon \beta \partial_t^2 (p^2),$$

where μ and β are positive constants.
(a) For small ε, show that a regular expansion for the solutions of the form

$$p(x, t) \sim p_0(kx - \omega t) + \varepsilon p_1(x, t) + \cdots$$

leads to the appearance of secular terms.
(b) Using multiple scales, find the reduced problem that must be solved to obtain a two-term expansion of the solution.

3.64. This problem concerns the nonlinear diffusion equation

$$u_t = \varepsilon u_{xx} + c u_x + f(u) \quad \text{for} \quad -\infty < x < \infty \text{ and } 0 < t,$$

where $u(x, 0) = g(x)$. Assume $f(u)$ satisfies the same conditions imposed on the nonlinearity in (3.132).
(a) After making the change of variables $r = x + ct$, $\tau = t$, find an asymptotic approximation of the solution that is valid for large τ. Express the result in x, t-coordinates.
(b) What does the approximate solution reduce to when $f(u) = u(1 - u)$?

3.65. Suppose the initial condition for the nonlinear diffusion problem in (3.132) is

$$g(x) = \frac{1}{1 + e^{\lambda(x - x_0)(x - x_1)}},$$

where $\lambda > 0$, x_0, and x_1 are constants.
(a) For $f(u) = u(1 - u)$, find a first-term approximation of the solution.
(b) Taking $\lambda = 1$, $\varepsilon = 10^{-2}$, and $x_0 = -x_1 = 5$, sketch the approximate solution for a few values of t. Comment on the differences in the solution with the that in Fig. 3.11.

3.66. In the nonlinear diffusion Eq. (3.132), suppose $f(u) = u(1 - u^n)$, where n is a positive integer.

(a) Find a first-term approximation of the solution that is valid for large t [use the initial condition in (3.134)].
(b) What does the approximate solution in part (a) reduce to when $n = 2$? Is the result a traveling wave? Also, compare this result with the solution shown in Fig. 3.11.

3.67. In the study of wave propagation along the human spine, one comes across the following problem (Murray and Tayler, 1970):

$$\partial_t^2 u = [1 + \varepsilon f(u_x)] \partial_x^2 u \quad \text{for } 0 < x < \infty \text{ and } 0 < t,$$

where $u(0, t) = g(t)$ and $u(x, 0) = \partial_t u(x, 0) = 0$. Also, f and g are smooth functions with $f(0) = g(0) = g'(0) = 0$. If we assume that $g(\theta) = 0$ if $\theta \leq 0$, then the solution of this problem is $u = g(t - x)$ when $\varepsilon = 0$.
(a) Based on what happens when $\varepsilon = 0$, suppose we assume an expansion of the form

$$u(x, t) \sim u_0(t - x) + \varepsilon u_1(x, t) + \cdots .$$

Show that this leads to secular terms.
(b) Using multiple scales, find an asymptotic approximation of the solution that is valid for large x.
(c) Suppose there is a positive value of s where $g''(s) f'(-g'(s)) < 0$. Explain why your approximate solution in part (b) holds for $x < x_c$ but may not hold for $x > x_c$, where

$$x_c = \frac{-2}{\varepsilon \min_{0 \leq s} [g''(s) f'(-g'(s))]} . \tag{3.149}$$

(d) The equation for a weakly nonlinear elastic string is

$$\partial_t^2 u = \frac{\partial}{\partial x} \left(\frac{u_x}{\sqrt{1 + \varepsilon (u_x)^2}} \right) \quad \text{for } 0 < x < \infty \text{ and } 0 < t.$$

Assuming the initial and boundary conditions are the same as those used in the spine problem, explain why the results from parts (a)–(c) apply to this problem.

3.68. The equation for the longitudinal motion of a nonlinear elastic bar is

$$\partial_t^2 s = \partial_x^2 s + \varepsilon f(s, s_x) \quad \text{for } -\infty < x < \infty \text{ and } 0 < t,$$

where $f(s, s_x) = 2 \partial_x(s \partial_x s)$. The initial conditions are $s(x, 0) = F(x)$ and $s_t(x, 0) = 0$, where $F(x)$ is a smooth bounded function. In this problem, $s(x, t)$ is the strain in the bar and is related to the displacement $u(x, t)$ through the equation $s = \partial_x u$. This problem is based on the papers of Lardner (1975) and Mortell and Varley (1970).
(a) Using multiple scales, find an asymptotic approximation of the strain $s(x, t)$ that is valid for large t.

(b) Explain why your approximate solution holds for $t < t_c$ but may not hold for $t > t_c$, where

$$t_c = \frac{2}{\varepsilon \max\limits_{-\infty < x < \infty} |F'(x)|}.$$

(c) How do your results in parts (a) and (b) change if $f(s, s_x) = n\lambda\partial_x(s^{n-1}\partial_x s)$, where λ is a nonzero constant and $n \geq 3$ is an integer?

3.69. The nonlinear progressive wave equation that is used to describe the density fluctuations due to an acoustic disturbance, for uniaxial motion, is (McDonald and Kuperman, 1985)

$$\partial_t \rho + \partial_x \left[\alpha(x)\rho + \frac{1}{2}\beta\rho^2 \right] = 0 \quad \text{for} \quad -\infty < x < \infty,$$

where $\rho(x, 0) = 1 + \varepsilon f(x)$. Here $\alpha(x) > 0$ is smooth and $\beta \geq 0$ is constant.

(a) Suppose α is constant and we are interested in finding an asymptotic approximation to the traveling-wave solutions. This is to be found by assuming an expansion of the form $\rho \sim 1 + \varepsilon\rho_0(x - \lambda t) + \varepsilon^2\rho_1(x, t) + \cdots$, where λ is a constant to be determined from the preceding equation. Show that this leads to secular terms.

(b) For the situation described in part (a), use multiple scales to prevent secularity in the $O(\varepsilon^2)$ term of the expansion.

(c) Redo part (b), but assume α is not a constant.

3.70. Consider the problem

$$\partial_t u + g(u)\partial_x u = 0 \quad \text{for} \quad -\infty < x < \infty \text{ and } 0 < t,$$

where $u(x, 0) = u_0 + \varepsilon v(x)$ for constant u_0. The function $g(u)$ is assumed smooth and is such that $g(u_0) \neq 0$ and $g'(u_0) > 0$. The function $v(x)$ is also assumed smooth and is zero outside a bounded interval. This differential equation is a scalar conservation law. One interesting feature of the equation is that shock (i.e., discontinuous) solutions will appear in finite time. The effect of this on the multiple-scale expansion developed in this problem can be found in DiPerna and Majda (1985).

(a) Show that a regular expansion of the solution leads to the appearance of secular terms.

(b) Using multiple scales, find the reduced problem that must be solved to obtain a two-term expansion of the solution.

(c) The exact solution of this problem is defined implicitly as $u = u_0 + \varepsilon v(x - tg(u))$. Derive your expansion in part (b) from this equation.

3.71. In this exercise some of the steps in the derivation of the asymptotic approximation of the nonlinear diffusion problem are worked out.

(a) Suppose that $t_1 + \theta \to \infty$ in (3.139). Explain why this requires $u_0 \to 1$. Assuming that u_0 is close to one, use Taylor's theorem on $f(r)$ to show that the first term approximation of (3.139) gives

$$u_0 = 1 - e^{f'(1)(t_1+\theta)}.$$

Use this to explain why a secular term does not appear in (3.141) in the case where $t_1 + \theta \to \infty$.

(b) Suppose that $t_1 + \theta \to -\infty$ in (3.139). Explain why this requires $u_0 \to 0$. Assuming that u_0 is close to zero, use Taylor's theorem on $f(r)$ to show that the first-term approximation of (3.139) gives

$$u_0 = e^{f'(0)(t_1+\theta)}.$$

Use this to explain why a secular term can appear in (3.141) in the case where $t_1 + \theta \to -\infty$.

(c) Explain why the improper integral in (3.145) is well defined if $h(x) \leq M e^{\alpha x^2}$ for $-\infty < x < \infty$, where α and M are constants with $\alpha < 1$. Use the ideas developed in parts (a) and (b) to show that for $x \to \infty$, $h \sim 1/g$, while for $x \to -\infty$, $h \sim (1-g)^\gamma$, where $\gamma = f'(0)/f'(1)$. Use this to explain why the initial condition in (3.134) means that the solution in (3.144) is well defined.

3.11 Difference Equations

An interesting application of multiple scales arises in their use in finding asymptotic solutions of difference equations. This is a relatively new area but one that is important because of the prominence of difference equations in problems in scientific computing, modeling, and other areas. Some pertinent results for solving difference equations are given in Appendix D.

3.11.1 Weakly Nonlinear Difference Equation

As we usually do, we will introduce the ideas using a relatively simple example. The one we will consider is the following weakly nonlinear, second-order difference equation:

$$y_{n+1} - 2y_n + y_{n-1} = -2\omega^2 \left(y_n + \varepsilon y_n^3 \right) \qquad \text{for } n = 1, 2, 3, \ldots , \qquad (3.150)$$

where y_0 and y_1 are prescribed. Also, ω is a constant that is assumed to satisfy $0 < \omega < 1$. In the case of small ω, this equation is associated with a finite-difference approximation of the nonlinear differential equation $y'' + y = \varepsilon y^3$.

Consequently, it might be expected that the methods we have developed for differential equations can be adapted in a relatively straightforward way to difference equations. This is not exactly true, because difference equations have some unusual features, but this analogy at least allows us to get started.

From what we have seen for differential equations, it is expected that for this problem the solution depends on n and on εn. The latter is the analog of the slow scale and is incorporated into the problem by assuming the solution has an expansion of the form

$$y_n \sim \overline{y}_0(n, s) + \varepsilon \overline{y}_1(n, s) + \cdots , \tag{3.151}$$

where $s = \varepsilon n$. With this,

$$y_{n\pm1} \sim \overline{y}_0(n \pm 1, s \pm \varepsilon) + \varepsilon \overline{y}_1(n \pm 1, s \pm \varepsilon) + \cdots$$
$$\sim \overline{y}_0(n \pm 1, s) + \varepsilon[\overline{y}_1(n \pm 1, s) \pm \partial_s \overline{y}_0(n \pm 1, s)] + \cdots . \tag{3.152}$$

Substituting (3.152) and (3.151) into (3.150), we obtain the following equation:

$O(1)$ $\overline{y}_0(n + 1, s) - 2\overline{y}_0(n, s) + \overline{y}_0(n - 1, s) = -2\omega^2 \overline{y}_0(n, s).$

This is a second-order, homogeneous, linear difference equation, and the general solution is (Appendix D)

$$\overline{y}_0 = A(s) \cos[n\phi + \theta(s)], \tag{3.153}$$

where $A(s)$ and $\theta(s)$ are arbitrary functions of the slow scale s and

$$\cos(\phi) = 1 - \omega^2. \tag{3.154}$$

$O(\varepsilon)$ $\overline{y}_1(n + 1, s) - 2\overline{y}_1(n, s) + \overline{y}_1(n - 1, s)$
$$= -2\omega^2 \overline{y}_1(n, s) - \partial_s \overline{y}_0(n + 1, s)$$
$$- 2\omega^2 \overline{y}_0(n, s)^3 + \partial_s \overline{y}_0(n - 1, s). \tag{3.155}$$

To determine how to remove the secular terms, note that a particular solution of the difference equation (Appendix D)

$$z_{n+1} - 2z_n + z_{n-1} = -2\omega^2 z_n + f_n \quad \text{for } n = 1, 2, 3, \ldots \tag{3.156}$$

is

$$z_n = \frac{1}{\sin\phi} \sum_{i=1}^{n} f_{n-i} \sin(i\phi). \tag{3.157}$$

When $f_i = \cos(i\phi)$, the solution can be written as

$$z_n = \frac{1}{2\sin\phi}\left[-n\cos(n\phi) + \cot(\phi)\sin(n\phi)\right].$$ (3.158)

Similarly, if $f_i = \sin(i\phi)$, then a particular solution is

$$z_n = \frac{1}{2}(n+1)\frac{\sin(n\phi)}{\sin\phi}.$$ (3.159)

In both (3.158) and (3.159) we are getting terms that grow with n. If either of these is allowed into our expansion, then we will end up with a secular term. To make sure this does not happen, the functions A and θ will be chosen to prevent the $\cos(n\phi + \theta)$ and $\sin(n\phi + \theta)$ terms from appearing in the right-hand side of (3.155). One finds, after using a few trigonometric identities, that this requirement yields

$$A' = 0$$

and

$$4A\theta'\sin\phi = 3\omega^2 A^3.$$

From this we get that A is constant and

$$\theta = \frac{3\omega A^2 s}{4\sqrt{2-\omega^2}} + \theta_0.$$

Therefore, a first-term approximation of the solution of (3.150) for small ε, that holds for large n, is

$$y_n \sim A\cos[(\phi + \varepsilon\alpha)n + \theta_0],$$ (3.160)

where

$$\alpha = \frac{3\omega A^2}{4\sqrt{2-\omega^2}}$$ (3.161)

and A and θ_0 are constants (independent of n and ε) and ϕ is defined in (3.154).

To compare our asymptotic approximation (3.160) with the exact solution of (3.150), suppose $y_0 = 0$, $y_1 = 1$, $\omega = 1/2$, and $\varepsilon = 0.01$. The resulting curves are shown in Fig. 3.15; clearly they are very close (at least over the interval shown).

The approach used here is based on some of the ideas developed in Mickens (1990), Hoppensteadt and Miranker (1977), and Torng (1960). More recent contributions to this area can be found in van Horssen and ter Brake (2009) and Rafei and Van Horssen (2010), but the application of multiple scales to difference equations still has many open problems.

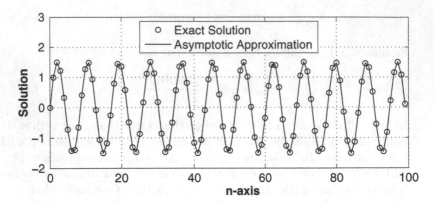

Figure 3.15 Graph of exact solution of difference equation in (3.150) and asymptotic approximation of solution given in (3.160)

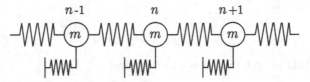

Figure 3.16 Chain of masses and springs

3.11.2 Chain of Oscillators

An interesting example involving difference equations arises when studying the motion of a chain of identical particles. Assuming nearest-neighbor interactions, the equations of motion are

$$my_n'' = V'(y_{n+1} - y_n) - V'(y_n - y_{n-1}) - W'(y_n) \text{ for } n = 0, \pm 1, \pm 2, \ldots,$$
$$(3.162)$$

where $y_n(t)$ is the displacement of the nth particle, m is the particle mass, $V(y)$ is the interaction potential, and $W(y)$ is called the on-site potential. This situation is illustrated in Fig. 3.16, where the springs between the masses all have potential V and the potential W is for each of the springs on the lower level.

The system in (3.162) arises in numerous applications and has various names. For example, in the case where $W = 0$ it reduces to the Fermi–Pasta–Ulam (FPU) chain, while if $V(y) = \alpha y^2$, one obtains the Klein–Gordon chain. The best-known recent application of this model involves a chain of atoms that are subject to nearest-neighbor interactions (Iooss and James, 2005). However, the problem goes back centuries, to at least Bernoulli (1728), and has been used in a wide spectrum of applications, including the interactions of stars (Voglis, 2003) and the modeling of swarming motion (Erdmann et al., 2005)

Mathematically, (3.162) is an example of a differential-difference equation. To relate it to some of the other problems we have studied in this chapter, consider the special case where $V(y) = \frac{1}{2}\varepsilon k_c y^2$ and $W(y) = \frac{1}{2}\varepsilon k y^2$. For these particular functions, (3.162) reduces to

$$my_n'' + ky_n = \varepsilon k_c(y_{n+1} - 2y_n + y_{n-1}),$$

which is the same problem given in (3.57)–(3.62) when $k_{ij} = k_c$, $c_{ij} = 0$, $m_i = m$, and $k_i = k$. Our interest now is not modal vibrations, as it was in Sect. 3.5, but the wavelike behavior seen with long chains of such oscillators. This leads us to a second example, when one takes $V(y) = \alpha y^2$ and $W(y) = \beta_0 y^2 + \beta_1 y^4$. In this case (3.162) is a finite-difference version of the nonlinear Klein–Gordon equation studied in Sect. 3.10.1. One of the goals here is the adaption of the multiple-scale method, as expressed in Sects. 3.9 and 3.10.1, to waves from the discrete system given in (3.162).

3.11.2.1 Example: Exact Solution

To understand the wavelike solutions obtained from differential-difference equations, consider

$$y_n'' = \mu^2(y_{n+1} - 2\lambda y_n + y_{n-1}) \text{ for } n = 0, \pm 1, \pm 2, \ldots, \tag{3.163}$$

where $y_n(0) = a_n$ and $y_n'(0) = b_n$. Using the Laplace transform one can show that the solution is (Pinney, 1958)

$$y_n(t) = a_n K(\sqrt{2}\mu t, 0) + \sum_{\ell=1}^{\infty}(a_{n+\ell} \mid a_{n-\ell})K(\sqrt{2}\mu t, \ell)$$

$$+ b_n L(\sqrt{2}\mu t, 0) + \sum_{\ell=1}^{\infty}(b_{n+\ell} + b_{n-\ell})L(\sqrt{2}\mu t, \ell), \tag{3.164}$$

where

$$K(t, \ell) = \frac{1}{\pi}\int_0^{\pi} \cos\left(t\sqrt{\lambda - \cos\theta}\right)\cos(\ell\theta)d\theta$$

and

$$L(t, \ell) = \frac{1}{\sqrt{2}\mu}\int_0^t K(\tau, \ell)d\tau.$$

In the special case where $\lambda = 1$, one finds that

$$K(t, \ell) = J_{2\ell}(\sqrt{2}t), \tag{3.165}$$

where $J_n(x)$ is the Bessel function of the first kind and

$$L(t, \ell) = \frac{1}{2\mu} \int_0^{\sqrt{2}t} J_{2n}(s)\mathrm{d}s$$

$$= L(t, 0) - \frac{1}{\mu} \sum_{m=0}^{\ell-1} J_{2m+1}(\sqrt{2}t). \qquad (3.166)$$

To get an idea of what the solution looks like, suppose the initial condition consists of plucking one of the oscillators, and so assume that the a_n and b_n are all zero except $a_0 = 1$. The preceding solution in this case reduces to

$$y_n(t) = J_{2n}(2\mu t). \qquad (3.167)$$

This function is plotted in Fig. 3.17 in the case where $\mu = 1$. It is evident that the disturbance spreads out as time passes. The edge of the disturbance is located at approximately $n = \pm\mu t$. Between the two fronts, for larger values of t, one can use the large-argument approximation for the Bessel function to obtain (Olver et al., 2010)

$$y_n(t) \sim \frac{1}{\sqrt{\pi\mu t}} \cos\left(2\mu t - n\pi - \frac{\pi}{4}\right). \qquad (3.168)$$

This approximation holds for $2n^2 << \mu t$. In this region the solution is a wave with phase velocity $2\mu/\pi$ and an amplitude that decays in time but is independent of n. What is somewhat surprising is that the solution is nonzero for large values of n for any $t > 0$. The reason why this might be unexpected is explained in Exercise 3.74(c). ∎

3.11.2.2 Example: Plane Wave Solution

Instead of solving an initial-value problem as in the last example, we now look for solutions of (3.163) of the form

$$y_n = \mathrm{e}^{\mathrm{i}(kn-\omega t)}. \qquad (3.169)$$

Taking $\mu = \lambda = 1$ and substituting the preceding solution into (3.163), the equation reduces to the dispersion relation

$$\omega = \pm 2\sin(k/2). \qquad (3.170)$$

Consequently, traveling waves are possible and propagate with velocity

$$v_{ph} = \pm\frac{2\sin(k/2)}{k}. \qquad (3.171)$$

The exception to the preceding statement arises when k is an even multiple of π, in which case one obtains a standing wave. Also note that the plane

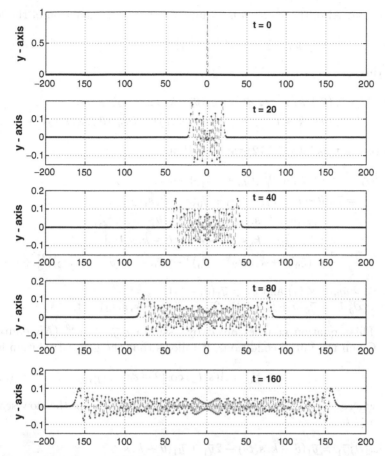

Figure 3.17 Solution of (3.163) in the case where $\mu = \lambda = 1$

waves match the velocity of the wave given in (3.168) only when $k = \pm\pi$. ∎

Similar to the approach taken in Sect. 3.10.1, we will investigate what the nonlinearity does to the preceding traveling wave. In particular, we will consider the equation

$$y_n'' = y_{n+1} - 2y_n + y_{n-1} + \varepsilon[(y_{n+1} - y_n)^3 - (y_n - y_{n-1})^3]. \qquad (3.172)$$

The analysis to follow is straightforward in the sense that it simply combines the ideas developed in Sects. 3.10.1 and 3.11.1 to construct an asymptotic approximation of the solution.

It is assumed that the initial condition is consistent with the linear ($\varepsilon = 0$) plane wave solution, $y_n = \alpha \cos(kn - \omega t)$, where $\omega = 2\sin(k/2)$ is assumed to be nonzero. The appropriate scales for this problem are $\theta = kn - \omega t$, $s = \varepsilon n$, and $t_2 = \varepsilon t$. Also, the solution is assumed to have an expansion of the form

$$y_n \sim \overline{y}_0(\theta, s, t_2) + \varepsilon\,\overline{y}_1(\theta, s, t_2) + \cdots . \tag{3.173}$$

With this,

$$y_{n\pm1} \sim \overline{y}_0(\theta \pm k, s \pm \varepsilon, t_2) + \varepsilon\,\overline{y}_1(\theta \pm k, s \pm \varepsilon, t_2) + \cdots$$
$$\sim \overline{y}_0(\theta \pm k, s, t_2) + \varepsilon\,[\overline{y}_1(\theta \pm k, s, t_2) \pm \partial_s\overline{y}_0(\theta \pm k, s, t_2)] + \cdots .$$

Substituting this into (3.172) we have that

$$(\omega^2\partial_\theta^2 - 2\varepsilon\omega\partial_\theta\partial_{t_2} + \partial_{t_2}^2)(\overline{y}_0 + \varepsilon\,\overline{y}_1 + \cdots)$$
$$= \overline{y}_0(\theta + k, s, t_2) - 2\overline{y}_0 + \overline{y}_0(\theta - k, s, t_2)$$
$$+ \varepsilon[\overline{y}_1(\theta + k, s, t_2) - 2\overline{y}_1 + \overline{y}_1(\theta - k, s, t_2)]$$
$$+ \varepsilon[\partial_s\overline{y}_0(\theta + k, s, t_2) - \partial_s\overline{y}_0(\theta - k, s, t_2)]$$
$$+ \varepsilon[(\overline{y}_0(\theta + k, s, t_2) - \overline{y}_0)^3 - (\overline{y}_0 - \overline{y}_0(\theta - k, s, t_2))^3] + \cdots ,$$

where $\overline{y}_0 = \overline{y}_0(\theta, s, t_2)$ and $\overline{y}_1 = \overline{y}_1(\theta, s, t_2)$.

$O(1)$ $\omega^2\partial_\theta^2\overline{y}_0 = \overline{y}_0(\theta + k, s, t_2) - 2\overline{y}_0 + \overline{y}_0(\theta - k, s, t_2)$.

This equation can be solved by assuming $\overline{y}_0 = e^{\lambda\theta}$ (Appendix E). Given the initial conditions, it follows that the general solution is

$$\overline{y}_0 = A(s, t_2)\cos[\theta + \phi(s, t_2)], \tag{3.174}$$

where $A(s, t_2)$ and $\phi(s, t_2)$ are arbitrary functions of the slow scales s and t_2.

$O(\varepsilon)$ $\omega^2\partial_\theta^2\overline{y}_1 = \overline{y}_1(\theta + k, s, t_2) - 2\overline{y}_1 + \overline{y}_1(\theta - k, s, t_2)$
$$+ 2\omega\partial_\theta\partial_{t_2}\overline{y}_0 + \partial_s\overline{y}_0(\theta + k, s, t_2) - \partial_s\overline{y}_0(\theta - k, s, t_2)$$
$$+ [\overline{y}_0(\theta + k, s, t_2) - \overline{y}_0]^3 - [\overline{y}_0 - \overline{y}_0(\theta - k, s, t_2)]^3.$$

To prevent secular terms we need to keep terms of the form $\cos[\theta + \phi(s, t_2)]$ and $\sin[\theta + \phi(s, t_2)]$ from appearing in the right-hand side of the preceding equation. Given (3.174), and using a few trigonometric identities, one finds that

$$(\partial_{t_2} + v\partial_s)A = 0$$

and

$$(\partial_{t_2} + v\partial_s)\phi = -3A^2\sin^3(k/2),$$

where $v = \sin(k)/\omega = \cos(k/2)$. The general solutions of these equations are $A = A(s - vt_2)$ and $\phi = -3A^2\sin^3(k/2)t_2 + \phi_0(s - vt_2)$.

Given that $\overline{y}_0 = \alpha \cos(kn)$ and $\partial_\theta \overline{y}_0 = -\alpha \sin(kn)$ at $t = 0$, it follows that $A = \alpha$ and $\phi_0 = 0$. With this, we have that a first-term approximation is

$$\overline{y}_0 = \alpha \cos(nk - \kappa \omega t), \qquad (3.175)$$

where $\kappa = 1 + \frac{3}{8} \varepsilon \alpha^2 \omega^2$.

Our analysis has produced a traveling-plane-wave solution, with a frequency modified by the nonlinearity in the equation. Other types of waves are also possible. As an example, the general solution obtained from our analysis is that

$$\overline{y}_0 = A(\xi) \cos(nk - \kappa \omega t + \phi_0(\xi)), \qquad (3.176)$$

where $\xi = \varepsilon(n - vt)$ and $\kappa = 1 + \frac{3}{8} \varepsilon A^2 \omega^2$. Consequently, the amplitude and phase can be traveling waves, but with velocity $v = \cos(k/2)$. The latter is the group velocity because it satisfies $v = \omega'(k)$. What is interesting is that A and ϕ can move in the opposite direction from the plane-wave solution. It is also worth pointing out that (3.176) is the basis for what is known as a bright breather, which is a time-periodic and spatially localized solution. A discussion of this can be found in Butt and Wattis (2007) and Iooss and James (2005).

Exercises

3.72. For small ε find a first-term approximation of the solution of each of the following equations that holds for large n.

(a) $y_{n+1} = y_n + \varepsilon(y_n + y_{n+1})$. Also, compare the approximation with the exact solution.

(b) $y_{n+1} - 2y_n + y_{n-1} + 2\omega^2 y_n + \alpha\varepsilon(y_{n+1} - y_{n-1}) = 0$.

(c) $y_{n+1} - 2y_n + y_{n-1} + 2\omega^2(y_n + \varepsilon y_n^3) = \varepsilon \cos[(1 + \varepsilon\omega_0)n]$.

3.73. This problem uses multiple scales to explore forced motion near resonance for a difference equation. The equation to solve is

$$y_{n+1} + ay_n + by_{n-1} = \cos(\Omega n) \qquad \text{for } n = 1, 2, 3, \ldots,$$

where $b \leq 1$, $-2 < a < 0$, and $a^2 - 4b < 0$. An equation of this form is obtained when using centered differences to find the numerical solution of $y'' + \varepsilon y' + y = c \cos(\bar{\Omega} t)$. This type of weakly damped oscillator was considered in Sect. 3.4. It also means that a and b depend on ε, and the exact dependence is specified below. Also, in what follows assume that $y_0 = y_1 = 0$.

(a) The undamped case corresponds to $b = 1$. Solve the resulting difference equation and explain why the solution is unbounded as a function of n if $\Omega = \theta$ (the definition of θ is given in Appendix D). In this context, θ is the resonant frequency for the undamped difference equation.

(b) When damping is included, the coefficients have the form $a = \alpha/(1+\varepsilon)$ and $b = (1-\varepsilon)/(1+\varepsilon)$, where $-2 < \alpha < 0$. Also, assume that $\Omega = \theta_0 + \varepsilon\omega$, where $\cos(\theta_0) = -\alpha/2$. Find a first-term approximation of the solution of the resulting difference equation.

3.74. This problem explores the differential-difference Eq. (3.163) and its solution given in (3.164).
(a) Derive (3.164).
(b) If $\lambda = -1$, then show that $K(t, \ell) = I_{2\ell}(\sqrt{2}t)$, where $I_n(x)$ is the modified Bessel function.
(c) Equation (3.163), when $\lambda = 1$, is obtained when using a centered-difference approximation on the wave equation $\partial_t^2 u = c^2 \partial_x^2 u$. In this case, $y_n(t)$ is the finite-difference approximation for $u(x_n, t)$ and $\alpha^2 = c^2/h^2$, where h is the spatial stepsize (so $h = x_{n+1} - x_n$). Assuming that the initial conditions are $u(x, 0) = f(x)$ and $\partial_t u(x, 0) = 0$, then the solution of the wave equation is

$$u(x, t) = \frac{1}{2}\left[f(x - ct) + f(x + ct)\right].$$

Comment on the differences and similarities of this solution with that shown in Fig. 3.17.

3.75. The Hamiltonian, or total energy, of the chain described by (3.162) is

$$H(t) = \sum_n \left[\frac{1}{2}m(y_n')^2 + V(y_n - y_{n-1}) + W(y_n)\right].$$

Show that H is constant and is therefore determined by the initial conditions.

3.76. Rayleigh (1909) considered a chain where the springs between the particles were replaced with elastic rods. In doing this he obtained the equation

$$my_n'' = -V'(y_{n+2}-2y_{n+1}+y_n)+2V'(y_{n+1}-2y_n+y_{n-1})-V'(y_n-2y_{n-1}+y_{n-2}),$$

where $V(y) = \frac{1}{2}\alpha y^2$.
(a) What is the resulting differential-difference equation?
(b) Verify that

$$y_n(t) = \frac{1}{\pi}\int_0^\pi \cos[s(n - v(s)t)]\,ds,$$

where $v(s) = 2a\sin^2(s/2)/s$ and $a = 2\sqrt{\alpha/m}$, is a particular solution of the equation. What initial conditions does this solution satisfy?
(c) Show that the solution in part (b) can be written as

$$y_n(t) = J_n(at)\cos\left(\frac{n\pi}{2} - at\right).$$

3.77. Consider the first-order difference equation

$$y_{n+1} = \alpha y_n + \varepsilon f(y_n, y_{n+1}) \quad \text{for} \ \ n = 0, 1, 2, 3, \ldots,$$

where α is a nonzero constant.

(a) Find a first-term approximation of the solution that holds for large n. In the derivation, you will need to assume that the function

$$g(x) = \lim_{n \to \infty} \frac{1}{n} \sum_{j=1}^{n} \frac{1}{\alpha^j} f(\alpha^{j-1} x, \alpha^j x)$$

is well defined.

(b) To find the numerical solution of the linear differential equation

$$y'(t) = \lambda y + \varepsilon y \quad \text{for} \ \ t > 0, \tag{3.177}$$

where λ is a nonzero constant, one can use Euler's method

$$y_{n+1} = (1 + h\lambda)y_n + \varepsilon h y_n \tag{3.178}$$

or the backward Euler method

$$(1 - h\lambda)y_{n+1} = y_n + \varepsilon h y_{n+1}. \tag{3.179}$$

Here y_n is the finite-difference approximation of $y(t_n)$, where $t_n = nh$ for a given value of the stepsize h. What does the result from part (a) reduce to for these two difference equations? Assume $1 \pm h\lambda \neq 0$ and $\lambda + \varepsilon \neq 0$.

(c) Find the exact solution of (3.177) and of the two difference Eqs. (3.178) and (3.179).

(d) Suppose $y(0) = 1$, $\lambda = -2$, $\varepsilon = 10^{-2}$, and $h = 10^{-1}$. On the same axes, plot the exact solution of (3.177), the exact solution of (3.178), and the asymptotic approximation of the solution of (3.178) for $0 \leq t \leq 2$. Also, plot them for $\lambda = -20$. Comment on the relative accuracy of the results.

(e) Redo part (d), but use (3.179) instead of (3.178).

3.78. This problem considers the accuracy of various finite-difference approximations that can be used to solve the nonlinear differential equation numerically:

$$y''(t) + y + \varepsilon y^3 = 0 \quad \text{for} \ \ t > 0. \tag{3.180}$$

(i) The first equation, which comes directly from using a centered-difference approximation of the derivative, is

$$y_{n+1} - 2y_n + y_{n-1} = -h^2(y_n + \varepsilon y_n^3). \tag{3.181}$$

This equation is basically the same as that given in (3.150). Note that y_n is the finite-difference approximation of $y(t_n)$, where $t_n = nh$ for some value of the stepsize h.

(ii) The second method, which uses the trapezoidal rule in its derivation, results in the difference equation

$$y_{n+1} - 2y_n + y_{n-1} = -\frac{h^2}{2}\left[y_{n+1} + y_{n-1} + \varepsilon\left(y_{n+1}^3 + y_{n-1}^3\right)\right]. \quad (3.182)$$

(iii) The third method, which uses Simpson's rule in its derivation, results in the difference equation

$$y_{n+1} - 2y_n + y_{n-1} = -\frac{h^2}{6}\left[y_{n+1} + 4y_n + y_{n-1} + \varepsilon\left(y_{n+1}^3 + 4y_n^3 + y_{n-1}^3\right)\right].$$

$$(3.183)$$

So we have three second-order difference equations from which to pick to solve (3.180).

(a) For small ε, find the first term in an expansion of the solution of (3.180) that is valid for large t.
(b) For small ε, find first-term expansions of the solutions of (3.181)–(3.183) that are valid for large n.
(c) Comparing your results in parts (a) and (b), comment on the relative accuracy of the finite-difference approximations. Of interest here is the solution when h is small. For example, does the phase cause the numerical solution to lead or follow the asymptotic solution? Also, what if one were to use the approximation given in (3.153) but the phase θ were found by adding together certain combinations of the phase determined from (i), (ii), and (iii)?

3.79. In the study of the structural changes in DNA, one finds a model for the motion of a harmonic lattice that involves localized oscillating nonlinear waves (Dauxois et al., 1992). These waves are associated with what are called "breathers." The problem is to find the function $u_n(t)$ that satisfies the equation

$$u_n'' + \omega_d^2\left(u_n - u_n^2\right) = u_{n+1} - 2u_n + u_{n-1} \quad \text{for} \quad n = 0, \pm 1, \pm 2, \pm 3, \ldots.$$

The initial conditions have the form $u_n(0) = \varepsilon f_n$ and $u_n'(0) = \varepsilon g_n$, where f_n and g_n are independent of ε.

(a) For small ε, suppose one looks for traveling-wave solutions using an expansion of the form

$$u_n \sim \varepsilon[A\cos(kn - \omega t) + \cdots], \quad (3.184)$$

where k and ω are positive constants. From the $O(\varepsilon)$ problem, find the relationship between ω and k so that the solution has the form assumed in (3.184). This equation for ω and k is the dispersion relation. Also, show that the form of expansion assumed in (3.184) leads to secular terms.

(b) Use multiple scales to find a first-term expansion of the traveling-wave solutions that is valid for large t. What form must f_n and g_n have to be able to generate this traveling-wave solution?

3.80. In a certain problem from population genetics, one is interested in the proportion g_n of a population in the nth generation with a particular genetic trait. The Fisher–Wright–Haldane model for this is (Hoppensteadt, 1982)

$$g_{n+1} = \frac{r_n g_n^2 + s_n g_n(1 - g_n)}{r_n g_n^2 + 2s_n g_n(1 - g_n) + t_n(1 - g_n)^2} \quad \text{for } n = 1, 2, 3, \ldots .$$

Assume that $r_n = 1 + \varepsilon \rho f(n)$, $s_n = 1 + \varepsilon \sigma f(n)$, and $t_n = 1 + \varepsilon \tau f(n)$, where ρ, σ, and τ are constants and $f(n)$ is a positive function with period N [i.e., $f(n + N) = f(n)$]. Also, because of how g_n is defined, it is required that $0 \leq g_n \leq 1$. This will happen if the start-off value satisfies $0 \leq g_1 \leq 1$.

(a) Using an expansion similar to that in (3.151), find a first-term approximation of g_n for small ε that holds for large n. Assume that at least one of the two constants $\alpha = \sigma - \tau$, $\beta = \rho - 2\sigma + \tau$ is nonzero.

(b) Using the result from part (a), find a first-term approximation for the possible steady states for g_n. Under what conditions on the constants α and β and the start-off value g_1 will the steady state g_∞ satisfy $0 < g_\infty < 1$?

Chapter 4
The WKB and Related Methods

4.1 Introduction

In the method of matched asymptotic expansions studied in Chap. 2, the dependence of the solution on the boundary-layer coordinate was determined by solving the boundary-layer problem. In a similar way, when using multiple scales the dependence on the fast time scale was found by solving a differential equation. This does not happen with the WKB method because one begins with the assumption that the dependence is exponential. This is a reasonable expectation since many of the problems we studied in Chap. 2 ended up having an exponential dependence on the boundary-layer coordinate. Also, with this assumption, the work necessary to find an asymptotic approximation of the solution can be reduced significantly.

The popularity of the WKB method can be traced back to the 1920s and the development of quantum mechanics. In particular, it was used to find approximate solutions of Schrödinger's equation. The name of the method is derived from three individuals who were part of this development, namely, Wentzel, Kramers, and Brillouin. However, as often happens in mathematics, the method was actually derived much earlier. Some refer to it as the method of Liouville and Green since they both published papers on the procedure in 1837. It appears that this too is historically incorrect since Carlini, in 1817, used a version of the approximation in studying elliptical orbits of planets. Given the multiple parenthood of the method it should not be unexpected that there are other names it goes by; these include the phase integral method, the WKBJ method ("J" standing for Jefferys), the geometrical acoustics approximation, and the geometrical optics approximation. The history of the method is surveyed very nicely by Heading (1962) and in somewhat more mathematical detail by Schlissel (1977b). A good, but dated, introduction to its application to quantum mechanics can be found in the book by Borowitz (1967) and to solid mechanics in the review article by Steele (1976). What

M.H. Holmes, *Introduction to Perturbation Methods*, Texts in Applied
Mathematics 20, DOI 10.1007/978-1-4614-5477-9_4,
© Springer Science+Business Media New York 2013

this all means is that the WKB approximation is probably a very good idea since so many have rediscovered it in a wide variety of disciplines.

4.2 Introductory Example

In the same manner as was done for boundary layers and multiple scales, the ideas underlying the WKB method will be developed by using it to solve an example problem. The one we begin with is the equation

$$\varepsilon^2 y'' - q(x)y = 0. \tag{4.1}$$

For the moment we will not restrict the function $q(x)$ other than to assume that it is smooth. Our intention is to construct an approximation of the general solution of this equation. To motivate the approach that will be used, suppose that the coefficient q is constant. In this case the general solution of (4.1) is

$$y(x) = a_0 e^{-x\sqrt{q}/\varepsilon} + b_0 e^{x\sqrt{q}/\varepsilon}. \tag{4.2}$$

The hypothesis made in the WKB method is that the exponential solution in (4.2) can be generalized to provide an approximate solution of (4.1). All that is necessary is to ensure the expansion is general enough so it can handle the variable coefficient in the equation. The specific assumption made when using the WKB method is that the asymptotic expansion of a solution is

$$y \sim e^{\theta(x)/\varepsilon^\alpha}[y_0(x) + \varepsilon^\alpha y_1(x) + \cdots]. \tag{4.3}$$

One of the distinctive features of the WKB method is that it is fairly specific on how the solution depends on the fast variation, namely, the dependence is assumed to be exponential. This limits the method but it can make it easier to use than either the multiple-scale or boundary-layer methods.

From (4.3) we get that

$$y' \sim \left(\varepsilon^{-\alpha}\theta' y_0 + y_0' + \theta' y_1 + \cdots\right)e^{\theta/\varepsilon^\alpha} \tag{4.4}$$

and

$$y'' \sim \left[\varepsilon^{-2\alpha}\theta_x^2 y_0 + \varepsilon^{-\alpha}(\theta_{xx}y_0 + 2\theta_x y_0' + \theta_x^2 y_1) + \cdots\right]e^{\theta/\varepsilon^\alpha}. \tag{4.5}$$

These are sufficient for the problem we are considering. However, we will occasionally be considering third- and fourth-order problems, and it is worth having the formulas in one place so they are easy to find when the time comes. So, for the record, we also have that

$$y''' \sim \left[\varepsilon^{-3\alpha}q_x^3 y_0 + \varepsilon^{-2\alpha}\theta_x(3\theta_x y_0' + 3\theta_{xx}y_0 + \theta_x^2 y_1) + \cdots\right]e^{\theta/\varepsilon^\alpha} \tag{4.6}$$

and

$$y^{(4)} \sim \left[\varepsilon^{-4\alpha}\theta_x^4 y_0 + \varepsilon^{-3\alpha}\theta_x^2(6\theta_{xx}y_0 + 4\theta_x y_0' + \theta_x^2 y_1) + \cdots\right]e^{\theta/\varepsilon^\alpha}. \quad (4.7)$$

The next step is to substitute (4.3) and (4.5) into (4.1). Doing this, one finds that

$$\varepsilon^2 \left\{ \frac{1}{\varepsilon^{2\alpha}}(\theta_x)^2 y_0 + \frac{1}{\varepsilon^\alpha}\left[\theta'' y_0 + 2\theta' y_0' + (\theta_x)^2 y_1\right] + \cdots \right\}$$
$$- q(x)(y_0 + \varepsilon^\alpha y_1 + \cdots) = 0. \quad (4.8)$$

Something important happened in this step, namely, the exponential dropped out. This would not have happened if the equation were nonlinear. This point will be raised again after the example has been completed. Now, balancing the terms in (4.8), one finds that $\alpha = 1$, and this leads to the following equations:

$O(1)$ $(\theta_x)^2 = q(x)$.

This is called the *eikonal equation* and its solutions are

$$\theta(x) = \pm \int^x \sqrt{q(s)}\,ds. \quad (4.9)$$

To obtain the first term in the expansion, we need to also find $y_0(x)$, and this is accomplished by looking at the $O(\varepsilon)$ problem.

$O(\varepsilon)$ $\theta'' y_0 + 2\theta' y_0' + (\theta_x)^2 y_1 = q(x)y_1$,

This is the *transport equation*. Since $\theta(x)$ satisfies the eikonal equation, it follows that $\theta'' y_0 + 2\theta' y_0' = 0$. The solution of this is

$$y_0(x) = \frac{c}{\sqrt{\theta_x}}, \quad (4.10)$$

where c is an arbitrary constant.

We have therefore found that a first-term approximation of the general solution of (4.1) is

$$y(x) \sim q(x)^{-1/4}\left[a_0 \exp\left(-\frac{1}{\varepsilon}\int^x \sqrt{q(s)}\,ds\right) + b_0 \exp\left(\frac{1}{\varepsilon}\int^x \sqrt{q(s)}\,ds\right)\right], \quad (4.11)$$

where a_0 and b_0 are arbitrary, possibly complex, constants. It is apparent from this that the function $q(x)$ must be nonzero. The values of x where $q(r)$ is zero are called *turning points*, and these will be discussed in the next section (also see Exercise 4.13).

The easiest way to check on the accuracy of the WKB approximation is to compare it with the exact solution. This is done in the following two examples.

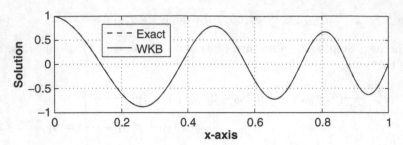

Figure 4.1 Comparison between exact solution (4.14) and WKB approximation given in (4.13). In the calculations, $\varepsilon = 0.1$, $a = 1$, and $b = 0$. The two curves are so close that they are indistinguishable from each other

Examples

1. Suppose $q(x) = -e^{2x}$. In this case (4.11) reduces to

$$y \sim e^{-x/2}\left(a_0 e^{-ie^x/\varepsilon} + b_0 e^{ie^x/\varepsilon}\right)$$
$$= e^{-x/2}[\alpha_0 \cos(\lambda e^x) + \beta_0 \sin(\lambda e^x)], \tag{4.12}$$

where $\lambda = \varepsilon^{-1}$. To determine the constants in (4.12), suppose the boundary conditions are $y(0) = a$ and $y(1) = b$. One finds that the WKB approximation is then

$$y \sim e^{-x/2}\left(\frac{be^{1/2}\sin\lambda(e^x - 1) - a\sin\lambda(e^x - e)}{\sin\lambda(e - 1)}\right). \tag{4.13}$$

For comparison, the exact solution is

$$y(x) = c_0 J_0(\lambda e^x) + c_1 Y_0(\lambda e^x), \tag{4.14}$$

where

$$c_0 = \frac{1}{d}[bY_0(\lambda) - aY_0(\lambda e)] \quad \text{and} \quad c_1 = \frac{1}{d}[aJ_0(\lambda e) - bJ_0(\lambda)]. \tag{4.15}$$

Here, J_0 and Y_0 are Bessel functions and $d = J_0(\lambda e)Y_0(\lambda) - Y_0(\lambda e)J_0(\lambda)$. The values obtained from the WKB approximation and the exact solution are shown in Fig. 4.1; it is clear that they are in very good agreement. ∎

2. Another use of the WKB method is to find approximations of large eigenvalues. To understand how this is done, suppose the problem is

$$y'' + \lambda^2 e^{2x}y = 0 \quad \text{for } 0 < x < 1, \tag{4.16}$$

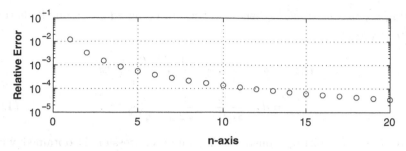

Figure 4.2 Relative error, in absolute value, between the WKB approximation (4.18) of the eigenvalues and the exact values obtained by solving (4.19) numerically. Here $\varepsilon = 10^{-1}$

where

$$y(0) = y(1) = 0. \qquad (4.17)$$

We are interested in what values of λ produce a nonzero solution. The solution in (4.13) will be nonzero and satisfy (4.17) so long as $\sin \lambda(e - 1) = 0$. Therefore, a first-term approximation of the large eigenvalues is

$$\lambda \sim \frac{n\pi}{e - 1} \quad \text{for integer } n. \qquad (4.18)$$

The exact value of each eigenvalue is determined from the denominator in (4.15), that is, they are solutions of the equation

$$J_0(\lambda e)Y_0(\lambda) - Y_0(\lambda e)J_0(\lambda) = 0. \qquad (4.19)$$

To see just how well (4.18) does in approximating the eigenvalues, the relative error is plotted in Fig. 4.2 for the first 20 eigenvalues. Clearly the approximation is very good and is much easier to use than (4.19). Moreover, the WKB approximation does well even for the first few eigenvalues. ∎

4.2.1 Second Term of Expansion

To obtain a measure of the error in using (4.11) to approximate the solution, we will calculate the second term y_1 in the expansion (4.3). This is not difficult, and carrying out the calculation it is found from (4.8) that the $O(\varepsilon^2)$ equation is

$$\theta'' y_1 + 2\theta' y_1' + (\theta_x)^2 y_2 + y_0'' = q(x)y_2.$$

Using the eikonal equation this can be reduced to

$$2\theta' y_1' + \theta'' y_1 = -y_0''.$$

Letting $y_1(x) = y_0(x)w(x)$, then from the preceding equation one finds that, for $\theta_x = \sqrt{q}$,

$$w(x) = d + \frac{1}{4}\frac{\theta_{xx}}{(\theta_x)^2} + \frac{1}{8}\int^x \frac{(\theta_{xx})^2}{(\theta_x)^3}\,dx$$

$$= d + \frac{1}{8}\frac{q_x}{q^{3/2}} + \frac{1}{32}\int^x \frac{q_x^2}{q^{5/2}}\,dx, \tag{4.20}$$

where d is an arbitrary constant. A similar expression is obtained when $\theta_x = -\sqrt{q}$.

Now, for small ε, the expansion in (4.3) will be well ordered so long as $\varepsilon y_1(x) \ll y_0(x)$. Given that $y_1(x) = y_0(x)w(x)$, then well-ordering is guaranteed if $|\varepsilon w(x)| \ll 1$. From the formula for $w(x)$ given in (4.20), we can express this requirement in terms of $q(x)$ and its first derivative. In particular, for $x_0 \le x \le x_1$ we will have an accurate approximation if

$$\varepsilon\left[|d| + \frac{1}{32}\left|\frac{q'}{q^{3/2}}\right|_\infty\left(4 + \int_{x_0}^{x_1}\left|\frac{q'}{q}\right|\,dx\right)\right] \ll 1 \tag{4.21}$$

holds, where $|h(x)|_\infty = \max_{x_0 \le x \le x_1}|h(x)|$. This holds if the interval does not contain a turning point, that is, a point where $q(x) = 0$. The preceding formula can also be used to determine just how close one can get to a turning point before the expansion becomes nonuniform, and this is explored in Exercise 4.11. For a more extensive investigation into the theory underlying the WKB method the book by Olver (1974) may be consulted.

The general solutions given in (4.11) and (4.20) contain constants that are determined from boundary conditions. This step is usually not thought provoking, but when using the WKB approximation we have to be a little more careful. The reason is that when attempting to satisfy the boundary conditions it is very possible that the constants depend on ε. It is therefore necessary to make sure this dependence does not interfere with the ordering assumed in the WKB expansion in (4.3). To illustrate this situation, consider the boundary-value problem

$$\varepsilon^2 y'' + e^{2\kappa x}y = 0 \quad \text{for } 0 < x < 1. \tag{4.22}$$

This is a generalization of the examples considered earlier (where $\kappa = 1$). Since $\theta_x = \pm i e^{\kappa x}$, we get, from (4.20), that

$$w(x) = d \mp \frac{i\kappa}{8}e^{-\kappa x}. \tag{4.23}$$

A two-term expansion of the general solution of (4.22) is therefore

$$y \sim e^{-\kappa x/2} \left\{ a_0 e^{-\chi} \left[1 + \varepsilon \left(a_1 - \frac{i\kappa}{8} e^{-\kappa x} \right) \right] + b_0 e^{\chi} \left[1 + \varepsilon \left(b_1 + \frac{i\kappa}{8} e^{-\kappa x} \right) \right] \right\},$$

$$(4.24)$$

where

$$\chi = \frac{i}{\kappa \varepsilon} e^{\kappa x}$$

and a_0 and b_0 are constants determined by the boundary conditions. As in the example shown in Fig. 4.1, we take $\kappa = 1$, $y(0) = 1$, and $y(1) = 0$. For these values

$$a_1 = -\frac{\cos \lambda (e - 1) - e^{-1-i(e-1)\lambda}}{8 \sin \lambda (e - 1)},$$

where $\lambda = \frac{1}{\varepsilon}$. One also finds that b_1 is the complex conjugate of a_1. This shows that for this example the second term in the combined WKB expansion in (4.24) is small except at, or near, the resonant points where $\lambda(e-1) = n\pi$. In particular, if $\varepsilon = 0.1$, then the maximum value of $|a_1 - \frac{i\kappa}{8} e^{-\kappa x}|$ for $0 \leq x \leq 1$ is approximately 0.08 . This explains why the curves in Fig. 4.1 are essentially indistinguishable from each other.

4.2.2 General Discussion

The most distinguished characteristic of the WKB approximation is the assumed exponential dependence on the fast variation. Making this assumption, we determine the first-term approximation of the solution by solving two first-order differential equations. The first one, the eikonal equation, is nonlinear, and it determines the fast variation in the solution. It was not difficult to solve the eikonal equation for this example, but this can be one of the more difficult steps in using the WKB method. The nonlinearity is important, however, because it means the solution of the eikonal equation is not unique, and this gives rise to the two independent solutions in (4.11). What can be done when there is a unique solution of the eikonal equation is explored in Exercises 4.4(c) and 4.10(b).

The second equation coming from the WKB approximation is the transport equation, and this determines the slow variation. Fortunately, the transport equation is linear, as are the equations for determining the higher-order terms in the expansion.

The WKB method is relatively easy to use and is capable of providing an approximation with less effort than other methods. However, with this advantage come limitations. The most significant is the requirement that the equation be linear. It is possible to find a few special nonlinear problems where it can be used in a limited form, and an example is considered in Exercise 4.12. A discussion of various other possibilities can be found in Carles (2008).

Another observation is that the WKB method works just as well if q is positive or negative. In contrast, the method of matched asymptotic expansions only works on (4.1) if q is positive (Exercise 2.4). However, unlike the WKB method, the method of matched asymptotic expansions is not limited to linear problems.

Finally, in the physics literature it is common to see the WKB method used by first making the change of variables $y = e^w$. This produces a Riccati equation, and examples are considered in Exercise 4.4. The direct approach, as embodied in (4.3), is used here as it is easier to generalize to other problems, such as those considered in Exercise 4.14 and in Sect. 4.5.

Exercises

4.1. Use the WKB method to find an approximate solution of the following problems:

(a) $\varepsilon y'' + 2y' + 2y = 0$ for $0 < x < 1$, where $y(0) = 0$ and $y(1) = 1$. Compare your answer with (i) the composite expansion (2.16) obtained using matched asymptotic expansions and (ii) the exact solution.

(b) $\varepsilon^2 y'' + \varepsilon x y' - y = -1$ for $0 < x < 1$, where $y(0) = 0$ and $y(1) = 3$. Compare your answer with the composite expansion obtained using matched asymptotic expansions.

(c) $\varepsilon y'' + y' + e^x y = 0$ for $0 < x < 1$, where $y(0) = 0$ and $y(1) = 1$. Compare your answer with the composite expansion obtained using matched asymptotic expansions.

(d) $\varepsilon y'' + (x - \frac{1}{2})y' + y = 0$ for $0 < x < 1$, where $y(0) = 2$ and $y(1) = 3$.

(e) $\varepsilon y^{(iv)} + e^{4x} y = 0$ for $0 < x < 1$, where $y(0) = y(1) = 1$ and $y''(0) = y''(1) = 0$.

4.2. Consider the following problem:

$$y'' + k^2(\varepsilon t)y = 0 \quad \text{for } 0 < t,$$

where $y(0) = a$ and $y'(0) = b$. Make the change of variables, $\tau = \varepsilon t$, and then use the WKB method to construct a first-term approximation of the solution. Compare this result with that obtained using multiple scales in (3.79).

4.3. This problem and the one that follows consider extensions of the WKB formula for the general solution given in (4.11).

(a) Suppose, rather than (4.1), the equation is

$$\varepsilon^2 y'' - q(x, \varepsilon)y = 0.$$

Assuming $q(x, \varepsilon) \sim q_0(x) + \varepsilon q_1(x)$, show that

$$y \sim q_0(x)^{-1/4} \left(a_0 e^{-\kappa(x,\varepsilon)/\varepsilon} + b_0 e^{\kappa(x,\varepsilon)/\varepsilon} \right),$$

where

$$\kappa(x,\varepsilon) = \int^x \sqrt{q_0(s)} \left[1 + \frac{\varepsilon}{2} \frac{q_1(s)}{q_0(s)} \right] ds.$$

In this expansion is it possible to replace κ with $K(x,\varepsilon) = \int^x \sqrt{q(s,\varepsilon)}ds$?

(b) Consider the equation

$$\varepsilon^2 y'' + p(x)y' + q(x)y = 0.$$

Assuming $p(x) \neq 0$, show that the WKB approximation of the general solution is

$$y \sim a_0 e^{-\int^x (q/p)ds} + \frac{b_0}{p(x)} e^{\int^x (q/p)ds - \varepsilon^{-2} \int^x p(s)ds}.$$

(c) Assuming $q(x) > 0$, show that the WKB approximation of the general solution of $\varepsilon^2 y'' - q(x)y = f(x)$ is

$$y \sim \int^x G(x,s,\varepsilon)f(s)ds + y_h,$$

where y_h is given in (4.11) and

$$G(x,s,\varepsilon) = \frac{\sinh\left[\frac{1}{\varepsilon} \int_s^x \sqrt{q(r)}dr\right]}{\varepsilon[q(x)q(s)]^{1/4}}.$$

What is $G(x,s,\varepsilon)$ in the case where $q(x) < 0$?

(d) Find the WKB approximation of the general solution of the equation

$$\varepsilon^2 y'' + \varepsilon^2 p(x)y' - q(x)y = 0,$$

where $q(x)$ is nonzero.

4.4. Another approach to using WKB is to first make the substitution $y(x) = e^{w(x)}$. This problem explores this idea using the equation

$$\varepsilon^2 y'' - q(x,\varepsilon)y = 0.$$

(a) What equation does w satisfy?

(b) Suppose that $q(x,\varepsilon) \sim q_0(x) + \varepsilon q_1(x)$, where q_0 is nonzero. Assuming $w \sim \varepsilon^{-\alpha}[w_0(x) + \varepsilon^\beta w_1(x) + \cdots]$, find the first two terms in the expansion. Show that this gives the result given in Exercise (4.3)(a).

(c) Suppose that $q(x,\varepsilon) \sim \varepsilon q_0(x) + \varepsilon^2 q_1(x)$, where q_0 is nonzero. Find the first two terms in the expansion for w, and then determine the resulting expansion for y.

(d) Discuss what happens to the change of variables at points where $y(x) = 0$.

4.5. Consider the eigenvalue problem

$$\frac{d}{dx}\left[p(x)\frac{dy}{dx}\right] - r(x)y = -\lambda^2 q(x)y \quad \text{for } 0 < x < 1,$$

where $y(0) = y(1) = 0$. Here $p(x)$, $q(x)$, and $r(x)$ are given, smooth, positive functions and $\lambda \geq 0$ is the eigenvalue.

(a) Show that if one lets $y(x) = h(x)w(x)$, where $h(x) = 1/\sqrt{p(x)}$, then

$$p(x)w'' + [\lambda^2 q(x) - f(x)]w = 0 \quad \text{for } 0 < x < 1,$$

where $f(x) = r - (ph')'/h$.

(b) For the problem in part (a), use a first-term WKB approximation to show that for the large eigenvalues $\lambda \sim \lambda_n$, where $\lambda_n = n\pi/\kappa$ and

$$\kappa = \int_0^1 \sqrt{\frac{q(x)}{p(x)}}dx.$$

What is the corresponding WKB approximation of the eigenfunctions?

(c) To find the second term in the expansion of the eigenvalue, one starts with the expansion $\lambda \sim \lambda_n + \frac{1}{n}\lambda_c + \cdots$. Extending your argument in part (b) show that

$$\lambda_c = \frac{1}{2\pi}\int_0^1 [-g(x)g''(x) + (pq)^{-1/2}f(x)]dx,$$

where $g(x) = (p/q)^{1/4}$.

(d) In the case where $p = q = 1$, explain why $\lambda \sim \lambda_n$ gives an accurate approximation to all the eigenvalues so long as r_M is small compared to $2\pi^2$, where $r_M = \max_{0 \leq x \leq 1} r(x)$.

(e) To investigate the accuracy of the WKB result consider the eigenvalue problem

$$y'' + \lambda^2 q(x)y = 0, \quad \text{for } 0 < x < 1,$$

where $y(0) = y(1) = 0$ and $q(x) = (x + 1)^4$. Calculate the two-term expansion for λ derived in parts (b) and (c), and then compare it with the numerically computed values given in Table 4.1.

4.6. Find a first-term approximation of

$$\varepsilon^2 y'' + \varepsilon^{3/2}p(x)y' + q(x)y = 0 \quad \text{for } 0 < x < 1,$$

λ_1^2	λ_2^2	λ_3^2	λ_4^2	λ_5^2	λ_6^2	λ_7^2
1.6767	7.0649	16.1056	28.7823	45.0897	65.0254	88.5882

Table 4.1 Eigenvalues for Exercise 4.5(e)

where $q(x)$ is nonzero. [Hint: Follow the approach that was used to determine (4.3) using (4.2).]

4.7. The equation for the transverse displacement $u(x,t)$ of an inhomogeneous Euler–Bernoulli beam is

$$\varepsilon \partial_x^2 [D(x)\partial_x^2 u] + \mu(x)\partial_t^2 u = 0 \quad \text{for } 0 < x < 1 \text{ and } 0 < t,$$

where $u = u_{xx} = 0$ at $x = 0, 1$, $u(x,0) = f(x)$, and $u_t(x,0) = 0$. Here $D(x)$ and $\mu(x)$ are smooth, positive functions.

(a) Consider the following eigenvalue problem:

$$\varepsilon \frac{d^2}{dx^2}\left[D(x)\frac{d^2 F}{dx^2}\right] = \lambda^2 \mu(x) F \quad \text{for } 0 < x < 1,$$

where $F = F'' = 0$ at $x = 0, 1$. Find a first-term WKB approximation of the eigenfunction $F(x)$ and associated eigenvalue λ (note that both depend on ε).

(b) Use the result from part (a) and separation of variables to obtain a first-term approximation of the solution of the beam problem.

4.8. Bessel's equation is $x^2 y'' + xy' + (x^2 - \nu^2)y = 0$.

(a) Find a two-term approximation of the general solution of this equation for $1 \ll \nu$. You can assume here that $a \leq x \leq b$, where a and b are positive constants.

(b) From your result in (a), the series definition of the Bessel function $J_\nu(x)$, and the Stirling series

$$\Gamma(\nu) \sim \left(\frac{\nu}{e}\right)^\nu \sqrt{\frac{2\pi}{\nu}}\left[1 + \frac{1}{12\nu} + \frac{1}{288\nu^2} + O(\nu^{-3})\right] \quad \text{for } 1 \ll \nu,$$

find a two-term expansion of $J_\nu(x)$ for large ν.

(c) Using the formula $\Gamma(\nu) = \pi \csc(\pi\nu)/\Gamma(1-\nu)$, find a two-term expansion of $J_{-\nu}(x)$ for $1 \ll \nu$ (assuming ν is not a positive integer). From this show that

$$Y_\nu(x) \sim -\left(\frac{2\nu}{ex}\right)^\nu \sqrt{\frac{2\pi}{\nu}}\left[1 + \frac{1}{12\nu}(1+3x^2)\right] \quad \text{for } 1 \ll \nu.$$

4.9. Consider the equation

$$\varepsilon^2 y'' + P(x,\varepsilon)y' + Q(x,\varepsilon)y = 0.$$

(a) Setting $y = we^{-\kappa}$, where $\kappa = \frac{1}{2}\varepsilon^{-2}\int^x P\,ds$, show that

$$\varepsilon^2 w'' - g(x,\varepsilon)w = 0,$$

where $g = -Q + \frac{1}{4}(2P' + \varepsilon^{-2}P^2)$.

(b) Derive the expansions in Exercises 4.3(a) and 4.3(b) using (4.11) and the result from part (a).

(c) Use the result from part (a) and Exercise 4.3(a) to find the WKB expansion of the solution of

$$\varepsilon^2 y'' + \varepsilon p(x) y' + q(x) y = 0.$$

Assume here that $4q \neq p^2$.

4.10. This problem considers what happens when there is only one solution of the eikonal equation. The equation to solve is

$$\varepsilon^2 y'' + \varepsilon p(x, \varepsilon) y' + q(x, \varepsilon) y = 0,$$

where $p(x, \varepsilon) \sim p_0(x) + \varepsilon p_1(x) + \cdots$ and $q(x, \varepsilon) \sim q_0(x) + \varepsilon q_1(x) + \cdots$. The critical assumption made here is that $p_0^2 = 4q_0$.

(a) Show that the WKB expansion in (4.3) yields only one solution to the problem.

(b) Using a reduction-of-order approach, let $y = e^{\theta(x)/\varepsilon^\alpha} u$, where $\theta(x)$ and α are determined from part (a). Find the resulting equation for u and then derive a WKB approximation of u. From this, write down a WKB approximation for the general solution of the equation for y. Assume here that $2q_1 \neq p_0' + p_0 p_1$.

(c) In part (b), what happens when $2q_1 = p_0' + p_0 p_1$?

4.11. This problem considers turning points and just how close you can get to one of them and expect an accurate approximation with the WKB formula.

(a) In (4.1) suppose $q(x) = ax$, where $a > 0$ is constant. In this case there is a simple turning point at $x = 0$. Find the first two terms in the WKB approximation, and then show that for the expansion to be well ordered it must be that $\varepsilon^{2/3} \ll |x|$.

(b) In (4.1) suppose that $q(x) \sim q_0(x - x_0)^\alpha$ for x close to x_0, where $\alpha > 0$ and $q_0 \neq 0$. Show the first-term WKB approximation holds for $\varepsilon^{2/(2+\alpha)} \ll |x - x_0|$.

4.12. It is possible to use the WKB method on certain types of nonlinear problems. As an example, consider solving

$$\varepsilon^2 y'' + y + \varepsilon |y|^2 y = 0 \quad \text{for } 0 < t < \infty,$$

where $y(0) = a$ and $y'(0) = 0$. This type of nonlinearity arises in studying the evolution of wavepackets.

(a) To use (4.3), with x replaced with t, assume that $\theta(t)$ is imaginary. With this, find and then solve the eikonal equation.

(b) Find, and then solve, the transport equation.

(c) What is the resulting WKB approximation? What, if any, difficulty is there in satisfying the initial conditions? Also, for a linear problem one can add solutions, as in (4.11). What about in this problem?

4.13. Consider the initial-value problem

$$\varepsilon^2 y'' + e^{-2t} y = 0 \quad \text{for } 0 < t < \infty,$$

where $y(0) = 0$ and $y'(0) = 1$. The exponential approaches zero as $t \to \infty$, raising the question of whether there is a turning point at infinity. This problem explores this possibility.
(a) Find a first-term WKB approximation of the solution.
(b) Determine the second term in the WKB expansion and then discuss the shortcomings of this approximation. Are any of these complications similar to what might be expected near a turning point?
(c) Find the exact solution. On the same axes plot the solution and the first-term expansion from part (a) for $0 \le t \le 10$ (take $\varepsilon = 10^{-2}$). Comment on the differences between the curves in conjunction with the results from part (b).

4.14. In the theory of collective ruin (Peters and Mangel, 1990) one comes across a variable $R(x)$, which is the probability of having resources available for emergencies (i.e., a risk reserve). It satisfies the integrodifferential equation

$$\varepsilon \beta(x) R' - R + \lambda \int_0^{x/\varepsilon} R(x - \varepsilon y) e^{-\lambda y} dy = 0 \quad \text{for } 0 < x < \infty,$$

where $R(x) \to 1$ as $x \to \infty$. Also, λ is a positive constant and $\beta(x)$ is a smooth, positive function.
(a) Find the exact solution in the case where β is a positive constant. Explain why it is necessary to require $\lambda\beta > 1$.
(b) Based on your observations from part (a), find the first-term in a WKB expansion of the solution when β is not constant. What conditions must be imposed on λ and $\beta(x)$?

4.15. In the theory of multiserver queues with a large number of servers and short interarrival times, the following problem appears (Knessl, 1991): find $P(x, \eta)$ that satisfies the first-order differential equation

$$\partial_\eta P(x, \eta) = \mu_0(x + \varepsilon) P(x + \varepsilon, \eta) - g(x, \eta) P(x, \eta) \quad \text{for } 0 < \eta < \infty,$$

where

$$P(x + \varepsilon, 0) = \int_0^\infty \lambda(s) P(x, s) ds$$

In this problem, $0 < x < 1$ and $g(x, \eta) = \mu_0 x + \lambda(\eta)$, where $\mu_0 > 0$ is constant and $\lambda(\eta) > 0$ is continuous with $\lambda \to \infty$ as $\eta \to \infty$. Find a first-term approximation of the solution of this problem for small ε.

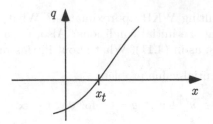

Figure 4.3 In the analysis of the turning point, $q(x)$ is assumed to have a simple zero at $x = x_t$ with $q'(x_t) > 0$. This will enable us to use a linear approximation for $q(x)$ near the turning point [see (4.30)]

4.3 Turning Points

We will introduce the analysis for turning points using the example from the previous section. The equation to solve is therefore

$$\varepsilon^2 y'' - q(x)y = 0. \tag{4.25}$$

As was pointed out in deriving the WKB approximation, we must stay away from the points where $q(x)$ is zero. To explain how to deal with these points, we will assume that $q(x)$ is smooth and has a simple zero at $x = x_t$. In other words, we will assume that $q(x_t) = 0$ and $q'(x_t) \neq 0$. To start, we will consider what happens when there is only one such point.

4.3.1 The Case Where $q'(x_t) > 0$

We are assuming here that there is a simple turning point at x_t, with $q(x) > 0$ if $x > x_t$ and $q(x) < 0$ if $x < x_t$ (Fig. 4.3). This means that the solution of (4.25) will be oscillatory if $x < x_t$ and exponential for $x > x_t$. The fact that the solution is oscillatory for negative $q(x)$ can be understood if one considers the constant coefficient equation $y'' + \lambda^2 y = 0$, where $\lambda > 0$. The general solution in this case is $y(x) = A_0 \cos(\lambda x + \phi_0)$, and this clearly is oscillatory. A similar explanation can be given for the exponential solutions for a positive coefficient.

We can use the WKB approximation on either side of the turning point. This gives the following approximation for the general solution (see Fig 4.6):

$$y \sim \begin{cases} y_L(x, x_t) & \text{if } x < x_t, \\ y_R(x, x_t) & \text{if } x_t < x, \end{cases} \tag{4.26}$$

where

$$y_R(x, x_t) = \frac{1}{q(x)^{1/4}} \left[a_R \exp\left(-\frac{1}{\varepsilon} \int_{x_t}^{x} \sqrt{q(s)}ds\right) + b_R \exp\left(\frac{1}{\varepsilon} \int_{x_t}^{x} \sqrt{q(s)}ds\right) \right]$$
(4.27)

and

$$y_L(x, x_t) = \frac{1}{q(x)^{1/4}} \left[a_L \exp\left(-\frac{1}{\varepsilon} \int_{x}^{x_t} \sqrt{q(s)}ds\right) + b_L \exp\left(\frac{1}{\varepsilon} \int_{x}^{x_t} \sqrt{q(s)}ds\right) \right].$$
(4.28)

These expressions come directly from (4.11), except that we have now fixed one of the endpoints in the integrals at the turning point. This particular choice is not mandatory, but it does simplify the formulas in the calculations to follow. It is important to recognize that the coefficients in (4.27) and (4.28) are not independent, and we must find out how they are connected by investigating what is happening in a transition layer centered at $x = x_t$. This is very similar to what happened when we were studying interior-layer problems in Chap. 2. When we finish, the approximation of the general solution will only contain two arbitrary constants rather than four as in (4.27) and (4.28).

4.3.1.1 Solution in Transition Layer

To determine the solution near the turning point, we introduce the transition layer coordinate

$$\bar{x} = \frac{x - x_t}{\varepsilon^\beta}$$
(4.29)

or, equivalently,

$$x = x_t + \varepsilon^\beta \bar{x}.$$

We know the point x_t, but we will have to determine the value for β from the analysis to follow. Now, to transform the differential equation, we first use

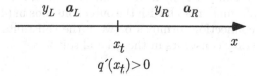

Figure 4.4 Schematic of the case of a simple turning point and the corresponding WKB approximations used on either side of x_t. Also shown are the respective coefficient vectors $\mathbf{a}_L = (a_L, b_L)^T$ and $\mathbf{a}_R = (a_R, b_R)^T$

Taylor's theorem to conclude

$$q(x_t + \varepsilon^\beta \bar{x}) \sim q(x_t) + \varepsilon^\beta \bar{x} q'(x_t) + \cdots$$
$$= \varepsilon^\beta \bar{x} q'(x_t) + \cdots . \tag{4.30}$$

We will assume $q(x)$ has a simple zero at x_t, so $q'(x_t) \neq 0$. With this, and letting $Y(\bar{x})$ denote the solution in this layer, we have that

$$\varepsilon^{2-2\beta} Y'' - (\varepsilon^\beta \bar{x} q'_t + \cdots) Y = 0, \tag{4.31}$$

where $q'_t = q'(x_t)$. For balancing we need $2 - 2\beta = \beta$, and so $\beta = \frac{2}{3}$. The appropriate expansion of the solution in this region is

$$Y \sim \varepsilon^\gamma Y_0(\bar{x}) + \cdots . \tag{4.32}$$

Introducing this into (4.31), we get the following equation to solve:

$$Y_0'' - \bar{x} q'_t Y_0 = 0 \quad \text{for} \ -\infty < \bar{x} < \infty. \tag{4.33}$$

This can be transformed into an equation whose solutions are known. Letting $s = (q_t')^{1/3} \bar{x}$, we get *Airy's equation*, which is

$$\frac{\mathrm{d}^2}{\mathrm{d}s^2} Y_0 - s Y_0 = 0, \quad \text{for} \ -\infty < s < \infty. \tag{4.34}$$

Because of its importance in applied mathematics, Airy's equation has been studied extensively. It can be solved using power series expansions or the Laplace transform. One finds that the general solution can be written as

$$Y_0 = a \, \mathrm{Ai}(s) + b \, \mathrm{Bi}(s),$$

where $\mathrm{Ai}(\cdot)$ and $\mathrm{Bi}(\cdot)$ are Airy functions of the first and second kinds, respectively, and a and b are arbitrary constants. The definitions and some of the properties of these functions are given in Appendix B. We are now able to write the general solution of the transition layer Eq. (4.33) as

$$Y_0(\bar{x}) = a \, \mathrm{Ai}[(q'_t)^{1/3} \bar{x}] + b \, \mathrm{Bi}[(q'_t)^{1/3} \bar{x}]. \tag{4.35}$$

From (4.27), (4.28), and (4.35) we have six undetermined constants. However, the solution in (4.35) must match with the outer solutions in (4.27) and (4.28). This will lead to connection formulas between the constants, and these will result in two arbitrary constants in the general solution.

4.3.1.2 Matching

The solution in the transition region must match with the outer solutions given in (4.27) and (4.28). In what follows, keep in mind that $q_t' > 0$. To do the matching, we will use the intermediate variable

$$x_\eta = \frac{x - x_t}{\varepsilon^\eta}, \tag{4.36}$$

where $0 < \eta < 2/3$. Before matching the solutions, note that the terms in the two outer solutions contain the following: for $x > x_t$

$$\int_{x_t}^x \sqrt{q(s)}ds = \int_{x_t}^{x_t + \varepsilon^\eta x_\eta} \sqrt{q(s)}ds$$
$$\sim \int_{x_t}^{x_t + \varepsilon^\eta x_\eta} \sqrt{(s - x_t)q_t'}ds$$
$$= \frac{2}{3}\varepsilon r^{3/2} \tag{4.37}$$

and

$$q(x)^{-1/4} \sim [q_t + (x - x_t)q_t']^{-1/4}$$
$$= \varepsilon^{-1/6}(q_t')^{-1/6}r^{-1/4}, \tag{4.38}$$

where $r = (q_t')^{1/3}\varepsilon^{\eta - 2/3}x_\eta$.

4.3.1.3 Matching for $x > x_t$

Using the asymptotic expansions for the Airy functions given in Appendix B, one finds that

$$Y \sim \varepsilon^\gamma Y_0(\varepsilon^{\eta - 2/3}x_\eta) + \cdots$$
$$\sim \frac{a\varepsilon^\gamma}{2\sqrt{\pi}r^{1/4}}e^{-\frac{2}{3}r^{3/2}} + \frac{b\varepsilon^\gamma}{\sqrt{\pi}r^{1/4}}e^{\frac{2}{3}r^{3/2}}, \tag{4.39}$$

and for the WKB solution in (4.27)

$$y_R \sim \frac{\varepsilon^{-1/6}}{(q_t')^{1/6}r^{1/4}}\left(a_R e^{-\frac{2}{3}r^{3/2}} + b_R e^{\frac{2}{3}r^{3/2}}\right). \tag{4.40}$$

For these expressions to match, we must have $\gamma = -\frac{1}{6}$,

$$a_R = \frac{a}{2\sqrt{\pi}}(q_t')^{1/6} \quad \text{and} \quad b_R = \frac{b}{\sqrt{\pi}}(q_t')^{1/6}. \tag{4.41}$$

4.3.1.4 Matching for $x < x_t$

The difference with this case is that $x_\eta < 0$, which introduces complex numbers into (4.28). As before, using the expansions for the Airy functions given in the Appendix B,

$$
\begin{aligned}
Y &\sim \varepsilon^\gamma Y_0(\varepsilon^{\eta-2/3} x_\eta) \\
&\sim \frac{a\varepsilon^\gamma}{\sqrt{\pi}|r|^{1/4}} \cos\left(\frac{2}{3}|r|^{3/2} - \frac{\pi}{4}\right) + \frac{b\varepsilon^\gamma}{\sqrt{\pi}|r|^{1/4}} \cos\left(\frac{2}{3}|r|^{3/2} + \frac{\pi}{4}\right) \\
&= \frac{\varepsilon^\gamma}{2\sqrt{\pi}|r|^{1/4}} \left[(ae^{-i\pi/4} + be^{i\pi/4})e^{i\zeta} + (ae^{i\pi/4} + be^{-i\pi/4})e^{-i\zeta}\right], \quad (4.42)
\end{aligned}
$$

where $\zeta = \frac{2}{3}|r|^{3/2}$. In the last step leading to (4.42), the identity $\cos\theta = \frac{1}{2}(e^{i\theta} + e^{-i\theta})$ was used. As for the WKB expansion, from (4.28),

$$
y_L \sim \frac{\varepsilon^{-1/6}}{(q_t')^{1/6}|r|^{1/4}} \left(a_L e^{i(\zeta - \frac{\pi}{4})} + b_L e^{-i(\zeta + \frac{\pi}{4})}\right). \quad (4.43)
$$

Matching (4.42) and (4.43) yields

$$
a_L = \frac{(q_t')^{1/6}}{2\sqrt{\pi}}(ia + b) \quad (4.44)
$$

and

$$
b_L = \frac{(q_t')^{1/6}}{2\sqrt{\pi}}(a + ib) = i\bar{a}_L, \quad (4.45)
$$

where $\bar{a}_L$ is the complex conjugate of a_L. Equations (4.41), (4.44), and (4.45) are known as *connection formulas*, and they constitute a system of equations that enable us to solve for four of the constants in terms of the remaining two.

4.3.1.5 Summary

Solving the connection formulas in (4.41), (4.44), and (4.45) we find that

$$
\mathbf{a}_L = \mathbf{M}\mathbf{a}_R, \quad (4.46)
$$

where

$$
\mathbf{a}_L = \begin{pmatrix} a_L \\ b_L \end{pmatrix}, \qquad \mathbf{a}_R = \begin{pmatrix} a_R \\ b_R \end{pmatrix}, \qquad \mathbf{M} = \begin{pmatrix} i & \frac{1}{2} \\ 1 & \frac{1}{2}i \end{pmatrix}. \quad (4.47)
$$

The resulting WKB approximation in (4.26) can be written as

$$y(x) \sim \begin{cases} \dfrac{1}{|q(x)|^{1/4}} \left[2a_R \cos\left(\dfrac{1}{\varepsilon}\theta(x) - \dfrac{\pi}{4}\right) + b_R \cos\left(\dfrac{1}{\varepsilon}\theta(x) + \dfrac{\pi}{4}\right) \right] & \text{if } x < x_t, \\[2ex] \dfrac{1}{q(x)^{1/4}} \left(a_R e^{-\kappa(x)/\varepsilon} + b_R e^{\kappa(x)/\varepsilon} \right) & \text{if } x_t < x, \end{cases}$$

$$\text{(4.48)}$$

where

$$\theta(x) = \int_x^{x_t} \sqrt{|q(s)|}\,ds \qquad\qquad (4.49)$$

and

$$\kappa(x) = \int_{x_t}^x \sqrt{|q(s)|}\,ds. \qquad\qquad (4.50)$$

It should be remembered that this expansion was derived under the assumption that $x = x_t$ is a simple turning point with $q'(x_t) > 0$. The accuracy of this approximation near the turning point depends on the specific problem. However, one can show that in general one must require $\varepsilon^{2/3} \ll |x - x_t|$ (Exercise 4.11). Also, as expected, we have ended up with an expansion for the solution of (4.25) that contains two arbitrary constants (a_R, b_R) .

Example

As an example of a turning point problem consider

$$\varepsilon^2 y'' = x(2 - x)y \quad \text{for } -1 < x < 1, \qquad\qquad (4.51)$$

where $y(-1) = y(1) = 1$. In this case, $q(x) = x(2 - x)$, and so there is a simple turning point at $x = 0$ with $q'(0) = 2$. The WKB approximation in (4.48) therefore applies where, from (4.49) and (4.50),

$$\theta(x) = \frac{1}{2}(1 - x)\sqrt{x(x - 2)} - \frac{1}{2}\ln[1 - x + \sqrt{x(x - 2)}]$$

and

$$\kappa(x) = \frac{1}{2}(x - 1)\sqrt{x(2 - x)} - \frac{1}{2}\arcsin(x - 1) + \frac{\pi}{4}.$$

The constants a_R and b_R are found from the two boundary conditions. For example, since $q(1) = 1$ and $\kappa(1) = \pi/4$, the condition $y(1) = 1$ leads to the equation $a_R \exp(-\frac{\pi}{4}\varepsilon) + b_R \exp(\frac{\pi}{4}\varepsilon) = 1$. The resulting approximation obtained from (4.48) is shown in Fig. 4.5a along with the approximation from the transition layer. The singularity in y_R and y_L at $x = 0$ is evident in this plot, as is the matching by the Airy functions from the transition layer. To demonstrate the accuracy of the WKB approximation even when a turning point is present, in Fig. 4.5b the numerical solution is plotted along with the left and right WKB approximations. Within their respective domains

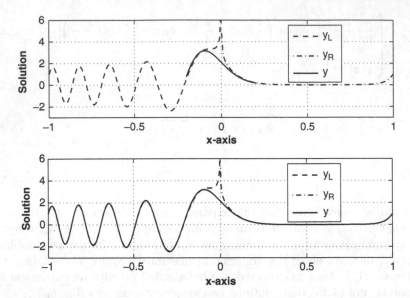

Figure 4.5 (a) Transition layer solution (4.32), and left and right WKB approximations given in (4.48), for the solution of (4.51). **(b)** Comparison between left and right WKB approximations and numerical solution of (4.51). In these calculations, $\varepsilon = \frac{1}{25}$

of applicability, the WKB approximations are essentially indistinguishable from the numerical solution. ■

4.3.2 The Case Where $q'(x_t) < 0$

This approximation derived for $q'(x_t) > 0$ can be used when $q'(x_t) < 0$ by simply making the change of variables $z = x_t - x$ (Exercise 4.17). The result is that

$$y \sim \begin{cases} Y_{\text{L}}(x, x_t) & \text{if} \quad x < x_t, \\ Y_{\text{R}}(x, x_t) & \text{if} \quad x_t < x, \end{cases} \tag{4.52}$$

where

$$Y_{\text{R}}(x, x_t) = \frac{1}{q(x)^{1/4}} \left[A_{\text{R}} \exp\left(\frac{1}{\varepsilon} \int_{x_t}^x \sqrt{q(s)} \, ds\right) + B_{\text{R}} \exp\left(-\frac{1}{\varepsilon} \int_{x_t}^x \sqrt{q(s)} \, ds\right) \right] \tag{4.53}$$

and

$$Y_L(x, x_t) = \frac{1}{q(x)^{1/4}} \left[A_L \exp\left(\frac{1}{\varepsilon} \int_x^{x_t} \sqrt{q(s)} ds \right) + B_L \exp\left(-\frac{1}{\varepsilon} \int_x^{x_t} \sqrt{q(s)} ds \right) \right].$$
(4.54)

The coefficients in these expressions satisfy the connection formula

$$\mathbf{A_R} = \mathbf{N A_L},$$
(4.55)

where

$$\mathbf{A_L} = \begin{pmatrix} A_L \\ B_L \end{pmatrix}, \qquad \mathbf{A_R} = \begin{pmatrix} A_R \\ B_R \end{pmatrix}, \qquad \mathbf{N} = \begin{pmatrix} \frac{1}{2}i & 1 \\ \frac{1}{2} & i \end{pmatrix}.$$
(4.56)

Making use of the fact that $q'(x_t) < 0$, one can rewrite the approximation in this case as

$$y(x) \sim \begin{cases} \dfrac{1}{q(x)^{1/4}} \left(A_L e^{\theta(x)/\varepsilon} + B_L e^{-\theta(x)/\varepsilon} \right) & \text{if } x < x_t, \\ \dfrac{1}{|q(x)|^{1/4}} \left[2B_L \cos\left(\dfrac{1}{\varepsilon}\kappa(x) - \dfrac{\pi}{4} \right) + A_L \cos\left(\dfrac{1}{\varepsilon}\kappa(x) + \dfrac{\pi}{4} \right) \right] & \text{if } x_t < x, \end{cases}$$
(4.57)

where $\theta(x)$ and $\kappa(x)$ are given in (4.49) and (4.50). Again, we have a solution that contains two arbitrary constants (A_L, B_L).

4.3.3 Multiple Turning Points

It is not particularly difficult to combine the preceding approximations to analyze problems with multiple turning points. To illustrate, suppose x_t and x_T, with $x_t < x_T$, are turning points with $q'(x_t) > 0$ and $q'(x_T) < 0$ (Fig. 4.6). Moreover, these are the only turning points. In this case, (4.26) holds for $x < x_T$. Similarly, (4.52) applies, with x_T replacing x_t, and this holds for $x > x_t$. What this means is that it is required that $y_R(x, x_t) = Y_L(x, x_T)$ for $x_t < x < x_T$. This holds if

$$\mathbf{a_R} = \mathbf{Q A_L},$$
(4.58)

where

$$\mathbf{Q} = \begin{pmatrix} e^\phi & 0 \\ 0 & e^{-\phi} \end{pmatrix}$$
(4.59)

and

$$\phi = \frac{1}{\varepsilon} \int_{x_t}^{x_T} \sqrt{q(s)} ds.$$
(4.60)

Therefore,

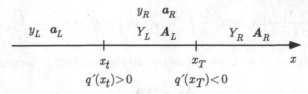

Figure 4.6 Schematic of case of two turning points and corresponding WKB approximations used in various intervals. Also shown are the respective coefficient vectors

$$y \sim \begin{cases} y_L(x, x_t) & \text{if } x < x_t, \\ Y_L(x, x_T) & \text{if } x_t < x < x_T, \\ Y_R(x, x_T) & \text{if } x_T < x, \end{cases} \qquad (4.61)$$

where $\mathbf{A_R} = \mathbf{N}\mathbf{A_L}$ and $\mathbf{a_L} = \mathbf{M}\mathbf{Q}\mathbf{A_L}$.

If, instead, $q'(x_t) < 0$ and $q'(x_T) > 0$, then a similar analysis (Exercise 4.17) shows that

$$y \sim \begin{cases} Y_L(x, x_t) & \text{if } x < x_t, \\ y_L(x, x_T) & \text{if } x_t < x < x_T, \\ y_R(x, x_T) & \text{if } x_T < x, \end{cases} \qquad (4.62)$$

where $\mathbf{a_R} = \mathbf{M}^{-1}\mathbf{a_L}$, $\mathbf{A_L} = \mathbf{N}^{-1}\mathbf{Q}^{-1}\mathbf{a_L}$, and $\mathbf{A_R} = \mathbf{Q}^{-1}\mathbf{a_L}$.

4.3.4 Uniform Approximation

In the case of a single turning point, with $q'(x_t) > 0$, the solution is in three pieces: the two WKB approximations (y_L, y_R) on either side of the turning point and the transition layer solution that is in between. The first-term approximation of the solution in the transition layer was found to be a solution of Airy's equation (4.33), which is the prototype equation for a simple turning point. In fact, it is possible to find a uniform first-term expansion of the solution in terms of Airy functions. This was first derived by Langer (1931), and the result is

$$y(x) \sim \varepsilon^{-1/6} \left(\frac{f(x)}{q(x)} \right)^{1/4} \left[a_0 \mathrm{Ai}\left(\varepsilon^{-2/3} f(x) \right) + b_0 \mathrm{Bi}\left(\varepsilon^{-2/3} f(x) \right) \right], \qquad (4.63)$$

where

$$f(x) = \begin{cases} \left[\dfrac{3}{2} \displaystyle\int_{x_t}^{x} \sqrt{q(s)}\, ds \right]^{2/3} & \text{if } x_t \le x \\[4mm] -\left[\dfrac{3}{2} \displaystyle\int_{x}^{x_t} \sqrt{-q(s)}\, ds \right]^{2/3} & \text{if } x \le x_t. \end{cases}$$

The derivation of (4.63) is outlined in Exercise 4.25. Also, in connection with the WKB approximations given in (4.48),

$$a_R = \frac{a_0}{2\sqrt{\pi}} \quad \text{and} \quad b_R = \frac{b_0}{2\sqrt{\pi}}.$$

Because of the value of having a composite expansion, there have been numerous generalizations of Langer's result. A discussion of some of these can be found in Nayfeh (1973).

In referring to the turning point as simple, it is meant that $q(x)$ has a simple zero at $x = x_t$, i.e., $q(x_t) = 0$ but $q'(x_t) \neq 0$. Higher-order turning points do arise, and an example of one of second order, at $x = 0$, is

$$\varepsilon^2 y'' - x^2 e^x y = 0. \tag{4.64}$$

The reason this is second order is simply that $q(x_t) = q'(x_t) = 0$ but $q''(x_t) \neq 0$. It is also possible to have one of fractional order (e.g., when $q(x) = x^{1/3} e^x$). The prototype equations in these cases are discussed in Appendix B. Other types do occur, such as logarithmic, though they generally are harder to analyze.

Another complication that can arise is that the position of the turning point may depend on ε. This can lead to coalescing turning points; an example of this is found in the equation

$$\varepsilon^2 y'' - (x - \varepsilon)(x + \varepsilon)y = 0. \tag{4.65}$$

This has simple turning points at $x_t = \pm \varepsilon$, but to the first-term approximation they look like one of second order at $x = 0$. Situations such as this are discussed in Steele (1976) and Dunster (1996).

Examples

1. Suppose the problem is

$$\varepsilon^2 y'' + \sin(x)y = 0 \quad \text{for } 0 < x < 2\pi,$$

where $y(0) = a$ and $y(2\pi) = b$. This has three turning points: $x_t = 0, \pi, 2\pi$. Because two of these are endpoints, there will be two outer WKB approximations, one for $0 < x < \pi$ and one for $\pi < x < 2\pi$. Since $\frac{d}{dx}\sin(x) \neq 0$ at $x = \pi$, the turning point $x_t = \pi$ is simple and is treated in much the same way as the one analyzed previously. The solution in the transition layer at $x_t = 0$ will be required to satisfy the boundary condition $y(0) = a$, and the solution in the layer at $x_t = 2\pi$ will satisfy $y(2\pi) = b$. ■

2. The equation

$$\varepsilon^2 y'' + p(x)y' + q(x)y = 0 \tag{4.66}$$

differs from that considered previously because of the first derivative term. The WKB approximation of the general solution is given in Exercise 4.3. In this case, the turning points now occur when $p(x) = 0$. It is not a coincidence that these are also the points that can give rise to an interior layer, and an example of this is given in Sect. 2.5.6. ∎

Exercises

4.16. Consider the boundary-value problem

$$\varepsilon^2 y'' + x(x+3)^2 y = 0 \quad \text{for } a < x < b,$$

where $y(a) = \alpha$ and $y(b) = \beta$. Find a first-term WKB expansion of the solution in cases where:
(a) $a = 0$, $\alpha = 0$, $b = 1$, $\beta = 1$;
(b) $a = -1$, $\alpha = 1$, $b = 0$, $\beta = 0$;
(c) $a = -1$, $\alpha = 0$, $b = 1$, $\beta = 1$.

4.17. This exercise concerns the derivation of the WKB approximation when one or more turning points are present.
(a) Making the change of variable $z = x_t - x$ in (4.25) derive (4.57) from (4.48).
(b) Derive (4.62).

4.18. In quantum mechanics one must solve the time-independent Schrödinger equation given as

$$\varepsilon^2 \psi'' - [V(x) - E]\psi = 0 \quad \text{for } -\infty < x < \infty, \tag{4.67}$$

where the wave function $\psi(x)$ must be bounded. Also, the potential $V(x)$ is a given, smooth function while the energy E is a nonnegative constant that must be solved for along with the wave function. Assume here that the situation is as depicted in Fig. 4.7, with $V_m < E$, where $V_m = \min_{-\infty < x < \infty} V(x)$. Also, $V(a) = V(b) = E$.
(a) Using (4.48) and (4.57) find a first-term approximation of the solution. From this conclude that for there to be a nonzero solution it must be that E satisfies

$$\int_a^b \sqrt{E - V(x)}\,dx = \varepsilon\pi\left(n + \frac{1}{2}\right) \quad \text{for } n = 0, 1, 2, 3, \ldots. \tag{4.68}$$

This is known as the WKB quantization condition.

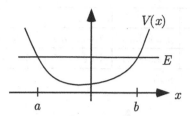

Figure 4.7 Potential used in Exercise 4.18

(b) For a harmonic oscillator the potential is $V(x) = \mu x^2$, where $\mu > 0$ is constant. Solve the quantization condition (4.68) in this case. Also describe, in terms of nodes and wavelengths, what the integer n signifies for the solution in the region $a < x < b$. What does n signify for the solution outside this region? It is worth pointing out that for this potential the WKB quantization condition gives the exact values for E.

(c) In deriving the WKB solution in part (a), it was assumed that a, b, and E were independent of ε. However, the quantization condition (4.68) means they must depend on ε for there to be a nontrivial solution. How does this dependence affect the derivation of the WKB approximation? Also, what affect, if any, does the dependence of E on the integer n (which may be quite large) do to the accuracy of the approximation?

(d) In the case where $V(x) = \mu x^m$, where $m = 2, 4, 6, \ldots$, one can show that $E^\gamma = \mu \varepsilon^m \chi$, where $\gamma = 1 + 2/m$ and χ satisfies the equation

$$\chi^{1/2} \sum_{k=0}^{\infty} a_k \chi^{-k} = 2n + 1, \qquad (4.69)$$

where a_k is independent of χ (Dunham, 1932; Bender et al., 1977). For example,

$$a_0 = \frac{2}{\sqrt{\pi}} \frac{\Gamma(1 + 1/m)}{\Gamma(3/2 + 1/m)} \quad \text{and} \quad a_1 = -\frac{m-1}{6\sqrt{\pi}} \frac{\cot(\pi/m)\Gamma(1/2 + 1/m)}{\Gamma(1/m)}.$$

Setting $N = 2n + 1$ and assuming $m \neq 2$, use (4.69) to find a two-term expansion of E for $N \gg 1$. How does the first term compare with your result from part (a)?

(e) The numerical values of $E/\varepsilon^{m/\gamma}$ calculated by Secrest et al. (1962) are given in Table 4.2 in the case where $V(x) = x^m$. Compare the results from parts (a) and (d) with these values. Make sure to comment on the accuracy as a function of n.

4.19. This problem concerns the Schrödinger equation in (4.67) and the potential shown in Fig. 4.8. Because $V(x)$ acts like a barrier, this example is commonly used to illustrate the phenomenon of tunneling.

m/n	0	2	4
4	1.0603621	7.4556979	16.261826
8	1.2258201	10.244947	25.809007

Table 4.2 For Exercise 4.18(e)

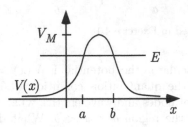

Figure 4.8 Potential used in Exercise 4.19

(a) Using (4.48) and (4.57) find a first-term approximation of the solution of (4.67) in the case of the barrier potential shown in Fig. 4.8. Assume E is given and satisfies $0 < E < V_M$, where $V_M = \max_{-\infty < x < \infty} V(x)$ and $V(-\infty) = V(\infty) = 0$.

(b) The time-dependent Schrödinger equation has the form

$$- \varepsilon^2 \partial_x^2 \Psi + V \Psi = i \partial_t \Psi.$$

In regard to the barrier potential, suppose a wave approaches from the left. Because of reflection from the barrier, the solution should then consist of left and right traveling waves in the region $x < a$. However, part of the incident wave will be transmitted through the barrier and result in a right-running wave for $x > b$ (this is the phenomenon of tunneling). Assuming $\Psi = \exp(-iEt)\psi(x)$, use the results from part (a) to find first-term approximations of the waves in these two regions.

4.20. The motion of planetary rings is described using the theory of self-gravitating annuli orbiting a central mass. For circular motion in the plane, with the planet at the origin, one ends up having to find the circumferential velocity $V(r, \theta, t) = v(r)e^{i(\omega t + m\theta)}$. The function $v(r)$ satisfies (Papaloizou and Pringle, 1987)

$$\frac{d}{dr}\left(r\frac{d}{dr}(rv)\right) = m^2(1 - \kappa^2 r^2)v \quad \text{for } 0 < r < \infty,$$

where

$$\kappa = \frac{\alpha + \beta m}{m}.$$

Here r is the radial coordinate and α and β are positive constants. The parameter m is positive and is a mode number. Find a first-term approximation of the solution for large m.

4.21. Find a first-term approximation of the solution of

$$\varepsilon^2 y'' + xy' - x(1+x)y = 0 \quad \text{for} \ -1 < x < 1,$$

where $y(-1) = 1$ and $y(1) = 3e^2$. (Hint: For large r,

$$\int_0^r e^{-\alpha s^2} ds \sim \frac{1}{2}\sqrt{\frac{\pi}{\alpha}} - \frac{1}{2\alpha r}e^{-\alpha r^2}\bigg).$$

4.22. Find a first-term approximation of the solution of $\varepsilon^2 y'' - q(x)y = 0$ for $-\infty < x < \infty$, where $q(x)$ is the even function shown in Fig. 4.9. Thus, $q(0) = q'(0) = 0$ but $q''(0) \neq 0$, and $q(\pm a) = 0$, with $q'(\pm a) \neq 0$.

4.23. Consider the problem

$$\varepsilon y'' + x^2 y' + y = 0 \quad \text{for} \ -1 < x < 1,$$

where $y(-1) = 1$ and $y(1) = -1$.
(a) The WKB/transition-layer analysis based on the coordinate transformation (4.29) does not work on this problem. Explain why. Also, explain why failure should be expected based on the balancing in the transition layer and its relationship to the terms responsible for the singularity in the WKB expansion.
(b) Letting $r = (x + \varepsilon^\beta x_0)/\varepsilon^\alpha$, find the transition regions that are needed to complete the approximations. Also state the equations to be solved in each region and the matching conditions to be imposed.
(c) Show that it is possible to resolve the difficulty by making the substitution

$$y(x) = u(x)e^{-x^2/(6\varepsilon)}.$$

4.24. For Schrödinger's equation it has been observed that in certain cases it is possible to improve on the WKB quantization condition given in (4.68). In the SWKB method, one introduces what is known as a supersymmetric potential $W(x) \equiv -\varepsilon \psi_0'/\psi_0$, where $\psi_0(x)$ is an eigenfunction corresponding to the lowest energy E_0. Under certain restrictions, it is then possible to transform (4.68) into the following equation:

$$\int_a^b \sqrt{E^* - W^2}dx = \varepsilon\pi n \quad \text{for} \ n = 0, 1, 2, 3, \ldots, \tag{4.70}$$

where $E^* = E - E_0$. It is significant, and somewhat surprising, that for many of the cases where the problem can be solved in closed form the condition in

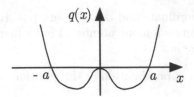

Figure 4.9 Potential used in Exercise 4.22

(4.70) actually yields the exact value for E (Comtet et al., 1985; Dutt et al., 1991). In comparison, (4.68) has been found to be exact for very few cases.

(a) Show that $W(x)$ satisfies the equation $\varepsilon W' - W^2 = -V^*$, where $V^*(x) = V(x) - E_0$. From this and the WKB quantization condition (4.68), derive (4.70). To do this, remember that a and b are turning points and assume that $W(a) = -W(b)$.

(b) As another approach, introduce the supersymmetric potential into (4.67) and then derive a first-term approximation of the solution. With this, obtain the SWKB quantization condition (4.70).

4.25. This problem concerns the derivation of the uniformly valid approximation in (4.63).

(a) Change variables by letting $s = \varepsilon^{-\alpha} f(x)$ and $y(x) = \Phi(x)Y(s)$. Show that by taking $\Phi = f_x^{-1/2}$, where f satisfies $f f_x^2 = q$, (4.1) can be transformed into an equation of the form $Y'' = [s + G(\varepsilon, s)]Y$.

(b) From the result in part (a) derive (4.63).

(c) Show that (4.63) reduces to the approximation in (4.48) when $x \gg x_t$ or when $x_t \ll x$. Does it reduce to (4.35) when $x \approx x_t$?

4.26. In the semiclassical description for what are called shape resonances (Combes et al., 1984), one finds the eigenvalue problem

$$\varepsilon^2 \psi'' - [V(x) - E]\psi = 0 \quad \text{for } 0 < x < \infty,$$

where $\psi(0) = \psi(\infty) = 0$. The potential here is $V(x) = (x - 1)^2 \exp(-x^2/4)$. Find a first-term approximation of the solution, and from this derive the quantization condition.

4.4 Wave Propagation and Energy Methods

The WKB method is quite useful for finding an asymptotic approximation of a traveling-wave solution of a linear partial differential equation. To illustrate this, we consider the problem

$$u_{xx} = \mu^2(x)u_{tt} + \alpha(x)u_t + \beta(x)u \quad \text{for} \quad \begin{cases} 0 < x < \infty \\ 0 < t, \end{cases} \qquad (4.71)$$

where

$$u(0,t) = \cos(\omega t). \qquad (4.72)$$

This is the equation for the displacement of a string that is damped (αu_t) and that has an elastic support (βu). Because the string is being forced periodically at the left end, the solution will develop into a wave that travels to the right. To find this solution, we will set $u(x,t) = e^{i\omega t}v(x)$ and then require that the function $v(x)$ be consistent with the observation that the wave moves to the right. Also, for the record, it is assumed that the functions in (4.71) are smooth with $\mu(x) > 0$, and $\alpha(x)$ and $\beta(x)$ nonnegative.

The equation is linear, but there is no obvious small parameter ε that can serve as the basis of the WKB approximation. To motivate what will be done, suppose $\alpha = \beta = 0$ and μ is constant. In this case, the plane wave solutions of (4.71), (4.72) are

$$u(x,t) = e^{i(\omega \tau - \kappa x)},$$

where $k = \pm \omega \mu$. This shows that for high frequencies the waves have a relatively short wavelength (that is, $|\frac{2\pi}{k}| \ll 1$). This observation will serve as the basis of our approximation, and in such a circumstance the small parameter is $\varepsilon = \frac{1}{\omega}$. This assumption is what underlies the application of the WKB method to many problems in optics, electromagnetics, and acoustics. For example, in the case of ultrasound, the waves can have frequencies on the order of 10^6 Hz with wavelengths of approximately 3×10^{-2} m (in air). Even more extreme are the waves for visible light, where the frequencies are in the neighborhood of 10^{15} Hz and the wavelengths are approximately 5×10^{-7} m.

Based on the preceding discussion, we will construct an asymptotic approximation of the traveling-wave solution of (4.71) in the case of a high frequency. This is done by assuming

$$u(x,t) \sim e^{i[\omega t - \omega^\gamma \theta(x)]} \left[u_0(x) + \frac{1}{\omega^\gamma} u_1(x) + \cdots \right]. \qquad (4.73)$$

Substituting this into (4.71) we obtain

$$-\omega^{2\gamma}\theta_x^2(u_0 + \omega^{-\gamma}u_1 + \cdots) + i\omega^\gamma \theta_x(u_0' + \cdots) + \frac{\mathrm{d}}{\mathrm{d}x}(i\omega^\gamma \theta_x u_0 + \cdots)$$
$$= -\mu^2\omega^2(u_0 + \omega^{-\gamma}u_1 + \cdots) - i\omega\alpha(u_0 + \cdots) + \beta(u_0 + \cdots).$$

In writing down the preceding equation, only those terms that might contribute to the first term in the expansion are written out explicitly. Balancing the first terms on each side of this equation yields $\gamma = 1$. Thus, we have the following problems:

$O(1)$ $(\theta_x)^2 = \mu^2(x)$.

The solutions of this eikonal equation are $\theta = \pm \int_0^x \mu(s)ds$. Because we are investigating waves that are moving to the right, we will only consider the $+$ sign in what follows.

$O(\frac{1}{\omega})$ $\theta'' u_0 + 2\theta' \partial_x u_0 = -\alpha u_0$.

The solution of this transport equation, for $\theta_x = \mu(x)$, is

$$u_0(x) = \frac{a_0}{\sqrt{\mu}} \exp\left(-\frac{1}{2} \int_0^x \frac{\alpha(s)}{\mu(s)} ds\right).$$

Putting the results together and imposing boundary condition (4.72), we have that the wave, which propagates to the right, has the expansion

$$u(x,t) \sim \sqrt{\frac{\mu(0)}{\mu(x)}} \exp\left(-\frac{1}{2} \int_0^x \frac{\alpha(s)}{\mu(s)} ds\right) \cos\left(\omega t - \omega \int_0^x \mu(s)ds\right). \quad (4.74)$$

Our solution is a traveling wave with an amplitude and phase that depend on x. The damping, as might be expected, causes the wave to be attenuated, while the coefficient $\mu(x)$ contributes both to the phase and amplitude. Interestingly, the approximation is independent of the coefficient $\beta(x)$ in (4.71).

4.4.1 Energy Methods

To connect what we have done with the energy methods that are also used to study wave problems, we first need to determine an energy equation associated with (4.71). One way to do this is to multiply (4.71) by the velocity u_t (also see Exercise 4.31). Doing this and collecting terms, one obtains an equation of the form

$$\partial_t E + \partial_x S = -\Phi, \quad (4.75)$$

where $E(x,t) = \frac{1}{2}\mu^2(\partial_t u)^2 + \frac{1}{2}(\partial_x u)^2 + \frac{1}{2}\beta u^2$, $S(x,t) = -\partial_t u \partial_x u$, and $\Phi(x,t) = \alpha(\partial_t u)^2$. As is always the case, the problem we are considering has been nondimensionalized. However, if we were to convert back to dimensional coordinates, we would find that E corresponds to the sum of the kinetic and potential energy densities, S is the energy flux, and Φ is the rate of energy dissipation (or, simply, the dissipation function). We will refer to E, S, and Φ in these physical terms, although it is understood that they are actually nondimensional.

We will be interested in the energy over spatial intervals of the form $[x_1(t), x_2(t)]$, where the endpoints $x_i(t)$ depend on time with $x_1 < x_2$.

Integrating (4.75) over this interval, and then using Leibniz's rule for differentiation under the integral sign, yields

$$\frac{d}{dt} \int_{x_1}^{x_2} E dx = E_2 x_2' - S_2 - E_1 x_1' + S_1 - \int_{x_1}^{x_2} \Phi dx, \qquad (4.76)$$

where $E_i = E(x_i, t)$ and $S_i = S(x_i, t)$ for $i = 1, 2$. The right-hand side is the time rate of change of the total energy in the interval. On the right-hand side, $E_i x_i'$ is the change of energy due to the motion of the endpoint, S_i is the flux of energy across the endpoint due to wave motion, and $-\int \Phi dx$ is the energy loss over the interval due to dissipation.

Because wave problems are of particular importance in most areas of application, numerous methods have been devised to find approximations of the solution. Of interest here are those based on using averages of the energy. The assumption made is that for a high frequency

$$u(x, t) \sim A(x) F[\omega t - \varphi(x)], \qquad (4.77)$$

where the amplitude A is slowly varying in comparison to the change in the phase φ (i.e., the wave is slowly modulated). It should be noted that the phase velocity v_{ph} for this wave is

$$v_{ph} \equiv \frac{\omega}{\varphi_x}. \qquad (4.78)$$

With this, curves $x = \bar{x}(t)$ in the x,t-plane that satisfy $\bar{x}' = v_{ph}$ are called phase lines. To help in understanding the next result, recall that, from the WKB expansion, the wavelength of the fast variation is $O(\omega^{-1})$.

Claim 1. From the eikonal equation for (4.71) it follows that the total energy remains constant (to the first term) between any two phase lines that are $O(\omega^{-1})$ apart.

The total energy $E_{tot}(t)$ between two phase lines is

$$E_{tot} = \int_{\bar{x}_1(t)}^{\bar{x}_2(t)} E(x, t) dx, \qquad (4.79)$$

where $\bar{x}_i' = v_{ph}(\bar{x}_i)$ with $\bar{x}_1 < \bar{x}_2$. With the assumption in (4.77), we have

$$E \sim \frac{1}{2} A^2 (\mu^2 \omega^2 + \varphi_x^2) \sin^2(\omega t - \varphi), \qquad (4.80)$$

$$S \sim \omega \varphi_x A^2 \sin^2(\omega t - \varphi), \qquad (4.81)$$

$$\Phi \sim \alpha \omega^2 A^2 \sin^2(\omega t - \varphi). \qquad (4.82)$$

Now, from (4.79), the energy Eq. (4.75), and the fact that $\bar{x}_2 - \bar{x}_1 = O(\omega^{-1})$ we have that

$$\frac{\mathrm{d}}{\mathrm{d}t} E_{\mathrm{tot}} = E_2 \bar{x}_2' - E_1 \bar{x}_1' + \int_{\bar{x}_1(t)}^{\bar{x}_2(t)} \partial_t E(x,t) \mathrm{d}x$$

$$= E_2 \bar{x}_2' - S_2 - E_1 \bar{x}_1' + S_1 - \int_{\bar{x}_1(t)}^{\bar{x}_2(t)} \Phi \mathrm{d}x,$$

where $E_i = E(\bar{x}_i, t)$ and $S_i = S(\bar{x}_i, t)$ for $i = 1, 2$. Using the expansions in (4.80) and (4.81) yields

$$E\bar{x}' - S \sim \frac{\omega A^2}{2\varphi_x} (\omega^2 \mu^2 - \varphi_x^2) \sin^2(\omega t - \varphi).$$

Recall that the eikonal equation for this problem is $\omega^2 \mu^2 = \varphi_x^2$. From this it follows that to the first term, $\frac{\mathrm{d}}{\mathrm{d}t} E_{\mathrm{tot}} = 0$.

Claim 2. The transport equation corresponds to the balance of energy over a period.

Given a function $g(t)$, the average to be used is

$$g_{\mathrm{avg}} = \frac{\omega}{2\pi} \int_0^{2\pi/\omega} g(t) \mathrm{d}t.$$

Averaging the energy Eq. (4.75) and using (4.80) one obtains

$$\partial_x S_{\mathrm{avg}} = -\Phi_{\mathrm{avg}}. \tag{4.83}$$

Using (4.81) and (4.82) yields

$$\partial_x (\varphi_x A^2) = -\alpha A^2.$$

The solution of this is the same as the solution of the transport equation.

The preceding results are interesting because they give a physical interpretation of the WKB approximation. However, these claims are not necessarily true for other problems, and so one must use caution in trying to use them to construct the WKB approximation. It is much better to rely on the original expansion. Also, the preceding discussion involved the phase velocity, but the group velocity plays an even more fundamental role in energy propagation. The interested reader may consult Whitham (1974) and Lighthill (1965), as well as Exercises 4.32 and 4.33.

Exercises

4.27. For flow of a gas in a long, thin duct, the equation for the velocity potential $\phi(x, t)$ is

$$\partial_x^2\phi + \partial_x(\ln A)\partial_x\phi = \partial_t^2\phi,$$

where $A(x)$ is the cross-sectional area of the duct. This is known as the Webster horn equation (Webster, 1919). Find a first-term approximation of the high-frequency waves. If there are turning points, make sure to identify where they are located.

4.28. In the kinetic theory of waves in plasmas, one comes across having to solve the Fokker–Planck equation,

$$\partial_t u = \partial_x(\alpha(x)u) + \partial_x^2(\beta(x)u) \quad \text{for } 0 < x < \infty,$$

where $u(0,t) = \cos(\omega t)$ and $u \to 0$ as $x \to \infty$. Here $\alpha(x)$ and $\beta(x)$ are positive functions known as the friction and diffusion coefficients, respectively. Find a first-term WKB approximation of the time-periodic solution in the case where ω is large.

4.29. In modeling the transduction of sound by the human ear, one finds the following equation for the pressure (Holmes, 1982):

$$\partial_x[H(x)\partial_x p] = K(x)\partial_t^2 p \quad \text{for } 0 < x < 1,$$

where $p(0,t) = \cos(\omega t)$ and $p(1,t) = 0$. Here $H(x) = A(x) - \delta P(x)/\sqrt{i\omega}$, where $A(x)$, $P(x)$, and $K(x)$ are smooth, positive functions. Also, $\delta > 0$ is constant and ω is the frequency of the sound signal. Find a first-term approximation of the time-periodic solution in the case where ω is large.

4.30. The equation for the vertical displacement $u(x,t)$ of an inhomogeneous Euler–Bernoulli beam is

$$\partial_x^2[D(x)\partial_x^2 u] + \mu(x)\partial_t^2 u = 0,$$

where $D(x)$ is the bending rigidity and $\mu(x)$ the mass density. Both of these functions are smooth and positive.
(a) Find a first-term approximation of the high-frequency traveling waves. If there are turning points, make sure to identify where they are located.
(b) Find E, S, and Φ in the energy Eq. (4.75).
(c) Do claims (1) and (2) hold for this equation?
(d) For a cantilevered beam, the boundary conditions are $u = u_x = 0$ at $x = 0$ and $u_{xx} = \partial_x(Du_{xx}) = 0$ at $x = 1$. Show that a first-term approximation of the high-frequency modes is

$$\omega \sim k_n^2 \left(\int_0^1 (\mu/D)^{1/4} d\tau \right)^{-1},$$

where the k_n are solutions of the equation $\cos(k_n)\cosh(k_n) = -1$.

4.31. Derive the energy Eq. (4.75) by multiplying (4.71) by u_t and then integrating the result over $x_0 \leq x \leq x_1$. Using integration by parts, and the fact that x_0 and x_1 are arbitrary, obtain (4.75).

4.32. The group velocity is defined as $v_g \equiv (\frac{d}{d\omega}\varphi_x)^{-1}$. Is Claim 1 true if "phase lines" is replaced with "group lines"?

4.33. The Lagrangian L is the difference between the kinetic and potential energy, and so, from (4.71), $L = \frac{1}{2}\mu^2(u_t)^2 - \frac{1}{2}(u_x)^2 - \frac{1}{2}\beta u^2$. The averaged Lagrangian L is obtained from this by assuming (4.77) and then averaging L over a period. This exercise introduces some of the ideas underlying the method of averaged Lagrangians developed in Whitham (1974).
(a) Show that $L \sim \frac{1}{4}A^2(\mu^2\omega^2 - \varphi_x^2)$.
(b) In the case where $\alpha = 0$, show that the eikonal and transport equations correspond to (i) $\partial_A L = 0$ and (ii) $\partial_t(\partial_\omega L) - \partial_x(\partial_{\varphi_x} L) = 0$.

4.34. The wave equation in three spatial dimensions is $\nabla^2 u = \mu^2 \partial_t^2 u$, where $\mu = \mu(\mathbf{x})$ is smooth and positive.
(a) Show that the energy equation has the form $\partial_t E + \nabla \cdot \mathbf{S} = 0$.
(b) Assume $u(\mathbf{x}, t) \sim A(\mathbf{x})\cos[\omega t - \varphi(\mathbf{x})]$, and introduce the average

$$\langle f \rangle = \frac{1}{2T}\int_{-T}^{T} f(\mathbf{x}, t)dt.$$

In this case show that, for $T \gg 2\pi/\omega$,

$$\langle E \rangle \sim \frac{1}{4}\left[\omega^2\mu^2 A^2 + \nabla(Ae^{-i\varphi}) \cdot \nabla(Ae^{i\varphi})\right]$$

and

$$\langle \mathbf{S} \rangle \sim -\frac{i\omega}{4}\left[Ae^{-i\varphi}\nabla(Ae^{i\varphi}) - Ae^{i\varphi}\nabla(Ae^{-i\varphi})\right].$$

4.5 Wave Propagation and Slender-Body Approximations

Another important situation where the WKB approximation can be used to obtain an asymptotic approximation of traveling-wave solutions arises when the geometry is long and thin. This sort of situation is referred to as slender-body theory. To understand this, consider the following problem:

$$\varepsilon^2 u_{xx} + u_{yy} = \mu^2(x)u_{tt} \quad \text{for} \quad \begin{cases} 0 < x < \infty, \\ -G(x) < y < G(x), \end{cases} \tag{4.84}$$

where

$$u(x, y, t) = 0 \quad \text{for } y = \pm G(x) \tag{4.85}$$

and

$$u(0, y, t) = f(y) \cos(\omega t). \tag{4.86}$$

This problem corresponds to finding the vertical displacement $u(x, y, t)$ of an elastic membrane that is much longer than it is wide. The membrane is fixed along its lateral sides ($y = \pm G$) and is forced periodically at the left end ($x = 0$). In effect, what this forcing is doing is generating waves at $x = 0$ that propagate down the membrane. We are interested in finding these waves and determining their basic properties. For example, once a wave is sent down the membrane, do we ever see it again? You might expect that because the membrane is infinitely long it just keeps traveling out away from the end at $x = 0$. However, as will be seen below, it is actually possible for the wave to travel a certain distance along the membrane and then turn around and come back to $x = 0$. It is not obvious why this should happen without help from the analysis to follow. In this example, the forcing frequency ω is not restricted, other than being positive.

To find the traveling-wave solutions for small ε, we use the WKB approximation

$$u(x, y, t) \sim e^{i[\omega t - \theta(x)/\varepsilon^\alpha]}[u_0(x, y) + \varepsilon^\alpha u_1(x, y) + \cdots]. \tag{4.87}$$

Substituting this into (4.84) yields

$$\varepsilon^{2-2\alpha}[-\theta_x^2 u_0 - i\varepsilon^\alpha(\theta_{xx}u_0 + 2\theta_x\partial_x u_0 + \theta_x^2 u_1) + \cdots] + \partial_y^2 u_0 + \varepsilon^\alpha \partial_y^2 u_1 + \cdots$$
$$= -\omega^2\mu^2(u_0 + \varepsilon^\alpha u_1 + \cdots). \tag{4.88}$$

Balancing the terms in this equation, it follows that $\alpha = 1$. This leads to the following problems:

$O(1)$ $\quad \partial_y^2 u_0 + (\omega^2\mu^2 - \theta_x^2)u_0 = 0,$
$\qquad u_0 = 0$ for $y = \pm G(x).$

This is the eikonal problem. It is actually an eigenvalue problem where θ_x^2 is the eigenvalue and u_0 is an eigenfunction. Setting $\lambda = \sqrt{\omega^2\mu^2 - \theta_x^2}$, the solution can be written as

$$u_0(x, y) = A(x) \sin[\lambda(y + G)], \tag{4.89}$$

where to satisfy the boundary condition (4.85) we must have $2\lambda G = n\pi$. Thereforo,

$$\theta_x = \pm\sqrt{\omega^2\mu^2 - \lambda_n^2} \quad \text{for } n = 1, 2, 3, \ldots, \tag{4.90}$$

where $\lambda_n(x) = n\pi/(2G)$. This means that the function θ_x depends on the mode number (n) as well as on the position (x). Also, as long as θ_x is real and nonzero, the wave propagates along the x-axis. However, the higher modes (where n is large enough that $\lambda_n > \omega\mu$) do not propagate since the wavenumber is imaginary.

$O(1)$ $\quad \partial_y^2 u_1 + \lambda_n^2 u_1 = i(\theta_{xx} u_0 + 2\theta_x \partial_x u_0),$
$\qquad u_1 = 0$ for $y = \pm G(x).$

This is the transport problem, and we will use it to complete the determination of u_0. Because it is an inhomogeneous version of the eikonal problem, we will have to impose a condition on the right-hand side of the transport equation to guarantee a solution. This situation is covered by the Fredholm alternative theorem. To obtain the condition, we multiply the equation by u_0 and then integrate with respect to y. Doing this one obtains

$$\int_{-G}^{G} u_0(\partial_y^2 u_1 + \lambda_n^2 u_1)dy = i \int_{-G}^{G} u_0(\theta_{xx} u_0 + 2\theta_x \partial_x u_0)dy.$$

Integrating by parts twice with the left integral and using the eikonal problem shows that the left-hand side is zero. Thus,

$$0 = \int_{-G}^{G} \partial_x(\theta_x u_0^2)dy. \tag{4.91}$$

This equation for u_0 is known as a *solvability condition*, and it must be satisfied for the transport problem to have a solution. To solve it, we use Leibniz's rule to rewrite the equation as

$$0 = \frac{d}{dx} \int_{-G}^{G} (\theta_x u_0^2)dy - \theta_x u_0^2(x, G)G' - \theta_x u_0^2(x, -G)G'.$$

Since $u_0 = 0$ for $y = \pm G$, we have that

$$\theta_x \int_{-G}^{G} u_0^2 dy = a,$$

where a is a constant. Thus, from (4.89),

$$A = \frac{a}{\sqrt{\theta_x G(x)}}.$$

Putting the results together, we have found that the WKB approximation for the modes has the form

$$u(x, y, t) \sim \frac{a}{\sqrt{\theta_x G(x)}} e^{i[\omega t \pm \theta(x)/\varepsilon]} \sin[\lambda_n(y + G)], \tag{4.92}$$

where

$$\theta(x) = \int^x \sqrt{\omega^2\mu^2 - \lambda_n^2} \, dx \qquad (4.93)$$

and

$$\lambda_n = \frac{n\pi}{2G(x)} \quad \text{for } n = 1, 2, 3, \ldots . \qquad (4.94)$$

The solution of the original problem is obtained by adding the modes together and then imposing the boundary condition at $x = 0$. However, before getting to that step, note that the preceding approximation breaks down at points where $\theta_x = 0$. These are the turning points for this problem, and they need to be investigated to complete the solution.

Before starting the analysis for the transition-layer region it is worth making a few observations. First, the eikonal problem differs significantly from that obtained in Sect. 4.4. However, even though it is a linear boundary-value problem, the equation that determines θ_x is nonlinear. In other words, it is still true that the equation that determines $\theta(x)$ is a nonlinear, first-order differential equation. Second, even though the next order problem differs from what we saw earlier, it is still true that the transport equation, which we referred to as a solvability condition for this problem, is a first-order linear differential equation.

4.5.1 Solution in Transition Region

The turning points in the WKB approximation (4.92) occur when $\theta_x = 0$. Using (4.90), these points satisfy

$$\mu(x_t) = \frac{n\pi}{2\omega G(x_t)} . \qquad (4.95)$$

We will consider the case when there is a single turning point $0 < x_t < \infty$. To guarantee that there is only one, it is assumed that $\frac{d}{dx}(\theta_x^2) < 0$ (i.e., θ_x^2 is a decreasing function of x, so if it does go through zero it does not do so again).

The first step in the procedure to find the solution in the transition region is to introduce the transition-layer coordinate

$$\bar{x} = \frac{x - x_t}{\varepsilon^\beta} . \qquad (4.96)$$

Letting $U(\bar{x}, y, t)$ denote the solution in this region, we assume that

$$U(\bar{x}, y, t) \sim \varepsilon^\gamma e^{i\omega t} \{U_0(\bar{x}) \sin[\lambda_n(y + G_t)] + \cdots \}, \qquad (4.97)$$

where $G_t = G(x_t)$. Substituting (4.96) and (4.97) into the differential Eq. (4.84) produces the following result:

$$\varepsilon^{2-2\beta}U_0'' + \theta_x^2(x_t + \varepsilon^\beta \bar{x})U_0 + \cdots = 0, \tag{4.98}$$

where

$$\theta_x^2(x_t + \varepsilon^\beta \bar{x}) \sim \theta_x^2(x_t) + \varepsilon^\beta \bar{x}\frac{d}{dx}\theta_x^2(x_t) + \cdots$$

$$\sim \varepsilon^\beta \bar{x}\frac{d}{dx}\theta_x^2(x_t) + \cdots .$$

For the terms to balance in (4.98), we need $2 - 2\beta = \beta$, and so $\beta = 2/3$. The equation to be solved is thus

$$U_0'' = \kappa \bar{x} U_0 \quad \text{for } -\infty < \bar{x} < \infty, \tag{4.99}$$

where

$$\kappa = -\frac{d}{dx}\theta_x^2(x_t) = -2\omega^2\mu_t^2\left(\frac{\mu_t'}{\mu_t} + \frac{G_t'}{G_t}\right) > 0. \tag{4.100}$$

This is an Airy equation, and the general solution is (Appendix B)

$$U_0 = a\,\mathrm{Ai}(\kappa^{1/3}\bar{x}) + b\,\mathrm{Bi}(\kappa^{1/3}\bar{x}). \tag{4.101}$$

4.5.2 Matching

What remains is the matching. We can use the WKB approximation of the nth mode on either side of the turning point. Thus,

$$u(x,y,t) \sim \begin{cases} u_\mathrm{L}(x,y,t) & \text{if } 0 \le x < x_t, \\ u_\mathrm{R}(x,y,t) & \text{if } x_t < x < \infty, \end{cases} \tag{4.102}$$

where

$$u_\mathrm{L} = \frac{1}{\sqrt{\theta_x G(x)}}\left[a_\mathrm{L}e^{i[\omega t - \theta(x)/\varepsilon]} + b_\mathrm{L}e^{i[\omega t + \theta(x)/\varepsilon]}\right]\sin[\lambda_n(y + G)] \tag{4.103}$$

and

$$u_\mathrm{R} = \frac{1}{\sqrt{|\theta_x|G(x)}}a_\mathrm{R}e^{i\omega t - \varphi/\varepsilon}\sin[\lambda_n(y + G)]. \tag{4.104}$$

Also,

$$\theta = \int_x^{x_t}\sqrt{\omega^2\mu^2 - \lambda_n^2}\,ds \quad \text{and} \quad \varphi = \int_{x_t}^x\sqrt{\lambda_n^2 - \omega^2\mu^2}\,ds. \tag{4.105}$$

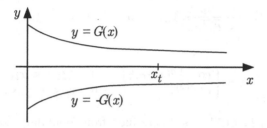

Figure 4.10 Tapered membrane that leads to the WKB approximation given in (4.110). It is assumed that for the nth mode there is a turning point at $x = x_t$

Before discussing the matching, note that for the transition-layer solution to be bounded we must have $b = 0$ in (4.101). It is for this same reason that the b_R term is not included in (4.104). Now, the matching is very similar to what was done in Sect. 4.3 to derive (4.44) and (4.45). One finds that in (4.97), $\gamma = -1/6$. Using the asymptotic properties of the Airy functions given in Appendix B, the connection formulas are found to be

$$a = \frac{2\sqrt{\pi}}{\kappa^{1/6}G_t^{1/2}}\,e^{-i\pi/4}a_L, \tag{4.106}$$

$$b_L = -ia_L \tag{4.107}$$

and

$$a_R = \frac{\kappa^{1/6}G_t^{1/2}}{2\sqrt{\pi}}\,a = e^{-i\pi/4}a_L. \tag{4.108}$$

With this, (4.103) takes the form

$$u_L = \frac{2a_L}{\sqrt{\theta_x G(x)}}\cos\left[\frac{1}{\varepsilon}\theta(x) + \frac{\pi}{4}\right]\sin[\lambda_n(y+G)]e^{i(\omega t - \pi/4)}. \tag{4.109}$$

Example

To see more clearly what we have derived, suppose the membrane is tapered and has a uniform density (Fig. 4.10). In particular, assume that $G = e^{-x}$, μ is constant, and in (4.86), $f(y) = \sin[\lambda_n(y+G)]$. Solving (4.95) we have that

$$x_t = \ln\left(\frac{2\omega\mu}{n\pi}\right).$$

It is assumed that the frequency is high enough that $0 < x_t < \infty$. With this, the WKB approximation of the solution is

$$u(x,y,t) \sim \frac{\alpha_n}{\sqrt{\theta_x G(x)}} V_n(x) \sin[\lambda_n(y+G)]\cos(\omega t), \qquad (4.110)$$

where

$$V_n(x) = \begin{cases} \cos\left[\frac{1}{\varepsilon}\theta(x) + \frac{\pi}{4}\right], & \text{if } 0 \le x < x_t, \\ e^{-\varphi(x)/\varepsilon} & \text{if } x_t < x < \infty. \end{cases} \qquad (4.111)$$

The constant α_n in (4.110) is determined from boundary condition (4.86), and the functions θ and φ are given in (4.105). ∎

We are now in a position to summarize the situation. From (4.103) it is seen that in the region $0 \le x < x_t$ the solution for the mode consists of the superposition of left and right traveling waves. As the right-running wave moves toward $x = x_t$, its amplitude grows since both G and θ_x are decreasing. Similarly, the wave that is reflected from the transition region decreases in amplitude as it travels away from the turning point. Together they result in a standing wave as described in (4.110) and (4.111). The portion of the disturbance that gets through the transition region ends up decaying exponentially, as shown in (4.111). This is known as the *evanescent region*.

There are several interesting observations that can be made using our solution. First, if the frequency is decreased, then the position of the turning point, which is a solution of the equation

$$\omega\mu = \frac{n\pi}{2G(x_t)},$$

moves to the left. This brings up the interesting possibility of being able to lower the frequency so the turning point moves to the boundary. To investigate this situation, suppose that there is a frequency $\omega = \omega_c$ for which the turning point is at $x_t = 0$. In this case, ω_c corresponds to a frequency below which waves do not propagate on the membrane without decaying exponentially. This is called a *cutoff frequency*. In this example, it is referred to as a low-frequency cutoff because frequencies below this value do not propagate. The second observation to be made here is that as the mode number (n) increases, the position of x_t moves to the left. Once n is large enough that this point moves off the membrane (i.e., $x_t \le 0$), the wave for that mode will not propagate. In other words, the higher modes do not result in propagating waves, and these are known as evanescent modes.

The conclusions stated here clearly depend on the problem that was solved. For example, if the membrane widens instead of tapers, then a completely different response is obtained. Also, in the preceding example, for each mode there is a possibility of resonance. This occurs when $\cos[\theta(0)/\varepsilon + \pi/4] = 0$, but the analysis of this situation will not be considered.

Exercises

4.35. The equation for a slender membrane with small damping is

$$\varepsilon^2 u_{xx} + u_{yy} = \mu^2(x)u_{tt} + \varepsilon\alpha(x)u_t \quad \text{for} \quad \begin{cases} 0 < x < \infty \\ 0 < y < 1, \end{cases}$$

where $u(x, 0, t) = u(x, 1, t) = 0$, $u(0, y, t) = f(y)\cos(\omega t)$ and $u(x, y, t)$ is bounded as $x \to \infty$. Assume $\mu(x)$ and $\alpha(x)$ are positive and $\mu' < 0$.
(a) Use the WKB method to find an approximation of the long-time solution outside of the transition region. You only need to find the general solution for each mode.
(b) Assuming the turning point $x_t > 0$, find a first-term approximation in the transition layer and match it to the solutions from part (a). Assume it is a simple turning point.

4.36. How does the WKB approximation of the mode given in (4.103) change if μ also depends on y?

4.37.(a) Use the results of this section to design a method to determine the width $2G(x)$ of the membrane using information from the point $x = 0$. Assume here that μ is constant.
(b) Does your method work to also determine μ if it depends on x?

4.38. Consider the equation for an elastic string

$$\partial_y^2 u = \mu^2 \partial_t^2 u \quad \text{for} \quad -G < y < G,$$

where $u = 0$ for $y = \pm G(x)$. This problem is obtained when setting $\varepsilon = 0$ in (4.84). Find the natural frequencies of the string and connect them to the cutoff frequency described previously.

4.39. The equation for the velocity potential $\phi(r, \theta, z, t)$ of a gas in a long narrow tube is

$$(\partial_r^2 + \frac{1}{r}\partial_r + \frac{1}{r^2}\partial_\theta^2 + \varepsilon^2\partial_z^2)\phi = \partial_t^2\phi \quad \text{for} \quad \begin{cases} 0 \le r \le \bar{r} \\ 0 \le \theta \le 2\pi. \end{cases}$$

Suppose the tube is axisymmetric and the cross-sectional radius $r = \bar{r}(z)$ of the tube depends on the longitudinal coordinate z. In this case, the boundary condition along the lateral wall of the tube is

$$(\partial_R - \varepsilon^2 \bar{r}_z \partial_z)\phi = 0 \quad \text{for } r = \bar{r}(z).$$

The parameter ε is the ratio of the characteristic radius of the tube to the tube's length, and it is assumed that ε is small.

(a) Use the WKB method to show that the waves propagating down the tube are linear combinations of the modes (Rabbitt and Holmes, 1988)

$$G_{nm}(r, \theta, z, t) = J_m[\gamma_{nm}(z)r] e^{i[\omega t - \varphi(z)/\varepsilon]} \cos(m\theta)$$

and

$$H_{nm}(r, \theta, z, t) = J_m[\gamma_{nm}(z)r] e^{i[\omega t - \varphi(z)/\varepsilon]} \sin(m\theta),$$

where γ_{nm} is determined from the boundary condition in the eikonal problem. (Hint: Leibniz's rule states

$$\frac{\mathrm{d}}{\mathrm{d}z} \int_{\Omega(z)} f(r, \theta, z) \mathrm{d}A_{r,\theta} = \int_{\partial\Omega(z)} \bar{r}_z f \mathrm{d}s + \int_{\Omega(z)} \frac{\partial f}{\partial z} \mathrm{d}A_{r,\theta}.\Big)$$

(b) Assuming $\bar{r}_z \neq 0$, find the turning points of the preceding modes and the transition-layer equation. Under what conditions will a mode be a propagating wave to the left of the turning point and evanescent to the right? Is this a situation of a low- or high-frequency cutoff?

4.6 Ray Methods

The extension of the WKB method to multidimensional problems is straightforward, although the equations are somewhat harder to solve. To understand what is involved, consider the wave equation

$$\nabla^2 u = \mu^2(\mathbf{x}) \partial_t^2 u, \tag{4.112}$$

where ∇^2 is the Laplacian and $\mathbf{x} \in \mathbb{R}^n$. We are interested in the time-periodic response, and so let

$$u(\mathbf{x}, t) = e^{-i\omega t} v(\mathbf{x}). \tag{4.113}$$

With this, (4.112) yields what is known as the reduced wave, or Helmholtz, equation, given as

$$\nabla^2 v + \omega^2 \mu^2(\mathbf{x}) v = 0. \tag{4.114}$$

It is possible to derive the WKB approximation without having the slightest idea of what the solution looks like. However, it is more instructive to have some understanding of what properties the solution has and how the WKB approximation takes advantage of them. It is for this reason that we first consider an example where the solution is known.

Example: Constant μ

Suppose, for $\mathbf{x} \in \mathbb{R}^2$ and μ constant, the problem is to solve (4.114) in the region exterior to the circle $\|\mathbf{x}\| = a$. Given the geometry, it is convenient to switch to polar coordinates $x = \rho \cos \varphi$, $y = \rho \sin \varphi$. The boundary conditions to be used are

$$v = f(\varphi) \quad \text{for} \quad \rho = a \qquad (4.115)$$

and

$$\sqrt{\rho}(\partial_\rho v - i\omega\mu v) = 0 \quad \text{for} \quad \rho \to \infty. \qquad (4.116)$$

This last condition needs some explanation. The assumption is that nothing is coming in from infinity. In other words, the solution is determined from the information supplied at $\rho = a$, and this produces waves propagating outward from the circle. The condition in (4.116) guarantees that this happens and is known as the Sommerfeld radiation condition. Using separation of variables to solve (4.114) it is found that the solution is

$$v(\rho, \varphi) = \sum_{n=-\infty}^{\infty} \alpha_n H_n^{(1)}(\omega\mu\rho)e^{-in\varphi}/H_n^{(1)}(\omega\mu a), \qquad (4.117)$$

where $H_n^{(1)}$ is the Hankel function of the first kind and the α_n are determined from (4.115) and are independent of ω. It is known that for large values of z (Abramowitz and Stegun, 1972)

$$H_n^{(1)}(z) \sim \sqrt{\frac{2}{\pi z}} \, e^{i(z-n\pi/2-\pi/4)}.$$

From this it follows that for large ω (4.117) reduces to

$$v(\rho, \varphi) \sim f(\varphi)\sqrt{\frac{a}{\rho}} \, e^{i\omega\mu(\rho-a)}. \qquad (4.118)$$

This is the WKB approximation for this example, and it is certainly more tractable than the exact solution given in (4.117). Our goal is to be able to derive it without having to first derive (4.117). Using the terminology that is introduced in the derivation of the WKB approximation, radial lines in this example are called *rays*. What we see from (4.118) is that along a ray, so φ is fixed, the solution consists of a highly oscillatory component that is multiplied by an amplitude $v_0 = f(\varphi)\sqrt{a/\rho}$ that decays as ρ increases. As will be shown subsequently, the amplitude is determined using the transport equation, while the phase $\theta = \mu(\rho - a)$ is determined from the eikonal equation. ∎

To complete the formulation of the problem, the domain and boundary conditions that will be used when solving (4.114) need to be specified.

To keep things simple, we will use a generalization of the previous example. Specifically, (4.114) is to be solved in the region exterior to a smooth surface S, where S encloses a bounded convex domain. This means that given a point $\mathbf{x}_0 \in S$, there is a well-defined unit outward normal $\mathbf{n}_0$. The boundary condition used is

$$v(\mathbf{x}_0) = f(\mathbf{x}_0) \quad \text{for} \ \ \mathbf{x}_0 \in S. \tag{4.119}$$

We will also concentrate on the outward propagating waves.

4.6.1 WKB Expansion

In a similar manner as was done in Sect. 4.4, for a high frequency, the WKB expansion for the solution of (4.114) is

$$v(\mathbf{x}) \sim e^{i\omega\theta(\mathbf{x})} \left[v_0(\mathbf{x}) + \frac{1}{\omega}v_1(\mathbf{x}) + \cdots \right]. \tag{4.120}$$

The expansions for the derivatives of this function can be obtained directly from the one-dimensional formulas in (4.4) and (4.5). The result is that (4.114) reduces to

$$(\nabla\theta \cdot \nabla\theta - \mu^2)v_0 + \frac{1}{\omega} \left[(\nabla\theta \cdot \nabla\theta - \mu^2)v_1 - i(\nabla^2\theta)v_0 - 2i\nabla\theta \cdot \nabla v_0 \right]$$

$$+ O\left(\frac{1}{\omega^2}\right) = 0. \tag{4.121}$$

The eikonal equation, which comes from the $O(1)$ term in (4.121), is

$$\nabla\theta \cdot \nabla\theta = \mu^2. \tag{4.122}$$

Similarly, the transport equation comes from the $O(1/\omega)$ term, and it is

$$2\nabla\theta \cdot \nabla v_0 + (\nabla^2\theta)v_0 = 0. \tag{4.123}$$

Note that, as usual, the nonlinearity of the eikonal equation means that the solution is not unique. Specifically, if θ is a solution of (4.122), then $-\theta$ is also a solution. In what follows, we will concentrate on outward propagating waves, so we will only consider the positive solutions of (4.122).

Everything is working out nicely, but we are about to run into a complication, namely, the eikonal equation is a nonlinear partial differential equation for θ, and this requires some effort to solve. The usual approach is to introduce characteristic coordinates. This means using curves that are orthogonal to the level surfaces of $\theta(\mathbf{x})$ (Fig. 4.9). Because of this, it is best to briefly review some of the salient features of the mathematical description of surfaces.

4.6.2 Surfaces and Wave Fronts

The level surfaces determined by $\theta(\mathbf{x})$ are called *wave fronts* or *phase fronts*. To explain why these play an important role in the solution, note that our WKB approximation of the solution of the wave Eq. (4.112) has the form

$$u(\mathbf{x}, t) \sim e^{i(\omega\theta - \omega t)} v_0(\mathbf{x}).$$

With this, we introduce the phase function $\Theta(\mathbf{x}, t) = \omega\theta(\mathbf{x}) - \omega t$. Suppose we start (at $t = 0$) with the surface S_c over which Θ is constant, namely, $\Theta(\mathbf{x}, 0) = \omega c$. As time increases, the points where $\Theta = \omega c$ change, and therefore the points forming S_c, move and form a new surface S_{c+t}. Because this new surface is where $\Theta(\mathbf{x}, t) = \omega c$, it is determined by the equation $\theta(\mathbf{x}) = c + t$ (Fig. 4.11). The path each point takes to get from S_c to S_{c+t} is obtained from the eikonal equation. One such path, a straight line, is shown in Fig. 4.11. In the WKB method these paths are called *rays*.

The evolution of the phase front gives us a natural coordinate system for this problem. This consists of the two coordinates α and β, which come from the parameterization of the phase front, and a coordinate s associated with the parameterization of the rays. Determining these coordinates turns out to be one of the critical steps in the derivation of the WKB approximation. The following example illustrates some of the issues that will arise in the derivation.

Example

Suppose that $\theta(\mathbf{x}) \equiv \mathbf{x} \cdot \mathbf{x}$. In this case, the surface S_{c+t} is the sphere $\mathbf{x} \cdot \mathbf{x} = c + t$. It is assumed that the rays are radial lines, so the points forming

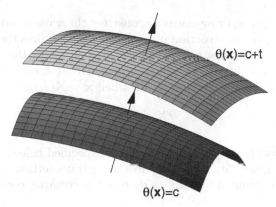

Figure 4.11 Schematic drawing illustrating the wave fronts (defined as the set of points where $\theta = c + t$) and the path followed by one of the points in the wave front

the original surface S_c move along radial lines to produce S_{c+t}. Because of the geometry and motion of these surfaces, spherical coordinates are a natural choice. However, it is more informative to consider a variation of spherical coordinates. In particular, we switch from (x, y, z) to (s, α, β) using the formula

$$(x, y, z) = \rho(s) \left(\sin \alpha \cos \beta, \sin \alpha \sin \beta, \cos \alpha \right),$$

where $0 \leq \alpha \leq \pi$, $0 \leq \beta < 2\pi$, and $0 \leq s$. The function $\rho(s)$ is required to be smooth and strictly increasing, and examples are $\rho = s$, $\rho = \ln(1 + s)$, and $\rho = e^s - 1$. This freedom in choosing the parameterization for the rays will be taken advantage of in the WKB approximation and is the reason for the arbitrary function λ in (4.124). A particularly important property of the preceding change of variables is that (s, α, β) forms an orthogonal coordinate system. Specifically, writing the change of variables as $\mathbf{x} = \mathbf{X}(s, \alpha, \beta)$, the preceding formula produces a vector $\partial_s \mathbf{X}$ tangent to the ray that is normal to the phase front S_{c+t}. ∎

In the preceding example we were given $\theta(\mathbf{x})$ and from this determined the appropriate change of variables $\mathbf{x} = \mathbf{X}(s, \alpha, \beta)$. The reverse occurs in the derivation of the WKB approximation, where conditions on $\mathbf{X}(s, \alpha, \beta)$ are specified so it is easy to find $\theta(\mathbf{x})$. In the construction process it is assumed that (s, α, β) forms an orthogonal coordinate system. This requires a result from multivariable calculus, which states that the gradient vector $\nabla \theta$ is normal to the level surface $\theta(\mathbf{x}) = c + t$. In what follows, whenever using this result, it is assumed that the gradient is nonzero. As will be seen subsequently, this assumption is equivalent to assuming the problem does not have a turning point.

4.6.3 Solution of Eikonal Equation

By following the paths, or rays, we can construct the wave at any point in time (this is the Huygens construction procedure). This requires the introduction of a change of coordinates $\mathbf{x} = \mathbf{X}(s, \alpha, \beta)$ that fits with the geometry and motion of the phase fronts. The requirement is that a ray's tangent vector $\partial_s \mathbf{X}$ points in the same direction as $\nabla \theta$ when $\mathbf{x} = \mathbf{X}(s, \alpha, \beta)$. Therefore, we will require

$$\frac{\partial \mathbf{X}}{\partial s} = \lambda \nabla \theta, \tag{4.124}$$

where λ is a smooth, positive function that is specified below. It is assumed, without loss of generality, that the rays are parameterized so that $0 \leq s$. It should also be pointed out that s does not necessarily correspond to arc length.

On a ray, $\partial_s\theta(\mathbf{X}) = \nabla\theta \cdot \partial_s\mathbf{X}$. With this, we can rewrite the eikonal Eq. (4.122) as

$$\frac{\partial\theta}{\partial s} = \lambda\mu^2. \tag{4.125}$$

This last equation can be integrated to yield

$$\theta = \theta(0, \alpha, \beta) + \int_0^s \lambda\mu^2 d\sigma. \tag{4.126}$$

The hitch here is that we have not yet completely determined the coordinates that lead to (4.125). To do so requires the solution of (4.124). This system of n equations is generally nonlinear and solved more often than not using numerical methods. However, we still have the freedom of choosing λ, and this will be discussed below.

4.6.4 Solution of Transport Equation

We have not completed the analysis for determining the first term of the WKB approximation. It remains to solve the transport Eq. (4.123). Since $\partial_s v_0 = \nabla v_0 \cdot \partial_s\mathbf{X}$, then using (4.124) we can rewrite (4.123) as

$$2\frac{\partial}{\partial s}v_0 + \lambda(\nabla^2\theta)v_0 = 0. \tag{4.127}$$

To express the second term of this equation in terms of the ray coordinates, we use the equation (Exercise 4.43)

$$\frac{\partial}{\partial s}\left(\frac{1}{\lambda}J\right) = J\nabla^2\theta,$$

where J is the Jacobian of the transformation $\mathbf{x} = \mathbf{X}(s, \alpha, \beta)$. With this, (4.127) can be written as

$$\frac{\partial}{\partial s}\left(\frac{1}{\lambda}Jv_0^2\right) = 0.$$

Therefore,

$$v_0(\mathbf{x}) = f(\mathbf{x}_0)\sqrt{\frac{\lambda(\mathbf{x})J(\mathbf{x}_0)}{\lambda(\mathbf{x}_0)J(\mathbf{x})}}. \tag{4.128}$$

In deriving the preceding formula, to satisfy (4.119), it is required that $\theta(0, \alpha, \beta) = 0$ in (4.126).

4.6.5 Ray Equations

The entire problem reduces to solving (4.124). To rewrite this equation so that it is independent of θ, let $\mathbf{X} = (X_1, X_2, X_3)$. Dividing (4.124) by λ, and then differentiating the result, yields

$$
\frac{\partial}{\partial s}\left(\frac{1}{\lambda}\frac{\partial}{\partial s}X_i\right) = \frac{\partial}{\partial s}\frac{\partial \theta}{\partial x_i}
$$

$$
= \left(\frac{\partial}{\partial x_1}\frac{\partial \theta}{\partial x_i}\right)\frac{\partial X_1}{\partial s} + \left(\frac{\partial}{\partial x_2}\frac{\partial \theta}{\partial x_i}\right)\frac{\partial X_2}{\partial s} + \left(\frac{\partial}{\partial x_3}\frac{\partial \theta}{\partial x_i}\right)\frac{\partial X_3}{\partial s}
$$

$$
= \left(\frac{\partial}{\partial x_i}\nabla\theta\right)\cdot(\lambda\nabla\theta) = \frac{1}{2}\lambda\frac{\partial}{\partial x_i}(\nabla\theta\cdot\nabla\theta)
$$

$$
= \frac{1}{2}\lambda\frac{\partial}{\partial x_i}\mu^2.
$$

Expressing this in vector form, we have that

$$
\frac{\partial}{\partial s}\left(\frac{1}{\lambda}\frac{\partial}{\partial s}\mathbf{X}\right) = \lambda\mu\nabla\mu. \tag{4.129}
$$

These are the ray equations, and together they form a second-order system of equations that is nonlinear except for special choices for μ.

To complete the formulation of the ray equation problem, we must specify, or determine, what happens at $s = 0$. Each ray starts on the boundary surface S. Therefore, given any point $\mathbf{x}_0 \in S$, its ray satisfies

$$
\mathbf{X}|_{s=0} = \mathbf{x}_0. \tag{4.130}
$$

The ray Eq. (4.129) is second order, and so we need a second boundary condition. The one used is (Exercise 4.44)

$$
\frac{\partial \mathbf{X}}{\partial s}\bigg|_{s=0} = \lambda_0\mu_0\,\mathbf{n}_0, \tag{4.131}
$$

where $\mathbf{n}_0$ is the unit outward normal at $\mathbf{x}_0$, and λ_0 and μ_0 are the values of the respective quantities at $\mathbf{x}_0$.

Another useful equation coming directly from (4.124) is

$$
\frac{\partial \mathbf{X}}{\partial s}\cdot\frac{\partial \mathbf{X}}{\partial s} = \lambda^2\mu^2. \tag{4.132}
$$

For example, if ℓ is arc length measured along the ray, then

$$\ell = \int_0^s \|\mathbf{X}_s\| \, ds$$

$$= \int_0^s \lambda\mu \, ds. \tag{4.133}$$

To go any further, we need to select λ, and the two most common choices are $\lambda = 1/\mu$ and $\lambda = 1$. The former is used in the remainder of this section, and the latter is considered in Exercise 4.41.

4.6.6 Summary

Letting $\lambda = 1/\mu$, (4.133) shows that the parameter s corresponds to arc length along the path. Also, from (4.129), the ray equations reduce to solving

$$\frac{\partial}{\partial s}\left(\mu\frac{\partial}{\partial s}\mathbf{X}\right) = \nabla\mu(\mathbf{X}), \tag{4.134}$$

where $\mathbf{X}|_{s=0} = \mathbf{x}_0$ and $\partial_s\mathbf{X}|_{s=0} = \mathbf{n}_0$. Once this is solved, the phase function (4.126) reduces to

$$\theta = \int_0^s \mu(\mathbf{X}) \, d\sigma, \tag{4.135}$$

and the amplitude is

$$v_0(\mathbf{x}) = f(\mathbf{x}_0)\sqrt{\frac{\mu(\mathbf{x}_0)J(\mathbf{x}_0)}{\mu(\mathbf{x})J(\mathbf{x})}}, \tag{4.136}$$

where $J(\mathbf{x})$ is the Jacobian of the transformation $\mathbf{x} = \mathbf{X}(s, \alpha, \beta)$. The resulting WKB approximation for the outward propagating wave is therefore

$$u(\mathbf{x}, t) \sim f(\mathbf{x}_0)\sqrt{\frac{\mu(\mathbf{x}_0)J(\mathbf{x}_0)}{\mu(\mathbf{x})J(\mathbf{x})}} \, \exp\left[i\omega\left(-t + \int_0^s \mu(\mathbf{X}(\sigma)) \, d\sigma\right)\right], \tag{4.137}$$

where s is the value for which the solution of (4.134) satisfies $\mathbf{X}(s) = \mathbf{x}$.

Example: Constant μ

When μ is constant, the solution of (4.134) that satisfies the stated initial conditions is $\mathbf{X} = \mathbf{x}_0 + s\mathbf{n}_0$, and (4.135) becomes $\theta = \mu s$. Given a point $\mathbf{x}$ on the ray, $s = \mathbf{n}_0 \cdot (\mathbf{x} - \mathbf{x}_0)$. Therefore, (4.137) reduces to

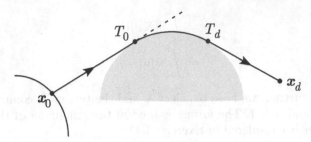

Figure 4.12 Schematic illustrating the wave along the surface of an obstacle and its subsequent path to a point in the shadow

$$u(\mathbf{x},t) \sim f(\mathbf{x}_0)\sqrt{\frac{J(\mathbf{x}_0)}{J(\mathbf{x})}}\ \exp[\mathrm{i}\,(\mathbf{k}\cdot(\mathbf{x}-\mathbf{x}_0)-\omega t)], \qquad (4.138)$$

where

$$\mathbf{k} = \mu\omega\mathbf{n}_0 \qquad (4.139)$$

is the wave vector for the ray. In $\mathbb{R}^2$, when the boundary surface is the circle $\|\mathbf{x}\| = a$, $\mathbf{n}_0$ is the outward normal to the circle and s is the distance from the circle. Switching to polar coordinates, so that $x = \rho\cos\varphi$ and $y = \rho\sin\varphi$, the Jacobian of the transformation from Cartesian to polar coordinates is ρ. With this, (4.138) becomes

$$u \sim f(\mathbf{x}_0)\sqrt{\frac{a}{\rho}}\,e^{\mathrm{i}\omega[\mu(\rho-a)-t]}.$$

This is exactly the same result as given in (4.118). ∎

It is important to consider what can go wrong with the solution given in (4.137). For example, it does not hold at any point where $\mu(\mathbf{x}) = 0$. These correspond to turning points and can be handled using a method similar to what was used in Sect. 4.3. However, for most application μ is positive. A more likely complication arises when $J = 0$, and the points where this occurs are called *caustics*. These arise when two or more rays intersect, which results in the breakdown of the characteristic coordinate system. If a ray passes through a caustic, then the approximation in (4.137) must be multiplied by $\exp(\mathrm{i}m\pi/2)$, where the integer m is determined from the rank of the Jacobian matrix at the caustic. These phase shifts give rise to what is known as the Keller–Maslov index, and this is addressed in Ludwig (1966), Lewis (1966), Bremmer and Lee (1984), and Fedoryuk (1999).

A less obvious complication with (4.137) arises with the requirement that s must be the value for which the solution of (4.134) satisfies $\mathbf{X}(s) = \mathbf{x}$. The fact is that it is possible that there is no such value of s. This happens with shadow regions, and an illustration is shown in Fig 4.12. Assuming μ is constant, so that the rays are straight lines, it is impossible to connect a point on the boundary surface S with $\mathbf{x}_d$. This is resolved by introducing the idea of a surface, or creeping, wave. Figure 4.12 shows a ray from S that is tangent to the given obstacle boundary. At the point of tangency, T_0, two rays are generated. One is simply the linear continuation of the original ray, but the other propagates along the boundary surface of the obstacle. This surface ray continues to T_d, at which point it splits, with one branch continuing along the boundary (not shown) and another that follows a straight line path to $\mathbf{x}_d$. The latter line is the one that is tangent to the obstacle surface at T_d. The derivation of the theory and approximations for surface waves can be found in Keller and Lewis (1995).

Example: Interface Conditions

Consider an incident wave of the form

$$u_I \sim e^{i(\omega\theta_I - \omega t)} u_0(\mathbf{x})$$

that strikes an interface across which $\mu(\mathbf{x})$ is discontinuous (Fig. 4.13). This happens, for example, when sunlight strikes the surface of a lake. The wave is partly reflected, producing a wave u_R, and partly transmitted (or refracted), producing a wave u_T. The latter two waves are determined from what happens when u_I strikes the interface, and our objective is to find the corresponding boundary conditions on u_R and u_T. Letting Q designate the interface, $u(\mathbf{x}, t)$ and its normal derivative are assumed to be continuous across Q. So the following conditions are imposed on Q:

$$u_I(\mathbf{x}, t) + u_R(\mathbf{x}, t) = u_T(\mathbf{x}, t) \tag{4.140}$$

and

$$\partial_n[u_I(\mathbf{x}, t) + u_R(\mathbf{x}, t)] = \partial_n u_T(\mathbf{x}, t). \tag{4.141}$$

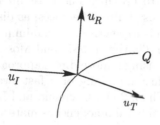

Figure 4.13 Schematic illustrating the reflection and transmission of an incident wave at an interface S across which the function $\mu(\mathbf{x})$ is discontinuous

As usual, the expansions for the reflected and transmitted waves are

$$u_R \sim e^{i(\omega_R \theta_R - \omega_R t)} v_0(\mathbf{x})$$

and

$$u_T \sim e^{i(\omega_T \theta_T - \omega_T t)} w_0(\mathbf{x}).$$

Given that the amplitude and phase are independent of frequency, and (4.140) and (4.141) hold for any value of t and ω, it must be that $\omega = \omega_R = \omega_T$ and $\theta_I(\mathbf{x}) = \theta_R(\mathbf{x}) = \theta_T(\mathbf{x})$ for $\mathbf{x} \in Q$. The interface conditions then give us that $u_0(\mathbf{x}) + v_0(\mathbf{x}) = w_0(\mathbf{x})$ and $u_0(\mathbf{x})\partial_n\theta_I + v_0(\mathbf{x})\partial_n\theta_R = w_0(\mathbf{x})\partial_n\theta_T$ for $\mathbf{x} \in Q$. Thus, for $\mathbf{x} \in Q$, $v_0 = Ru_0$ and $w_0 = Tu_0$, where the reflection (R) and transmission (T) coefficients are

$$R(\mathbf{x}) = \frac{\partial_n\theta_I - \partial_n\theta_T}{\partial_n\theta_T - \partial_n\theta_R} \quad \text{and} \quad T(\mathbf{x}) = \frac{\partial_n\theta_I - \partial_n\theta_R}{\partial_n\theta_T - \partial_n\theta_R} .$$

These expressions can be written in terms of the incoming phase θ_I using the eikonal equation. To do this, let μ_+ be the limiting value of μ when approaching Q from the incident side and μ_- the limiting value coming from the transmitted side. From the continuity of the phase functions across the interface and the eikonal equation, it follows that $\partial_n\theta_R = -\partial_n\theta_I$. For the same reasons,

$$\partial_n\theta_T = \pm\sqrt{\mu_-^2 - \mu_+^2 + (\partial_n\theta_I)^2},$$

where the plus sign is used if $\partial_n\theta_I > 0$ and the minus sign is used if $\partial_n\theta_I < 0$. With this, R and T are determined entirely in terms of the incident wave. Note, however, that for larger values of μ_+, $\partial_n\theta_T$ is not real-valued. This situation is explored in Exercise 4.52. ∎

Ray methods have been used to find approximate solutions to some very challenging problems in, for example, electromagnetics, acoustics, geophysics, and solid mechanics. Interested readers may consult Keller and Lewis (1995), Born and Wolf (1999), Cerveny (2001), Broutman et al. (2004), and Kravtsov (2005). In many of these real-world applications it is not possible to solve (4.129) in closed form, and it is necessary to find the solution numerically. There are different ways to do this. The most direct route for solving the ray tracing problem is to use finite differences, as discussed in Pereyra et al. (1980). Another approach is to rewrite it as a minimization problem; this has been pursued by Um and Thurber (1987) and Moser et al. (1992). Recently there has been interest in using a more direct approach to solving the eikonal equation using either a fast marching or a fast sweeping method (Sethian and Vladimirsky, 2000; Zhao, 2005; Gremaud and Kuster, 2006). Finally, it is possible to consider weakly inhomogeneous materials, which is explored in Exercise 4.51.

Exercises

4.40. Assuming a high frequency, find a first-term approximation of the time-periodic response of
(a) $\nabla^2 u = \partial_t^2 u + \beta(\mathbf{x})\partial_t u$ where $\beta(\mathbf{x})$ is smooth and positive, and
(b) $\nabla^2 u + \mathbf{a}\cdot\nabla u = \partial_t^2 u$ where $\mathbf{a} = \mathbf{a}(\mathbf{x})$ is smooth.

4.41. In this problem, the consequences of letting $\lambda = 1$ in (4.124) are explored.
(a) Show that, instead of (4.134), one solves

$$\frac{\partial^2 \mathbf{X}}{\partial s^2} = \frac{1}{2}\nabla(\mu^2).$$

(b) Explain why (4.132) shows that s does not correspond to arc length. If ℓ is arc length, then show that $d\ell = \mu ds$.
(c) Show that (4.135) changes, but (4.137) still applies.
(d) Even though there are various ways to select λ, they all produce the same answer. Demonstrate this by working out the constant μ example when $\lambda = 1$.

4.42. Many textbooks on partial differential equations consider the problem of solving the nonlinear first-order equation $F(x, y, u, p, q) = 0$ for $u(x, y)$, where $u = u_0$ on a prescribed boundary curve. In this equation, $p = \partial_x u$ and $q = \partial_y u$. Using the method of characteristics, the conventional method is to rewrite the problem as (Debnath, 2012; Guenther and Lee, 1996)

$$\partial_s x = \partial_p F,$$
$$\partial_s y = \partial_q F,$$
$$\partial_s u = p\partial_p F + q\partial_q F,$$
$$\partial_s p = -\partial_x F - p\partial_u F,$$
$$\partial_s q = -\partial_y F - q\partial_u F.$$

Assuming the boundary curve can be parameterized as $(x, y) = (x_0(\tau), y_0(\tau))$, at $s = 0$ the solution of the preceding system satisfies $(x, y, u, p, q) = (x_0(\tau), y_0(\tau), u_0(\tau), p_0(\tau), q_0(\tau))$, where p_0 and q_0 are determined by solving $F(x_0, y_0, u_0, p_0, q_0) = 0$ and $p_0 x_0' + q_0 y_0' = u_0'$.
(a) Assuming $\mathbf{x} \in \mathbb{R}^2$, what is u and what is F for the eikonal Eq. (4.122)?
(b) Show that the preceding system reduces to (4.129) with $\lambda = 2$.
(c) Show that the preceding initial conditions reduce to (4.130) and (4.131).

4.43. In the case where $\mathbf{x} \subset \mathbb{R}^2$, the ray coordinates are given as $\mathbf{x} = \mathbf{X}(s, \alpha)$. If J is the Jacobian of the transformation, then

$$J = \left|\frac{\partial(x, y)}{\partial(s, \alpha)}\right|.$$

Note that the formulas derived in this exercise also hold when $\mathbf{x} \in \mathbb{R}^3$.
(a) Show that

$$\partial_s J = J \nabla \cdot (\lambda \nabla \theta).$$

(b) Given a smooth function $q(\mathbf{x})$, show that

$$\partial_s (qJ) = J \nabla \cdot (q \lambda \nabla \theta).$$

4.44. The objective of this exercise is to derive (4.131). It is assumed that the boundary surface S is parameterized as $\mathbf{x} = \mathbf{X}(0, \alpha, \beta)$, where the tangent vectors $\mathbf{t}_\alpha = \partial_\alpha \mathbf{X}(0, \alpha, \beta)$ and $\mathbf{t}_\beta = \partial_\beta \mathbf{X}(0, \alpha, \beta)$ are orthonormal. Also, $\mathbf{n}$ is the unit outward normal to S. You can assume that S, f, and λ are independent of ω.
(a) Given that $\nabla v = (\nabla v \cdot \mathbf{n})\mathbf{n} + (\nabla v \cdot \mathbf{t}_\alpha)\mathbf{t}_\alpha + (\nabla v \cdot \mathbf{t}_\beta)\mathbf{t}_\beta$, show that

$$(\nabla v \cdot \mathbf{n})\mathbf{n} + (\nabla f \cdot \mathbf{t}_\alpha)\mathbf{t}_\alpha + (\nabla f \cdot \mathbf{t}_\beta)\mathbf{t}_\beta \sim \frac{i\omega}{\lambda} f(\mathbf{x}) \partial_s \mathbf{X}|_{s=0} + \cdots.$$

Explain why this means that $\partial_s \mathbf{X}|_{s=0}$ is a scalar multiple of $\mathbf{n}$.
(b) Given that $\partial_s \mathbf{X}|_{s=0} = \kappa \mathbf{n}$, show that $\kappa = \lambda \mu$.

4.45. This problem examines the situation where there is spherical symmetry, in which case $\mu = \mu(r)$, where $r = ||\mathbf{x}||$ is the radial distance from the origin for $\mathbf{x} \in \mathbb{R}^3$. This is a common assumption when studying optics of lenses or wave propagation through the Earth's atmosphere.
(a) Use the ray equation in (4.134) to show that the vector $\mathbf{p} = \mathbf{X} \times (\partial_s \mathbf{X}/\lambda)$ is independent of s.
(b) Taking $\lambda = 1$, explain why the result from part (a) shows that each ray lies in a plane that contains the origin. Moreover, show that $\mu ||\mathbf{X}|| \sin(\chi)$ is constant along a ray, where χ is the angle between $\mathbf{X}$ and $\partial_s \mathbf{X}$.
(c) Let ρ, φ be polar coordinates in the plane of the ray. In calculus, it is shown that for a polar curve $\rho = \rho(\varphi)$ the angle χ between the position and tangent vectors satisfies

$$\sin(\chi) = \frac{\rho}{\sqrt{\rho^2 + (\partial_\varphi \rho)^2}}.$$

Assuming $\partial_\varphi \rho \neq 0$, use the preceding result and part (b) to show that

$$\varphi = \varphi_0 + \kappa \int_{\rho_0}^{\rho} \frac{dr}{r\sqrt{\mu^2 r^2 - \kappa^2}},$$

where ρ_0, φ_0, and κ are constants. What happens if $\partial_\varphi \rho = 0$?
(d) Using (4.133), show that for a polar curve $\mu ds = \sqrt{\rho^2 + (\partial_\varphi \rho)^2} d\varphi$. With this and the result from part (c), show that

$$\theta = \frac{1}{\kappa} \int_{\varphi_0}^{\varphi} \mu^2 \rho^2 d\varphi.$$

4.46. For Maxwell's fish-eye lens it is assumed that

$$\mu(r) = \frac{1}{1 + r^2},$$

where r is the radial distance from the origin. In this problem x, y designate Cartesian coordinates in the plane of the ray (Exercise 4.45). It is of interest to note that this lens was thought, perhaps mistakenly, to provide a solution of the "perfect lens problem," which means one that is capable of infinite resolution (Leonhardt and Philbin, 2010; Merlin, 2011).

(a) Using Exercise 4.45(c), show that each ray lies on a circle and its equation has the form

$$(x - x_0)^2 + (y - y_0)^2 = 1 + x_0^2 + y_0^2.$$

[Hint: Find

$$\frac{d}{dr} \arcsin\left(\frac{r^2 - 1}{2\alpha r}\right).]$$

(b) Show that the solution of the eikonal equation has the form

$$\theta = \theta_0 + \frac{1}{2}\arccos\left(\frac{2x}{x^2 + y^2 + z^2 + 1}\right).$$

(c) Find the solution of the transport equation for this problem.

4.47. This problem examines the situation where $\mu = \mu(z)$ for $\mathbf{x} \in \mathbb{R}^3$. This is a common assumption in seismology and oceanography, where the medium is vertically stratified and μ depends on the depth coordinate z. In this problem the waves will be assumed to be propagating upward.

(a) Show that $\mathbf{p} = \mathbf{e}_z \times (\partial_s \mathbf{X}/\lambda)$ is independent of s, where $\mathbf{e}_z$ is the unit vector pointing in the z-direction. What does this mean for the value of $\mu \sin(\chi)$, where χ is the angle between $\mathbf{X}$ and the z-axis?

(b) Taking $\lambda = 1/\mu$ show that on a ray, given a value of s, the z-coordinate is found by solving

$$s = \int_{z_0}^{z} \frac{\mu}{\sqrt{\mu^2 - \gamma^2}} dz,$$

where $\gamma^2 = \mu_0^2(1 - n_z^2)$, $\mathbf{x}_0 = (x_0, y_0, z_0)$, and $\mathbf{n}_0 = (n_x, n_y, n_z)$. It is assumed here that $\mu > \gamma$. What happens when $\mu = \gamma$ is considered in Ahluwalia and Keller (1977), Stickler et al. (1984). [Hint: (4.132) is useful here.]

(c) Show that the other two coordinates of the ray are given as

$$x = x_0 + \int_{z_0}^{z} \frac{\alpha}{\sqrt{\mu^2 - \gamma^2}} dz$$

and

$$y = y_0 + \int_{z_0}^{z} \frac{\beta}{\sqrt{\mu^2 - \gamma^2}} dz,$$

where $\alpha = \mu_0 n_x$ and $\beta = \mu_0 n_y$.
(d) Using the separation-of-variables assumption that $\theta = F(x)+G(y)+H(z)$, show

$$\theta = \alpha(x - x_0) + \beta(y - y_0) + \int_{z_0}^{z} \sqrt{\mu^2 - \gamma^2} dz.$$

(e) After solving the transport equation, show that the WKB approximation takes the form

$$v(\mathbf{x}) \sim f(\mathbf{x}_0) \left(\frac{\mu_0^2 - \gamma^2}{\mu^2 - \gamma^2} \right)^{1/4} e^{-i\omega\theta},$$

where θ is given in part (c). Given $\mathbf{x}$, explain how the solution is calculated using the preceding formula.

4.48. The equation for the vertical displacement $u(x, y, t)$ of an Euler–Bernoulli plate is

$$\partial_x^2 M_x + 2\partial_x\partial_y M_{xy} + \partial_y^2 M_y + \mu(x, y)\partial_t^2 u = 0,$$

where $M_x = D(\partial_x^2 u + \nu\partial_y^2 u)$, $M_{xy} = (1 - \nu)D\partial_x\partial_y u$, $M_y = D(\partial_y^2 u + \nu\partial_x^2 u)$. Also, D is a positive constant, and $\mu(x, y)$ is a smooth, positive function. Further, ν is constant and satisfies $0 < \nu < \frac{1}{2}$.
(a) For the time-periodic response, in the case of a high frequency, show that the eikonal equation is

$$D \left[(\partial_x\theta)^2 + (\partial_y\theta)^2 \right]^2 = \mu.$$

What is the transport equation?
(b) Explain why if there is one solution of the eikonal equation then there can be, in fact, four solutions.

4.49. In the study of acoustic wave propagation in an ocean with a slowly varying bottom, one comes across the following problem (Burridge and Weinberg, 1977):

$$\varepsilon^2 \phi_{xx} + \varepsilon^2 \phi_{yy} + \phi_{zz} + k^2\phi = 0 \quad \text{for } 0 < z < y,$$

where $\phi = 0$ on $z = 0$ and $\partial_z\phi = \partial_y\phi$ on $z = y$. Also, $k = \omega\mu$ is a positive (fixed) constant. It is assumed that a sound source is located at $x = 0$, $y = 1$. In this problem, we are interested in constructing a first-term WKB approximation of the solution for small ε.
(a) Assuming the phase is independent of z, show that the eikonal equation is

$$\theta_x^2 + \theta_y^2 = \kappa_n^2,$$

where $\kappa_n^2 = k^2 - (n\pi/y)^2$ and n is a positive integer.
(b) Setting $\mathbf{X} = (X, Y)$, show, using (4.124) with $\lambda = 1/\kappa_n$, that $X_s^2 + Y_s^2 = 1$. Thus, we can take $X_s = \cos(\xi)$, where ξ is the angle between the ray and the x-axis.

(c) From the ray equations, show that $\kappa_n X_s = \alpha$, where $\alpha = \sqrt{k^2 - (n\pi)^2} \cos(\xi_0)$ and ξ_0 is the angle ξ at the source. Also, show that

$$X = \int_1^Y \frac{\alpha}{\sqrt{\kappa_n^2 - \alpha^2}} dy \quad \text{and} \quad \theta = \int_1^Y \frac{\kappa_n^2}{\sqrt{\kappa_n^2 - \alpha^2}} dy.$$

(d) Find, and solve, the transport equation.

4.50. One can show that the solution of the transport Eq. (4.123), when $\lambda = 1/\mu$, can be written as (Kline, 1961; Bremmer and Lee, 1984)

$$v_0 = f(\mathbf{x}_0) \exp\left(-\frac{1}{2} \int_0^s \left(\frac{1}{R_1} + \frac{1}{R_2}\right) ds\right) \sqrt{\frac{\mu(\mathbf{x}_0)}{\mu(\mathbf{x})}},$$

where $R_1(s)$ and $R_2(s)$ are the principal radii of curvature of the wave front at s. Here R_i is taken to be positive if the corresponding normal section bends away from the direction of propagation. Note that focal points, which are where either R_1 or R_2 goes through zero, require special consideration and are discussed in Bremmer and Lee (1984).

(a) In the case where μ is constant, show that

$$v_0 = f(\mathbf{x}_0) \sqrt{\frac{\rho_1 \rho_2}{(\rho_1 + s)(\rho_2 + s)}},$$

where ρ_1 and ρ_2 are the principal radii of curvature of the wave front at $s = 0$. What does this formula reduce to in $\mathbb{R}^2$?

(b) Show that the result in part (a) can be written as

$$v_0 = f(\mathbf{x}_0) \sqrt{\frac{G(s)}{G(0)}},$$

where $G(s)$ is the Gaussian curvature of the wave front.

4.51. More often than not, for an inhomogeneous material it is necessary to solve the ray equations numerically. One alternative to this is to consider the medium to be weakly inhomogeneous, that is, to assume $\mu(\mathbf{x}) = 1 + \varepsilon \mu_1(\mathbf{x})$. This problem explores some of the consequences of this assumption.

(a) Find the first two terms in the expansion of $\mathbf{X}$ for small ε. To do this assume that the initial values $\mathbf{X}(0)$ and $\mathbf{X}'(0)$ are independent of ε.

(b) Suppose $\mu_1(\mathbf{x}) = \sin(\mathbf{k} \cdot \mathbf{x})$, where $\mathbf{k}$ is a constant vector. Are there limitations on s for the expansion to be valid? Would multiple scales help here to extend the range of the expansion of $\mathbf{x}$?

(c) Light rays are bent as they pass through the atmosphere, and this makes determining the exact location of a star difficult. One approach that has been taken to account for the variation of the index of refraction in the atmosphere is to let $\mu(r) = 1 + \varepsilon e^{-\kappa r}$, where r is the (nondimensional)

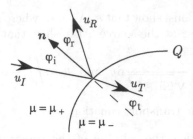

Figure 4.14 Reflection and transmission of an incident wave at an interface Q as assumed in Exercise 4.52

radial distance measured from the surface of the Earth (Park, 1990). One finds that ε is small and, in fact, $\varepsilon \approx 0.032$ (Green, 1985). On the basis of this index of refraction, what are the first two terms in the expansion of $\mathbf{X}$ for small ε?

4.52. This problem examines what happens when μ is discontinuous across a smooth interface Q. As shown in Fig. 4.14, the incident wave makes an angle φ_i with $\mathbf{n}$, where $\mathbf{n}$ is the unit normal to Q pointing to the incident side. It is assumed that $0 < \varphi_i < \frac{\pi}{2}$, and the resulting plane formed by $\mathbf{n}$ and the ray for u_I determine what is called the plane of incidence. Also note that each wave has its own set of ray coordinates, which will be designated as $\mathbf{X}_I(s)$, $\mathbf{X}_R(s)$, and $\mathbf{X}_T(s)$. Assuming that the incident ray strikes Q at $s = s_1$, then $\mathbf{X}_I(s_1) = \mathbf{X}_R(s_1) = \mathbf{X}_T(s_1)$. Moreover, the tangent vectors $\partial_s \mathbf{X}_I(s_1)$, $\partial_s \mathbf{X}_R(s_1)$, and $\partial_s \mathbf{X}_T(s_1)$, which are assumed to be nonzero, determine the angle of the respective wave with the normal to the surface. Finally, assume that Q can be parameterized as $\mathbf{x} = \mathbf{r}(\alpha, \beta)$, where the tangent vectors $\mathbf{t}_\alpha = \partial_\alpha \mathbf{r}$ and $\mathbf{t}_\beta = \partial_\beta \mathbf{r}$ are orthonormal and $\mathbf{n} = \mathbf{t}_\alpha \times \mathbf{t}_\beta$.

(a) Use the fact that $\theta_I(\mathbf{r}) = \theta_R(\mathbf{r})$ to show that $\nabla\theta_I - (\partial_n \theta_I)\mathbf{n} = \nabla\theta_R - (\partial_n \theta_R)\mathbf{n}$. With this and the eikonal equation show that $\partial_n \theta_R = -\partial_n \theta_I$.

(b) The vector $\mathbf{n} \times \partial_s \mathbf{X}_I(s_1)$ is normal to the plane of incidence. Use this and the fact that $\theta_I(\mathbf{r}) = \theta_R(\mathbf{r})$ to show that $\partial_s \mathbf{X}_R(s_1)$ is in the plane of incidence. Moreover, show that $\cos \varphi_i = \cos \varphi_r$, and therefore $\varphi_i = \varphi_r$.

(c) Using an argument similar to that used in part (b), show that the ray for u_T is in the plane of incidence and $\mu_+ \sin \varphi_i = \mu_- \sin \varphi_t$. This is known as Snell's law of refraction. Also, μ_+ is the limiting value of μ when approaching S from the incident side and μ_- the limiting value coming from the transmitted side.

(d) As shown in part (c), the transmitted angle is determined from the equation $\sin \varphi_t = (\mu_+/\mu_-) \sin \varphi_i$. If $\mu_+ > \mu_-$, then there are incident angles such that $(\mu_+/\mu_-) \sin \varphi_i > 1$, and this means there is no real-valued solution for the transmitted angle. In this case, show that $R = e^{-i\delta}$, where $0 < \delta < \pi$. This produces a phase shift in the reflected wave that is associated with what is known as the Goos–Hänchen effect (Goos and Hanchen,

1947; de Micheli and Viano, 2006). Also show that the transmitted wave decays exponentially with distance from the interface; for this reason it it is called an evanescent wave.

(e) If μ is constant, then the WKB approximation for the incident wave is given in (4.138). It is possible to write the reflected and transmitted waves in a similar form. Determine the reflected and transmitted wave vectors $\mathbf{k}_R$ and $\mathbf{k}_T$ in terms of the incident wave.

4.7 Parabolic Approximations

In using the WKB method to solve wave propagation problems, one assumes there is a relatively high frequency. This applies to many problems, but there are situations where this assumption is not applicable or where the procedure involved in ray tracing is too prohibitive to be of practical use. An interesting example of the latter situation arises in the use of sound waves in the ocean to measure the effects of global warming (or cooling). This operation gives rise to what is known as an acoustic thermometer. It uses the fact that a $1°C$ increase in the ocean temperature increases the speed of sound by approximately $4.6\,\text{m/s}$. The change in ocean temperature due to climatic changes is small, and to measure it, one must have the sound signal propagate over a long distance. An experiment used to test the feasibility of this idea was undertaken by placing underwater sound sources near Heard Island, located in the southern Indian Ocean (Baggeroer and Munk, 1992; Munk et al., 1992). Measuring devices to detect the signals were placed off the coasts of Seattle and Cape Cod, which means the waves traveled up to $18,000\,\text{km}$ through the ocean! To achieve such long-range sound propagation in seawater, it is necessary to employ a low-frequency signal, and in this experiment tones centered at $57\,\text{Hz}$ were used. What we would like to do here is develop an approximation method that will work for such situations, and the one to be considered is known as the parabolic wave approximation. This appears to have been first used by Leontovich and Fock (1946) to study long-wave radio-wave propagation in the troposphere. The popularity of the method, however, is due to Tappert (1977). The ideas developed below are based on his paper as well as the paper of Siegmann et al. (1985). Readers interested in whether the ideas underlying the Heard Island experiment actually work, and how the test might affect marine mammals, should consult Frisk et al. (2003) and Dushaw et al. (2009).

To explain how this method works, consider the wave equation

$$\nabla^2 u = \mu^2(\mathbf{x})\partial_t^2 u, \qquad (4.142)$$

where $\mathbf{x} \in \mathbb{R}^3$. We are interested in the time-periodic response, and so let $u(\mathbf{x}, t) = e^{-i\omega t}v(\mathbf{x})$. With this, (4.142) yields the reduced wave, or Helmholtz,

equation
$$\nabla^2 v + \omega^2 \mu^2 v = 0. \tag{4.143}$$

To simplify the situation further, we assume the problem is cylindrically symmetric. In this case, using cylindrical coordinates, (4.143) takes the form

$$v_{rr} + \frac{1}{r} v_r + v_{zz} + \omega^2 \mu^2(r, z) v = 0. \tag{4.144}$$

As stated earlier, the parabolic approximation is widely used on acoustic wave propagation problems. With this in mind, we assume there is a sound source located at $r = r_0$, so the domain we consider is $r_0 < r < \infty$ and $0 < z < 1$. The boundary conditions are that $\partial_z v = 0$ for $z = 0$, $v = 0$ for $z = 1$ and $v = f(z)$ at $r = r_0$. Also, because the stimulus generates waves that propagate away from the source, we also impose the Sommerfeld radiation condition
$$\lim_{r \to \infty} (v_r - i\omega v) = 0. \tag{4.145}$$

This condition means there is no energy coming in from infinity. In other words, the waves propagate out from $r = r_0$ and not toward $r = r_0$.

To gain some insight into the behavior of the solutions of (4.144), suppose for the moment that $\mu = \mu_0$ is constant and v is independent of z. In this case, the solution of (4.144) that satisfies (4.145) is

$$v = \alpha_0 H_0^{(1)}(kr), \tag{4.146}$$

where $H_0^{(1)}$ is the zeroth order Hankel function of the first kind and $k = \omega\mu_0$. For large r this function is oscillatory. In fact, one can show that the solution in (4.146) has the form (Abramowitz and Stegun, 1972)

$$v \sim \alpha_0 \sqrt{\frac{2}{\pi k r}} \, e^{i(kr - \pi/4)} \text{ for } kr \gg 1. \tag{4.147}$$

This result is similar to the approximation given in (4.118). In this sense, the WKB approximation is based on the observation that (4.118) and (4.147) hold for any nonzero r (or ρ) so long as k (or ω) is large. In contrast, the parabolic approximation is based on the observation that (4.118) and (4.147) hold for any nonzero k (or ω) so long as r (or ρ) is large enough. Consequently, the parabolic approximation method works for any frequency, small or large, but assumes that the distance from the source is large.

4.7.1 Heuristic Derivation

Before deriving the parabolic approximation using multiple scales, we outline the argument that was used in the original derivation of the method. The fundamental assumption made is that the solution of (4.144) can be written as

$$v(r, z) = w(r, z)h(r), \qquad (4.148)$$

where $h(r)$ represents a rapidly varying portion of the solution and $w(r, z)$ is its modulation. This is similar to the assumption made for the WKB method, although we are not assuming that h is an exponential function. If we substitute this into (4.144), then we find that

$$\left[w_{rr} + \left(\frac{1}{r} + \frac{2}{h}h_r \right)w_r + w_{zz} + \omega^2\mu^2 w \right]h + \left(h_{rr} + \frac{1}{r}h_r \right)w = 0. \quad (4.149)$$

The choice for the function $h(r)$ is based on the solution of (4.144) when $\mu = \mu_0$. In this approximation, we are primarily interested in the far field, that is, the behavior of the solution for large r. Therefore, we will use the first-term expansion in (4.147) as our choice for $h(r)$. Dropping terms of order $O(\frac{1}{r})$ and smaller, we obtain

$$w_{rr} + 2ikw_r + w_{zz} + (\omega^2\mu^2 - k^2)w = 0. \qquad (4.150)$$

The last step in the reduction is based on the assumption that $w(r, z)$ is supposed to describe the slow modulation of the wave. It seems plausible in this case that the w_{rr} term in the preceding equation is negligible compared to the others and can therefore be omitted. This is called the paraxial approximation, and the result is the standard two-dimensional parabolic equation (PE) given as

$$w_r = \frac{i}{2k}(\omega^2\mu^2 - k^2)w + \frac{i}{2k}w_{zz}. \qquad (4.151)$$

There is no doubt that the derivation of this parabolic equation leaves much to be desired, and this issue will be addressed subsequently. However, before doing so, note that the value of k has not been specified. This is known as the separation constant, and it is reasonable to assume that it is a reference or characteristic value of $\omega\mu$.

4.7.2 Multiple-Scale Derivation

The PE in (4.151) can be obtained directly using multiple scales; the first step is to reconsider the scaling of the physical problem. If one nondimensionalizes the radial coordinate using a length scale associated with a typical wavelength of the wave, then (4.144) takes the form (Exercise 4.53)

$$v_{rr} + \frac{1}{r}v_r + \varepsilon v_{zz} + \mu^2 v = 0 \quad \text{for} \quad \begin{cases} 0 < r < \infty, \\ 0 < z < 1, \end{cases} \qquad (4.152)$$

where
$$\mu = 1 + \varepsilon\mu_1(\varepsilon r, z). \tag{4.153}$$

The boundary conditions we will use for this problem are the radiation condition (4.145) along with

$$v(0, z) = f(z) \quad \text{for } 0 < z < 1 \tag{4.154}$$

and

$$v(r, 1) = 0, \ v_z(r, 0) = 0 \quad \text{for } 0 < r < \infty. \tag{4.155}$$

Here $f(z)$ is a given function. In (4.153) we have made an important assumption about how μ depends on the spatial variables. It is assumed that μ is a small perturbation from a constant and the perturbation is a slowly varying function of r. This leads to the PE in (4.151), but other variations are of interest; one is examined in Exercise 4.55.

To find a first-term approximation of the solution of (4.152)–(4.155), we use multiple scales and introduce the slowly varying radial coordinate $R = \varepsilon r$. In this case the radial derivative transforms as $\partial_r \to \partial_r + \varepsilon\partial_R$, and so (4.152) becomes

$$\left(\partial_r^2 + \frac{1}{r}\partial_r + 1 + 2\varepsilon\mu_1 + 2\varepsilon\partial_r\partial_R + \frac{\varepsilon}{r}\partial_R + \varepsilon\partial_z^2 + \cdots\right)v = 0. \tag{4.156}$$

The appropriate expansion of the solution in this case is

$$v \sim v_0(r, R, z) + \varepsilon v_1(r, R, z) + \cdots. \tag{4.157}$$

Introducing this into (4.156), one finds that the $O(1)$ equation is

$$(\partial_r^2 + \frac{1}{r}\partial_r + 1)v_0 = 0.$$

The general solution of this is $v_0 = V(R, z)H_0^{(1)}(r)$. With this, the $O(\varepsilon)$ equation obtained from (4.156) is

$$\left(\partial_r^2 + \frac{1}{r}\partial_r + 1\right)v_1 = -2V_R\partial_r H_0^{(1)} - \frac{1}{r}V_R H_0^{(1)} - V_{zz}H_0^{(1)} - 2\mu_1 V H_0^{(1)}. \tag{4.158}$$

Using variation of parameters, one finds that the general solution of this equation is

$$v_1 = aH_0^{(1)}(r) + bH_0^{(2)}(r)$$
$$+ \alpha\int_0^r \left[H_0^{(1)}(s)H_0^{(2)}(r) - H_0^{(1)}(r)H_0^{(2)}(s)\right]sF(s, R, z)ds, \tag{4.159}$$

where F designates the right-hand side of (4.158), $H_0^{(2)}$ is the zeroth-order Hankel function of the second kind, and α is a constant. The Hankel functions in (4.159) generate secular terms (this is shown in Exercise 4.54a). To prevent them from appearing, we require that $V(r, z)$ satisfies (Exercise 4.54b)

$$\partial_R V = \frac{i}{2} \partial_z^2 V + i\mu_1 V \quad \text{for } 0 < R < \infty \text{ and } 0 < z < 1. \tag{4.160}$$

This is, in effect, the PE given in (4.151) in the case where $k = 1$. The boundary conditions that go along with this equation are given in (4.154), (4.155). After solving this problem the first-order approximation of the solution of (4.152) is

$$v \sim V(\varepsilon r, z) H_0^{(1)}(r), \tag{4.161}$$

which is valid for $0 \le z \le 1$ and $0 \le r \le O(\varepsilon^{-1})$.

The PE we have derived here is sometimes referred to as the standard small-angle PE. Other PEs are applicable in other situations. The reader is referred to the papers of Siegmann et al. (1985) and Tappert (1977) for discussions of these.

Exercises

4.53. In dimensional variables, (4.144) takes the form

$$\frac{\partial^2}{\partial(r^*)^2} v^* + \frac{1}{r^*} \frac{\partial}{\partial(r^*)} v^* + \frac{\partial^2}{\partial(z^*)^2} v^* + (\omega^*)^2 (\mu^*)^2 v^* = 0.$$

Assume that $\mu^* = \mu_0^*(1 + \varepsilon\mu_1)$, where μ_0^* is a positive constant and $\mu_1 = \mu_1(\sqrt{\varepsilon} r^*/h, z^*/h)$. Also, the region in which this equation holds is $0 < r^* < \infty$ and $0 < z^* < h$. The asterisks indicate dimensional variables.
(a) Find the scaling that leads to (4.152)–(4.155).
(b) It was stated that (4.152)–(4.155) are obtained if one nondimensionalizes the radial coordinate using a length scale r_c associated with a typical wavelength of the wave. Comment on this. Also comment on what assumption is made on the height of the channel compared to r_c.

4.54. For large r the Hankel functions have the expansions (Abramowitz and Stegun, 1972)

$$H_\nu^{(1)}(r) \sim \sqrt{\frac{2}{\pi r}} \, \mathrm{e}^{i(r - \pi/4 - \pi\nu/2)} \quad \text{and} \quad H_\nu^{(2)}(r) \sim \sqrt{\frac{2}{\pi r}} \, \mathrm{e}^{-i(r - \pi/4 - \pi\nu/2)}$$

(a) Use these expansions to show that, for $1 \ll r$,

$$\int_0^r s H_0^{(1)}(s) H_0^{(2)}(s) \mathrm{d}s \sim \frac{2r}{\pi} \quad \text{and} \quad \int_0^r s H_1^{(1)}(s) H_0^{(2)}(s) \mathrm{d}s \sim -\frac{2ir}{\pi}.$$

(b) Using the ideas developed in part (a) and the fact that $\frac{\mathrm{d}}{\mathrm{d}s} H_0^{(j)} = -H_1^{(j)}(s)$, show that to prevent secular terms from appearing in the expansion for v, it is necessary that (4.160) holds.

4.55. The assumption made on the function μ in Exercise 4.53 is fairly restrictive. Instead, suppose it is assumed that $\mu^* = \mu_0^*(\varepsilon r, z)$.
(a) Find the equation satisfied by $v(r, z)$ and explain the differences between the assumption made here and that used in Exercise 4.53.
(b) Using the ideas developed in Sect. 3.6 find a first-term approximation for v that is valid for large r.

4.56. Suppose (4.152) is rewritten as

$$v_{rr} + \frac{\varepsilon}{R} v_r + \varepsilon v_{zz} + [1 + \varepsilon \mu_1(\varepsilon r, z)]^2 v = 0.$$

(a) Find a first-term approximation for v that is valid for large r.
(b) How does the approximation you found in part (a) differ from that given in (4.161)? In answering this, include a comment or two on which approximation should be more accurate.

4.8 Discrete WKB Method

The ideas underlying the WKB method can be extended to difference equations without much difficulty. To show how, and in view of the first example considered in Sect. 4.2, our starting point is the second-order difference equation

$$y_{n+1} - 2y_n + y_{n-1} = q_n y_n \quad \text{for } n = 0, \pm 1, \pm 2, \ldots, \tag{4.162}$$

where $q_n = q(\varepsilon n)$. This equation arises, for example, from a finite-difference approximation of (4.1). It also comes from the equation for an elastic string on which point masses are placed at uniform intervals.

The form of the discrete WKB approximation we will use here is

$$y_n \sim e^{\theta(\varepsilon n)/\varepsilon^\alpha} \left[\bar{y}_0(\varepsilon n) + \varepsilon^\beta \bar{y}_1(\varepsilon n) + \cdots \right]. \tag{4.163}$$

The functions $\theta(\nu)$, $\bar{y}_0(\nu)$, $\bar{y}_1(\nu)$, ... are assumed smooth. In preparation for substituting this expansion into (4.162), we first apply Taylor's theorem to obtain

$$\theta(\varepsilon(n \pm 1)) \sim \theta(\varepsilon n) \pm \varepsilon \theta'(\varepsilon n) + \frac{1}{2} \varepsilon^2 \theta''(\varepsilon n) \pm \cdots.$$

Doing the same for $\bar{y}_0$ and $\bar{y}_1$ we have, from (4.163), that

$$y_{n\pm1} \sim \exp\left[\varepsilon^{-\alpha}\left(\theta_n \pm \varepsilon\theta'_n + \frac{1}{2}\varepsilon^2\theta''_n\right)\right]\left[\bar{y}_{0_n} \pm \varepsilon\bar{y}'_{0_n} + \varepsilon^\beta\bar{y}_{1_n} + \cdots\right]. \quad (4.164)$$

In this equation, $\theta_n = \theta(\varepsilon n)$, $\bar{y}_{0_n} = \bar{y}_0(\varepsilon n)$, etc.

For the terms in (4.162) to balance, we need to take $\alpha = \beta = 1$. In this case, (4.162) takes the form

$$e^{\theta'_n}\left(1 + \frac{1}{2}\varepsilon\theta''_n + \cdots\right)\left[\bar{y}_{0_n} + \varepsilon\bar{y}'_{0_n} + \varepsilon\bar{y}_{1_n} + \cdots\right]$$

$$+ e^{-\theta'_n}\left(1 + \frac{1}{2}\varepsilon\theta''_n + \cdots\right)\left[\bar{y}_{0_n} - \varepsilon\bar{y}'_{0_n} + \varepsilon\bar{y}_{1_n} + \cdots\right]$$

$$= (2 + q_n)[\bar{y}_{0_n} + \varepsilon\bar{y}_{1_n} + \cdots]. \quad (4.165)$$

This leads to

$$O(1) \quad e^{\theta'_n} + e^{-\theta'_n} = 2 + q_n.$$

This is the eikonal equation. Multiplying by $e^{\theta'_n}$ and then solving the resulting quadratic yields

$$\theta'_n = \ln\left[\frac{1}{2}\left(2 + q_n \pm \sqrt{q_n(q_n + 4)}\right)\right]. \quad (4.166)$$

It is a simple matter to integrate this last equation to obtain $\theta_n(x)$, and so our next task is to find the equation for $\bar{y}_{0_n}$.

$$O(\varepsilon) \quad \left(e^{\theta'_n} - e^{-\theta'_n}\right)\bar{y}'_{0_n} + \frac{1}{2}\theta''_n\,\bar{y}_{0_n}\left(e^{\theta'_n} + e^{-\theta'_n}\right) = 0.$$

This is the transport equation, and it can be solved by noting that it can be rewritten as

$$\bar{y}'_{0_n} + \frac{1}{2}\left[\ln\left(e^{\theta'_n} - e^{-\theta'_n}\right)\right]'\bar{y}_{0_n} = 0.$$

Integrating this yields

$$\bar{y}_{0_n}\sqrt{e^{\theta'_n} - e^{-\theta'_n}} = A_0, \quad (4.167)$$

where A_0 is a constant. Thus, from (4.166) we have

$$\bar{y}_{0_n} = \frac{A}{[q_n(q_n + 4)]^{1/4}}, \quad (4.168)$$

where A is constant.

Adding the two solutions together, a first-term discrete WKB approximation of the general solution of (4.162) in the case of small ε is

$$\bar{y}_n \sim \frac{1}{[q_n(q_n+4)]^{1/4}} \left(a_0 e^{\theta_+/\varepsilon} + b_0 e^{\theta_-/\varepsilon} \right), \qquad (4.169)$$

where

$$\theta_\pm(\varepsilon n) = \int^{\varepsilon n} \ln\left[\frac{1}{2}\left(2 + q(\nu) \pm \sqrt{q(\nu)(q(\nu)+4)}\right)\right] d\nu. \qquad (4.170)$$

Using the properties of inverse functions, the preceding expression for $\theta_\pm(x)$ can be simplified to

$$\theta_\pm(\varepsilon n) = \begin{cases} \pm i \int^{\varepsilon n} \cos^{-1}\left[1 + \frac{1}{2}q(\nu)\right] d\nu & \text{if } -4 < q < 0, \\ \pm \int^{\varepsilon n} \cosh^{-1}\left[1 + \frac{1}{2}q(\nu)\right] d\nu & \text{if } 0 < q, \\ i\pi\varepsilon n \pm \int^{\varepsilon n} \cosh^{-1}\left|1 + \frac{1}{2}q(\nu)\right| d\nu & \text{if } q < -4. \end{cases} \qquad (4.171)$$

The similarities between this result and the WKB approximation given in (4.11) are evident. However, there are differences, and perhaps one of the more important differences concerns the turning points. From (4.11) these occur when $q(x) = 0$. For the discrete WKB formula in (4.169), these include the values of n, where $q_n = 0, -4$. However, there is also a problem with (4.169) if $q(\nu)$, or $q(\nu) + 4$, changes sign between $\nu = \varepsilon n$ and $\nu = \varepsilon(n+1)$. This will be referred to as a turning-point interval. The resolution of the solution near a turning point is carried out in a manner similar to what was done for differential equations, and this is demonstrated below for a simple turning point.

Example

To illustrate the behavior of the solution when a turning point is present, consider the difference equation

$$y_{n+1} - 2y_n + y_{n-1} = \frac{2n - N}{2N} y_n \quad \text{for } n = -N+1, \dots, N-1, \qquad (4.172)$$

where $y_{-N} = y_N = 1$. In this problem, $\varepsilon = 1/N$ and $q(\nu) = (2\nu - 1)/2$. Also, note that if N is even, then there is a turning point at $n = N/2$ and there is a turning-point interval when N is odd. We will assume N is even and the solution in the case where $N = 100$ is shown in Fig. 4.15. The oscillatory nature of the solution for $n < 50$ and the exponential character of the solution for $n > 50$ are evident in this figure. These are contained in our WKB approximation (4.170), which, for this problem, reduces to

$$y_n \sim \begin{cases} d_n \left(a_L e^{-iN\theta_n} + b_L e^{iN\theta_n} \right) & \text{if } -N \leq n < N/2, \\ d_n \left(a_R e^{N\kappa_n} + b_R e^{-N\kappa_n} \right) & \text{if } N/2 < n \leq N, \end{cases} \qquad (4.173)$$

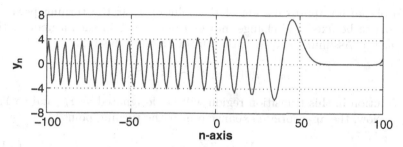

Figure 4.15 Solution of difference Eq. (4.172) in the case where $N = 100$. To draw this graph, the individual points were connected by a smooth curve. This was done, as opposed to showing the individual data points, because of the large number of points included in the plot

where $d_n = [(\frac{n}{N} - \frac{1}{2})(\frac{n}{N} + \frac{7}{2})]^{-1/4}$. From (4.171) one finds that

$$\kappa_n = 2\sqrt{1 - \eta^2} - 2\eta \cos^{-1}(\eta) \tag{4.174}$$

and

$$\theta_n = 2\eta \cosh^{-1}(\eta) - 2\sqrt{\eta^2 - 1}, \tag{4.175}$$

where $\eta = \frac{1}{4}(3 + 2n/N)$. The solution in (4.173) must satisfy the conditions at $n = \pm N$. That still leaves two undetermined constants, and these are found by matching (4.173) with the solution from the turning-point region.

4.8.1 Turning Points

We will now investigate the behavior of the solution of (4.162) for n near n_0, where $\nu_0 = \varepsilon n_0$ is a simple turning point. In what follows, it is assumed that $q(\nu_0) = 0$ with $q'(\nu_0) > 0$. It is also assumed that $q(\nu) + 4 > 0$. In this case the WKB approximation in (4.169) can be written as

$$y_n \sim \begin{cases} \alpha_n \left(a_L e^{-i\theta_n/\varepsilon} + b_L e^{i\theta_n/\varepsilon}\right) & \text{if } n < n_0, \\ \alpha_n \left(a_R e^{\kappa_n/\varepsilon} + b_R e^{-\kappa_n/\varepsilon}\right) & \text{if } n_0 < n, \end{cases} \tag{4.176}$$

where

$$\alpha_n = \frac{1}{[q_n(q_n + 4)]^{1/4}},$$

$$\theta_n = \int_{\varepsilon n}^{\varepsilon n_0} \cos^{-1}\left[1 + \frac{1}{2}q(\nu)\right] d\nu, \tag{4.177}$$

and

$$\kappa_n = \int_{\varepsilon n_0}^{\varepsilon n} \cosh^{-1}\left|1 + \frac{1}{2}q(\nu)\right| d\nu. \tag{4.178}$$

To obtain an approximation of the solution near the turning point, we take εn_0 to be fixed and change indices from n to k in such a manner that $n = n_0 + k$. Assuming $\varepsilon|k| \ll 1$, then from Taylor's theorem,

$$q_n = q(\varepsilon(n_0 + k)) \sim \varepsilon k q'(\varepsilon n_0).$$

The solution in this transition region will be designated as Y_k, and so $Y_k = y_{n_0+k}$. Now, the appropriate expansion near the turning point is

$$Y_k \sim \varepsilon^\gamma \overline{Y}_k + \cdots . \tag{4.179}$$

Substituting this into (4.162) one finds that

$$\overline{Y}_{k+1} - 2\overline{Y}_k + \overline{Y}_{k-1} = \varepsilon k Q_0' \overline{Y}_k, \tag{4.180}$$

where $Q_0' = q'(\varepsilon n_0)$. The general solution of this second-order difference equation involves Bessel functions (Exercise 4.67). However, we will approach the problem directly, and a standard method for solving difference equations like this one is to introduce the transformation (Carrier et al., 1966)

$$\overline{Y}_k = \int_C e^{ikz} f(z) dz, \tag{4.181}$$

where the contour C is in the complex plane. Both C and the function $f(z)$ are determined from (4.180). Substituting this into (4.180) yields

$$2 \int_C e^{ikz} [\cos(z) - 1] f(z) dz = \mu \int_C k e^{ikz} f(z) dz, \tag{4.182}$$

where $\mu = \varepsilon Q_0'$. The next step is to use integration by parts on the integral on the right-hand side of (4.182) to obtain

$$\int_C e^{ikz} \{2[\cos(z) - 1] f(z) - i\mu f'(z)\} dz = 0. \tag{4.183}$$

In this step, it was assumed that the contour was chosen such that the contribution from the endpoints is zero. Later, once C has been determined, we will have to come back to check that this does indeed hold. What we have accomplished by integrating by parts is to get the terms in the braces { } in (4.183) to be independent of k. So for (4.183) to be satisfied for all k we require that

$$2[\cos(z) - 1] f(z) - i\mu f'(z) = 0. \tag{4.184}$$

Solving this equation yields

$$f(z) = B_0 \exp\left[-\frac{2i}{\mu} (\sin(z) - z)\right], \tag{4.185}$$

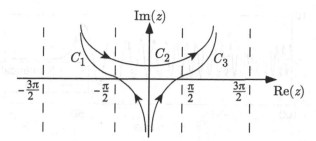

Figure 4.16 Three possible contours that can be used in the WKB approximation (4.186). Note that C_1 has vertical asymptotes $\mathrm{Re}(z) = -\pi, 0$, while C_2 has asymptotes $\mathrm{Re}(z) = \pm\pi$, and C_3 has asymptotes $\mathrm{Re}(z) = 0, \pi$

where B_0 is a constant. With this, and the original transformation given in (4.181), we have that

$$\overline{Y}_k = B_0 \int_C e^{ikz - 2i(\sin(z) - z)/\mu} dz. \qquad (4.186)$$

We must now specify the contour C so all this makes sense. Actually, we are looking for two contours because the general solution of (4.180) consists of the superposition of two independent solutions. The requirements on these contours are that they produce a convergent integral in (4.186), that (4.186) is obtained after integration by parts, and the solution from this region matches with the outer solutions. The first two requirements are easily satisfied if we make use of the exponential decay of the integrand. There are many possibilities, three of which are shown in Fig. 4.16. From this and (4.186) we have that the WKB approximation of the general solution of (4.177) can be written as

$$\overline{Y}_k = \alpha_0 \int_{C_1} e^{ikz - 2i(\sin(z) - z)/\mu} dz + \beta_0 \int_{C_3} e^{ikz - 2i(\sin(z) - z)/\mu} dz, \qquad (4.187)$$

where α_0 and β_0 are arbitrary constants.

It remains to match (4.187) with the outer solutions given in (4.176). Determining the behavior of the functions in (4.176) near the turning point is relatively easy and follows the steps used in Sect. 4.3. Finding out how $\overline{Y}_k$ behaves as $k \to \pm\infty$, however, requires extensive use of the method of steepest descents (Murray, 1984). The details of the calculations are outlined in Exercise 4.66. One finds that the connection formulas relating the coefficients in the three regions are

$$b_{\mathrm{L}} = b_{\mathrm{R}} + \frac{1}{2} i a_{\mathrm{R}}, \quad a_{\mathrm{L}} = \frac{1}{2} a_{\mathrm{R}} + i b_{\mathrm{R}} \qquad (4.188)$$

and

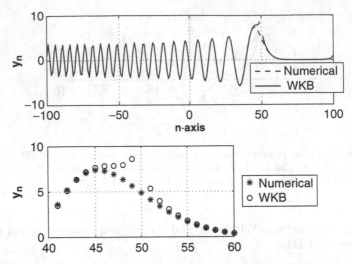

Figure 4.17 Plot of WKB approximation (4.190) for solution of difference Eq. (4.172) in the case where $N = 100$. For comparison the numerical solution is also shown. The *lower plot* shows the solution in the region near $n = 50$

$$\alpha_0 = \frac{b_L}{\sqrt{2\pi Q_0'}}, \qquad \beta_0 = \frac{-ia_L}{\sqrt{2\pi Q_0'}}, \qquad (4.189)$$

where $\gamma = -1/2$.

Putting our results together, we have from (4.176) and (4.188) that the WKB approximation of the solution of (4.162) is

$$y_n \sim \begin{cases} A_n \left(a_R \cos\left(\frac{1}{\varepsilon}\theta_n + \frac{\pi}{4}\right) + 2b_R \cos\left(\frac{1}{\varepsilon}\theta_n - \frac{\pi}{4}\right) \right) & \text{if } n < n_0, \\ A_n \left(a_R e^{\kappa_n/\varepsilon} + b_R e^{-\kappa_n/\varepsilon} \right) & \text{if } n_0 < n, \end{cases} \qquad (4.190)$$

where $A_n = [|q_n|(q_n + 4)]^{-1/4}$, θ_n is given in (4.177), and κ_n is given in (4.178).

To illustrate the accuracy of this approximation, we return to the example in (4.172). The coefficients a_R and b_R in (4.190) are determined from the boundary conditions $y_{-N} = y_N = 1$. The resulting approximation is shown in Fig. 4.17 along with the numerical solution. It is seen that in their respective regions of applicability the discrete WKB approximations are almost indistinguishable from the numerical solution.

The discrete WKB approximation developed here is based, in part, on the ideas introduced by Dingle and Morgan (1967a,b) and Wilmott (1985). A somewhat different approach can be found in Braun (1979) and Geronimo and Smith (1992). Those interested in a more theoretical analysis of the discrete WKB method may consult Costin and Costin (1996).

Exercises

4.57. This problem concerns the equation

$$p_{n+1}y_{n+1} + q_n y_n + p_n y_{n-1} = 0 \quad \text{for } n = 0, 1, 2, \ldots,$$

where $q_n = q(\varepsilon n)$ and $p_n = p(\varepsilon n)$. It is assumed here that $p(\nu)$ and $q(\nu)$ are smooth functions of ν, and $p(\nu) > 0$.
(a) Find a first-term approximation of the solution of this equation.
(b) Identify the turning points, and simplify your answer from part (a) if a turning point is not present [as is done in (4.171) and (4.173)].
(c) A nonzero solution is said to be oscillatory if, given any n, there is a $k \geq n$ such that $y_k y_{k+1} \leq 0$. In Patula (1979) it is proved that if $q(\nu) \geq 0$, then all nontrival solutions of the preceding difference equation are oscillatory. Is your approximation consistent with this result? In the case of $q(\nu) < 0$, Hooker and Patula (1981) show that all nontrival solutions are oscillatory if $(4 - \delta)p_n^2 \geq q_n q_{n+1}$ for some $\delta > 0$. Is your approximation consistent with this result?

4.58. Find the general solution of the turning point equation when:
(a) εn_0 is a simple turning point where $q_n = 0$ and $\mu < 0$,
(b) εn_0 is a simple turning point where $q_n = -4$ and $\mu > 0$,
(c) εn_0 is a simple turning point where $q_n = -4$ and $\mu < 0$.

4.59. The development of the discrete WKB method assumed that the variable coefficient in (4.162) could be described with a continuous function $q(\nu)$. When starting out with a difference equation, there will inevitably be numerous choices that can be made for $q(\nu)$. It is therefore a natural question to ask what effect, if any, this choice will have on the approximation. To investigate this, suppose the difference equation is

$$y_{n+1} - 2y_n + y_{n-1} = 2y_n \quad \text{for } n = 0, \pm 1, \pm 2, \ldots.$$

(a) Assuming $y_n = r^n$, find the general solution of this equation.
(b) Now the question is, what should we choose for $q(\nu)$? For example, we can take (i) $q(\nu) = 2$, (ii) $q(\nu) = 2 + \sin(\nu\pi)$, or (iii) $q(\nu) = 2 + \sin(20\nu\pi)$. Discuss the effect of these choices on the accuracy of the WKB approximation. You might do this by either investigating the second term in the expansion or else examining the effects on the value of the function $\theta_\pm$ as given in (4.170).
(c) Based on your findings in part (b), discuss the case where $q(\nu) = 2 + 4\sin(\nu\pi)$.

4.60. Find a discrete WKB approximation of the solution of the following problem. If turning points, or turning-point intervals, are present, then identify where they are located.

(a) $y_{n+2} - 4y_{n+1} + 6y_n - 4y_{n-1} + y_{n-2} = q_n y_n$,

(b) $\left(1 + \frac{1}{2}p_n\right) y_{n+1} + (q_n - 2)y_n + \left(1 - \frac{1}{2}p_n\right) y_{n-1} = 0$,

(c) $\left(1 + \frac{1}{2}\sqrt{\varepsilon}p_n\right) y_{n+1} + (q_n - 2)y_n + \left(1 - \frac{1}{2}\sqrt{\varepsilon}p_n\right) y_{n-1} = 0$. [Hint: Assume $y_n \sim \exp(\varepsilon^{-\alpha}\theta_0 + \varepsilon^{-\beta}\theta_1)(y_0 + \varepsilon^\gamma y_1 + \cdots).$]

4.61. Consider the eigenvalue problem

$$y_{n+1} - 2y_n + y_{n-1} + \lambda^2 \varepsilon^{-2} q_n y_n = 0 \quad \text{for } n = 1, 2, \ldots, N,$$

where $y_0 = y_{N+1} = 0$. Also, $q_n = q(\varepsilon n)$, where $\varepsilon = 1/(N+1)$, and $q(v)$ is a smooth, positive function for $0 \le v \le 1$.

(a) Assuming $\lambda \sim \varepsilon^\gamma(\lambda_0 + \varepsilon^\alpha \lambda_1 + \cdots)$, use the discrete WKB approximation to determine the first term in the expansion for the eigenvalue λ (i.e., find λ_0).

(b) In the case where $q(v) = \kappa^2$, where κ is a positive constant, the eigenvalues for the problem are $\lambda = \frac{2}{\kappa}(N+1)\sin(\frac{j\pi}{2(N+1)})$ for $j = 1, 2, \ldots, N$. How does the WKB approximation of the eigenvalues compare with this exact result?

(c) The eigenvalue problem comes from a finite-difference approximation of the differential equation

$$y'' + \lambda^2 q(x)y = 0 \quad \text{for } 0 < x < 1,$$

where $y(0) = y(1) = 0$. In this case, y_n is the approximation of $y(x_n)$, where $x_n = nh$ and $h = 1/(N+1)$. In (4.18), the WKB approximation of the large eigenvalues for this problem is given in the case where $q(x) = e^{2x}$. Compare this result with the values obtained from the discrete WKB approximation obtained in part (a) in the case where $q(v) = e^{2v}$.

4.62. Consider the second-order difference equation

$$y_{n+1} + y_{n-1} = q_n y_n \quad \text{for } n = 0, \pm 1, \pm 2, \ldots, \tag{4.191}$$

where $q_n = q(\varepsilon n)$.

(a) Find a WKB approximation of the general solution of this equation for small ε.

(b) Setting $z_n = \alpha_n y_n$, show how the equation

$$a_{n+1}z_{n+1} + b_n z_n + c_{n-1}z_{n-1} = 0 \quad \text{for } n = 0, \pm 1, \pm 2, \ldots$$

can be transformed into the one in (4.191). It is assumed here that $a_n = a(\varepsilon n)$ and $c_n = c(\varepsilon n)$ are nonzero. From this and the result from part (a), write down a WKB approximation of the general solution of this equation.

4.63. The Bessel function $J_n(x)$ satisfies the second-order difference equation

$$y_{n+1} + y_{n-1} = \frac{2n}{x} y_n, \quad \text{for } n = 1, 2, 3, \ldots.$$

(a) For large x find a WKB approximation of the general solution of this equation.

(b) Your solution in part (a) should contain two arbitrary constants. If, for large x,

$$y_0 \sim \sqrt{\frac{2}{\pi x}} \cos\left(x - \frac{\pi}{4}\right) \quad \text{and} \quad y_1 \sim \sqrt{\frac{2}{\pi x}} \cos\left(x - \frac{3\pi}{4}\right),$$

then find the resulting expansion for y_n.

(c) For large x, one can show that

$$J_n(x) \sim \sqrt{\frac{2}{\pi x}} \cos\left(x - \frac{n\pi}{2} - \frac{\pi}{4}\right).$$

How does this compare with your result in part (b)?

4.64. Consider the difference equation

$$q_n z_{n+1} - (1 + 2q_n - \varepsilon\beta) z_n + (q_n - \varepsilon\beta) z_{n-1} = 0 \quad \text{for } n = 1, 2, 3, \ldots,$$

where $q(\nu) = \nu$ and β is a positive constant.

(a) For small ε find a WKB approximation of the general solution of this equation.

(b) If $z_1 \sim \Gamma(\beta)\varepsilon^{\beta-1} e^{1/\varepsilon}$ and $z_2 \sim z_1/\varepsilon$, then find the resulting expansion for z_n.

(c) Compare the result from part (b) with the expansion of Kummer's function $M(n, \beta, x)$ for large x. What difference equation does $Z_n = M(n, \beta, x)$ satisfy?

4.65. In combinatorics, it is often of interest to find the asymptotic behavior of a sequence that is defined by a recursion equation. An example of this arises with Stirling numbers of the second kind, $S(n, k)$. Given a set of n elements, $S(n, k)$ is the number of ways to partition it into k nonempty subsets. From this definition one finds that these numbers satisfy the recursion equation

$$S(n + 1, k) = S(n, k - 1) + kS(n, k) \quad \text{for } k = 1, 2, \ldots, n,$$

where $S(n, 0) = 0$, $S(n, n) = 1$, and $S(n, k) = 0$ for $k > n$. One can show that the solution of this can be written in series form as

$$S(n, k) = \frac{1}{k!} \sum_{j=0}^{k} (-1)^{k-j} \binom{k}{j} j^n.$$

This exercise derives an approximation of $S(n, k)$ for n large and is based on the paper of Knessl and Keller (1991b).

(a) For large n and fixed k show that $S(n+1,k) \sim kS(n,k)$. From this conclude $S(n,k) \sim k^n/k!$.

(b) To find an approximation for $1 \ll k < n$, let $k = \kappa/\varepsilon$, where $\varepsilon = 1/n$. Letting $S(n,k) = \varepsilon^\gamma R(n,\kappa)$, show

$$R(n+1,\kappa) = R(n,\kappa-\varepsilon) + \kappa R(n,\kappa).$$

Using a WKB expansion, find a first-term expansion of the solution of this equation for small ε. To determine the solution uniquely, it is necessary to match with the expansion in part (a). You do not need to do this, but the steps involved can be found in Knessl and Keller (1991b).

4.66. This exercise concerns the transition-layer solution (4.187) and its matching with the WKB approximation in (4.176).

(a) In regard to the two contour integrals in (4.187), use Cauchy's theorem to relate them to the integral over C_2 (Fig. 4.16).

(b) Consider the limit $k \to \infty$. Setting $g(z) = \frac{2i}{\mu k}(\sin(z) - (1 + \frac{1}{2}\mu k)z)$, find the stationary points for $g(z)$ and the paths of steepest descent for the contour integrals over C_2 and C_3. Assuming $k \to \infty$, with $\varepsilon k \ll 1$, use the method of steepest descents to show that

$$\overline{Y}_k \sim \left(\frac{\pi^2\mu}{k}\right)^{1/4} \left[i(\alpha_0 + \beta_0)\Lambda + \frac{1}{2}(\beta_0 - \alpha_0)\Lambda^{-1}\right],$$

where $\Lambda = \exp(\frac{2}{3}\mu^{1/2}k^{3/2})$.

(c) Suppose $k \to -\infty$, with $|\varepsilon k| \ll 1$. Setting $h(z) = \frac{2i}{\mu\lambda}(\sin(z) - (1 - \frac{1}{2}\mu\lambda)z)$, where $\lambda = -k$, use the method of steepest descents to show that

$$\overline{Y}_k \sim \left(\frac{\pi^2\mu}{\lambda}\right)^{1/4}(-\alpha_0\Theta + \beta_0\Theta^{-1}),$$

where $\Theta = \exp[i(\frac{2}{3}\mu^{1/2}\lambda^{3/2} - \pi/4)]$.

(d) Find first-term approximations of the WKB approximations in (4.176) for n near n_0.

(e) Matching the results from parts (b)–(d), derive the connection formulas in (4.188) and (4.189).

4.67. This problem develops another approach to matching the transition-layer solution with the WKB expansion in (4.176).

(a) Show that the general solution of (4.180) is

$$\overline{Y}_k = AJ_{k+2/\mu}\left(\frac{2}{\mu}\right) + BY_{k+2/\mu}\left(\frac{2}{\mu}\right),$$

where $\mu = \varepsilon Q_0'$.

(b) Using the general solution in part (a), derive the corresponding connection formulas. What assumptions do you need to make approximately εk?

Chapter 5
The Method of Homogenization

5.1 Introduction

It is common in engineering and scientific problems to have to deal with materials that are formed from multiple constituents. Some examples are shown in Fig. 5.1 and include laminated wood, a fluid-filled porous solid, an emulsion, and a fiber-reinforced composite. Solving a mathematical problem that includes such variations in the structure can be very difficult. It is therefore natural to try to find simpler equations that effectively smooth out whatever substructure variations there may be. An example of this situation occurs when describing the motion of a fluid or solid. One usually does not consider them as composites of discrete interacting molecules. Instead, one uses a continuum approximation that assumes the material to be continuously distributed. Using this approximation, material parameters, such as the mass density, are assumed to represent an average.

In this chapter, we investigate one approach that can be used to smooth out substructure variations that arise with spatially heterogeneous materials. In this sense, our objective differs from what we have done in the other chapters. Usually we have been interested in deriving an approximate solution, but in this chapter our primary goal is to derive an approximate problem. The substructural geometric and material variations will not be present in the reduced problem, but they will be used to determine the coefficients of the problem. The procedure we will be developing is called homogenization. It goes by other names, including effective media theory, bulk property theory, and the two-space scale method.

5.2 Introductory Example

To introduce the method of homogenization, we will examine the boundary-value problem

M.H. Holmes, *Introduction to Perturbation Methods*, Texts in Applied
Mathematics 20, DOI 10.1007/978-1-4614-5477-9_5,
© Springer Science+Business Media New York 2013

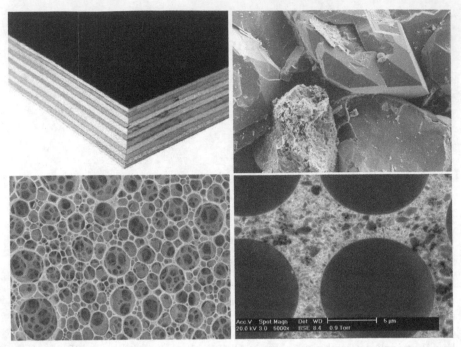

Figure 5.1 Examples of composite systems. This includes plywood, a 400-μm view of quartz grains with water-filled pores (Haszeldine, 2010), an 80-μm view of an oil and water emulsion (Akay, 2010), and a 20-μm view of a fiber-reinforced ceramic composite (MEIAF, 2010)

$$\frac{\mathrm{d}}{\mathrm{d}x}\left(D\frac{\mathrm{d}u}{\mathrm{d}x}\right) = f(x) \quad \text{for} \;\; 0 < x < 1, \tag{5.1}$$

where

$$u(0) = a \;\; \text{and} \;\; u(1) = b. \tag{5.2}$$

Of interest here is when the function D includes a relatively slow variation in x as well as a faster variation over a length scale that is $O(\varepsilon)$. Two examples of this are illustrated in Fig. 5.2. The function in Fig. 5.2a is an example of the type of variation that might be found in a laminated structure such as plywood since D has jump discontinuities but is constant over intervals that are $O(\varepsilon)$. The function in Fig. 5.2b has a similar variation to that in Fig. 5.2a, but it is continuous. We will concentrate on examples of the latter type, but the discontinuous problem is not that much harder (Exercise 5.9). It is worth pointing out that a problem with rapidly varying coefficients, as in (5.1), is not easy to solve, even numerically. This makes the smoothing process underlying homogenization quite valuable when dealing with such situations.

We need to express the sort of variation seen in Fig. 5.2b mathematically, and the way this assumption is incorporated into the problem is to assume $D = D(x, x/\varepsilon)$. As an example, setting $y = x/\varepsilon$ we could have

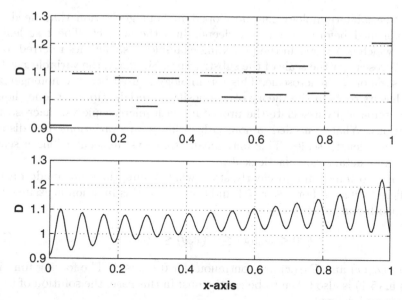

Figure 5.2 Examples illustrating the coefficient D in (5.1) and how it can vary over a length scale that is small in comparison to the length of the interval. In (a) the function is a piecewise constant function, and in (b) the dependence is continuous

$$D(x,y) = \frac{1}{1 + \alpha x + \beta g(x)\cos(y)}. \tag{5.3}$$

This function happens to be periodic in the fast variable (y), but this is not required for the derivation that follows. It is of interest to note that this function is plotted in Fig. 5.2b for the case of $\alpha = -0.1$, $\beta = 0.1$, $g(x) = e^{4x(x-1)}$, and $\varepsilon = 0.01$.

The question we are going to ask is whether or not it is possible to smooth out $D(x, x/\varepsilon)$ and replace it with an averaged value that is independent of ε. One might try to argue that this is possible because the variations of D are over such a small length scale that a first-term approximation that uses an averaged D is reasonable. If so, then the question arises as to what average should be used. One possibility is to simply average over the fast variable. Doing this and using a first-term approximation we get

$$\langle D \rangle_\infty = \lim_{y \to \infty} \frac{1}{y} \int_0^y D(x,r)dr. \tag{5.4}$$

For example, if $D(x, y)$ is given in (5.3), then one finds that

$$\langle D \rangle_\infty = \frac{1}{\sqrt{(1 + \alpha x)^2 - [\beta g(x)]^2}}. \tag{5.5}$$

As we will see below, it is possible to approximate D with an average, but $\langle D \rangle_\infty$ is not the one that should be used. Any ideas what it might be?

A special terminology has developed in homogenization that needs to be explained before getting any deeper into the subject. The fast length scale, which is $y = x/\varepsilon$ in the preceding example, is sometimes referred to as the microscale or the scale of the substructure. Similarly, the variable x is the slow scale or the macroscale. This terminology is used because homogenization has its roots in the theory of composite materials. However, one should not automatically associate the procedure with microscopic vs. macroscopic variations. What is needed, among other things, are variations over disproportionate length scales. This may mean atomic vs. molecular length scales or even planetary vs. galactic scales.

It is also important to clearly state what assumptions are made on the coefficient $D(x, y)$. For $0 \leq x \leq 1$ and $0 < y < \infty$ this function is assumed to be smooth and satisfies

$$0 < D_m(x) \leq D(x, y) \leq D_M(x), \tag{5.6}$$

where $D_m(x)$ and $D_M(x)$ are continuous for $0 \leq x \leq 1$. The forcing function $f(x)$ in (5.1) is also taken to be continuous. In this case, the solution of (5.1), (5.2) is well defined.

Given the disparity of the two length scales in this problem, it is natural to use multiple scales to find an approximation of the solution. Normally we would introduce the scales $x_1 = x/\varepsilon$ and $x_2 = x$. However, with the objective of keeping the notation relatively simple, we will only introduce new notation for the scale $y = x/\varepsilon$ and designate the slow scale simply as x. In this case, the derivative transforms into

$$\frac{d}{dx} \to \frac{1}{\varepsilon}\partial_y + \partial_x.$$

The differential Eq. (5.1) now takes the form

$$(\partial_y + \varepsilon\partial_x)[D(x, y)(\partial_y + \varepsilon\partial_x)u] = \varepsilon^2 f(x). \tag{5.7}$$

A regular multiple-scale expansion is appropriate for this problem, and so we take

$$u \sim u_0(x, y) + \varepsilon u_1(x, y) + \varepsilon^2 u_2(x, y) + \cdots . \tag{5.8}$$

Because the solution of the problem is well behaved, we will assume that each term in the expansion is smooth and a bounded function of y. Now, substituting our expansion in (5.8) into (5.7) leads to the following equations:

$O(1)$ $\partial_y[D(x, y)\partial_y u_0] = 0.$

The general solution of this equation is

$$u_0(x, y) = c_1(x) + c_0(x) \int_{y_0}^{y} \frac{ds}{D(x, s)}, \tag{5.9}$$

where y_0 is any given fixed value of y. It turns out that $c_0 = 0$, and the reason for this is that the preceding integral is not a bounded function of y. This can be shown using the upper bound in (5.6). If $y > y_0$, then

$$\int_{y_0}^{y} \frac{ds}{D_M(x)} \leq \int_{y_0}^{y} \frac{ds}{D(x,s)},$$

and so

$$\frac{y - y_0}{D_M(x)} \leq \int_{y_0}^{y} \frac{ds}{D(x,s)}. \qquad (5.10)$$

The left-hand side of this last inequality becomes infinite as $y \to \infty$. Therefore, if u_0 is to be bounded, it must be that $c_0 = 0$. In what follows, we will write this dependence simply as $u_0 = u_0(x)$ and forgo the use of c_1.

Before proceeding to the $O(\varepsilon)$ equation, we need to make another observation about the integral in (5.9). Using the inequality $D_m(x) \leq D(x,y)$, we can extend the result in (5.10) to

$$\frac{y - y_0}{D_M(x)} \leq \int_{y_0}^{y} \frac{ds}{D(x,s)} \leq \frac{y - y_0}{D_m(x)}. \qquad (5.11)$$

Therefore, the integral is unbounded, but its growth is confined by linear functions in y as $y \to \infty$. This information will be needed subsequently when dealing with the secular terms that appear in the expansion of the solution. We are now ready to go on to the next order equation.

$O(\varepsilon) \quad \partial_y[D(x,y)\partial_y u_1] = -\partial_x u_0 \, \partial_y D.$

The general solution of this equation is

$$u_1(x,y) = b_1(x) + b_0(x) \int_{y_0}^{y} \frac{ds}{D(x,s)} - y\partial_x u_0. \qquad (5.12)$$

Again, the integral becomes unbounded as y increases, but so does the last term. Moreover, from (5.11), they are both $O(y)$. To prevent this from occurring, we will require that these two terms cancel each other as y increases. This is done by imposing the condition

$$\lim_{y \to \infty} \frac{1}{y} \left[b_0(x) \int_{y_0}^{y} \frac{ds}{D(x,s)} - y\partial_x u_0 \right] = 0.$$

This can be rewritten as

$$\partial_x u_0(x) = \langle D^{-1} \rangle_{\infty} b_0(x), \qquad (5.13)$$

where

$$\langle D^{-1} \rangle_\infty \equiv \lim_{y \to \infty} \frac{1}{y} \int_{y_0}^{y} \frac{ds}{D(x, s)}. \tag{5.14}$$

There are several mathematical questions that need to be considered in connection with this average, but we will continue the derivation of the asymptotic approximation and return to these questions later. Our immediate objective is to find $u_0(x)$, and this takes us to the next, and final, equation.

$$O(\varepsilon^2) \; \partial_y[D(x, y)\partial_y u_2] = f(x) - b_0' - \partial_y(D\partial_x u_1).$$

The general solution of this is

$$u_2 = d_1(x) + d_0(x) \int_{y_0}^{y} \frac{ds}{D(x, s)} - \int_{y_0}^{y} \partial_x u_1(x, s) ds + (f - b_0') \int_{y_0}^{y} \frac{s \, ds}{D(x, s)}.$$

The last integral is $O(y^2)$ for large y. There are no other terms in the expression that can cancel this growth, and it is therefore necessary to require

$$b_0'(x) = f(x). \tag{5.15}$$

This is the equation we have been looking for.

After all the preceding work, we are now in a position to state how to determine the first-term approximation of the solution of (5.1), (5.2). From (5.13) and (5.15) we have that the approximation is the solution of the boundary-value problem

$$\frac{d}{dx}\left(\overline{D}\frac{d}{dx}u_0\right) = f(x) \quad \text{for } 0 < x < 1, \tag{5.16}$$

where

$$u_0(0) = a \quad \text{and} \quad u_0(1) = b. \tag{5.17}$$

The coefficient here is

$$\overline{D}(x) \equiv \lim_{y \to \infty} \frac{y}{\int_{y_0}^{y} \frac{ds}{D(x,s)}} \tag{5.18}$$

or, equivalently, $\overline{D}(x) = \langle D^{-1} \rangle_\infty^{-1}$.

It may seem like we have accomplished very little since (5.16) is so similar to the original equation given in (5.1). However, what is significant is that the equation no longer explicitly contains the fast scale. The result is a *homogenized differential equation* with what is called a *homogenized*, or *effective, coefficient*. The fast variation still contributes, albeit indirectly, through this averaged coefficient. The average that appears in (5.18) is the *harmonic mean* of D (this average was first introduced by Archytas, circa 400 BC, who was one of the last warrior mathematicians). The relationship of $\overline{D}(x)$ to the arithmetic, and geometric, mean is given in Exercise 5.8.

Examples

1. If $D(x,y)$ is given in (5.3), then from (5.18),

$$\langle D^{-1}\rangle_\infty = \lim_{y\to\infty} \frac{1}{y}\int_{y_0}^{y}(1+\alpha x + \beta g(x)\cos(s))\mathrm{d}s = 1 + \alpha x. \qquad (5.19)$$

Taking $\alpha = 0$, the homogenized coefficient is $\overline{D} = 1$. Also, letting $f(x) = 0$, $a = 0$, and $b = 1$, the solution of the homogenized problem in (5.16) is

$$u_0(x) = x. \qquad (5.20)$$

The problem, using the $D(x,y)$ given in (5.3), can be solved exactly, and the solution is

$$u(x) = \frac{x + \beta\varepsilon\sin(x/\varepsilon)}{1 + \beta\varepsilon\sin(1/\varepsilon)}. \qquad (5.21)$$

A comparison between these two functions is given in Fig. 5.3. The exact solution does contain a $O(\varepsilon)$ variation in x, but the amplitude of the variation is also $O(\varepsilon)$. Consequently, as ε decreases, the exact solution is a rapidly oscillatory function, but the amplitude of the oscillations goes to zero with ε. In contrast, the coefficient D has a $O(\varepsilon)$ variation in x, but the amplitude of the variation is $O(1)$ (Fig. 5.4). ■

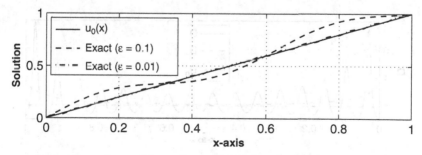

Figure 5.3 Comparison between first-term homogenization approximation given in (5.20) and the exact solution, (5.21), for two values of ε. For $\varepsilon = 10^{-2}$ the asymptotic approximation is essentially indistinguishable from the exact solution. In these calculations $\alpha = 0$, $\beta = 0.83$, and $g(x) = e^{4x(x-1)}$

2. Earlier the question was asked if a good approximation to the problem
   would result if the arithmetic average given in (5.4) was used. In the case
   where D is given in (5.3), the arithmetic average is

$$\langle D \rangle_\infty = \frac{1}{\sqrt{(1 + \alpha x)^2 - [\beta g(x)]^2}}. \tag{5.22}$$

This shows just how bad our first guess was since $\overline{D}(x) = (1 + \alpha x)^{-1}$,
which clearly is independent of β or the function $g(x)$. An illustration of
how these functions can differ is given in Fig. 5.4. ∎

3. As another example, suppose

$$D(x, y) = \frac{\sqrt{1 + y}}{1 + \sqrt{1 + y}}. \tag{5.23}$$

In this case,

$$\int_0^y \frac{ds}{D(x, s)} = y + 2\sqrt{1 + y} - 2, \tag{5.24}$$

and so $\langle D^{-1} \rangle_\infty = 1$. The first term from the homogenization procedure is
just fine. However, note that the approximation for u_1 is

$$u_1 = b_1(x) + 2b_0(x)\left(\sqrt{1 + y} - 1\right). \tag{5.25}$$

It is clear that this function is not bounded. The problem here is that our
assumption of a regular power series, as given in (5.8), is incorrect in this
case. For this problem we should use an expansion in powers of $\varepsilon^{1/2}$. As it
turns out, this modification does not affect the first-term approximation
(Exercise 5.4). ∎

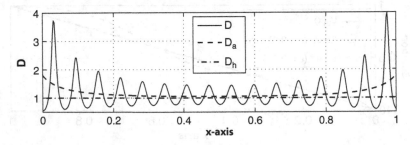

Figure 5.4 Shown are the values of (1) D obtained from (5.3), (2) the averaged
values $D_a = \langle D \rangle_\infty$ as given in (5.22), and (3) the harmonic average as given in
(5.14). In these calculations $\alpha = 0$, $\beta = 0.83$, $g(x) = e^{4x(x-1)}$, and $\varepsilon = 10^{-2}$

5.2.1 Properties of the Averaging Procedure

Because of the central role the harmonic mean plays in the homogenization procedure, it is worth spending some time going over its properties. As stated earlier, it is assumed that $D(x, y)$ is smooth and satisfies

$$0 < D_m(x) \le D(x, y) \le D_M(x). \tag{5.26}$$

In this case, it is not hard to show that $\overline{D}(x)$ is a smooth, positive function that is independent of the lower endpoint y_0. One can also show that (Exercise 5.5)

$$0 < D_m(x) \le \overline{D}(x) \le D_M(x). \tag{5.27}$$

The connection of the harmonic mean with the arithmetic and geometric means is given in Exercise 5.8.

Of particular interest for us is how to determine $\overline{D}(x)$. Besides using the definition given in (5.14) and (5.18), there are also the following two special cases:

1. If $\lim_{y \to \infty} D(x, y) = D_\infty(x)$, then from l'Hospital's rule applied to (5.14) it follows that

$$\overline{D}(x) = D_\infty(x). \tag{5.28}$$

   As an example, if $D(x, y) = (3 + y^2)/(1 + y^2)$, then $\overline{D}(x) = 3$.
2. If $D(x, y)$ is periodic in y, then there is a positive y_p such that $D(x, y + y_p) = D(x, y)$ for all x and y that are under consideration. In this case, $\overline{D}(x) = 1/d(x)$, where

$$d(x) = \frac{1}{y_p} \int_0^{y_p} \frac{ds}{D(x, y_0 + s)}. \tag{5.29}$$

This can be applied to (5.3), where $y_p = 2\pi$, and the result is that $d(x) = 1 + \alpha x$ and $\overline{D}(x) = 1/(1 + \alpha x)$. Note that y_0 in (5.29) is arbitrary, except that it must be a point within the domain for y. From this result, we see that $\overline{D}$ is directly related to the average of D over a period in y. This sort of averaging will play a prominent role in the multidimensional problems examined in later sections.

5.2.2 Summary

The method of homogenization is in many ways simply an extension of multiple scales. One difference is that it appears to be necessary to consider the equation for the third term in the expansion even though two scales are used.

In our example, this was not a problem, but in multidimensional situations it can be very difficult to find the correct secularity condition. This is the primary reason why many who use the method assume that the substructure is periodic. Another characteristic of homogenization is that it generally produces a problem in the slow variable where the coefficients are determined from averages over the fast scale. This situation is discussed in detail in Sect. 5.4.

The procedure can be applied without a great deal of modification to more complicated problems. For example, if we had started with the heat equation

$$\partial_x(D\partial_x u) = \partial_t u + f(x), \tag{5.30}$$

then we would have ended up with the homogenized equation

$$\partial_x(\overline{D}\partial_x u_0) = \partial_t u_0 + f(x). \tag{5.31}$$

It is assumed here that, as before, $D = D(x, y)$. The complication in this case is that the time scale $\tau = t/\varepsilon^2$ appears, and this represents the diffusion time associated with the microscale.

As stated earlier, the homogenization procedure works just as well on problems where the coefficients are discontinuous. An example of this is given in Exercise 5.9. A more interesting situation arises when the equations change from layer to layer (rather than just the coefficients). This occurs, for example, for materials made up of a fluid and solid (like a fluid-saturated sponge) or a bubbly fluid. Problems like these are investigated in Sect. 5.4.

Where the method runs into difficulty is with nonlinear equations. For example, if $D = D(x, y, u)$ in (5.1) or in (5.30), then it is not at all clear how useful the method is because the effective coefficient now depends on the unknown u_0. However, not all nonlinear equations cause complications, as demonstrated in Exercises 5.3, 5.19, and 5.20.

The theory underlying homogenization is extensive. Little has been said about the theory in our development, but a good introduction can be found in Pavliotis and Stuart (2008) and Tartar (2009). The theoretical aspects are also important when incorporating the ideas of homogenization into a numerical algorithm. A discussion of numerical methods for homogenization can be found in Allaire and Brizzi (2005) and Engquist and Souganidis (2008).

Exercises

5.1. For each of the following modifications of (5.1), (5.2) determine the homogenized problem:
(a) $f = f(x, \varepsilon)$, where f has a regular expansion in ε.
(b) $D = D(x, y, \varepsilon)$, where D has a regular expansion in ε.
(c) $D = D(x, g(x)/\varepsilon)$, where $g(x)$ is smooth and $g' > 0$.

5.2. This exercise considers homogenization applied to

$$\frac{d}{dx}\left(D\frac{du}{dx}\right) + p\frac{du}{dx} + qu = f \quad \text{for } 0 < x < 1,$$

where

$$u(0) = a \quad \text{and} \quad u(1) = b.$$

In this problem, $D = D(x, x/\varepsilon)$, $p = p(x, x/\varepsilon)$, $q = q(x, x/\varepsilon)$ are smooth bounded functions, with D positive.
(a) Assuming $f = f(x, x/\varepsilon)$, determine the resulting homogenized problem.
(b) How does the homogenized problem change if $f = g(x, x/\varepsilon)u^n$?
(c) Explain why homogenization does not work so well in the case where $f = g(x, x/\varepsilon, u)$.

5.3. Consider the problem

$$\partial_x(D\partial_x u) + g(u) = \partial_t u + f(x, x/\varepsilon) \quad \text{for } 0 < x < 1,$$

where $u = 0$ when $x = 0, 1$ or $t = 0$. Assume $D = D(x, x/\varepsilon)$.
(a) Find the homogenized problem for the steady state. Make sure to point out what assumptions you need to impose on f so the procedure can be used.
(b) Find the homogenized problem for the time-dependent problem.

5.4. Find the exact solution when D is given in (5.23). Also, derive the homogenized problem and find its solution.

5.5. Assume $D(x, y)$ is continuous.
(a) Show $\overline{D}(x)$ is independent of y_0.
(b) If $D(x, y)$ satisfies (5.26), then show that $\overline{D}(x)$ satisfies (5.27).

5.6. What differences, if any, are there between $\overline{D}(x)$ and D_{avg} for the special cases in (5.28) and (5.29)?

5.7. Suppose $D(x, y)$ is periodic in y. Assuming the terms in the expansion (5.8) are periodic in y (with the same period as D), derive the homogenized problem. Note the $O(\varepsilon^2)$ equation does not have to be solved in this case as the homogenized equation can be obtained from a solvability condition.

5.8. The arithmetic mean of $D(x, y)$, with respect to y, is $A(D) = \langle D \rangle_\infty$, and the geometric mean is

$$G(D) \equiv \lim_{y \to \infty} \exp\left[\frac{1}{y}\int_{y_0}^{y} \ln(D(x, s))ds\right].$$

Show that $\overline{D}(x) \le G(D) \le A(D)$. (Hint: Jensen's inequality is useful here.)

5.9. Consider a layered material where the equation for the ith layer is

$$\frac{d}{dx}\left(D_i\frac{du}{dx}\right) = f(x) \quad \text{for } x_{i-1} < x < x_i.$$

The layers are such that $0 = x_0 < x_1 < x_2 < \cdots < x_n < x_{n+1} = 1$, where $x_{i+1} - x_i = O(\varepsilon)$ and $\varepsilon = 1/n$. Each layer has its own coefficient $D_i(x)$, as illustrated in Fig. 5.2a (in the figure the D_i are constants, but this is not assumed in this problem). At the interface between layers, it is required that

$$u(x_i^-) = u(x_i^+) \quad \text{and} \quad \lim_{x\uparrow x_i}(D_iu_x) = \lim_{x\downarrow x_i}(D_{i+1}u_x).$$

Also, each $D_i(x)$ satisfies $0 < D_m(x) \le D_i(x) \le D_M(x)$.

(a) Let $y = (x - \bar{x})/\varepsilon$, where $\bar{x}$ is any fixed point satisfying $0 < \bar{x} < 1$. Assuming $\bar{x}$ is in the i_0th layer, so $x_{i_0-1} \le \bar{x} \le x_{i_0}$, then the neighboring layer interfaces are located at $y_{i_0}, y_{i_0\pm1}, y_{i_0\pm2}, \ldots$. Setting

$$S_j(x) = \sum_{k=1}^{j} \frac{y_{i_0+k} - y_{i_0+k-1}}{D_{i_0+k}},$$

explain why

$$\frac{y_{i_0+j} - y_{i_0}}{D_M(x)} \le S_j(x) \le \frac{y_{i_0+j} - y_{i_0}}{D_m(x)}.$$

This inequality is the discrete version of (5.11).

(b) Using multiple scales, show that in the (i_0+j)th layer the first term in the expansion is $u_0 = a_j(x) + yb_j(x)$. Use the interface conditions to conclude that a_j is independent of j and $b_j = 0$. Thus, we can write $u_0 = u_0(x)$.

(c) Show that in the $(i_0 + j)$th layer the second term in the expansion is $u_1 = c_j(x) + yd_j(x)$. Use the interface conditions to conclude that

$$\partial_x u_0 = \langle\!\langle D^{-1}\rangle\!\rangle_\infty D_{i_0+1}[d_1(x) + \partial_x u_0],$$

where

$$\langle\!\langle D^{-1}\rangle\!\rangle_\infty \equiv \lim_{j\to\infty} \frac{1}{y_{i_0+j}} S_j(x).$$

Comment on what needs to be assumed for the limit of $j \to -\infty$.

(d) From the third term, show that (5.16), (5.17) still holds but the averaging is as defined in part (c).

(e) Suppose the system is biphasic. In particular, for i even the layers have width α and $D_i = D_\alpha$, while if i is odd the layers have width β and $D_i = D_\beta$ (where D_α and D_β are constants). Explain why $\phi_\alpha = \alpha/(\alpha+\beta)$ might be called the volume fraction for the α material and $\phi_\beta = \beta/(\alpha+\beta)$ the volume fraction for the β material. Express the effective coefficient for D in terms of these volume fractions. When is the effective coefficient the harmonic mean of D_α and D_β?

(f) In materials engineering of composites, a rule of mixtures is used, and, applied to the biphasic system in part (e), it states that the effective coefficient is $\overline{D} = \phi_\alpha D_\alpha + \phi_\beta D_\beta$ (Kollar and Springer, 2003). How well does this compare with your answer in (e)?

5.3 Multidimensional Problem: Periodic Substructure

The method of homogenization produces some interesting results in multidimensional problems. To illustrate the ideas involved with using the procedure, consider the inhomogeneous Dirichlet problem

$$\nabla \cdot (D\nabla u) = f(\mathbf{x}) \quad \text{for } \mathbf{x} \in \Omega, \tag{5.32}$$

where

$$u = g(\mathbf{x}), \quad \text{for } \mathbf{x} \in \partial\Omega. \tag{5.33}$$

As before, the coefficient $D = D(\mathbf{x}, \mathbf{x}/\varepsilon)$ is assumed to be positive and smooth. What is different is that we also assume that it is periodic in the fast coordinate. Specifically, there is a vector $\mathbf{y}_p$ with positive entries so that $D(\mathbf{x}, \mathbf{y} + \mathbf{y}_p) = D(\mathbf{x}, \mathbf{y})$ for all $\mathbf{x}$ and $\mathbf{y}$ under consideration. The consequences of this periodicity will be discussed below. It is also necessary to say something about the domain Ω. In what follows it is assumed to be an open connected set with a smooth enough boundary $\partial\Omega$ so that the solution of the preceding problem is well defined.

5.3.1 Implications of Periodicity

The assumption of periodicity plays a pivotal role in the homogenization procedure that is developed subsequently. It is therefore worthwhile to spend a moment or two to consider some of its consequences. In this discussion, we will consider the problem to be in $\mathbb{R}^2$.

An example of the function D that appears in (5.32) is

$$D = 6 + \cos(2y_1 - 3y_2).$$

In this case, the period vector is $\mathbf{y}_p = (\pi, 2\pi/3)$. The periodicity of this function means that if we know its values in the rectangle $\alpha_0 \le y_1 \le \alpha_0 + \pi$, $\beta_0 \le y_2 \le \beta_0 + 2\pi/3$, where α_0 and β_0 are arbitrary, then we can determine it anywhere (Fig. 5.5). This observation motivates our introduction of a *cell*. This is denoted by Ω_p, and it is a rectangle consisting of the points that satisfy $\alpha_0 \leq y_1 \leq \alpha_0 + p_1$ and $\beta_0 \leq y_2 \leq \beta_0 + p_2$, where $\mathbf{y}_p = (p_1, p_2)$ is a period vector. In this definition the numbers α_0 and β_0 are arbitrary, but they are assumed to be given and they must be consistent with Ω. For example, if Ω is the square $0 < x_1 < 1$ and $0 < x_2 < 1$, then we must have $\alpha_0, \beta_0 \geq 0$. Also note that in $\mathbb{R}^3$ a cell Ω_p will have the form of a parallelepiped.

It is possible for the period $\mathbf{y}_p$ to depend on the slow variable $\mathbf{x}$. For example, if $D = 6 + \cos(y_1 e^{x_2} + 4y_2)$, then $\mathbf{y}_p = (2\pi e^{-x_2}, \pi/2)$. However, in the derivation of the homogenization problem that follows, it is assumed that $\mathbf{y}_p$ is independent of $\mathbf{x}$.

An important consequence of periodicity arises when determining what values a function can take on the boundary of a cell. To explain, consider opposing boundary points $\mathbf{y}_\ell$ and $\mathbf{y}_r$, as illustrated in Fig. 5.5b. A one-dimensional version of this situation is shown in Fig. 5.6. If a function, say $w = w(\mathbf{y})$, is to be periodic, then it is necessary that $w(\mathbf{y}_r) = w(\mathbf{y}_\ell)$. Moreover, if it is to be smooth, then it must also be that $\nabla_y w(\mathbf{y}_r) = \nabla_y w(\mathbf{y}_\ell)$. In fact, if it is to be C^2, then we must have $\partial_{y_i} \partial_{y_j} w(\mathbf{y}_r) = \partial_{y_i} \partial_{y_j} w(\mathbf{y}_\ell) \, \forall i, j$. These conditions must hold at all opposing boundary points of a cell.

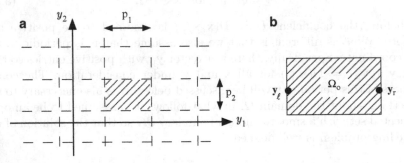

Figure 5.5 (a) Cells in a plane. They are rectangular regions with sides of length p_1 and p_2. (b) A single cell illustrating opposing boundary points $\mathbf{y}_\ell$ and $\mathbf{y}_r$

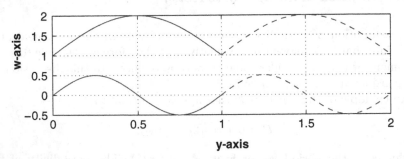

Figure 5.6 Graphs of two functions that are periodic with period $y_p = 1$. In the *top curve* the periodic extension, from $0 \leq y \leq 1$ to $1 \leq y \leq 2$, results in a function that is not differentiable at $y = 1$. In contrast, the extension for the *bottom curve* is differentiable. Thus, for opposing boundary points we require $\partial_y w(y_\ell) = \partial_y w(y_r)$

The primary reason for assuming the problem is periodic can be demonstrated with the simple equation $w''(y) = \kappa$. The general solution of this is $w(y) = \frac{1}{2}\kappa y^2 + c_0 y + c_1$. If $w(y)$ is periodic, it must be that $\kappa = 0$. This conclusion can be obtained without solving the differential equation. All we have to do is to integrate the differential equation over a period $(0 \leq y \leq y_p)$ and use the periodicity to conclude that $\kappa = 0$. This labor-saving observation is used in the homogenization procedure to avoid having to solve the $O(\varepsilon^2)$ problem.

It is also worth pointing out that periodicity is a modeling assumption that is often used in applications. For example, the four composite systems shown in Fig. 5.1 are not exactly periodic, but, given the complexity of the substructure, assuming periodicity provides a reasonable approximation.

5.3.2 Homogenization Procedure

The procedure to be used is essentially the same as in the previous section. Thus, letting $\mathbf{y} = \mathbf{x}/\varepsilon$, the derivatives transform as $\nabla \to \nabla_x + \varepsilon^{-1}\nabla_y$, where the subscript indicates the variables involved in the differentiation. In this case, (5.32) takes the form

$$(\nabla_y + \varepsilon\nabla_x) \cdot [D(\nabla_y + \varepsilon\nabla_x)u] = \varepsilon^2 f(\mathbf{x}). \qquad (5.34)$$

A regular multiple-scale expansion is appropriate here, and so

$$u \sim u_0(\mathbf{x}, \mathbf{y}) + \varepsilon u_1(\mathbf{x}, \mathbf{y}) + \varepsilon^2 u_2(\mathbf{x}, \mathbf{y}) + \cdots . \qquad (5.35)$$

Because of the assumption on D, we will assume that the terms in this expansion are periodic in $\mathbf{y}$ with period $\mathbf{y}_p$.

With the expansion given in (5.35), the following equations are obtained:

$O(1)$ $\nabla_y \cdot (D\nabla_y u_0) = 0.$

> The general solution of this equation, which is bounded, is $u_0 = u_0(\mathbf{x})$. Readers interested in a plausible explanation of why this holds may recall Liouville's theorem from complex variables. It states that the only bounded solutions of Laplace's equation, over $\mathbb{R}^2$, are constants. The proof for the case where D is not constant, or when considering $\mathbb{R}^3$, requires more function theory (Pavliotis and Stuart, 2008).

$O(\varepsilon)$ $\nabla_y \cdot (D\nabla_y u_1) = -\nabla_y D \cdot \nabla_x u_0.$

> Because u_1 is periodic in $\mathbf{y}$, it suffices to solve this equation in a cell Ω_p and then simply extend the solution using periodicity. Two other important observations are that u_0 does not depend on $\mathbf{y}$ and the

problem for u_1 is linear. Therefore, the general solution can be written as

$$u_1(\mathbf{x}, \mathbf{y}) = \mathbf{a} \cdot \nabla_x u_0 + c(\mathbf{x}), \qquad (5.36)$$

where the components $a_i(\mathbf{x}, \mathbf{y})$ of the vector $\mathbf{a}(\mathbf{x}, \mathbf{y})$ are periodic functions in $\mathbf{y}$ that satisfy the equation

$$\nabla_y \cdot (D\nabla_y a_i) = -\partial_{y_i} D \text{ for } \mathbf{y} \in \Omega_p.$$

It is not possible to determine the a_i until the coefficient function D is specified. Also, the function $c(\mathbf{x})$ in (5.36) is arbitrary except for being independent of $\mathbf{y}$.

$$O(\varepsilon^2) \ \nabla_y \cdot [D(\nabla_y u_2 + \nabla_x u_1)] + \nabla_x \cdot [D(\nabla_y u_1 + \nabla_x u_0)] = f(\mathbf{x}).$$

The periodicity assumption will be used to derive the homogenized equation for u_0. To do this, we introduce the average of a function $v(\mathbf{x}, \mathbf{y})$ over a cell Ω_p. This is defined as

$$\langle v \rangle_p \equiv \frac{1}{|\Omega_p|} \int_{\Omega_p} v(\mathbf{x}, \mathbf{y}) dV_y, \qquad (5.37)$$

where $|\Omega_p|$ is the volume (or the area if in $\mathbb{R}^2$) of the cell. We will average the $O(\varepsilon^2)$ equation. First note that, using the divergence theorem,

$$\langle \nabla_y \cdot [D(\nabla_y u_2 + \nabla_x u_1)] \rangle_p = \frac{1}{|\Omega_p|} \int_{\partial\Omega_p} D\mathbf{n} \cdot (\nabla_y u_2 + \nabla_x u_1) \, dS_y$$

$$= 0.$$

The fact that the preceding integral around the boundary is zero comes from the periodicity. Also, from (5.36),

$$\langle D\partial_{y_i} u_1 \rangle_p = \langle D\partial_{y_i} (\mathbf{a} \cdot \nabla_x u_0) \rangle_p$$

$$= \langle D\partial_{y_i} \mathbf{a} \rangle_p \cdot \nabla_x u_0. \qquad (5.38)$$

The other terms in the $O(\varepsilon^2)$ equation average to produce $\langle f \rangle_p = f$ and $\langle \nabla_x \cdot (D\nabla_x u_0) \rangle_p = \nabla_x \cdot (\langle D \rangle_p \nabla_x u_0)$.

The resulting homogenized problem therefore has the form

$$\nabla_x \cdot (\mathbf{D}\nabla_x u_0) = f(\mathbf{x}) \quad \text{for } \mathbf{x} \in \Omega, \qquad (5.39)$$

where

$$u_0 = g(\mathbf{x}), \quad \text{for } \mathbf{x} \in \partial\Omega. \qquad (5.40)$$

In $\mathbb{R}^2$ the homogenized coefficients are

$$D = \begin{pmatrix} \langle D \rangle_p + \langle D \partial_{y_1} a_1 \rangle_p & \langle D \partial_{y_1} a_2 \rangle_p \\ \langle D \partial_{y_2} a_1 \rangle_p & \langle D \rangle_p + \langle D \partial_{y_2} a_2 \rangle_p \end{pmatrix}. \tag{5.41}$$

The functions a_j are smooth periodic solutions of the cell problem

$$\nabla_y \cdot (D \nabla_y a_j) = -\partial_{y_j} D \quad \text{for } \mathbf{y} \in \Omega_p. \tag{5.42}$$

At this point, there are two options. We can take the point of view that (5.39), (5.40) is the problem to solve and essentially ignore where it comes from. In this case, the coefficients D_{ij} are assumed to be given. This can be described as a macro viewpoint and is typically the approach used in elementary mathematical physics textbooks.

Another option open to us is to determine exactly how the homogenized coefficients depend on the substructure. This means the D_{ij} are not given, and to find them it is necessary to solve the cell problems in (5.42). The most common way to do this is numerically, but it is also possible to find examples where analytical solutions can be obtained.

Example

Suppose, in $\mathbb{R}^2$, the cell Ω_p is the rectangular region $0 \le y_1 \le a, 0 \le y_2 \le b$. To determine the homogenized coefficients in (5.41), it is necessary to solve the cell problem. In this example, we will take

$$D(x_1, x_2, y_1, y_2) = D_0(x_1, x_2)e^{\alpha(y_1)+\beta(y_2)}, \tag{5.43}$$

where $\alpha(y_1)$ and $\beta(y_2)$ are periodic. For example, one might have $\alpha(y_1) = 5\sin(2\pi y_1/a)$ and $\beta(y_2) = \cos(6\pi y_2/b)$. The form of the coefficient given in (5.43) results in a cell problem that is solvable in closed form because D is separable in terms of its dependence on the components of $\mathbf{y}$. With (5.43), the solution for u_1 given in (5.36) is

$$u_1(\mathbf{x}, \mathbf{y}) = c(\mathbf{x}) + \mathbf{a} \cdot \nabla_x u_0,$$

where

$$a_1 = -y_1 + \kappa_1 \int_0^{y_1} e^{-\alpha(s)} ds \quad \text{and} \quad a_2 = -y_2 + \kappa_2 \int_0^{y_2} e^{-\beta(s)} ds.$$

The constants κ_1 and κ_2 are determined from the requirement that u_1 must be periodic. To have $a_1(0) = a_1(a)$ and $a_2(0) = a_2(b)$, one finds that

$$\kappa_1 = a \left(\int_0^a e^{-\alpha(s)} ds \right)^{-1} \quad \text{and} \quad \kappa_2 = b \left(\int_0^b e^{-\beta(s)} ds \right)^{-1}.$$

With this it follows that $\partial_{y_1} a_2 = \partial_{y_2} a_1 = 0$, and so, from (5.41), we have that $D_{12} = D_{21} = 0$. Also,

$$\langle D\partial_{y_1} a_1 \rangle_p = \frac{1}{ab} \int_0^a \int_0^b (-1 + \kappa_1 e^{-\alpha}) D_0 e^{\alpha+\beta} dy_1 dy_2$$

$$= -\langle D \rangle_p + D_0 \left(\frac{1}{b} \int_0^b e^{\beta(s)} ds \right) \left(\frac{1}{a} \int_0^a e^{-\alpha(s)} ds \right)^{-1}$$

and

$$\langle D\partial_{y_2} a_2 \rangle_p = -\langle D \rangle_p + D_0 \left(\frac{1}{a} \int_0^a e^{\alpha(s)} ds \right) \left(\frac{1}{b} \int_0^b e^{-\beta(s)} ds \right)^{-1}.$$

Therefore, the homogenized differential Eq. (5.39) is

$$\partial_{x_1}(D_1 \partial_{x_1} u_0) + \partial_{x_2}(D_2 \partial_{x_2} u_0) = f(\mathbf{x}), \qquad (5.44)$$

where $D_i(\mathbf{x}) = \lambda_i D_0(\mathbf{x})$ for

$$\lambda_1 = \left(\frac{1}{b} \int_0^b e^{\beta(s)} ds \right) \left(\frac{1}{a} \int_0^a e^{-\alpha(s)} ds \right)^{-1} \qquad (5.45)$$

and

$$\lambda_2 = \left(\frac{1}{a} \int_0^a e^{\alpha(s)} ds \right) \left(\frac{1}{b} \int_0^b e^{-\beta(s)} ds \right)^{-1}. \qquad (5.46)$$

Interestingly, in this multidimensional example we do not get the harmonic mean. Rather, for each D_i we get the harmonic mean of one of the subscale components of D multiplied by the arithmetic mean in the other subscale component. ■

Several observations should be made at this point. First, if we consider (5.32) as a steady-state heat equation, then we have started out with a material that is inhomogeneous and isotropic. What comes out of the homogenization procedure is a steady-state heat equation for an inhomogeneous and anisotropic material. This means that a material that is isotropic on the microscale can be anisotropic on the macroscale. The degree of anisotropy depends on the coefficient D. Also, note that the contribution of the substructure is entirely through the homogenized coefficients D_{ij}. To find these terms, it is necessary to determine the solution of the cell equation given in (5.42). This cell problem needs to be solved only once, as it does not depend on the boundary conditions on $\partial\Omega$ or the forcing function. This situation, that the cell problem is not explicit in the homogenized equation, makes it similar to what are called hidden variables in thermoviscoelasticity (Fung, 1965). Numerous mathematical questions are also associated with the foregoing analysis. For one thing, it is not obvious that the homogenized problem is well posed. Related to this is the question of whether or not the coefficients D_{ij} are positive. The reader is referred to Pavliotis and Stuart (2008) and Tartar (2009), where some of these questions are addressed.

Exercises

5.10. In the study of the diffusion of a disease through a population one finds the following problem:

$$\nabla \cdot (D\nabla S) - \beta SI = 0,$$
$$\nabla \cdot (D\nabla I) + \beta SI - \lambda I = 0,$$

where β and λ are positive constants. The variables S and I are the densities of susceptible and infected populations, respectively. Also, the function $D = D(\mathbf{x}, \mathbf{x}/\varepsilon)$ is assumed to be positive, smooth, and periodic in the fast coordinate. Find the homogenized version of this problem. Note that a somewhat different version of this problem is analyzed in Garlick et al. (2011).

5.11. The homogenized Eq. (5.39) should equal the original (5.32) if D is independent of $\mathbf{y}$. For example, this happens in (5.44)–(5.46) if $\alpha = \beta = 0$. Is this true in general?

5.12. This exercise examines some of the properties of the homogenized coefficients.
(a) Suppose D is independent of y_2, that is, $D = D(x_1, x_2, y_1)$. What does (5.39)–(5.41) reduce to in this case?
(b) The Voigt–Reuss bound states that the eigenvalues λ of $\mathbf{D}$ satisfy $\langle D^{-1} \rangle_{\mathrm{p}}^{-1} \leq \lambda \leq \langle D \rangle_{\mathrm{p}}$ (i.e., the eigenvalues are bounded by the harmonic and arithmetic averages of D). Does your matrix in part (a) satisfy this bound?
(c) Does the $\mathbf{D}$ in (5.44) satisfy the Voigt–Reuss bound?

5.13. In $\mathbb{R}^2$ suppose $D(x_1, x_2, y_1, y_2) = D(x_2, x_1, y_2, y_1)$. Because of this symmetry, you might expect that $D_{11} = D_{22}$ and $D_{12} = D_{21}$. Is this the case?

5.14. Find the homogenized equation in the case where the period $\mathbf{y}_{\mathrm{p}}$ depends on $\mathbf{x}$. You can assume that the problem is in $\mathbb{R}^2$.

5.15. Suppose Green's function can be found for the cell problem. Explain why the solution of (5.42) can, in this case, be written as $a_i = G[\partial_{s_i} D]$. Also, show that the homogenized coefficients can now be written as

$$D_{ij} = \langle D \rangle_{\mathrm{p}} \delta_{ij} + \langle (\partial_{y_i} D) G[\partial_{s_j} D] \rangle_{\mathrm{p}}.$$

5.16. Consider the wave equation

$$\partial_x(D\partial_x u) = \partial_t^2 u \quad \text{for} \quad -\infty < x < \infty \quad \text{and} \quad 0 < t,$$

where $u(x,0) = f(x)$ and $\partial_t u(x,0) = 0$. In this problem, assume $D = D(x/\varepsilon)$, where $D(y)$ is smooth, positive, and periodic.

(a) Introducing the scales x, $y = x/\varepsilon$, $t_1 = t/\varepsilon$, and $t_2 = \varepsilon t$, show that the first term in the expansion has the form

$$u_0(x, y, t_1, t_2) = \sum A_n(x, t_2) F_n(y) \sin(\lambda_n t_1).$$

The substructure modes F_n can be determined once the coefficient D has been specified. This form of the solution is known as a Bloch expansion (Santosa and Symes, 1991; Conca and Vanninathan, 1997).

(b) Use a similar modal expansion to determine u_1. Is it necessary to restrict D to guarantee that secular terms will not be present in u_1?

(c) Using a modal expansion, find u_0 from the $O(\varepsilon^2)$ problem.

(d) The first-term approximation for the solution of the original problem depends on the substructure scale. Suppose we decide that it is sufficient to have an approximation obtained by averaging the solution over a spatial interval that is between the microscale $O(\varepsilon)$ and the macroscale $O(1)$. Explain why this leads to the following approximation:

$$u(x,t) \sim \frac{1}{y_p} \int_0^{y_p} u_0(x, y, t/\varepsilon, \varepsilon t) dy.$$

Carry out this average using your result from part (c) to obtain an approximation of the solution.

5.4 Porous Flow

One of the more interesting applications of homogenization is in the study of materials made up of multiple constituents. To illustrate the ideas, we will consider the flow of a viscous fluid through a porous solid. This particular problem is important in a wide variety of disciplines, such as in geophysics when studying the motion of a fluid, like water or oil, through sand. An example of the substructure geometry that is being considered is given in Fig. 5.7. The essential assumption concerning the geometry is that the length scale ℓ for the pores and other substructure components is small compared to the macroscopic length scale L. This disparity in length scales is what provides us with our expansion parameter $\varepsilon = \ell/L$.

The solid region is taken to be rigid, and the fluid is assumed to be incompressible and viscous. Letting R_f designate the fluid region, in R_f the following equations hold:

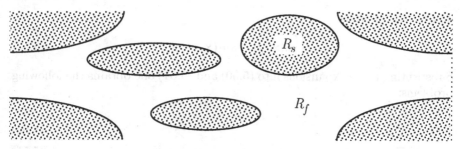

Figure 5.7 Magnified view of substructure of porous medium. It is assumed that the substructure length scales for the fluid (R_f) and solid (R_s) regions are much smaller than the macroscale. This is an idealistic version of the quartz/water configuration shown in Fig. 5.1

$$\varepsilon\nabla^2\mathbf{v} = \nabla p, \tag{5.47}$$

$$\nabla\cdot\mathbf{v} = 0, \tag{5.48}$$

where $\mathbf{v}(\mathbf{x},t)$ is the fluid velocity vector and $p(\mathbf{x},t)$ the fluid pressure. At the fluid–solid interface, ∂R_{fs}, the no-slip condition applies. Thus, at this interface

$$\mathbf{v} = \mathbf{0} \text{ for } \mathbf{x}\in\partial R_{\mathrm{fs}}. \tag{5.49}$$

What makes the problem difficult is the complexity of the region R_f. A closed-form solution is essentially unachievable, and even a numerical solution would be nearly impossible to calculate except for the simplest substructure geometries. Therefore, a procedure like homogenization is a particularly valuable tool for such a problem. It should be kept in mind that our objective is not to solve the problem. Rather, we will derive a set of equations that hold over a much simpler domain. The contributions of the substructure will come through the coefficients of the homogenized equation.

5.4.1 Reduction Using Homogenization

To describe the substructure near $\mathbf{x}_0$, we use the variable $\mathbf{y} = (\mathbf{x} - \mathbf{x}_0)/\varepsilon$. With this, (5.47) and (5.48) take the form

$$\varepsilon(\nabla_y + \varepsilon\nabla_x)^2\mathbf{v} = (\nabla_y + \varepsilon\nabla_x)p, \tag{5.50}$$

$$(\nabla_y + \varepsilon\nabla_x)\cdot\mathbf{v} = 0. \tag{5.51}$$

For small ε the appropriate expansion of the fluid velocity is

$$\mathbf{v} \sim \mathbf{v}_0(\mathbf{x},\mathbf{y}) + \varepsilon\mathbf{v}_1(\mathbf{x},\mathbf{y}) + \cdots, \tag{5.52}$$

and for the pressure

$$p \sim p_0(\mathbf{x}, \mathbf{y}) + \varepsilon p_1(\mathbf{x}, \mathbf{y}) + \cdots . \tag{5.53}$$

Substituting these expansions into (5.50) and (5.51) one obtains the following problems:

$$O(1) \quad \nabla_y p_0 = \mathbf{0}, \tag{5.54}$$
$$\nabla_y \cdot \mathbf{v}_0 = 0. \tag{5.55}$$

From (5.49), the interfacial boundary condition is

$$\mathbf{v}_0 = \mathbf{0} \text{ for } \mathbf{x} \in \partial R_{\text{fs}}. \tag{5.56}$$

It follows from (5.54) that

$$p_0 = p_0(\mathbf{x}). \tag{5.57}$$

It remains to solve (5.55), and this will be done in conjunction with solving the $O(\varepsilon)$ equation for the pressure.

$$O(\varepsilon) \quad \nabla_y p_1 + \nabla_x p_0 = \nabla_y^2 \mathbf{v}_0, \tag{5.58}$$
$$\nabla_y \cdot \mathbf{v}_1 + \nabla_x \cdot \mathbf{v}_0 = 0. \tag{5.59}$$

Unlike the other examples we have examined in this chapter, the first term in the expansion for the velocity depends on the substructure variable $\mathbf{y}$. This dependence is determined by solving (5.55), (5.56), and (5.58) for p_1 and $\mathbf{v}_0$ in terms of p_0. This is no small task, as these equations constitute a forced Stokes problem that must be solved in the substructure regions occupied by the fluid. What helps us significantly is that $\nabla_x p_0$ is independent of $\mathbf{y}$. With this, and the linearity of the problem, we can write p_1 and each component of $\mathbf{v}_0$ as linear combinations of the components of $\nabla_x p_0$. Therefore, we have that

$$\mathbf{v}_0 = \mathbf{B} \nabla_x p_0 \tag{5.60}$$

and

$$p_1 = \mathbf{g} \cdot \nabla_x p_0, \tag{5.61}$$

where the matrix $\mathbf{B}(\mathbf{x}, \mathbf{y})$ and vector $\mathbf{g}(\mathbf{x}, \mathbf{y})$ are determined by substituting (5.60) and (5.61) into (5.54) and (5.59). Their exact forms are not needed to complete the homogenization procedure, and a discussion of their properties can be found in Exercise 5.18.

5.4.2 *Averaging*

The next, and final, step in the homogenization process is to average the velocity over a length scale that is between the microscopic and macroscopic scales. This step was not used in the problems considered earlier in this chapter because the first terms in the expansions were always independent of the substructure variable $\mathbf{y}$. This is not the case for the porous flow problem.

To simplify the averaging, we will now assume the substructure is periodic with period cell Ω_p. In conjunction with this, we also assume the terms of the expansions in (5.52) and (5.53) are periodic with period cell Ω_p. In this case, the average of a dependent variable $\zeta(\mathbf{x})$ at $\mathbf{x}_0$ is defined as

$$\langle \zeta \rangle \equiv \frac{1}{|B_\varepsilon|} \int_{B_\varepsilon} \zeta(\mathbf{s}) dV_s, \qquad (5.62)$$

where $B_\varepsilon = \{\mathbf{s} : |\mathbf{s} - \mathbf{x}_0| < \sqrt{\varepsilon}\}$ and $|B_\varepsilon|$ is the volume of B_ε. Letting $\zeta_0(\mathbf{x}_0, \mathbf{y})$ be a first-term, multiple-scale approximation of ζ for small ε,

$$\langle \zeta \rangle \sim \frac{(\text{number of cells in } B_\varepsilon)}{|B_\varepsilon|} \int_{B_\varepsilon \cap \Omega_p} \zeta_0\left(\mathbf{x}_0, \frac{\mathbf{s} - \mathbf{x}_0}{\varepsilon}\right) dV_s$$

$$= \frac{(\text{number of cells in } B_\varepsilon)\varepsilon^3}{|B_\varepsilon|} \int_{\Omega_f} \zeta_0(\mathbf{x}_0, \mathbf{y}) dV_y$$

$$\sim \frac{1}{|\Omega_p|} \int_{\Omega_f} \zeta_0(\mathbf{x}_0, \mathbf{y}) dV_y, \qquad (5.63)$$

where $|\Omega_p|$ is the volume of Ω_p and Ω_f is the region occupied by the fluid in Ω_p. A variable of particular importance associated with this average is the volume fraction ϕ_f of the fluid. This comes from the average of the characteristic function of the fluid, that is,

$$\zeta = \begin{cases} 1 & \text{if } \mathbf{x} \in R_f, \\ 0 & \text{otherwise.} \end{cases}$$

In this case, $\langle \zeta \rangle \sim \phi_f$, where

$$\phi_f(\mathbf{x}_0) = \frac{|\Omega_f|}{|\Omega_p|} \qquad (5.64)$$

and $|\Omega_f|$ is the volume of Ω_f. In what follows, it is assumed that Ω_p is constant, but Ω_f may depend on $\mathbf{x}$.

Before going any further, it is necessary to state some of the properties of the boundary of the fluid domain Ω_f in a period cell. In particular, $\partial \Omega_f$ consists of two parts. One is the fluid–solid interface within Ω_p, which is denoted by $\partial \Omega_{fs}$, and the other is the portion that lies on $\partial \Omega_p$ (Fig. 5.7). It is assumed that these surfaces are smooth enough that the divergence theorem can be used.

To carry out the averaging, we will use Reynold's transport theorem, which states (Holmes, 2009)

$$\partial_{x_i} \int_{\Omega_f} \zeta(\mathbf{x}, \mathbf{y}) dV_y = \int_{\Omega_f} \partial_{x_i} \zeta(\mathbf{x}, \mathbf{y}) dV_y + \int_{\partial \Omega_f} \zeta \left(\frac{\partial \mathbf{r}}{\partial x_i} \right) \cdot \mathbf{n} \, dS_y, \qquad (5.65)$$

where $\mathbf{r} = \mathbf{r}(\mathbf{x}, \mathbf{y})$ describes $\partial \Omega_{fs}$ and $\mathbf{n}$ is the unit outward normal to $\partial \Omega_f$. Note that the surface integral in (5.65) does not contain any points on $\partial \Omega_p$ since this surface has been assumed to be independent of $\mathbf{x}$. It follows from (5.65) and (5.64) that

$$\partial_{x_i} \phi_f = \frac{1}{|\Omega_p|} \int_{\partial \Omega_{fs}} \left(\frac{\partial \mathbf{r}}{\partial x_i} \right) \cdot \mathbf{n} \, dS_y. \qquad (5.66)$$

Because p_0 is independent of $\mathbf{y}$, the easiest equation to average is (5.60). One finds that

$$\langle \mathbf{v}_0 \rangle = -\mathbf{L} \nabla_x p_0, \qquad (5.67)$$

where

$$\mathbf{L}(\mathbf{x}) = -\frac{1}{|\Omega_p|} \int_{\Omega_f} \mathbf{B} \, dV_y. \qquad (5.68)$$

The homogenized equation for the velocity comes from (5.55). Using the divergence theorem, Reynold's transport theorem (5.65), and the interfacial condition (5.56), the following is obtained:

$$\begin{aligned} 0 &= \int_{\Omega_f} (\nabla_y \cdot \mathbf{v}_1 + \nabla_x \cdot \mathbf{v}_0) dV_y \\ &= -\int_{\partial \Omega_f} \mathbf{v}_1 \cdot \mathbf{n} \, dS_y + \nabla_x \cdot \int_{\Omega_f} \mathbf{v}_0 \, dV_y - \sum_{i=1}^{3} \int_{\partial \Omega f} v_{0_i} \left(\frac{\partial \mathbf{r}}{\partial x_i} \right) \cdot \mathbf{n} \, dS_y \\ &= \nabla_x \cdot \int_{\Omega_f} \mathbf{v}_0 \, dV_y. \end{aligned} \qquad (5.69)$$

The first-term approximation of the averaged velocity therefore satisfies the equation

$$\nabla_x \cdot \langle \mathbf{v}_0 \rangle = 0. \qquad (5.70)$$

5.4.3 Homogenized Problem

Suppose the porous material occupies a (macroscopic) domain Ω made up of the solid (R_s) and fluid (R_f) regions. From the homogenization procedure we have derived a fluid velocity $\mathbf{v}_h(\mathbf{x})$ and pressure $p_h(\mathbf{x})$ that are defined throughout Ω. From (5.67) and (5.70), these variables satisfy the homogenized equations

$$\mathbf{v}_h = -\mathbf{L}\nabla_x p_h \qquad (5.71)$$

and

$$\nabla_x \cdot \mathbf{v}_h = 0. \qquad (5.72)$$

Note that (5.72) is the standard continuity equation for an incompressible fluid. The relationship in (5.71) is known as Darcy's law, and the matrix $\mathbf{L}$ is called the permeability tensor (Scheidegger, 1974; Mei and Vernescu, 2010). It is this homogenized coefficient tensor that contains the contributions of the substructure.

The ideas developed for the porous flow problem can also be used to analyze other composite materials. Examples can be found in the exercises as well as in Sanchez-Palencia (1980). His approach, like ours, assumes periodicity of the substructure. However, it is not necessary to make such an assumption, and examples of this are given in Sect. 5.4. An extension to multidimensional problems can be found in Burridge and Keller (1981).

Exercises

5.17. There are ways to smooth out the substructure to describe the flow in a porous solid that do not use homogenization. One of these is mixture theory, and for the case of a rigid solid the mixture equations are (Holmes, 1985)

$$\nabla_x \boldsymbol{\sigma}_M = -\boldsymbol{\pi}_M$$

and

$$\nabla_x \cdot (\phi_f \mathbf{v}_M) = 0.$$

Here $\boldsymbol{\sigma}_M$ is the stress in the fluid, $\mathbf{v}_M$ is the fluid velocity, and the vector $\boldsymbol{\pi}_M$ is assumed to have the form $\boldsymbol{\pi}_M = p_M \nabla_x \phi_f - \mathbf{K}\mathbf{v}_M$.
(a) Determine the mixture quantities ($\boldsymbol{\sigma}_M$, $\mathbf{v}_M$, p_M, and $\mathbf{K}$) in terms of the entries that make up the homogenized Eqs. (5.71) and (5.72).
(b) The averaging used in (5.63) is known as a phase average since it is based on the total volume of the cell. An alternative is the intrinsic phase average, which uses the true volume occupied by the constituent [i.e., $|\Omega_f|$ is used in (5.63) rather than $|\Omega_p|$]. Based on the results of part (a), comment on these averages and the mixture formulation.

5.18. This problem examines some of the properties of the elements of the matrix $\mathbf{B}$ in (5.60) and the matrix $\mathbf{L}$ in (5.68). It is assumed here that the substructure is periodic.

(a) Let $q_j(x, y)$ be the jth column vector from the matrix B. Show that

$$\nabla_y^2 q_j = \nabla_y g_j - e_j$$

and

$$\nabla_y \cdot q_j = 0$$

for $y \in \Omega_f$. What are the boundary conditions? Here $g(x, y)$ is the vector in (5.61) and e_j is the jth unit coordinate vector.

(b) Let $q_j = (\alpha_j, \beta_j, \gamma_j)$ and note, from part (a), that $e_j = \nabla_y g_j - \nabla_y^2 q_j$. For $j = 2$, take the dot product of this equation with $(0, \beta_1, 0)$ and then integrate over Ω_f to find a formula for L_{21}. Similarly, using $(\alpha_2, 0, 0)$ and $j = 1$, find L_{12}. With these formulas, and using the continuity equation, show $L_{12} = L_{21}$. This argument is easily extended to prove that L is symmetric.

(c) Darcy's law is usually applied to macroscopically isotropic porous materials. In such situations, the matrix L is replaced with a scalar. Under what conditions is L, as defined in (5.68), a scalar multiple of the identity?

5.19. This exercise investigates a problem that arises in the study of the diffusion of gases in polymers. In what is known as a dual-mode sorption model, the polymer is a composite with a substructure much like the one shown in Fig. 5.7, except now the fluid regions are called microvoids. The concentration of gas in the two regions is governed by the following equations (Sangani, 1986):

$$\partial_t C_s = D_s \nabla^2 C_s \text{ in } R_s \quad \text{and} \quad \partial_t C_f = D_f \nabla^2 C_f \text{ in } R_f,$$

where C_s and C_f are the concentrations in the respective regions. At the interface ∂R_s (or, equivalently, at ∂R_f) it is required that

$$(D_s \nabla C_s - D_f \nabla C_f) \cdot n = 0 \quad \text{and} \quad C_f = \frac{\alpha C_s}{1 + \varepsilon^2 \beta C_s}.$$

Here n is a normal to the interface and D_s, D_f, α, and β are positive constants (that are independent of ε). It is assumed that the substructure is periodic and does not vary with x.

(a) Assuming that $C_s \sim C_{s_0}(x, y, t) + \varepsilon C_{s_1}(x, y, t) + \cdots$ and $C_f \sim C_{f_0}(x, y, t) + \varepsilon C_{f_1}(x, y, t) + \cdots$, show that C_{s_0} and C_{f_0} are independent of y.

(b) Show that the solution of the $O(\varepsilon^2)$ problem can be written as

$$C_{s_1}(x, y, t) = A_s(x, t) - w_s \cdot \nabla_x C_{s_0},$$

where $w_s(x, y)$ and $A_s(x, t)$ are independent of the concentrations. You should also find a similar expression for $C_{f_1}(x, y, t)$.

(c) From the $O(\varepsilon^3)$ problem, show that $\partial_t C_{so} = \nabla_x \cdot (\mathbf{D}_s \nabla_x C_{so})$, where the diffusivity tensor $\mathbf{D}_s$ is given as

$$\mathbf{D}_s = \frac{1}{|\Omega_s| + \alpha|\Omega_f|} \left[\chi \mathbf{I} - D_s \int_{\Omega_s} (\nabla_y \mathbf{w}_s)^T dV_y - D_f \int_{\Omega_f} (\nabla_y \mathbf{w}_f)^T dV_y \right]$$

for $\chi = D_s |\Omega_s| + \alpha D_f |\Omega_f|$.

(d) What do the results from parts (a)–(c) reduce to in the one-dimensional case? Assume that the period cell is $0 < y < 1$, where the "solid" occupies the interval $0 < y < r$ and the "fluid" region is $r < y < 1$.

5.20. An interesting problem in electrochemistry concerns the diffusion of ions through a solution containing charged molecules. In reference to Fig. 5.7, the molecules form the solid region, and the solution makes up the fluid region. In R_f, (Holmes, 1990)

$$\varepsilon^2 \nabla^2 \phi = z\alpha S e^{z\phi(\mathbf{x},t)}$$

and

$$\partial_t S + z S \partial_t \phi = \nabla^2 S + z \nabla S \cdot \nabla \phi$$

hold, where $\varepsilon \partial_n \phi = \sigma$ and $\partial_n S = 0$ on ∂R_f. Here $\phi(\mathbf{x},t)$ is the electrostatic potential in the solution and $S(\mathbf{x},t)$ is related to the concentration $C(\mathbf{x},t)$ of diffusing ions through the relation $S(\mathbf{x},t) = C(\mathbf{x},t) \exp(-z\phi(\mathbf{x},t))$. This problem arises in the study of the physiology of tissue and in the conformational properties of polynucleotides like DNA. One of the questions of interest is how the geometry and charge density of the charged molecules affect the diffusion of ions.

(a) Using the multiple-scale expansions $\phi(\mathbf{x},t) \sim \phi_0(\mathbf{x},\mathbf{y},t) + \varepsilon \phi_1(\mathbf{x},\mathbf{y},t) + \cdots$ and $S(\mathbf{x},t) \sim 1 + \varepsilon S_1(\mathbf{x},\mathbf{y},t) + \cdots$, show that S_1 is independent of $\mathbf{y}$, and find the problem satisfied by ϕ_0.

(b) Show that the solution of the $O(\varepsilon^2)$ problem can be written as $S_2(\mathbf{x},\mathbf{y},t) = R_2(\mathbf{x},t) - \mathbf{w} \cdot \nabla_x S_1$ and $\phi_1(\mathbf{x},\mathbf{y},t) = S_1(\mathbf{x},t) f(\mathbf{y})$, where $R_2(\mathbf{x},t)$, $\mathbf{w}(\mathbf{y})$, and $f(\mathbf{y})$ are independent of S_1.

(c) From the $O(\varepsilon^3)$ problem, and assuming the substructure is periodic, show that

$$\partial_t S_1 \int_{\Omega_f} e^{z\phi_0} dV_y = \nabla_x \cdot \left[\nabla_x S_1 \int_{\Omega_f} (\mathbf{I} - \nabla_y \mathbf{w}) e^{z\phi_0} dV_y \right].$$

(d) Of particular interest is the concentration $C = S(\mathbf{x},t) e^{-z\phi(\mathbf{x},t)}$. Using the results from parts (a)–(c), show $\langle C \rangle \sim \langle C \rangle_0 + \varepsilon \langle C \rangle_1$, where $\langle C \rangle_0$ is independent of $\mathbf{x}_0$ and t, and

$$\langle C \rangle_1 = S_1(\mathbf{x}_0,t) \frac{\phi_f}{|\Omega_f|} \int_{\Omega_f} [1 + f(\mathbf{y})] e^{z\phi_0(\mathbf{y})} dV_y.$$

(e) Using the two-term expansion in part (d), show that the average concentration satisfies the linear diffusion equation $\partial_t C = \nabla_x \cdot (\mathbf{D} \nabla_x C)$, where the diffusivity tensor $\mathbf{D}$ is given as

$$\mathbf{D} = \mathbf{I} - \frac{\alpha z}{\sigma |\partial \Omega_f|} \int_{\Omega_f} e^{z\phi} (\nabla_y \mathbf{w})^{\mathrm{T}} dV_y.$$

Also, state what problem the potential $\phi(\mathbf{y})$ satisfies.

Chapter 6
Introduction to Bifurcation and Stability

6.1 Introduction

On several occasions when working out examples in the earlier chapters, we came across problems that had more than one solution. Such situations are not uncommon when studying nonlinear problems, and we are now going to examine them in detail. The first step is to determine when multiple solutions appear. Once the solutions are found, the next step is to determine if they are stable. Thus, we will focus our attention on what is known as linear stability theory. In terms of perturbation methods, almost all the tools we need were developed in earlier chapters. For example, the analysis of steady-state bifurcation uses only regular expansions (Chap. 1), and the stability arguments will use regular and multiple-scale expansions (Chap. 3). On certain examples, such as when studying relaxation dynamics, we will use matched asymptotic expansions (Chap. 2).

It is worth making a comment or two about the approach used in this chapter compared to most other textbooks on this subject. The conventional method is to first express the problem as a dynamical system and then prove results related to stability using the eigenvalues of a Jacobian matrix. The approach here concentrates more on the construction of an approximation of the solution, and stability considerations are a consequence of this analysis. Also, the objective is to develop methods that are applicable to a wide variety of problems, including nonlinear partial differential equations and integral equations. This does not mean that tidy little formulas related to the stability of particular problems are avoided, but they will not play a prominent role in the presentation.

M.H. Holmes, *Introduction to Perturbation Methods*, Texts in Applied Mathematics 20, DOI 10.1007/978-1-4614-5477-9_6, © Springer Science+Business Media New York 2013

6.2 Introductory Example

Before presenting the general formulation, it is worth while to briefly examine a relatively simple example. One that has a number of interesting characteristics is the nonlinear oscillator equation

$$y'' + 2\beta y' - \lambda y + y^3 = 0 \quad \text{for} \quad < t. \tag{6.1}$$

This is a Duffing-type equation, and we have come across these cubically nonlinear oscillators before. We begin by investigating the steady states. From (6.1) we see that they satisfy

$$\lambda y - y^3 = 0. \tag{6.2}$$

The number of solutions of this equation depends on the parameter λ. That is, $y = 0$ is a solution for all values of λ, and $y = \pm\sqrt{\lambda}$ are solutions if $\lambda \geq 0$. These steady states are shown in Fig. 6.1.

Given the existence of multiple steady states, it is natural to ask if any of them are achievable. In other words, will the solution of (6.1) approach one of the three steady states as $t \to \infty$? A partial answer to this question can be inferred from Fig. 6.2, where the numerical solution of (6.1) is shown for different initial values. When starting at $(y, y') = (0, 1)$, the solution spirals into the steady state $y_s = 1$, while the initial point $(y, y') = (0, 2)$ results in the solution spiraling into the steady state $y_s = -1$. In fact, one finds that when $\lambda > 0$, it is easy to find nonzero initial conditions that produce $y_s = 1$ or $y_s = 1$ but none that produces $y_s = 0$. For $\lambda < 0$ just the opposite occurs. The implication from these numerical experiments is that the stability of $y_s = 0$ changes at $\lambda = 0$. How to determine analytically the stability, or instability, of a steady state will be explained in the next section.

It is convenient to introduce some terminology that will be used throughout this chapter. The point $(\lambda_b, y_b) = (0, 0)$ where the solutions intersect is called a *bifurcation point*, and λ is the *bifurcation parameter*. Also, because of the way the solution curves come together, this is known as a pitchfork bifurcation (presumably, the possibility of calling it a trident bifurcation was also considered but it lost out to pitchfork). In any case, we will say that the

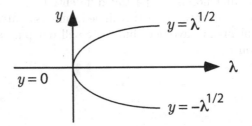

Figure 6.1 Bifurcation diagram for steady-state solutions of (6.1)

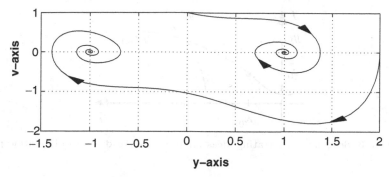

Figure 6.2 Numerical solution of (6.1) using the initial conditions (1) $y(0) = 0$, $y'(0) = 1$ and (2) $y(0) = 2$, $y'(0) = 0$. In these calculations, $\beta = 1/4$ and $\lambda = 1$. The possible steady states in this case are $0, \pm 1$

solutions $y = \pm\sqrt{\lambda}$ bifurcate from the solution $y = 0$, and we will refer to each of these curves as a branch of the solution. It is also important to point out that inherent in identifying $(\lambda_b, y_b) = (0, 0)$ as a bifurcation point is the observation that the stability of one of the solutions changes at this point.

6.3 Analysis of a Bifurcation Point

The steady-state Eq. (6.2) for the oscillator equation in (6.1) is simple enough that all the solutions can be found in closed form. However, this is not possible for an equation such as

$$y'' + 2\beta y' + \lambda y + \sin^2(y) = 0 \quad \text{for } 0 < t. \tag{6.3}$$

For situations like this, we will employ asymptotic approximations for finding the steady states and determining their stability.

To describe our method, suppose we want to find the solutions of the equation

$$F(\lambda, y) = 0, \tag{6.4}$$

where F is a smooth function of λ and y. To start, we are interested in the situation where we know one solution and are interested if there is a second solution that intersects it. This is shown in Fig. 6.3, where $y_s(\lambda)$ is taken as the known solution. The conditions needed to prevent such an intersection are contained in the implicit function theorem. This result, which is given below, determines what happens in a neighborhood of a point (λ_0, y_0) that satisfies (6.4).

Theorem 6.1. *Given a function $F \in C^1$, assume that $F(\lambda_0, y_0) = 0$ and $\partial_y F(\lambda_0, y_0) \neq 0$. In this case, for λ near λ_0,*

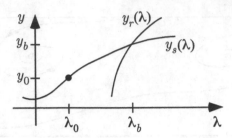

Figure 6.3 Schematic illustrating ideas underlying a steady-state bifurcation point

1. There is a unique solution $y = y_s(\lambda)$ of (6.4) that satisfies $y_s(\lambda_0) = y_0$;
2. If $F \in C^k$, where k is a positive integer, then $y_s \in C^k$.

This means that at a bifurcation point (λ_b, y_b) as shown in Fig. 6.3, the following two equations must hold:

$$F(\lambda_b, y_b) = 0, \qquad (6.5)$$
$$\partial_y F(\lambda_b, y_b) = 0. \qquad (6.6)$$

Examples

1. For the steady states of (6.3), $F(\lambda, y) = \lambda y + \sin^2(y)$. One solution of this is $y_s = 0$, and in this case $\partial_y F(\lambda, 0) = \lambda$. This means that the only possible bifurcation point associated with this solution occurs when $\lambda = 0$. ∎

2. As another example, suppose $F = y^2$. Taking the known solution to be $y_s = 0$, then $\partial_y F(\lambda, 0) = 0$. However, even though (6.5) and (6.5) are satisfied, there is no bifurcation point anywhere along the known solution. ∎

As the preceding examples illustrate, (6.5) and (6.6) are, by themselves, not enough to determine a bifurcation point. Even more importantly, even in the cases where they are sufficient, they provide no indication about what the bifurcating solutions look like. Nevertheless, the implicit function theorem has played an important role in the more theoretical aspects of bifurcation problems, and those interested in this may consult Crandall and Rabinowitz (1980).

There are two steps in analyzing a bifurcation point as shown in Fig. 6.3. First, given $y_s(\lambda)$, one must determine the intersection point and then find the solutions that branch from it. We will do this using what is known as the Lyapunov–Schmidt method. Once this is done, it is then necessary to determine the stability properties of the various solutions, and for this we will use a linearized stability analysis.

6.3.1 *Lyapunov–Schmidt Method*

We start out with the situation shown in Fig. 6.3. In particular, it is assumed we know y_s, and we will use this to determine the possible bifurcation points and the solutions that intersect y_s. The approach we will be using is known as the *Lyapunov–Schmidt method*, and it constructs the solution(s) near a bifurcation value $\lambda = \lambda_b$. This means we are going to look at what happens to the solution when $\varepsilon = \lambda - \lambda_b$ is small. Also, the derivation for the case of a general F is not pretty, and so we will assume that

$$F(\lambda, y) \equiv \lambda y + f(y), \tag{6.7}$$

where $f(y)$ is smooth, $f(0) = 0$, and $f'(0) = 0$. For example, we could have $f(y) = y^2$ or $f(y) = y^3$. This means that $y_s = 0$ is a solution of $F(\lambda, y) = 0$. Moreover, since $\partial_y F(0,0) = 0$, then $(\lambda_b, y_b) = (0,0)$ is a possible bifurcation point.

The first step is to expand $F(\lambda, y)$ in a Taylor series about the point $(\lambda_b, y_b) = (0,0)$. To do this, let n be an integer where $f(0) = f'(0) = f''(0) = \cdots = f^{(n-1)}(0) = 0$ but $f^{(n)}(0) \neq 0$. With this,

$$F(\lambda, y) = \lambda y + f(0) + y f'(0) + \frac{1}{2} y^2 f''(0) + \frac{1}{3!} y^3 f'''(0) + \cdots$$

$$= \lambda y + \frac{1}{n!} y^n f^{(n)}(0) + \cdots .$$

We are looking for the nonzero solutions of $F(\lambda, y) = 0$, and so we require that

$$\lambda + \frac{1}{n!} y^{n-1} f^{(n)}(0) + \cdots = 0. \tag{6.8}$$

Setting

$$\varepsilon = \lambda - \lambda_b \tag{6.9}$$

and assuming that

$$y \sim y_b + \varepsilon^\alpha y_1, \tag{6.10}$$

where $\alpha > 0$ and $y_1 \neq 0$, (6.8) reduces to

$$\varepsilon + \frac{1}{n!} y_1^{n-1} \varepsilon^{(n-1)\alpha} f^{(n)}(0) + \cdots = 0. \tag{6.11}$$

Balancing the terms, we get $(n-1)\alpha = 1$. Thus,

$$\alpha = \frac{1}{n-1} \quad \text{and} \quad y_1^{n-1} = \frac{-n!}{f^{(n)}(0)}. \tag{6.12}$$

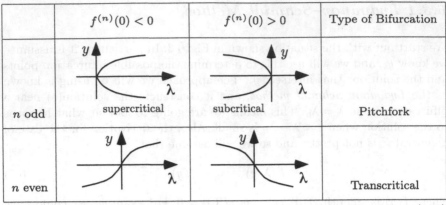

	$f^{(n)}(0) < 0$	$f^{(n)}(0) > 0$	Type of Bifurcation
n odd	supercritical	subcritical	Pitchfork
n even			Transcritical

Table 6.1 Schematic of structure of bifurcating solutions as determined from (6.13) and (6.14)

In writing down the answer it should be remembered that the solution is real-valued, and ε can be positive or negative. With this, the conclusions are as follows:

1. For n even,

$$y \sim y_b + \left(\frac{-n!}{f^{(n)}(0)} (\lambda - \lambda_b) \right)^{\frac{1}{n-1}}. \tag{6.13}$$

2. For n odd,

$$y \sim \begin{cases} y_b \pm \left(\dfrac{-n!}{f^{(n)}(0)} (\lambda - \lambda_b) \right)^{\frac{1}{n-1}} & \text{for } \lambda \geq \lambda_b \text{ if } f^{(n)}(0) < 0, \\[4mm] y_b \pm \left(\dfrac{-n!}{f^{(n)}(0)} (\lambda_b - \lambda) \right)^{\frac{1}{n-1}} & \text{for } \lambda \leq \lambda_b \text{ if } f^{(n)}(0) > 0. \end{cases} \tag{6.14}$$

Sketches of these solutions are given in Table 6.1. For the pitchfork bifurcation, *supercritical* means that the branches that bifurcate from $y_s = 0$ are to the right of the bifurcation value $\lambda = 0$, and *subcritical* means they are to the left.

There are two other types of bifurcation point that appear frequently. One is a Hopf bifurcation, and this will be discussed later (Sect. 6.5). The other is a *saddle-node*, or *tangent*, *bifurcation*. In this case, no solutions bifurcate from one side of the bifurcation point and (at least) two on the other side. An example of this occurs with the function $F \equiv \lambda - y^2$, which has $(0,0)$ as a saddle-node bifurcation point. In fact, it is a supercritical saddle node since the branches occur for $\lambda \geq 0$. Another, somewhat more substantial, example is considered in Sect. 6.3.2.

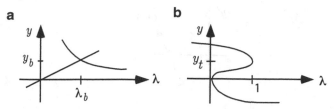

Figure 6.4 Bifurcation diagrams illustrating secondary bifurcation. In each case, $y_s = 0$ is a solution. In (**a**) there are two transcritical bifurcations, one at $(0,0)$, the other at (λ_b, y_b). In (**b**) there is a supercritical pitchfork bifurcation at $(0,0)$ followed by a subcritical saddle node at $(1, y_t)$

It needs to be emphasized that the analysis presented here is local in the sense that it only gives us the solutions in the immediate neighborhood of the bifurcation point. It is very possible that after a branch $y = y_r(\lambda)$ bifurcates from $y_s(\lambda)$, another solution bifurcates from $y_r(\lambda)$, a situation called *secondary bifurcation*. An example of this occurs with the function $F = y(y - \lambda)(y - e^{-\lambda})$. As illustrated in Fig. 6.4a, the branch that bifurcates transcritically from $y_s = 0$ at $(0,0)$ has a transcritical bifurcation at $\lambda_b > 0$. For a similar reason, in the case of a supercritical pitchfork bifurcation one should not expect the branches to be defined for all $\lambda > \lambda_b$, as might seem to be implied from the bifurcation diagram in Fig. 6.1. An example of this situation occurs with the function $F = y[\lambda + y^2(2y - 3)]$, and the bifurcation diagram for this is shown in Fig. 6.4b. There is a supercritical pitchfork bifurcation at $(0,0)$, but the only interval where there are three solutions is $0 < \lambda < 1$.

6.3.2 Linearized Stability

Once the equilibrium solutions are determined, the next question to answer is whether or not they are achievable. In other words, if we start the solution close to a steady state, will it then converge to that steady state as $t \to \infty$ or will it at least stay close to the steady state as $t \to \infty$? This is determined from the stability properties of the solution. In many textbooks in dynamical systems, this question is reduced to determining the eigenvalues of a Jacobian matrix, and we will consider that result later (Sect. 6.6). Our approach here is to derive the result from scratch for this particular example. The reason is that the ideas used in the derivation are applicable to other types of problems, and an example is the delay equation considered in the next section. Other examples include partial differential equations and integral equations, and these are amply represented in the exercises and following sections in this chapter.

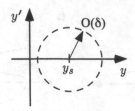

Figure 6.5 Initial conditions used in linearized stability analysis are taken to be from within a $O(\delta)$ region around the steady-state solution

To determine the stability properties of a steady state, we return to the damped oscillator equation

$$y'' + 2\beta y' + F(\lambda, y) = 0 \quad \text{for} \ 0 < t, \tag{6.15}$$

where $\beta > 0$. Assuming $y = y_s$ is a steady state, then the initial conditions we will use to investigate the stability of this solution are

$$y(0) = y_s + \alpha_0 \delta \quad \text{and} \quad y'(0) = \beta_0 \delta. \tag{6.16}$$

The idea here is that we are starting the solution close to the steady state (Fig. 6.5), and so we assume δ is small. In this context, α_0 and β_0 are arbitrary constants. We will also assume, in this discussion, that the value of λ is not such that (λ, y_s) is a bifurcation point. In other words, we will assume that $\partial_y F(\lambda, y_s) \neq 0$.

The appropriate expansion of the solution for small δ is

$$y(t) \sim y_s + \delta y_1(t) + \cdots. \tag{6.17}$$

Substituting this into (6.15) and using Taylor's theorem, we obtain

$$\delta y_1'' + \cdots + 2\beta \delta y_1' + \cdots + F(\lambda, y_s) + \delta y_1 F_y(\lambda, y_s) + \cdots = 0. \tag{6.18}$$

Now, since y_s is an equilibrium solution, $F(\lambda, y_s) = 0$. Thus, the $O(\delta)$ problem we get from (6.18) and (6.16) is

$$y_1'' + 2\beta y_1' + F_y(\lambda, y_s) y_1 = 0, \tag{6.19}$$

where

$$y_1(0) = \alpha_0 \quad \text{and} \quad y_1'(0) = \beta_0. \tag{6.20}$$

The general solution of the equation depends on the value of $F_y(\lambda, y_s)$. One finds that

$$y_1(t) = \begin{cases} a_0 e^{r_+ t} + a_1 e^{r_- t} & \text{if } F_y(\lambda, y_s) \neq \beta^2, \\ (a_0 + a_1 t) e^{r_+ t} & \text{if } F_y(\lambda, y_s) = \beta^2, \end{cases} \tag{6.21}$$

where a_0 and a_1 are constants determined by the initial conditions and

$$r_\pm = -\beta \pm \sqrt{\beta^2 - F_y(\lambda, y_s)}. \qquad (6.22)$$

We are only interested here in whether or not the steady state is stable. From our expansion in (6.17), and our solution (6.21), this can be determined from the signs of the exponents $r_\pm$. The easiest one to determine is r_- since $\mathrm{Re}(r_-) < 0$ [here $\mathrm{Re}(r)$ designates the real part of r]. So if $\mathrm{Re}(r_+) < 0$, then $y_1(t) \to 0$ as $t \to \infty$. From this condition we make the following conclusions:

(a) If $F_y(\lambda, y_s) > 0$, then $y_1(t) \to 0$ as $t \to \infty$ irrespective of the values of the coefficients a_0 and a_1. In this case, the equilibrium solution is said to be *asymptotically stable* (to small perturbations).

(b) If $F_y(\lambda, y_s) < 0$, then $y_1(t)$ remains bounded as $t \to \infty$ only so long as $a_0 = 0$ [assuming $F_y(\lambda, y_s) \neq \beta^2$]. This happens only if the initial conditions in (6.16) satisfy $\beta_0 = \alpha_0 r_-$. Because of this, the steady state is said to be *unstable*.

It should be observed that the cases of unstable and asymptotically stable are separated by those points that are bifurcation points. Also, our conclusions are based on the assumption that the expansion remains well ordered as $t \to \infty$. In certain circumstances, when this is not the case, it is necessary to use something like multiple scales to determine stability and instability.

Examples

1. If $F(\lambda, y) = \lambda y$, then $y = 0$ is an equilibrium solution and $F_y(\lambda, 0) = \lambda$. For this problem, the exact solution has the form given in (6.21). Therefore, we need not limit the initial conditions to a small $O(\delta)$ neighborhood of the equilibrium point to determine stability. What we conclude is that the solution is globally asymptotically stable if $\lambda > 0$ and unstable if $\lambda < 0$. By *globally asymptotically stable* is meant that any initial conditions will result in $y(t) \to y_s$ as $t \to \infty$. ∎

2. If $F(\lambda, y) = -\lambda y + y^3$, then the bifurcation diagram is as shown in Fig. 6.1. Since $F_y(\lambda, 0) = -\lambda$ and $F_y(\lambda, \pm\sqrt{\lambda}) = 2\lambda$, the stablity/instability of each branch is as indicated in Fig. 6.6. Note that there is an *exchange of stability* at $(0, 0)$. ∎

3. Suppose $F(\lambda, y) = \lambda y - e^y$ for $0 < \lambda$. Although λ is the bifurcation parameter, to determine the steady states, it is easier to solve for λ and sketch the solution as a function of y. From this it is seen that there are no steady states for $0 < \lambda < e$ and two for $e < \lambda$. The bifurcation point in this case

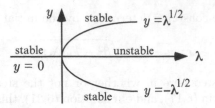

Figure 6.6 Bifurcation diagram for steady-state solutions of (6.1) and their stability. There is an exchange of stability at the bifurcation point

is $(\lambda_b, y_b) = (e, 1)$, and the solutions bifurcating from this point can be found by assuming $y \sim 1 + \varepsilon^\alpha y_1$. From this one finds that

$$y_\pm \sim 1 \pm \sqrt{\frac{2}{e}(\lambda - e)} \quad \text{for } 0 < \lambda - e \ll 1.$$

As for stability, $F_y(\lambda, y_\pm) \sim \mp\sqrt{2e(\lambda - e)}$, which means y_+ is unstable and y_- is asymptotically stable. ∎

The preceding discussion of stability was based on the assumption that the damping parameter β in (6.1) and (6.15) is positive. The case where $\beta < 0$, something known as *negative damping*, produces only unstable solutions, as is evident from (6.22). The undamped case, which occurs when $\beta = 0$, is somewhat harder to analyze. In this case, the exponents in (6.21) are imaginary, and the resulting solution is therefore periodic. This means that the expansion in (6.17) remains well ordered, or at least seems to from what we have derived. The reason for the hesitation is that from the Taylor expansion of the function $F(\lambda, y)$, the function $y_1(t)$ is going to appear in the higher-order problems. As demonstrated in Chap. 3, this may generate secular terms. Since we are trying to determine the long-time behavior of the solution, this development is important, and to deal with this one can use multiple scales. Multiple scales may also be needed in problems with weak damping; an example of this can be found in Exercises 6.1(g),(h).

6.3.3 Example: Delay Equation

The equation $y' = \alpha y$ is often used in population modeling based on the assumption that a population increases at a rate proportional to the current population. This produces exponential growth and, given the limits on food and space, it is expected that the birth rate will decrease for larger population levels. This gives rise to an equation such as $y' = \alpha y - \beta y^3$. A criticism of this comes from the observation that the birth rate should actually depend on the population that existed T time units ago, where T is the gestation period. With this, we have the first-order delay equation

$$y'(t) = \alpha y(t - T) - \beta y^3(t - T) \quad \text{for } t > 0. \tag{6.23}$$

It is assumed that $y(t) = a$ for $-T \le t \le 0$. It is also assumed that α, β, and a are positive, and $y(t) \ge 0$. Delay equations, such as the one above, can be very difficult to solve, even numerically. One of the objectives of this example is to determine if including a delay is worth the effort.

We first assume no delay, so $T = 0$. In this case the steady states are $y_s = 0$ and $y_s = \sqrt{\alpha/\beta}$. From the direction field for $y' = \alpha y - \beta y^3$ it is easy to demonstrate that $y_s = 0$ is unstable and $y_s = \sqrt{\alpha/\beta}$ is asymptotically stable. This is also evident from the exact solution, which is

$$y(t) = \sqrt{\frac{\alpha}{\beta + ce^{-2\alpha t}}},$$

where $c = -\beta + \alpha/a^2$.

When a delay is included, so that $T > 0$, one obtains the same steady states as for $T = 0$. To determine if they are stable, we use the initial condition

$$y(t) = y_s + \alpha_0 \delta \quad \text{for } -T \le t \le 0,$$

where y_s is one of the given steady states. The appropriate expansion of the solution for small δ is $y(t) \sim y_s + \delta y_1(t) + \cdots$. Substituting this into (6.23) and using Taylor's theorem, we obtain

$$\delta y_1'(t) + \cdots = \delta(\alpha - 3\beta y_s^2)y_1(t - T) + \cdots. \tag{6.24}$$

We will concentrate, from this point on, on the steady state $y_s = \sqrt{\alpha/\beta}$. In this case, the $O(\delta)$ equation we have is

$$y_1'(t) = -2\alpha y_1(t - T). \tag{6.25}$$

The general solution of this equation can be written as the superposition of solutions of the form $y = e^{rt}$ (Appendix E). Substituting this into (6.25) yields

$$r = -2\alpha e^{-rT}, \tag{6.26}$$

which is known as the characteristic equation for (6.25). With this, and assuming the roots are simple, the general solution of (6.25) has the form

$$y_1(t) = \sum_n a_n e^{r_n t}, \tag{6.27}$$

where the sum is over the roots from (6.26) and the a_n are determined from the initial condition. Therefore, if we find that all roots satisfy $\text{Re}(r_n) < 0$, then the steady state is asymptotically stable, and if even one root has $\text{Re}(r_n) > 0$, then it is unstable. Note that if a root is not simple, then (6.26) must be modified, as explained in Appendix E. However, this does not change our conclusions about stability.

For simplicity, assume that $T = 1$. With this, writing $r = A + iB$, and then equating real and imaginary parts in (6.26) we have that

$$A = -2\alpha e^{-A} \cos(B), \tag{6.28}$$

$$B = 2\alpha e^{-A} \sin(B). \tag{6.29}$$

It is not difficult to show that for small values of α all of the solutions of the preceding equations satisfy $A < 0$, which means the steady state is asymptotically stable. However, for larger values of α there is one or more solutions with $A > 0$, which implies instability. To determine where the switch occurs, and hence the location of the bifurcation point, we consider the case where $A = 0$. From (6.28), $B = \pm\frac{\pi}{2}(2k + 1)$, where $k = 0, 1, 2, \ldots$. From (6.29) it follows that the smallest value of α that yields $A = 0$ is $\alpha = \pi/4$. Consequently, if $0 < \alpha < \pi/4$, then the steady-state behavior of the delay equation is the same as for the equation with no delay. The implication here is that the delay does not need to be included for smaller values of the birth-rate coefficient. However, for larger values it makes a significant difference. What is also interesting is how the solution goes unstable. For the damped oscillator instability occurred because one root, r_+, became a problem. For the delay equation two roots cross over from $\text{Re}(r) < 0$ to $\text{Re}(r) > 0$, and they are complex conjugates. In particular, they cross over when $A + iB = \pm\frac{\pi}{2}i$. As we will see in Sect. 6.5, what is happening is that a periodic solution appears once α increases past $\pi/4$.

Exercises

6.1. For the following equations, describe the steady states, classify the bifurcation points, and determine the stability of each branch. Also, sketch the bifurcation diagram.
(a) $y'' + y' + y(y - \lambda^3 + 3\lambda^2 - 4) = 0$.
(b) $y'' + y' + 9y^2 + 4\lambda^2 = 1$.
(c) $y'' + y' + y[2y^2 + \lambda(\lambda - 2)] = 0$.
(d) $y'' + y' + y(\lambda - 3y + 2y^2) = 0$.
(e) $y'' + y' + \lambda + \alpha y - y^3 = 0$, where α is a given constant with $-\infty < \alpha < \infty$.
(f) $y'' + y' + e^{\lambda y} - 1 - (y - 1)^2 = 0$.
(g) $y'' + yy' + \lambda y + y^3 = 0$.
(h) $y'' + (y')^n + \lambda y - y^3 = 0$, where n is a positive integer.

6.2. In the study of chemical kinetics, one finds the equation

$$\frac{dc}{dt} = \frac{1}{\kappa}(1 - c) - c(1 + \beta - c)^2,$$

where $0 \le c \le 1$, and κ and β are positive constants with $0 < \beta < \frac{1}{8}$. This equation comes from the theory for a continuously fed, well-stirred tank reactor that involves the cubic autocatalytic reaction $C + 2B \to 3B$ (Gray and Scott, 1994). In this setting $c(t)$ is the (scaled) concentration of C.
(a) Sketch the steady states as a function of κ.
(b) As a function of κ, classify the bifurcation points and determine the stability of each branch.

6.3. In the theory for the buckling of an initially straight rod subjected to an axial load λ, the following problem arises (Euler, 1774):

$$\theta'' + \lambda \sin \theta = 0 \quad \text{for} \ \ 0 < x < 1,$$

where $\theta'(0) = \theta'(1) = 0$. The variable $\theta(x)$ is the angle the tangent to the rod makes with the horizontal at the spatial position x. This variable is related to the vertical displacement $w(x)$ of the rod through the relation $w' = \sin(\theta)$, where $w(0) = w(1) = 0$.
(a) Find the solutions that bifurcate from the equilibrium solution $\theta_s = 0$. The values of λ where bifurcation occurs are called buckling loads.
(b) In mechanics, the principle of minimum potential energy states that a system will move into a state with the minimum potential energy. For this problem, the potential energy is proportional to

$$V = \int_0^1 [\theta_x^2 + 2\lambda(\cos \theta - 1)] dx.$$

Use this principle to determine the preferred configuration of the rod near the first bifurcation point.
(c) Setting $\lambda_n = (n\pi)^2$, find a first-term approximation of V for the solutions that bifurcate from (λ_n, θ_s). Comparing these with the value of V when $\theta_s = 0$, comment on what configuration the rod might take when $\lambda_n < \lambda < \lambda_{n+1}$. Also, explain why this might not actually be the configuration.

6.4. Consider the following nonlinear eigenvalue problem:

$$y'' + \alpha y = 0 \quad \text{for} \ \ 0 < x < 1,$$

where $y(0) = y(1) = 0$. Also,

$$\alpha = \lambda - \int_0^1 y^2(x) dx,$$

where λ is a constant.

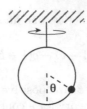

Figure 6.7 Bead on a rotating hoop, as described in Exercise 6.6

(a) Find the (exact) solutions of this problem, and classify the bifurcation points using λ as the bifurcation parameter.
(b) Use the principle of minimum energy (Exercise 6.3) to determine the stability of the solutions. To do this, let

$$V = \int_0^1 (y_x^2 - \alpha y^2)\mathrm{d}x.$$

6.5. The Wazewska–Czyzewska and Lasota model for the survivability of red blood cells is (Wazewska-Czyzewska and Lasota, 1976)

$$y'(t) = -\mu y(t) + e^{-y(t-T)} \quad \text{for } 0 < t.$$

Here $y(t)$ is the number of cells at time t and T is the time required to produce a red blood cell. The purpose of this exercise is to see if the delay in cell production can produce an instability in the system, and for this reason T is the bifurcation parameter. Assume that μ is a positive constant and $y(t) \geq 0$.
(a) In the case where there is no delay, so that $T = 0$, show that there is a nonzero steady state y_s and show that it is asymptotically stable. Also, explain what happens to the value of y_s as μ varies from 0 to ∞.
(b) The steady state in part (a) is also a steady state when $T > 0$. Show that it is asymptotically stable if $\mu > 1/e$.
(c) Assume that $0 < \mu < 1/e$. Show that y_s is asymptotically stable if $0 < T < T_c$, where

$$T_c = \frac{\pi - \arctan(\sqrt{y_s^2 - 1})}{\mu\sqrt{y_s^2 - 1}},$$

and it is unstable if $T > T_c$. What happens to the solution near $T = T_c$ is considered in Exercise 6.21.

6.6. Consider a wire hoop suspended from the ceiling (Fig. 6.7). Suppose the hoop is rotating and there is a bead that is able to slide around the hoop. The equation of motion for the bead, in nondimensional form, is

$$\theta'' + \kappa\theta' + \sin(\theta) = \omega^2 \sin(\theta)\cos(\theta) \quad \text{for } 0 < t,$$

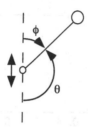

Figure 6.8 Inverted pendulum, as studied in Exercise 6.7

where κ and ω are positive constants. Here κ comes from the friction of the bead on the wire and ω comes from the frequency of rotation. Also, $\theta(t)$ is the angular displacement of the bead measured from the bottom of the hoop; you can assume that $-\pi \leq \theta \leq \pi$.

(a) Find the steady states and determine if they are stable. Also, sketch the bifurcation diagram and classify the bifurcation points.

(b) What happens to the equilibrium position of the bead as $\omega \to \infty$?

6.7. The simple pendulum has two steady states; one is stable ($\theta_s = 0$) and the other is unstable ($\theta_s = \pm\pi$). What is interesting is that if the support point of the pendulum is driven periodically, then it is possible for the inverted position to also be stable (did you ever try to balance an inverted broom on your hand?). In the case where the support is given a vertical periodic forcing, the equation of motion in terms of the polar angle $\phi(t)$ is (Fig. 6.8)

$$\phi'' - (\kappa - \varepsilon \cos t)\sin\phi = 0 \quad \text{for } 0 < t.$$

Here κ is a positive constant (it is inversely proportional to the square of the driving frequency Ω) and ε is a positive constant associated with the amplitude of the forcing.

(a) What are the steady states? For the inverted case, show that the linearized stability argument leads to having to solve Mathieu's equation, given as

$$\varphi'' - [\kappa - \varepsilon \cos(t)]\varphi = 0 \quad \text{for } 0 < t.$$

What are the initial conditions?

(b) From this point on, we will assume ε is small. Expanding in terms of ε, show that the inverted position is unstable if κ is independent of ε.

(c) Now assume $\kappa \sim \kappa_0\varepsilon + \kappa_1\varepsilon^2 + \cdots$. Find conditions on κ_0 and κ_1 such that the inverted position is stable (it does not have to be asymptotically stable). Recall that the initial conditions are independent of ε.

(d) Shown in Fig. 6.9 are the experimentally measured values for the frequency of oscillation of the inverted pendulum as a function of the amplitude of the forcing (Ness, 1967). Is your result from part (c) consistent with these values? Make sure to comment on (i) the approximate linearity

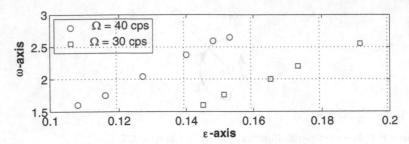

Figure 6.9 Experimentally determined values of frequency (ω) of oscillation of inverted pendulum as a function of the amplitude (ε) of the forcing (Ness, 1967)

of each data set, (ii) the change in slope when Ω is changed, and (iii) the increase in ω with Ω. Also, explain why the data indicate that there are at least two time scales in the problem. Note that t is nondimensional, and in physical coordinates $t = 2\pi\Omega t^*$, where t^* is measured in seconds and Ω is measured in cycles per second (cps).

6.8. Have you ever ridden in a swing? To get the swing higher, you move a small amount up and down in the seat with a period that is related to the period of the swing. The swing in this case can be modeled as a pendulum. Also, because the mass at the end is moving up and down, it is a pendulum with a variable length. The equation in this case is

$$\theta'' + 2\frac{\ell'(t)}{\ell(t)}\theta' + \sin(\theta) = 0 \quad \text{for} \ \ 0 < t.$$

Here $\ell(t) = 1 + \varepsilon\cos(\omega t)$ is the length of the swing, where ε and ω are the amplitude and frequency of the up and down motion in the seat, respectively.

(a) What are the steady states? For $\theta_s = 0$ show that the linearized stability argument leads to having to solve

$$v'' + 2\frac{\ell'(t)}{\ell(t)}v' + v = 0 \quad \text{for} \ \ 0 < t.$$

What are the initial conditions?

(b) Show that if the swinger does nothing (so that $\omega = 0$), then the steady state $\theta_s = 0$ is stable.

(c) What the swinger must do is find a frequency ω that makes the steady state unstable (so the amplitude grows and the swing goes higher). Find such a frequency in the case where ε is small. Do this by constructing a first-term expansion of v that is valid for large t.

6.4 Quasi-Steady States and Relaxation

One of the more interesting aspects of equilibrium points is their effect on the dynamic response of the solution. This can be quite dramatic. As an example of such a situation we will consider the initial-value problem

$$\varepsilon\frac{dy}{dt} = y - \frac{1}{3}y^3 - f(t), \quad \text{for } 0 < t, \tag{6.30}$$

where $y(0) = \alpha$. This equation does not arise from any particular physical problem. The reason for using it as our first example is that it gives a nice illustration of what is known as "fast dynamics" and how this works with the steady states to control the motion. After this, more physically relevant problems are easier to understand.

To explain what is distinctive about (6.30), suppose $f(t)$ is constant. In fact, let $f(t) = \lambda$. In this case, as shown in Fig. 6.10, the steady-state solution has two (saddle-node) bifurcation points, one at $(\lambda_b, y_b) = (-2/3, -1)$ and another at $(\lambda_b, y_b) = (2/3, 1)$. Using the methods developed in the previous section, it is not hard to show that the upper and lower branches are asymptotically stable but the middle branch (which connects the two bifurcation points) is unstable. Now, if we happen to pick the initial condition to be within the basin of attraction of the upper branch, then the solution will approach this branch very rapidly. This is because the ε multiplying the highest derivative in (6.30) will cause the solution to change on the time scale $\bar{t} = t/\varepsilon$. Consequently, there will be an initial time interval, of length $O(\varepsilon)$, during which the solution will approach the equilibrium solution exponentially. This is what is referred to as "fast dynamics." The second point to be made here is that this behavior will occur even if $f(t)$ depends on t. This will happen so long as $f(t)$ does not change too quickly, that is, it changes slowly over time intervals of length $O(\varepsilon)$. Assuming this is the case, then the question arises as to what the solution does if $f(t)$ causes the solution to pass

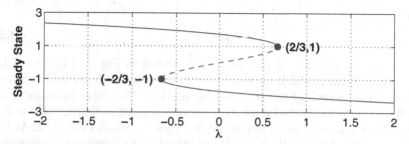

Figure 6.10 Steady-state solutions obtained from (6.30) in the case of when $f(t) = \lambda$. The two bifurcation points are also shown. The *dashed potion* of the curve is unstable, and the *solid portions* are stable

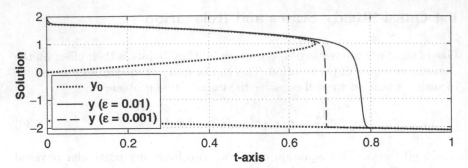

Figure 6.11 Numerical solution of (6.31), (6.32) for two values of ε. Also shown is the *curve* obtained from the outer solution y_0, which is the cubic shown in Fig. 6.10

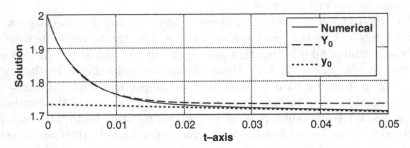

Figure 6.12 Comparison between numerical solution of (6.31), (6.32) and initial-layer approximation given in (6.35). Also shown is the *curve* obtained from the outer solution y_0. In this calculation, $\varepsilon = 10^{-2}$

near one of the bifurcation points. Answering this question, and making more sense out of this discussion, are some of the objectives of what follows.

To make the problem definite, suppose we want to solve

$$\varepsilon \frac{dy}{dt} = y - \frac{1}{3}y^3 - t \quad \text{for } 0 < t, \tag{6.31}$$

where

$$y(0) = 2. \tag{6.32}$$

The numerical solution of this problem is shown in Fig. 6.11 for two values of ε. There is an initial layer where the solution drops from $y(0) = 2$ down to the upper branch of the equilibrium solution. This layer occupies such a small time interval that the solution curve is almost indistinguishable from the vertical axis, and so a more magnified view is given in Fig. 6.12. As time increases, the solution follows the upper branch until it passes the bifurcation point, after which it drops quickly to the lower branch. This part of the curve is something like what occurs in cartoons, where the coyote (or some other poor soul) runs off a cliff but it takes a moment or two before he drops to the new equilibrium level. We are now going to analyze this situation using matched asymptotic expansions (Chap. 2).

6.4.1 Outer Expansion

The easiest expansion to determine is what we will call the outer solution. This is obtained from a regular expansion of the form

$$y(t) \sim y_0(t) + \varepsilon y_1(t) + \cdots . \tag{6.33}$$

Substituting this into (6.31), one finds that

$$y_0 - \frac{1}{3}y_0^3 = t. \tag{6.34}$$

The solutions of this equation generate the curve shown in Fig. 6.10 (where t replaces λ). Which branch of this curve we should use will be determined from matching. As we will see, (6.34) determines the solution everywhere except near $t = 0$ and near $t = 2/3$.

Because y_0 satisfies the equation obtained by setting $y' = 0$ in (6.31), it is referred to as a quasi-steady-state approximation (QSSA). It is *quasi* because it depends on t, and a *steady state* because it satisfies the equation used to determine a steady state. The QSSA is often used to reduce dynamical systems, and what we see is that it is effectively equivalent to using an outer approximation of the solution.

6.4.2 Initial Layer Expansion

The solution in (6.34) does not satisfy the initial condition. This is dealt with in the usual manner of introducing the initial-layer coordinate $\bar{t} = t/\varepsilon$. The appropriate expansion of the solution $y = Y(\bar{t})$ in this region is $Y(\bar{t}) \sim Y_0(\bar{t}) + \cdots$. Problems like this were studied extensively in Chap. 2, so only the results of this procedure will be given. One finds that

$$Y \sim 2\frac{\sqrt{12 - 3e^{-2\bar{t}}}}{4 - e^{-2\bar{t}}}. \tag{6.35}$$

By matching this expansion with the one from the outer region, we can determine which branch must be used from (6.34). In particular, since $Y_0(\infty) = 3^{1/2}$, then $y_0(t)$ is determined by the upper branch. More precisely, the upper branch determines $y_0(t)$ but only so long as $0 < t < 2/3$ (which is in agreement with the numerical solutions shown in Fig. 6.11).

Thus we have found an approximation of the solution up to the bifurcation point, where $(t, y) = (2/3, 1)$. Our next task is to determine how the solution drops down to the lower branch of the cubic. It does this in two steps. First, it must turn the corner and head toward the lower curve, and this portion will give rise to a corner layer. Second, it must make the transition onto the lower curve, and this will be described using an interior layer.

6.4.3 Corner-Layer Expansion

Now, as is evident from Fig. 6.10, the upper branch stops at $t = 2/3$. To determine what the solution does after this point, we introduce the coordinate

$$\tilde{t} = \frac{t - 2/3}{\varepsilon^\alpha}. \tag{6.36}$$

The solution in this region will be denoted by $y = \widetilde{Y}(\tilde{t})$. We are interested in the behavior in the immediate vicinity of the bifurcation point (where $y = 1$), and so the appropriate expansion is

$$\widetilde{Y} \sim 1 + \varepsilon^\gamma \widetilde{Y}_1 + \cdots. \tag{6.37}$$

Substituting (6.37) and (6.36) into (6.31), we get

$$\varepsilon^{1-\alpha+\gamma} \frac{\mathrm{d}}{\mathrm{d}\tilde{t}} \widetilde{Y}_1 + \cdots = -\varepsilon^{2\gamma} \widetilde{Y}_1^2 + \cdots - \varepsilon^\alpha \tilde{t}. \tag{6.38}$$

Balancing the terms in this equation gives us that $\alpha = 2/3$ and $\gamma = 1/3$. The resulting equation for $\widetilde{Y}_1$ is

$$\frac{\mathrm{d}}{\mathrm{d}\tilde{t}} \widetilde{Y}_1 + \widetilde{Y}_1^2 + \tilde{t} = 0. \tag{6.39}$$

This is a Riccati equation, and it is possible to transform it into a linear problem (Holmes and Stein, 1976). This is accomplished by making the change of variables $\widetilde{Y}_1 = w'/w$. The result is an Airy equation for $w(\tilde{t})$, and this yields (Appendix B)

$$\widetilde{Y}_1 = -\frac{a_0 \mathrm{Ai}'(-\tilde{t}) + a_1 \mathrm{Bi}'(-\tilde{t})}{a_0 \mathrm{Ai}(-\tilde{t}) + a_1 \mathrm{Bi}(-\tilde{t})}. \tag{6.40}$$

To determine the constants a_0 and a_1, it is necessary to match (6.40) with the outer solution (6.34) by considering the limits of $t \uparrow 2/3$ and $\tilde{t} \to -\infty$. With this in mind, we introduce the intermediate variable $t_\eta = (t - 2/3)/\varepsilon^\eta$, where $0 < \eta < 2/3$. From (6.40) we get that

$$\widetilde{Y} \sim \begin{cases} 1 + \varepsilon^{\eta/2}(-t_\eta)^{1/2} & \text{if } a_1 = 0, \\ 1 - \varepsilon^{\eta/2}(-t_\eta)^{1/2} & \text{if } a_1 \neq 0, \end{cases}$$

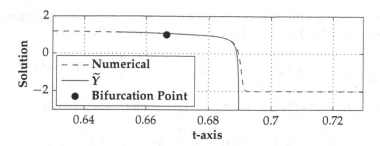

Figure 6.13 Comparison between numerical solution of (6.31), (6.32) and corner-layer approximation given in (6.41). In this calculation, $\varepsilon = 10^{-3}$

and from (6.34) $y \sim 1 + \varepsilon^{\eta/2}(-t_\eta)^{1/2}$ (Exercise 6.16). For these two expansions to match it is required that $a_1 = 0$. The corner-layer expansion is therefore

$$\widetilde{Y} \sim 1 - \varepsilon^{1/3}\frac{\text{Ai}'(-\tilde{t})}{\text{Ai}(-\tilde{t})}. \tag{6.41}$$

This function is plotted in Fig. 6.13, along with the numerical solution. It is seen that $\widetilde{Y}$ does indeed describe the solution as it leaves the top branch and starts heading downward toward the lower branch. However, rather than approach the lower branch asymptotically, $\widetilde{Y}$ shoots right by it and continues to head south. This is a rather strong indication that there is another layer, or transition region, that provides for the asymptotic approach onto the lower branch, and determining how this happens is objective of the next topic.

6.4.4 Interior-Layer Expansion

The major difficulty with the result in (6.41) is that the Airy function is zero for certain negative values of its argument. This can be seen in the graph of Ai(x) given in Appendix B. The first zero $x = x_0$ of this function occurs when $x_0 = -2.33811 \ldots$. This means that the expansion given in (6.41) is essentially meaningless when $\tilde{t} = 2.33811 \ldots$, and we have a situation reminiscent of what occurs at a turning point for the WKB expansion (Sect. 4.3). In terms of the original time variable, this singularity occurs when $t = t_0$, where $t_0 = 2/3 + \varepsilon^{2/3}\tilde{t}_0$ for $\tilde{t}_0 = 2.3381\ldots$. For example, in Fig. 6.13, $t_0 \approx 0.69$, and this corresponds to the t-value where the approximation for $\widetilde{Y}$ goes bad.

To determine what is happening near this point, we introduce the layer coordinate

$$t^* = \frac{t - t_0}{\varepsilon^\kappa}. \tag{6.42}$$

In this layer, we will designate the solution as $Y^*(t^*)$, and the appropriate expansion is $Y^* \sim Y_0^* + \cdots$. Introducing this into (6.31) yields

$$\varepsilon^{1-\kappa} \frac{\mathrm{d}}{\mathrm{d}t^*} Y_0^* + \cdots = Y_0^* + \cdots - \frac{1}{3}(Y_0^*)^3 + \cdots - \left(\frac{2}{3} + \varepsilon^{2/3}\tilde{t}_0 + \varepsilon^\kappa t^*\right). \quad (6.43)$$

Balancing terms yields $\kappa = 1$, and the $O(1)$ equation is given as

$$\frac{\mathrm{d}}{\mathrm{d}t^*} Y_0^* = Y_0^* - \frac{1}{3}(Y_0^*)^3 - \frac{2}{3} \quad \text{for} \quad -\infty < t^* < \infty. \quad (6.44)$$

It is possible to solve this equation by noting that the right-hand side can be factored to produce

$$\frac{\mathrm{d}}{\mathrm{d}t^*} Y_0^* = -\frac{1}{3}(Y_0^* + 2)(Y_0^* - 1)^2.$$

Separating variables, and integrating, one obtains

$$\frac{1}{3} \ln \left| \frac{2 + Y_0^*}{1 - Y_0^*} \right| + \frac{1}{1 - Y_0^*} = -t^* - A_0, \quad (6.45)$$

where A_0 is an arbitrary constant. After a little manipulation, the preceding solution can be written as

$$Y_0^* = \frac{-2 + q}{1 + q}, \quad (6.46)$$

where q satisfies

$$qe^q = e^{-3(t^* + A_0) - 1}. \quad (6.47)$$

From the graph of $h(q) = qe^q$ it is evident that the solution of (6.47) is unbounded as $t^* \to -\infty$, and it approachs zero as $t^* \to \infty$. From (6.46) it follows that $Y_0^* \to 1$ as $t^* \to -\infty$, and $Y_0^* \to -2$ as $t^* \to \infty$. This shows that (6.46) provides a transition from the upper bifurcation point down to the lower branch of the cubic. The details of the actual matching, using intermediate variables, are considered in Exercise 6.16. Note, however, that the value of A_0 is undetermined. A discussion of possible ways to find A_0 can be found in Sect. 2.5, but exactly what value it has is an open question. A comparison of the interior-layer solution and the numerical solution is given in Fig. 6.14. As expected, the interior-layer solution picks up where the corner layer approximation leaves off, namely, it completes the turn and then makes the transition onto the lower branch of the outer solution.

We have completed the analysis of the problem. However, it is worth commenting on the steps we had to take to obtain the solution. The major effort was to determine how the solution moves from one branch to the other. This takes place in two steps. The first takes the solution past the bifurcation point and lets the solution drop a short distance below $y_b = 1$. Once this happened,

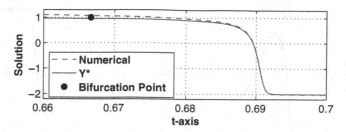

Figure 6.14 Comparison between numerical solution of (6.31), (6.32) and interior-layer approximation given in (6.46). In this calculation, $\varepsilon = 10^{-3}$ and $A_0 = -3/4$

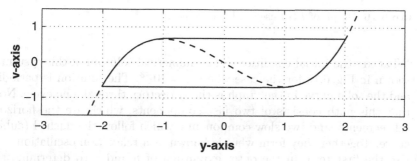

Figure 6.15 Diagram for Exercise 6.10

the balancing in the equation changed, and this was the reason for the interior layer. The drop in the interior-layer solution occurs very quickly, over a time scale of $O(\varepsilon)$, and this connects the upper branch with the lower.

Exercises

6.9. Suppose

$$\varepsilon \frac{dy}{dt} = \frac{1}{3}y^3 - y - t \quad \text{for } 0 < t,$$

where $y(0) = 0$. Find a first-term approximation of the solution.

6.10. The van der Pol equation is

$$\varepsilon y'' - (1 - y^2)y' + y = 0 \quad \text{for } t > 0.$$

Assume that $y(0) = \sqrt{3}$ and $y'(0) = 0$.
(a) Letting $y = -v'$, show that the equation can be written as the first-order system

$$v' = -y,$$

$$\varepsilon y' = v + y - \frac{1}{3}y^3.$$

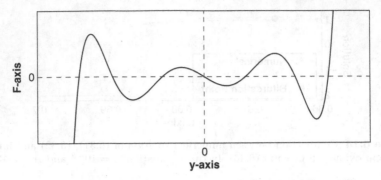

Figure 6.16 Graph of $F(y)$ used in Exercise 6.11

What are the initial conditions? The numerical solution of this system is shown in Fig. 6.16 for the case where $\varepsilon = 10^{-3}$. The solution is periodic, and the solid curve in the graph is the trajectory through one cycle. Note that this path consists of two fast components, which are the horizontal segments, and two slow components, which follow the dashed (cubic) curve. Together they form what is known as a relaxation oscillation.

(b) Use the first term in the outer expansion of y and v to determine the equation of the cubic curve shown in Fig. 6.16. Also, use it to determine where the first corner layer is located (give the value of t, y, and v).

(c) Find the first term in the corner-layer expansion of y and v. With this, determine the location of the first transition layer. The analysis of the transition layers is rather involved and can be found in MacGillivray (1990).

(d) In each transition layer, the solution very quickly switches from one branch of the cubic to another. Because of this, the time the solution spends on the slow component of the cycle serves as a first-term approximation of the period. Use this to find an approximation of the period of the oscillation. For comparison, the period found numerically is $T = 1.91$ when $\varepsilon = 10^{-2}$ and $T = 1.68$ when $\varepsilon = 10^{-3}$).

(e) Use the idea from part (d) to sketch the first-term approximation of $y(t)$ for $0 \le t \le 3T$.

6.11. The Liènard equation has the form

$$\varepsilon y'' + \Phi(y)y' + y = 0 \quad \text{for } t > 0.$$

The initial conditions are $y(0) = a$ and $y'(0) = b$. Assume that Φ is a smooth and even function of y, with $\Phi(0) \neq 0$.

(a) Is the steady state asymptotically stable or unstable? In answering this question, assume that ε is fixed and positive.

(b) Letting $y = -v'$, show that the equation can be written as the first-order system

$$v' = -y,$$
$$\varepsilon y' = v - F(y),$$

where $F' = \Phi$ and $F(0) = 0$. What are the initial conditions? Describe the location of the initial conditions in the (y, v)-plane, relative to the curve $v = F(y)$ for $\varepsilon \ll 1$.

(c) Suppose F is as shown in Fig. 6.16. Based on your answer from part (a), is the steady state asymptotically stable? Also, sketch the direction field in the (y, v)-plane, making use of the assumption that $\varepsilon \ll 1$.

(d) Using the first term in the outer expansion of y and v, as well as part (c), explain why there are two possible limit cycles in this problem.

6.12. Rayleigh (1883) undertook a study of what he called "maintained vibrations" that occur in organ pipes, singing flames, finger glasses, and other such systems. The equation he used to model such vibrations was

$$\varepsilon y'' - [1 - \frac{1}{3}(y')^2]y' + y = 0 \quad \text{for } t > 0.$$

Assume that $y(0) = 0$ and $y'(0) = -\sqrt{3}$.

(a) Letting $v = y'$, write the problem as a first-order system. Make sure to give the initial conditions.

(b) Find the first term in the outer expansion of y and v. From this determine where the first corner layer is located (give the value of t, y, and v).

(c) Find the first term in the corner-layer expansion of y and v. With this, determine the location of the first transition layer.

(d) In the transition layers, the solution very quickly switches from one branch of a cubic to another. Use this observation to find a first-term approximation of the period of the oscillation. Use this same idea to sketch the first-term approximation of $y(t)$ for $0 \le t \le 3T$.

6.13. In the study of pattern formation, a model for the activation of a gene by a chemical signal is (Lewis et al., 1977)

$$g' = \lambda - g + \frac{kg^2}{1 + g^2} \quad \text{for } 0 < t,$$

where $\lambda > 0$ and $k > 2$ are constants (to simplify things, take $k = 4$). Also note that because $g(t)$ designates a concentration, it must be nonnegative.

(a) Sketch the bifurcation diagram. Also, find and then classify the bifurcation point.

(b) Determine which of the three steady-state branches are stable and which are unstable.

(c) Suppose λ increases slowly in time. In particular, let $\lambda = \varepsilon t$, where $0 < \varepsilon \ll 1$. What problem does g satisfy if you change variables to $\tau = \varepsilon t$ and assume $g(0) = 0$? From this, find a first-term approximation of the solution for $0 \le \tau < \infty$.

(d) For the problem in part (c) give a qualitative description of the solution for $0 \leq t < \infty$. Why do you think this equation is used as a model of a *biological switching mechanism*? Also, discuss the limiting value of the solution (as $t \to \infty$) in the case where $\lambda = \varepsilon t e^{-\alpha \varepsilon t}$, where α is a positive constant.

6.14. A model for the bistable iodate–arsenous acid reaction gives rise to the equation (Ganapathisubramanian and Showalter, 1986)

$$\frac{dy}{dt} = y(1 + \alpha y)(\beta - y) + \lambda(1 - y),$$

where $\lambda \geq 0$. In this equation $y(t)$ designates a concentration, and it therefore must be nonnegative. Also, λ is the bifurcation parameter and α, β are assumed to be the fixed constants $\alpha = 5$ and $\beta = 15$.
(a) Sketch the bifurcation diagram and classify the two bifurcation points. Designate the values of λ where bifurcation occurs as λ_0 and λ_1, where $\lambda_0 < \lambda_1$.
(b) Determine which steady states are stable and which are unstable (assume $\lambda \neq \lambda_0, \lambda_1$).
(c) Suppose λ increases slowly in time. In particular, let $\lambda = \varepsilon t$, where $0 < \varepsilon \ll 1$. (i) What problem does y satisfy if you change variables to $\tau = \varepsilon t$ and assume $y(0) = 15$? (ii) Find a first-term approximation of the solution for $0 \leq \tau < \infty$.

6.15. This problem examines the behavior of the nonlinear oscillator

$$y'' + \varepsilon \kappa y' + \lambda y - 3y^2 + 2y^3 = \varepsilon F \cos(\omega t),$$

where κ, λ, and ω are positive constants.
(a) Consider the case of no forcing (i.e., $F = 0$). Sketch the bifurcation diagram, classify the two bifurcation points, and determine which steady states are stable and which are unstable.
(b) Letting y_s designate a steady-state solution of the unforced problem, suppose the forced problem is started out so that $y(0) = y_s$ and $y'(0) = 0$. In this case, given the amplitude of the forcing function, it might be expected that the forcing contributes to the $O(\varepsilon)$ term in the expansion of the solution. For resonance, however, one would expect it to contribute earlier, perhaps to the $O(1)$ or the $O(\varepsilon^{1/2})$ term (Sect. 3.4). For what values of the driving frequency ω do you expect there to be a resonant response?
(c) Suppose $y_s = \frac{1}{4}(3 + \sqrt{9 - 8\lambda})$, where $8\lambda < 9$ and $\omega^2 = \lambda - 6y_s + 6y_s^2$. Find a two-term approximation of $y(t)$ that is valid for large t. If you are not able to solve the problem that determines the t_2 dependence, then find the possible steady states (if any) for the amplitude.

6.16. To match the solution from the interior layer with those from the outer and corner regions, introduce the intermediate variable

$$\tau = \frac{t - (2/3 + \varepsilon^{2/3}\tilde{t}_0)}{\varepsilon^\nu},$$

where $0 < \nu < 1$.

(a) Assuming $\tau < 0$, show that $Y^* \sim 1 + \varepsilon^{1-\nu}/\tau$. Making the additional assumption that $\frac{2}{3} < \nu < 1$, show $\tilde{Y} \sim 1 + \varepsilon^{1-\nu}/\tau$. In other words, the two expansions match.

(b) For the case of $\tau > 0$, show that $Y^* \sim -2 + 3\exp(-3(\varepsilon^{\nu-1}\tau + A_0) - 1)$. Explain why this shows that the interior-layer solution matches with the outer solution.

6.5 Bifurcation of Periodic Solutions

One of the more interesting responses that can come from a nonlinear problem is a periodic motion known as a self-sustained oscillation. To explain what this is, we begin with van der Pol's equation, given as

$$y'' - \lambda(1 - y^2)y' + y = 0 \quad \text{for } 0 < t. \tag{6.48}$$

The nonlinearity in the equation is in the damping. What is important is that the coefficient can be both positive and negative, depending on the value of the solution. Assuming $\lambda > 0$, then for small values of $y(t)$ the coefficient of the damping term is negative, a situation that can lead to unbounded solutions. However, for large values of $y(t)$ the coefficient is positive, which should help stabilize the problem. As we will see below, this give-and-take in the damping will lead to a self-sustained oscillation.

The equation in (6.48) was studied by van der Pol (1926) in connection with the triode circuits that were used in some of the early radios. It is possible to build one of these circuits relatively cheaply, and readers interested in trying this may consult Horwitz and Hill (1989) or Keener (1983). This equation is the classic example in the field of relaxation oscillations. It has applicability in mechanics, electronics, and other fields, and an overview of the subject can be found in Strogatz (2001).

The first step in analyzing (6.48) is to consider the steady states. The only one we get is $y_s = 0$, and to determine if it is stable, we use the initial conditions

$$y(0) = \alpha_0\delta \quad \text{and} \quad y'(0) = \beta_0\delta. \tag{6.49}$$

Now, the appropriate expansion of the solution for small δ is $y \sim y_s + \delta y_1(t) + \cdots$. Substituting this into (6.48), one finds that

$$y_1(t) = a_0 e^{r+t} + a_1 e^{r-t}, \tag{6.50}$$

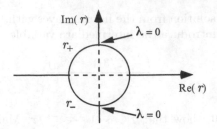

Figure 6.17 Solution of van der Pol equation showing the approach to the limit cycle starting at $y(0) = 1$, $y'(0) = 0$ and taking $\lambda = 0.1$. Shown are the trajectories obtained from the asymptotic approximations in (6.58) and (6.59) and from the numerical solution of (6.48). Also included is the limit cycle obtained from the asymptotic solution. Note $v = y'$

where a_0 and a_1 are constants and $2r_\pm = \lambda \pm \sqrt{\lambda^2 - 4}$. The values of these exponents are shown in Fig. 6.17, and it is seen that $\mathrm{Re}(r_\pm) < 0$ if $\lambda < 0$ and $\mathrm{Re}(r_\pm) > 0$ if $\lambda > 0$. From this it follows that $y_s = 0$ is asymptotically stable if $\lambda < 0$ and unstable if $0 < \lambda$.

The question now is, what happens to the solution when $\lambda > 0$? A hint of the answer to this can be found in the way the zero solution becomes unstable. As λ increases in (6.50) and goes from negative to positive, the exponents $r_\pm$ are complex conjugates and their real parts change sign. This differs from what we found earlier when a steady-state solution went unstable. For example, in (6.21), the solution becomes unstable because one (and not both) of the exponents goes from negative to positive. Moreover, the exponents are real.

To see exactly what is happening as λ goes from negative to positive, let

$$\varepsilon = \lambda - \lambda_b, \qquad (6.51)$$

where $\lambda_b = 0$. In this case (6.48) becomes

$$y'' - \varepsilon(1 - y^2)y' + y = 0 \quad \text{for } 0 < t. \qquad (6.52)$$

The equation is weakly damped for small ε, and so, based on our experience from Chap. 4, we will use the time scales $t_1 = t$ and $t_2 = \varepsilon t$. Introducing these into (6.52) yields

$$(\partial_{t_1}^2 + 2\varepsilon\partial_{t_1}\partial_{t_2} + \cdots)y - \varepsilon(1 - y^2)(\partial_{t_1} + \varepsilon\partial_{t_2})y + y = 0. \qquad (6.53)$$

The appropriate expansion for the solution is

$$y \sim y_0(t_1, t_2) + \varepsilon y_1(t_1, t_2) + \cdots. \qquad (6.54)$$

Introducing this into (6.53) we get the following problems:

$O(1)$ $\quad (\partial_{t_1}^2 + 1)y_0 = 0.$

The general solution of this problem is

$$y_0 = A(t_2) \cos[t_1 + \phi(t_2)]. \tag{6.55}$$

We will assume, without loss of generality, that $A \geq 0$.

$O(\varepsilon)$ $\quad (\partial_{t_1}^2 + 1)y_1 = (1 - y_0^2)\partial_{t_1} y_0 - 2\partial_{t_1}\partial_{t_2} y_0.$

Setting $\eta = t_1 + \phi(t_2)$, then

$$
\begin{aligned}
(1 - y_0^2)&\partial_{t_1} y_0 - 2\partial_{t_1}\partial_{t_2} y_0 \\
&= -[1 - A^2(1 - \sin^2(\eta)]A\sin\eta + 2(A'\sin\eta + A\phi'\cos\eta) \\
&= \left[2A' - A(1 - \frac{1}{4}A^2)\right]\sin\eta + 2A\phi'\cos\eta + \frac{1}{4}A^3\sin(3\eta).
\end{aligned}
$$

To prevent secular terms in the expansion, we set $A\phi' = 0$ and

$$2A' = A\left(1 - \frac{1}{4}A^2\right). \tag{6.56}$$

Thus, ϕ is constant and the nonzero solutions of (6.56) are

$$A(t_2) = \frac{2}{\sqrt{1 + ce^{-t_2}}}, \tag{6.57}$$

where c is a constant determined from the initial conditions.

To summarize, we have found that a first-term approximation of the solution of the initial-value problem, for λ close to the bifurcation value $\lambda_b = 0$, is

$$y(t) \sim \frac{2}{\sqrt{1 + ce^{-\lambda t}}}\cos(t + \phi_0) \tag{6.58}$$

and

$$y'(t) \sim -\frac{2}{\sqrt{1 + ce^{-\lambda t}}}\sin(t + \phi_0), \tag{6.59}$$

where c and ϕ_0 are constants determined from the initial conditions (Exercise 6.23). It is assumed here that the initial conditions are such that $y^2(0) + y_t^2(0) \neq 0$. From this we conclude that as $t \to \infty$, the solution $y(t) \to 0$ if $\lambda < 0$. This is expected because we have shown that the zero solution is an asymptotically stable steady state when λ is negative. However, when $\lambda > 0$, the limiting solution is $y(t) = 2\cos(t + \phi_0)$.

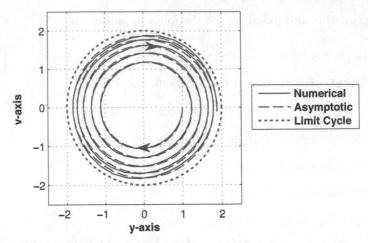

Figure 6.18 The values of the exponents $r_\pm$ in (6.50). If $\lambda \leq -2$, then both are negative and satisfy $r_\pm \leq -1$. When $\lambda \geq 2$, then both exponents are positive and satisfy $r_\pm \geq 1$. In the case of $-2 < \lambda < 2$, the exponents are complex conjugates with $|r_\pm| = 1$. The point at which $r_\pm = \pm i$ occurs when $\lambda = 0$

To describe the situation in more detail, it is convenient to use the phase plane. From (6.58) and (6.59) we have $y^2 + y_t^2 \sim 4/(1 + ce^{-\lambda t})$. Therefore, if $0 < \lambda \ll 1$, then $y^2 + y_t^2 \to 4$ as $t \to \infty$. In this case, the limit $(y, y') = (2\cos(t + \phi_0), -2\sin(t + \phi_0))$ is a periodic solution and there are no other periodic solutions near it. In other words, it is a limit cycle. The limit cycle and the asymptotic approximations (6.58) and (6.59) are shown in Fig. 6.17 in the cases $y(0) = 1.0$ and $y'(0) = 0$. For comparison, the numerical solution of the problem is also shown, and the two are seen to be in close agreement. The circular nature of the limit cycle is lost as λ increases. The solution in this case becomes more like a relaxation oscillation, and the analysis of this situation can be carried out using the ideas developed in Sect. 6.4 (Exercise 6.10).

We have found that the steady state $y_s = 0$ loses stability at $\lambda_b = 0$ and a stable limit cycle appears for $0 < \lambda - \lambda_b \ll 1$. In the vernacular of bifurcation theory, this is classified as a degenerate supercritical *Hopf bifurcation*. At a Hopf bifurcation point there is an exchange of stability, from a steady state to a limit cycle. This happens in this example because the two exponents $r_\pm$ in (6.50) cross the imaginary axis (Fig. 6.18). They are complex conjugates and they cross over with "nonzero velocity," in other words, $\frac{d}{d\lambda}\mathrm{Re}\,[r_\pm] \neq 0$ for $\lambda = \lambda_b$. The latter is known as a *transversality condition*.

Besides being supercritical, the bifurcation is said to be degenerate. What this means is that the steady state $y_s = 0$, for $\lambda < 0$, is not continuous with the limit cycle that is present for $\lambda > 0$. This discontinuity has an effect on the form of the expansion. If the limit cycle is continuous with the steady-state solution, then an example of an appropriate multiple-scale expansion is

$$y \sim y_b + \varepsilon^\alpha y_0(t_1, t_2) + \varepsilon^\beta y_1(t_1, t_2) + \cdots, \qquad (6.60)$$

where $0 < \alpha < \beta$. In writing this, it is assumed that two time scales are sufficient to determine the behavior of the solution for large t.

Exercises

6.17. This problem deals with the nonlinear oscillator

$$y'' + (y^2 - \lambda)y' + (y + 1)(y^2 + y - \lambda) = 0 \quad \text{for } 0 < t.$$

(a) Find the steady states and determine the values of λ at which they are stable. Sketch the bifurcation diagram and classify the three bifurcation points.
(b) Find a first-term approximation of the limit cycle that appears at the Hopf bifurcation point.

6.18. Consider the modified Rayleigh equation

$$y'' - \lambda[1 - \alpha(y')^{2n}]y' + y = 0 \quad \text{for } 0 < t,$$

where α is a positive constant and n a positive integer.
(a) Find the steady state and determine the values of λ at which it is stable.
(b) Find a first-term approximation of the limit cycle that appears at the Hopf bifurcation point.

6.19. The equation for the Hartley oscillator is (Hester, 1968)

$$y'' - \mu\left(4e^{2y'} - e^{4y'}\right) + y' + y = 0 \quad \text{for } 0 < t,$$

where μ is a positive constant.
(a) Find the steady state and determine the values of μ at which it is stable.
(b) Find a first-term approximation of the limit cycle that appears at the Hopf bifurcation point.

6.20. The logistic equation with delay, which is used in population modeling, is (Hutchinson, 1948)

$$y'(t) = ry(t)\left[1 - \frac{y(t - T)}{Y}\right] \quad \text{for } 0 < t,$$

where $y(t) = a$ for $-\infty < t \leq 0$. Without delay, so that $T = 0$, the steady state $y = Y$ is asymptotically stable. The purpose of this exercise is to investigate whether the delay can cause this to change, and for this reason T is the bifurcation parameter. Assume here that T, r, a, and Y are positive constants and $y(t) \geq 0$.

(a) Show that the steady state $y = 0$ is unstable.
(b) Show that the steady state $y = Y$ is stable if $0 < rT < \frac{\pi}{2}$ and unstable if $\frac{\pi}{2} < rT$. Note that the general solution of $y'(t) = c\,y(t-T)$ can be written as the superposition of solutions of the form $y = e^{\lambda t}$ (Appendix E). Also, recall that $2x/\pi < \sin x$ if $0 < x < \pi/2$.
(c) Letting $r = Y = 1$, use an expansion similar to (6.60) to determine what happens near $T = \frac{\pi}{2}$. Note that it is sufficient to concentrate only on the unstable mode you found in part (b). Also, you should derive the equations that determine the t_2 dependence of y_0, but you do not need to solve these equations.
(d) Express y_0 as in (6.55). Assuming that A approaches a nonzero constant as $t_2 \to \infty$, show that the constant is

$$A = 2\sqrt{\frac{10}{3\pi - 2}}$$

and the period of the oscillation approaches $2\pi/(1 + \varepsilon\kappa)$, where $\kappa = -6/(3\pi - 2)$.

6.21. The Wazewska–Czyzewska and Lasota model for the survivability of red blood cells is (Wazewska-Czyzewska and Lasota, 1976)

$$y'(t) = -\mu y(t) + e^{-y(t-T)} \quad \text{for } 0 < t.$$

Here $y(t)$ is the number of cells at time t and T is the time required to produce a red blood cell. Also, it is assumed that $0 < \mu < 1/e$. This problem is a continuation of Exercise 6.5.
(a) Determine what happens near $T = T_c$. Note that it is sufficient to concentrate only on the unstable mode you found in Exercise 6.5(c). Also, you should derive the equations that determine the t_2 dependence of y_0, but you do not need to solve these equations.
(b) Express y_0 as in (6.55). Assuming that A approaches a nonzero constant as $t_2 \to \infty$, show that

$$A^2 = \frac{8\mu\kappa(y_s + 1)(5y_s - 4)}{(2y_s - 1)(3y_s - 2)(1 + \mu y_s^2 T_c) - 2\kappa(y_s - 1)},$$

where $\kappa = \sqrt{y_s^2 - 1}$. What, if anything, does the period of the oscillation approach?

6.22. Suppose one uses the multiple-scale expansion in (6.60) to solve the van der Pol equation in (6.52). In this case $y_s = 0$.
(a) Show that the general solution of the $O(\varepsilon^\alpha)$ equation is the same as given in (6.55). With this, then show that one choice for the balancing is $\beta = 2$ and $\alpha = 1$.
(b) Show that removing secular terms results in the conclusion that $A = A_0 e^{t_2/2}$.

(c) The first-term approximation is oscillatory but with a growing amplitude. Explain why this violates the assumptions made in the balancing used to determine β in (a). What is the proper balance for this equation?

6.23. This problem examines some other aspects of the derivation of (6.58) and (6.59).

(a) The multiple-scale expansion assumes that ε is nonnegative. Show that (6.58) and (6.59) hold in the case of negative ε.

(b) Depending on the value of the constant c in (6.58) and (6.59), it would appear that there is a possibility that the amplitude has a singularity. Assuming $y(0) = a_0$ and $y'(0) = 0$, show that this does not happen.

(c) Describe the solution when $\lambda = 0$. Explain why it is not a limit cycle.

6.6 Systems of Ordinary Differential Equations

The ideas developed for scalar problems can be extended to systems of equations. To explain how, suppose the system is

$$\frac{d}{dt}\mathbf{y} = \mathbf{f}(\lambda, \mathbf{y}) \tag{6.61}$$

or, in component form,

$$\frac{d}{dt}y_1 = f_1(\lambda, y_1, y_2, \ldots, y_n),$$

$$\frac{d}{dt}y_2 = f_2(\lambda, y_1, y_2, \ldots, y_n),$$

$$\vdots \qquad \vdots$$

$$\frac{d}{dt}y_n = f_n(\lambda, y_1, y_2, \ldots, y_n).$$

It is assumed that $\mathbf{f}$ does not depend explicitly on t, but it does depend on a scalar λ. It is how the solution depends on this parameter that forms the basis of the bifurcation analysis that follows.

6.6.1 Linearized Stability Analysis

Given a steady-state solution $\mathbf{y}_s(\lambda)$, its stability is determined by solving (6.61) with the initial condition

$$\mathbf{y}(0) = \mathbf{y}_s + \delta\mathbf{a}. \tag{6.62}$$

As in the scalar case, the assumption is that we are starting the solution close to the steady state, and so we assume δ is small. The expansion of the solution in this case is assumed to have the form

$$y(t) \sim y_s + \delta v(t) + \cdots. \tag{6.63}$$

Substituting this into (6.61), and using Taylor's theorem, one finds that

$$\frac{d}{dt}(y_s + \delta v + \cdots) = f(\lambda, y_s + \delta v + \cdots)$$

$$= f(\lambda, y_s) + A(\delta v + \cdots) + \cdots, \tag{6.64}$$

where $A = \nabla f(\lambda, y_s)$. Here $\nabla f(\lambda, y)$ is the Jacobian matrix of $f(\lambda, y)$, in other words, it is the $n \times n$ matrix

$$\nabla f = \begin{pmatrix} \dfrac{\partial f_1}{\partial y_1} & \dfrac{\partial f_1}{\partial y_2} & \cdots & \dfrac{\partial f_1}{\partial y_n} \\[2mm] \dfrac{\partial f_2}{\partial y_1} & \dfrac{\partial f_2}{\partial y_2} & \cdots & \dfrac{\partial f_2}{\partial y_n} \\[2mm] \vdots & \vdots & & \vdots \\[2mm] \dfrac{\partial f_n}{\partial y_1} & \dfrac{\partial f_n}{\partial y_2} & \cdots & \dfrac{\partial f_n}{\partial y_n} \end{pmatrix}. \tag{6.65}$$

From the $O(\delta)$ terms in (6.64) and (6.62), it follows that

$$\frac{d}{dt}v = Av, \tag{6.66}$$

where $v(0) = a$. The solution of this problem is found by assuming that $v = xe^{rt}$. Substituting this into (6.66), the problem reduces to solving

$$Ax = rx. \tag{6.67}$$

This is an eigenvalue problem, where r is the eigenvalue and x the associated eigenvector. With this, the values for r are determined by solving the characteristic equation

$$\det(A - rI) = 0, \tag{6.68}$$

where I is the identity matrix. Given a value of r, the eigenvector is then determined by solving $(A - rI)x = 0$.

In the case where A has n distinct eigenvalues $r_1, r_2, \ldots, r_n$, with corresponding eigenvectors $x_1, x_2, \ldots, x_n$, the general solution of (6.66) has the form

$$v = \alpha_1 x_1 e^{r_1 t} + \alpha_2 x_2 e^{r_2 t} + \cdots + \alpha_n x_n e^{r_n t}, \tag{6.69}$$

where $\alpha_1, \alpha_2, \ldots, \alpha_n$ are arbitrary constants. The latter are determined from the initial condition. However, it is not necessary to calculate their values as we are only interested in the time dependence of the solution. In particular, from (6.69) it follows that $\mathbf{v} \to \mathbf{0}$ as $t \to \infty$ if $Re(r_i) < 0$, $\forall i$. However, if even one eigenvalue has $Re(r_i) > 0$, then it is possible to find values for $\mathbf{a}$ so that $\mathbf{v}$ is unbounded as $t \to \infty$.

If $\mathbf{A}$ does not have n distinct eigenvalues, then the general solution contains e^{rt} terms as well as those of the form $t^k e^{rt}$, where k is a positive integer. Consequently, the conclusion is the same, which is that $\mathbf{v} \to \mathbf{0}$ as $t \to \infty$ if $Re(r_i) < 0$, $\forall i$. Moreover, if there is an eigenvalue with $Re(r_i) > 0$, then $\mathbf{v}$ can become unbounded as $t \to \infty$. These conclusions have been reached without actually writing down the solution, which is possible because we only need to know the time dependence of the solution. Those interested in the exact formula, in the case where there are not n distinct eigenvalues, may consult Braun (1993).

The discussion in the previous paragraphs gives rise to the following result.

Theorem 6.2. *The steady state* $\mathbf{y}_s$ *is asymptotically stable if all of the eigenvalues of* $\mathbf{A}$ *satisfy* $Re(r) < 0$*, and it is unstable if even one eigenvalue has* $Re(r) > 0$*.*

The usefulness of the preceding theorem comes down to how difficult it is to determine the location of the eigenvalues of $\mathbf{A}$. For example, if $\mathbf{A}$ is triangular, then the eigenvalues are just the diagonal entries in the matrix. It is also easy to find the eigenvalues if $\mathbf{A}$ is 2×2. For more general matrices one might be able to use Gerschgorin's theorem to determine if the eigenvalues satisfy the given inequality (Horn and Johnson, 1990).

There are also tests using the characteristic polynomial. For these, note that (6.68) can be written as

$$r^n + a_1 r^{n-1} + \cdots + a_{n-1}r + a_n = 0. \tag{6.70}$$

In some cases it is possible to use Descartes' rule of signs to determine if the steady state is unstable. For example, it is unstable if one of the a_i is negative and the other coefficients satisfy $a_i \geq 0$. To determine if it is stable, one can use the Routh–Hurwitz criterion. To state this result, there are n associated Hurwitz matrices, and they are

$$\mathbf{H}_1 = \begin{pmatrix} a_1 \end{pmatrix}, \quad \mathbf{H}_2 = \begin{pmatrix} a_1 & 1 \\ a_3 & a_2 \end{pmatrix}, \quad \mathbf{H}_3 = \begin{pmatrix} a_1 & 1 & 0 \\ a_3 & a_2 & a_1 \\ a_5 & a_4 & a_3 \end{pmatrix}, \ldots,$$

$$\mathbf{H}_n = \begin{pmatrix} a_1 & 1 & 0 & 0 & 0 & 0 & \cdots \\ a_3 & a_2 & a_1 & 1 & 0 & 0 & \cdots \\ a_5 & a_4 & a_3 & a_2 & a_1 & 1 & \cdots \\ \cdot & \cdot & \cdot & \cdot & \cdot & \cdot & \\ a_{2n-1} & a_{2n-2} & \cdot & \cdot & \cdot & a_n \end{pmatrix}.$$

In these matrices, $a_k = 0$ if $k > n$. The Routh–Hurwitz criterion is given in the next theorem (Leipholz, 1987).

Theorem 6.3.

(i) If $det(\mathbf{H}_j) > 0$ for $j = 1, 2, \ldots, n$, then the solutions of (6.70) satisfy $Re(r) < 0$.

(ii) If there is a value of j so that $det(\mathbf{H}_j) < 0$, then there is a solution of (6.70) that satisfies $Re(r) > 0$.

Because this result requires the calculation of the determinant of several matrices, it is practical only for smaller values of n. The explicit requirements for $n \leq 4$ are given below.

Corollary 6.1.   *Assume that the characteristic Eq. (6.68) is written in the form given in (6.70).*

(i) $n = 2$: A steady state is asymptotically stable if $a_1 > 0$ and $a_2 > 0$ or, equivalently, if $tr(\mathbf{A}) < 0$ and $det(\mathbf{A}) > 0$. It is unstable if any of these inequalities is reversed.

(ii) $n = 3$: A steady state is asymptotically stable if $a_1 > 0$, $a_1 a_2 > a_3$, and $a_3 > 0$. It is unstable if any of these inequalities is reversed.

(iii) $n = 4$: A steady state is asymptotically stable if $a_1 > 0$, $a_3 > 0$, $a_4 > 0$, and $a_1 a_2 a_3 > a_1^2 a_4 + a_3^2$. It is unstable if any of these inequalities is reversed.

Given Theorem 6.2, the stability boundary occurs where one or more eigenvalues have $Re(r) = 0$ and all the others have $Re(r) < 0$. Assume that this occurs for $\lambda = \lambda_b$. If one of the $Re(r) = 0$ eigenvalues also has $Im(r) = 0$, then $r = 0$ is an eigenvalue of $\mathbf{A} = \nabla \mathbf{f}(\lambda_b, \mathbf{y}_s)$, and this means that $\mathbf{A}$ is singular when $\lambda = \lambda_b$. For this to happen, λ must satisfy

$$det(\mathbf{A}) = 0. \tag{6.71}$$

This is the multivariable version of (6.6). It is tempting to use this to find the bifurcation points and thereby avoid the stability analysis. However, like its one-dimensional version, this equation is not sufficient. Also, it is not necessary because it misses those eigenvalues with $Im(r) \neq 0$. As a final comment, determining the stability or instability of the steady state at $\lambda = \lambda_b$ can be challenging because the linear approximation used in (6.64) is insufficient. For this reason, in the exercises below we will be mostly interested in what happens when $\lambda < \lambda_b$ and when $\lambda > \lambda_b$.

Examples

1. Suppose the system to solve is

$$\frac{du}{dt} = v,$$

$$\frac{dv}{dt} = -v - \lambda u(1 - u).$$

In this case

$$\nabla \mathbf{f} = \begin{pmatrix} 0 & 1 \\ \lambda(2u - 1) & -1 \end{pmatrix}.$$

Steady states : There are two steady states – $(u_s, v_s) = (0, 0)$ and $(u_s, v_s) = (1, 0)$.

Linearized stability : Since $\mathbf{A} = \nabla \mathbf{f}(\lambda, \mathbf{y}_s)$, then $\text{tr}(\mathbf{A}) = -1$ and $\det(\mathbf{A}) = \lambda(1 - 2u_s)$. From Corollary (6.1) it follows that the steady state $(u_s, v_s) = (0, 0)$ is asymptotically stable if $\lambda > 0$ and unstable if $\lambda < 0$. Similarly, the steady state $(u_s, v_s) = (1, 0)$ is asymptotically stable if $\lambda < 0$ and unstable if $\lambda > 0$. ■

2. Consider the system

$$y' = v - y[v^2 + y^2 - \lambda(1 - \lambda)], \tag{6.72}$$

$$v' = -y - v[v^2 + y^2 - \lambda(1 - \lambda)]. \tag{6.73}$$

Steady states : There is only one steady state: $(y_s, v_s) = (0, 0)$. Setting $f(y, v) = v - y[v^2 + y^2 - \lambda(1 - \lambda)]$ and $g(y, v) = -y - v[v^2 + y^2 - \lambda(1 - \lambda)]$, the Jacobian matrix evaluated at this steady state is

$$\nabla \mathbf{f} = \begin{pmatrix} f_y(0, 0) & f_v(0, 0) \\ g_y(0, 0) & g_v(0, 0) \end{pmatrix}$$

$$= \begin{pmatrix} \lambda(1 - \lambda) & 1 \\ -1 & \lambda(1 - \lambda) \end{pmatrix}.$$

Linearized stability : Solving $\det(\mathbf{A} - r\mathbf{I}) = 0$ gives the eigenvalues $r_{\pm} = \lambda(1 - \lambda) \pm i$. This says that the steady state is stable if $\lambda < 0$ or if $\lambda > 1$ and unstable for $0 < \lambda < 1$. ■

6.6.2 Limit Cycles

The last example is interesting because it is unclear what is happening for $0 < \lambda < 1$. To determine what happens near $\lambda_b = 0$, set $\varepsilon = \lambda - \lambda_b$. We will use multiple scales, and so let $t_1 = t$ and $t_2 = \varepsilon t$. The appropriate expansion for the solution in this case is

$$\mathbf{y} \sim \mathbf{y}_s + \varepsilon^\alpha \mathbf{y}_0(t_1, t_2) + \varepsilon^\beta \mathbf{y}_1(t_1, t_2) + \cdots, \tag{6.74}$$

where $\mathbf{y}_s = \mathbf{0}$ and $\mathbf{y}_i = (y_i, v_i)$. Substituting this expansion into (6.72), (6.73) yields

$$(\partial_{t_1} + \varepsilon \partial_{t_2})(\varepsilon^\alpha y_0 + \varepsilon^\beta y_1 + \cdots)$$
$$= \varepsilon^\alpha v_0 + \varepsilon^\beta v_1 - \varepsilon^\alpha y_0 [\varepsilon^{2\alpha}(v_0^2 + y_0^2) - \varepsilon + \varepsilon^2] + \cdots,$$
$$(\partial_{t_1} + \varepsilon \partial_{t_2})(\varepsilon^\alpha v_0 + \varepsilon^\beta v_1 + \cdots)$$
$$= -\varepsilon^\alpha y_0 - \varepsilon^\beta y_1 - \varepsilon^\alpha v_0 [\varepsilon^{2\alpha}(v_0^2 + y_0^2) - \varepsilon + \varepsilon^2] + \cdots.$$

From this we get the following:

$O(\varepsilon^\alpha)$ $\partial_{t_1} y_0 = v_0$,
   $\partial_{t_1} v_0 = -y_0$.

The general solution of this system is $y_0 = A(t_2) \sin[t_1 + \theta(t_2)]$ and $v_0 = A(t_2) \cos[t_1 + \theta(t_2)]$. Also, from balancing, we have $\beta = \alpha + 1$ and $2\alpha = 1$.

$O(\varepsilon^\beta)$ $\partial_{t_1} y_1 = v_1 - y_0[(v_0^2 + y_0^2) - 1] - \partial_{t_2} y_0$,
   $\partial_{t_1} v_1 = -y_1 - v_0[(v_0^2 + y_0^2) - 1] - \partial_{t_2} v_0$.

One can solve the first equation for v_1 and reduce this system to a single second-order equation for y_1. Doing so one finds that to remove the secular terms, $\theta = \theta_0$ is constant and $A(t_2)$ satisfies $A' = A(1 - A^2)$. Solving this we obtain the nonzero solution

$$A(t_2) = \frac{1}{\sqrt{1 + ce^{-2t_2}}}, \tag{6.75}$$

where c is a nonnegative constant determined from the initial conditions.

Assuming that $\lambda > 0$, we have found that

$$\mathbf{y} \sim \sqrt{\frac{\lambda}{1 + ce^{-2t_2}}} \left(\sin(t + \theta_0), \cos(t + \theta_0) \right). \tag{6.76}$$

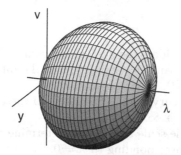

Figure 6.19 Surface formed from limit cycles of (6.72), (6.73). An approximation of this surface when $0 \leq \lambda \ll 1$ can be obtained directly from (6.77), and the general equation for the surface is derived in Exercise 6.24

Therefore, as $t \to \infty$,

$$y \sim \begin{cases} 0 & \text{if } \lambda < 0, \\ \sqrt{\lambda}\,(\sin(t + \theta_0), \cos(t + \theta_0)) & \text{if } 0 < \lambda \ll 1. \end{cases} \tag{6.77}$$

The zero solution is expected when $\lambda < 0$ since it is stable in that region. For positive values of λ we get that the limiting solution is periodic, and this means there is a supercritical Hopf bifurcation at the point $(\lambda_\beta, y_\beta) = (0, \mathbf{0})$. Using the same arguments, one can show that there is a subcritical Hopf bifurcation at $\lambda = 1$. As shown in Exercise 6.24, for $0 < \lambda < 1$ the solution approaches the curve $y^2 + v^2 = \lambda(1 - \lambda)$. The resulting bifurcation diagram is shown in Fig. 6.19.

It is worth pointing out what properties of the eigenvalues of $\nabla \mathbf{f}(\lambda, y_s)$ played an important role in the appearance, and disappearance, of the periodic solution. Recall that we found $r_\pm = \lambda(1 - \lambda) \pm i$. Therefore:

1. A single pair of eigenvalues crosses the imaginary axis to change the stability of the steady state. They moved into the right half-plane at $\lambda = 0$ and moved back into the left half-plane when $\lambda = 1$.

2. When the eigenvalues cross the imaginary axis, they satisfy the transversality condition. This condition states that $\frac{\mathrm{d}}{\mathrm{d}\lambda}\mathrm{Re}(r) \neq 0$ when $\mathrm{Re}(r) = 0$. In the preceding example, $\frac{\mathrm{d}}{\mathrm{d}\lambda}\mathrm{Re}(r) = 1$ when $\lambda = 0$ and $\frac{\mathrm{d}}{\mathrm{d}\lambda}\mathrm{Re}(r) = -1$ when $\lambda = 1$.

3. When the eigenvalues cross the imaginary axis, they are simple.

These conditions are the basis of what is known as the Poincaré–Andronov–Hopf bifurcation theorem. Readers interested in pursuing an investigation of the theory underlying this material may consult the books by Hale and Kocak (1991) and Guckenheimer and Holmes (1983).

Exercises

6.24. This problem considers the system given in (6.72), (6.73).
(a) Letting $y = r\cos(\theta)$ and $v = r\sin(\theta)$, find the differential equations satisfied by $r(t)$ and $\theta(t)$.
(b) Sketch the integral curves for $r(t)$, and from these comment on the limit of the solution as $t \to \infty$.
(c) Carry out a multiple-scale analysis to determine what happens near the bifurcation point corresponding to $\lambda = 0$.
(d) Find the exact solution and compare its behavior with what you found in parts (b) and (c).

6.25. In modeling the populations of a predator and prey, one finds the equations (Collings and Wollkind, 1990)

$$y' = y(1 - y) - \frac{\alpha y v}{\beta + y},$$

$$v' = \lambda v\left(1 - \frac{v}{y}\right),$$

where α, β, and λ are positive constants and $\beta < 1$. As usual, λ is the bifurcation parameter. Also, since $y(t)$ and $v(t)$ represent population densities of prey and predator, respectively, they are assumed to be positive.
(a) Find the steady state and show it is stable if

$$\lambda > \frac{(\alpha - \mu)(1 - \alpha - \beta + \mu)}{1 - \alpha + \beta + \mu},$$

where $\mu = \sqrt{(1 - \alpha - \beta)^2 + 4\beta}$.
(b) Describe what happens near the point where the steady state changes stability.
(c) Assuming $\alpha = 3/4$ and $\beta = 1/4$, use multiple scales to find the solution near the bifurcation point.

6.26. The Rosenzweig–MacArthur prey–predator model consists of the following two equations: (Collings and Wollkind, 1990)

$$y' = \lambda y - \mu y - \nu y^2 - \frac{\alpha y v}{1 + \alpha h y},$$

$$v' = \frac{\alpha y v}{1 + \alpha h y} - \gamma v,$$

where α, β, γ, λ, μ, ν, and h are positive constants. Also, since $y(t)$ and $v(t)$ represent population densities of prey and predator, respectively, they are assumed to be nonnegative.

(a) Find the three steady-state solutions. For each steady state, what conditions must the constants satisfy so the densities are nonnegative?

(b) On the same graph, sketch the three steady-state values for y as a function of λ. Do the same for v (on another graph). Assume here that $\beta > \alpha \gamma h$.

(c) There is one asymptotically stable steady state for $0 < \lambda < \mu$. Which one is it?

(d) There is one asymptotically stable steady state for $\lambda_c < \lambda$, where $\lambda_c = \mu + \nu \gamma / (\beta - \alpha \gamma h)$. Which one is it?

(e) Determine what happens near $\lambda = \mu$.

(f) Determine what happens near $\lambda = \lambda_c$.

6.27. The Sel'kov model for the enzymatic driven reactions that result in the conversion of ATP into ADP is (Sel'kov, 1968)

$$y' = 1 - yv^\gamma,$$
$$v' = \lambda v(yv^{\gamma-1} - 1),$$

where $\gamma > 1$ and $\lambda > 0$ are constants. In these equations, y and v designate the concentrations of ATP and ADP, respectively.

(a) Find the steady state and determine the values of λ at which it is stable.

(b) Is there a Hopf bifurcation in this problem?

(c) Assuming $\gamma = 2$, use multiple scales to find the solution near the bifurcation point.

6.28. In the description of the dynamics of excitons in semiconductor physics, one finds the following problem (Lee, 1992):

$$n' = \gamma - \alpha n^2 x,$$
$$x' = \alpha n^2 x - \frac{x}{1 + x},$$

where α and γ are constants that satisfy $0 < \gamma < 1$ and $\alpha > 0$. Also, $x(t)$ and $n(t)$ are densities and are therefore nonnegative.

(a) Find the steady state and determine the values of γ and α at which it is stable.

(b) Is there a Hopf bifurcation in this problem?

(c) Assuming $\gamma = 1/2$, use multiple scales to find the solution near the bifurcation point.

6.29. In the study of chemical reactions that involve thermal feedback, one finds the following problem (Gray and Scott, 1994):

$$c' = \mu - \kappa c e^T,$$
$$T' = ce^T - T,$$

where $c(t)$ is the concentration of chemical and $T(t)$ is the temperature rise due to self-heating. Also, μ and κ are positive constants.

(a) Find the steady state and determine the values of μ and κ at which it is stable. Explain why the steady state is stable if $\kappa > e^{-2}$.

(b) Suppose the value of κ is fixed and satisfies $0 < \kappa < e^{-2}$. Using μ as the bifurcation parameter, sketch the bifurcation diagrams for the steady states of c and T. Let μ_ℓ and μ_r, where $\mu_\ell < \mu_r$, and identify the two values of μ where c changes stability. Classify these two bifurcation points.

(c) Determine what happens near $\mu = \mu_\ell$. Also, comment on what you expect to occur near $\mu = \mu_r$.

(d) Use the smallness of κ to find the first three terms in expansions for μ_ℓ and μ_r.

6.30. A classic model in the study of oscillatory systems is the Brusselator. The equations in this case are (Gray et al., 1988)

$$x' = \mu - (1 + \alpha)x + x^2 y,$$
$$y' = x - x^2 y.$$

Here μ and α are positive constants.

(a) Find the steady state and determine its stability in the case where $\alpha \geq 1$.

(b) Suppose $0 < \alpha < 1$ and μ is the bifurcation parameter. Determine the stability of the steady state and describe what happens near the point where the steady state loses stability.

(c) Assuming $\alpha = 3/4$, use multiple scales to find the solution near the bifurcation point.

6.31. The Colpitts oscillator consists of the following equations (Maggio et al., 2004):

$$x' = \frac{\lambda}{1 - k}[1 + z - e^y],$$
$$y' = \frac{\lambda}{k} z,$$
$$z' = -\lambda(x + y) - z,$$

where $0 < k < 1$ and λ is positive.

(a) Find the steady state and determine its stability.

(b) Determine what happens near the point where the steady state loses stability.

6.32. Dynamic friction models are used to describe the interaction of surfaces as they slide past each other. An example is (Batista and Carlson, 1998)

$$y'' + \delta y' + y = \gamma t - \kappa - \alpha\phi,$$
$$\beta\phi' = \phi(1 - \phi) - \phi y'.$$

Here $y(t)$ is the distance the upper surface has moved relative to the lower surface, and $\phi(t)$ is a state variable related to the material properties of the interface. Also, κ, α, β, δ, and γ are positive constants.

(a) A steady sliding solution is of the form $y = t + a$ and $\phi = b$, where a and b are constants. There are two such solutions for this system. What are they?

(b) One of the solutions in part (a) has $b = 0$. Show that it is asymptotically stable if $\gamma > 1$ and unstable if $\gamma < 1$.

(c) One of the solutions in part (a) has $b \neq 0$. Show that it is unstable if $\gamma > 1$. Also, assuming $\gamma < 1$, show that it is asymptotically stable if $\delta \geq \alpha$, but if $\delta < \alpha$, then asymptotically stability requires that $\gamma_c < \gamma$, where

$$\gamma_c = 1 - \frac{\delta\beta}{2}\left(-1 + \sqrt{1 + \frac{4}{\delta(\alpha - \delta)}}\right).$$

(d) To explore what happens when the $b \neq 0$ solution loses stability near $\gamma = \gamma_c$, let $\delta = 4$, $\alpha = 5$, and $\beta = 1$. In this case, $\gamma_c = 3 - 2\sqrt{2}$. Using multiple scales and an expansion similar to that in (6.74), determine what happens to the solution for γ near γ_c.

6.33. A Volterra delay equation that comes up in population modeling is (Cushing, 1977)

$$y'(t) = ry(t)\left[Y - \int_{-\infty}^{t} K(t - \tau)y(\tau)d\tau\right],$$

where $K(t) = te^{-t}$. In this equation, r and Y are positive constants and $y(t) \geq 0$.

(a) Show that the steady state $y = 0$ is unstable.

(b) Using Appendix E, show that the delay equation can be written as a first-order system.

(c) Show that the steady state $y = Y$ is asymptotically stable if $rY < 2$ and unstable if $rY > 2$. Given how it goes unstable, what do you expect the solution is for $rY > 2$?

(d) Letting $Y = 1$, determine what happens near $r = 2$. Note that it is sufficient to concentrate only on the unstable mode you found in part (c). Also, you should derive the equations that determine the t_2 dependence of y_0, but you do not need to solve these equations.

(e) Express your solution for y_0 from part (d) as in (6.55). Assuming that A approaches a nonzero constant as $t_2 \to \infty$, find the constant and also determine the resulting period of oscillation.

6.34. An elastic pendulum is a pendulum where the rod contains an elastic spring as shown in Fig. 6.20. The (nondimensional) equations of motion for this system are (Minorsky, 1947; Heinbockel and Struble, 1963)

$$(1 + z)\theta'' + 2z'\theta' + \sin\theta = 0,$$
$$z'' + \kappa^2 z + 1 - \cos\theta - (1 + z)\theta_t^2 = 0,$$

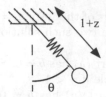

Figure 6.20 Elastic pendulum studied in Exercise 6.34

where $\theta(t)$ is the usual angular coordinate and $z(t)$ the stretch in the spring. Also, κ is a positive constant, and for physical reasons it is required that $-1 < z$.

(a) What is the steady state for this system?

(b) Suppose the initial conditions are $\theta(0) = \theta'(0) = 0$, $z(0) = \varepsilon \alpha_0$, and $z'(0) = 0$, where α_0 is a given nonzero constant. Describe what these conditions correspond to in physical (or geometric) terms. Use this to find the exact solution of the problem and describe the motion.

(c) To investigate the stability of the solution you found in part (b), suppose $\theta(0) = \varepsilon a_0$, $\theta'(0) = \varepsilon b_0$, $z(0) = \varepsilon \alpha_0$, and $z'(0) = 0$. For small ε, find an asymptotic approximation of the solution that is valid for large t. You should find the explicit t_1 dependence, but you only need to find the equations that determine the t_2 dependence. It is of interest to note that what is occurring in this problem is known as autoparametric self-excitation. This means that there is a transfer of energy from one mode, or type of motion, to another.

6.35. It is possible for traveling-wave solutions to bifurcate from a steady state. An example of this occurs with the system of equations

$$\partial_t u = [\lambda - f(r)]u - v + \partial_x^2 u,$$
$$\partial_t v = u + [\lambda - f(r)]v + \partial_x^2 v,$$

where $r^2 = u^2 + v^2$ and $f(r)$ is smooth with $f(0) = f'(0) = 0$, but $f''(0) > 0$. This is a prototypical problem that comes up in the study of pattern formation (Murray, 2003). In this context, u and v represent deviations of chemical concentrations from equilibrium and can be positive or negative.

(a) At what values of λ is the steady state $u = v = 0$ stable? Explain why one would classify the point where there is a change in the stability as a Hopf bifurcation point.

(b) Letting $u = r \cos \theta$ and $v = r \sin \theta$, show that the problem becomes

$$\partial_t r = [\lambda - f(r)]r - r\theta_x^2 + \partial_x^2 r,$$
$$\partial_t \theta = 1 + r^{-2}\partial_x(r^2 \theta_x).$$

With this, find the traveling-wave solutions that depend on $\theta = kx - \omega t$ and $r = r_0$.

(c) The multiple-scale analysis of the traveling wave will be replaced with a simpler approach. In particular, assume $\theta \sim kx - \omega t + \varepsilon^\alpha \phi(x,t)$ and $r \sim r_0 + \varepsilon^\alpha \rho(x,t)$, where $\phi = \phi_0 e^{(st+i\xi x)}$ and $\rho = \rho_0 e^{(st+i\xi x)}$. For stability, we need that $\text{Re}(s) < 0$ for all values of the constant ξ. Show that this results in the conclusion that the stable traveling waves are those that satisfy $0 < \lambda - f(r_0) < \frac{1}{4} r_0 f'(r_0)$.

6.36. Consider the system of equations

$$y' = v - y(v^2 + y^2 - \lambda),$$
$$v' = -y - v(v^2 + y^2 - \mu).$$

(a) In what region in the λ, μ-plane is the steady-state solution $y = v = 0$ stable? What portion of the boundary of this region gives rise to a Hopf bifurcation?

(b) For points along the boundary where a Hopf bifurcation occurs, find the solution bifurcating from the steady-state solution. You do not have to consider cases where $\mu = \pm 1$.

6.7 Weakly Coupled Nonlinear Oscillators

We return to a question first introduced in Sect. 3.5, which is what happens when oscillators are weakly coupled. Our earlier analysis, as illustrated in Fig. 3.9, dealt with linear oscillators that are coupled using springs and dashpots. The problem studied here is a bit more challenging and consists of the following system of four equations:

Oscillator 1:

$$\frac{dy_1}{dt} = v_1 - y_1(v_1^2 + y_1^2 - \kappa_1^2) + \varepsilon f_1(y_2 - y_1, v_2 - v_1), \qquad (6.78)$$

$$\frac{dv_1}{dt} = -y_1 - v_1(v_1^2 + y_1^2 - \kappa_1^2) + \varepsilon g_1(y_2 - y_1, v_2 - v_1); \qquad (6.79)$$

Oscillator 2:

$$\frac{dy_2}{dt} = v_2 - y_2(v_2^2 + y_2^2 - \kappa_2^2) + \varepsilon f_2(y_2 - y_1, v_2 - v_1), \qquad (6.80)$$

$$\frac{dv_2}{dt} = -y_2 - v_2(v_2^2 + y_2^2 - \kappa_2^2) + \varepsilon g_2(y_2 - y_1, v_2 - v_1). \qquad (6.81)$$

Although the preceding equations are not trivial, the underlying idea on which they are based is relatively simple. The reduced system, which is obtained when $\varepsilon = 0$, consists of two oscillators of the form considered in (6.72), (6.73). Each oscillator, by itself, will evolve into a limit cycle, as shown in (6.77). This is assuming, of course, that the κ_i are positive. The coupling, which is assumed to be weak, is contained in the ε terms. What is distinctive is that these terms depend on the differences $y_2 - y_1$ and $v_2 - v_1$. This is the same type of coupling that arose for the oscillator system in (3.57)–(3.58). As with that earlier problem, we will eventually assume the oscillators interact using an equal but opposite rule (so $f_1 = -f_2$ and $g_1 = -g_2$), and the coupling functions are linear. However, for the moment, it is only assumed that these functions are continuous.

Based on the preceding observations, what we are going to investigate is how a limit cycle solution is affected when coupled to another limit cycle. This might seem to be a question cooked up by applied mathematicians with too much time on their hands, but there are numerous applications of this material. For example, this type of system is the basis for a proposed neurocomputer (Hoppensteadt and Izhikevich, 1999; Gilli et al., 2005), and it is fundamental to understanding oscillator death and synchronization (Ermentrout, 1990; Pikovsky et al., 2001). It is also the basis of the Huygens clock problem described in Sect. 3.5, and a variation of this involving metronomes is considered in Exercise 6.38.

The preceding system of equations can be simplified by changing to polar coordinates. In particular, letting

$$y_1 = r_1 \cos(\theta_1), \qquad v_1 = r_1 \sin(\theta_1),$$
$$y_2 = r_2 \cos(\theta_2), \qquad v_2 = r_2 \sin(\theta_2);$$

then (6.78)–(6.81) reduce to

$$r_1' = -r_1(r_1^2 - \kappa_1^2) + \varepsilon F_1, \tag{6.82}$$
$$\theta_1' = -1 + \varepsilon G_1, \tag{6.83}$$
$$r_2' = -r_2(r_2^2 - \kappa_2^2) + \varepsilon F_2, \tag{6.84}$$
$$\theta_2' = -1 + \varepsilon G_2, \tag{6.85}$$

where

$$F_i = f_i(r_2 \cos\theta_2 - r_1 \cos\theta_1, r_2 \sin\theta_2 - r_1 \sin\theta_1) \cos\theta_i$$
$$+ g_i(r_2 \cos\theta_2 - r_1 \cos\theta_1, r_2 \sin\theta_2 - r_1 \sin\theta_1) \sin\theta_i, \tag{6.86}$$
$$r_i G_i = g_i(r_2 \cos\theta_2 - r_1 \cos\theta_1, r_2 \sin\theta_2 - r_1 \sin\theta_1) \cos\theta_i$$
$$- f_i(r_2 \cos\theta_2 - r_1 \cos\theta_1, r_2 \sin\theta_2 - r_1 \sin\theta_1) \sin\theta_i.$$

It is worth commenting about limit cycles and the polar coordinate variables introduced here. Given the solution in (6.76), the expectation is that both r_1 and r_2 will approach constants as $t \to \infty$, while θ_1 and θ_2 will approach monotonic functions of t. Also, in what follows it is assumed that both $r_1(0)$ and $r_2(0)$ are positive.

The derivation of the multiple-scale approximation will follow the same sequence of steps used in Sect. 6.6.2. Thus, letting $t_1 = t$ and $t_2 = \varepsilon t$, assume that

$$r_1 \sim r_{10}(t_1, t_2) + \varepsilon r_{11}(t_1, t_2), \quad \theta_1 \sim \theta_{10}(t_1, t_2) + \varepsilon \theta_{11}(t_1, t_2),$$
$$r_2 \sim r_{20}(t_1, t_2) + \varepsilon r_{21}(t_1, t_2), \quad \theta_2 \sim \theta_{20}(t_1, t_2) + \varepsilon \theta_{21}(t_1, t_2).$$

The problem for the $O(1)$ terms can be obtained by setting $\varepsilon = 0$ in (6.82)–(6.85). Solving these equations, one finds that the nonzero solutions are

$$r_{i0} = \frac{\kappa_i}{\sqrt{1 - a_i(t_2) \exp(-2\kappa_i^2 t_1)}}, \tag{6.87}$$

$$\theta_{i0} = -t_1 + \phi_i(t_2), \tag{6.88}$$

where a_i and ϕ_i are arbitrary functions of t_2.

To determine the t_2 dependence, we need to specify the coupling functions f_i and g_i. We will use linear functions, similarly to Sect. 3.5. In particular, it is assumed that

$$f_1(y, v) = -ay - bv, \qquad g_1(y, v) = by - av, \tag{6.89}$$
$$f_2(y, v) = ay + bv, \qquad g_2(y, v) = -by + av, \tag{6.90}$$

where a and b are constants. Looking at the various coefficients in the preceding expressions it is apparent that we are being very selective in what linear functions are used in the problem. Note that the assumption that $f_2 = -f_1$ and $g_2 = -g_1$ is consistent with Newton's third law (Sect. 3.5). The relationship between the coefficients of f_1 and g_1 is made to simplify the analysis.

The $O(\varepsilon)$ problem coming from (6.82)–(6.85) is

$$\partial_{t_1} r_{i1} + \partial_{t_2} r_{i0} = -(3r_{i0}^2 - \kappa_i^2) r_{i1} + F_{i0} \text{ for } i = 1, 2, \tag{6.91}$$

$$\partial_{t_1} \theta_{11} + \partial_{t_2} \theta_{10} = -b\left(1 - \frac{r_{20}}{r_{10}} \cos(\phi_2 - \phi_1)\right) + a\frac{r_{20}}{r_{10}} \sin(\phi_1 - \phi_2), \tag{6.92}$$

$$\partial_{t_1} \theta_{21} + \partial_{t_2} \theta_{20} = -b\left(1 - \frac{r_{10}}{r_{20}} \cos(\phi_2 - \phi_1)\right) - a\frac{r_{10}}{r_{20}} \sin(\phi_1 - \phi_2), \tag{6.93}$$

where F_{i0} is (6.86) evaluated at $(r_i, \theta_i) = (r_{i0}, \theta_{i0})$.

Because the problem for r_{i1} is linear, it is possible to solve it using an integrating factor. From this it is found that a_i in (6.87) is constant (Exercise 6.41). Therefore, $r_{i0} \to \kappa_i$ as $t \to \infty$. Also, in this limit, (6.92) and (6.93) take the forms

$$\partial_{t_1} \theta_{11} + \phi_1' = -b\left(1 - \frac{\kappa_2}{\kappa_1} \cos(\phi_2 - \phi_1)\right) + a\frac{\kappa_2}{\kappa_1} \sin(\phi_1 - \phi_2),$$

$$\partial_{t_1} \theta_{21} + \phi_2' = -b\left(1 - \frac{\kappa_1}{\kappa_2} \cos(\phi_2 - \phi_1)\right) - a\frac{\kappa_1}{\kappa_2} \sin(\phi_1 - \phi_2).$$

To minimize the error (Sect. 3.3), we conclude that

$$\phi_1' = -b\left(1 - \frac{\kappa_2}{\kappa_1}\cos(\phi_2 - \phi_1)\right) + a\frac{\kappa_2}{\kappa_1}\sin(\phi_1 - \phi_2), \qquad (6.94)$$

$$\phi_2' = -b\left(1 - \frac{\kappa_1}{\kappa_2}\cos(\phi_2 - \phi_1)\right) - a\frac{\kappa_1}{\kappa_2}\sin(\phi_1 - \phi_2). \qquad (6.95)$$

These equations can be solved by introducing the phase difference

$$\phi = \phi_2 - \phi_1. \qquad (6.96)$$

Subtracting (6.94) and (6.95), it follows that

$$\phi' = b\left(\frac{\kappa_1}{\kappa_2} - \frac{\kappa_2}{\kappa_1}\right)\cos\phi + a\left(\frac{\kappa_2}{\kappa_1} + \frac{\kappa_1}{\kappa_2}\right)\sin\phi. \qquad (6.97)$$

Making the final simplifying assumption that $\kappa_2 = \kappa_1 = \kappa$, the preceding equation reduces to

$$\phi' = 2a\sin\phi. \qquad (6.98)$$

It is evident that there are two possible steady states, $\phi = 0$ and $\phi = \pi$ (modulo integer multiples of 2π). Although it is possible to carry out a linear stability analysis for these steady states, it is actually easier to just solve the problem. Doing so one finds that

$$\phi = 2\arctan\left(c_0 e^{2at_2}\right), \qquad (6.99)$$

where c_0 is a constant. From (6.94) and (6.98) it follows that

$$\phi_1 = \phi_0 - \arctan(c_0 e^{2at_2}),$$

and from (6.96)

$$\phi_2 = \phi_0 + \arctan(c_0 e^{2at_2}).$$

As a final comment, the solution of (6.98) is not unique. In particular, if ϕ is a solution, then $\phi + 2\pi n$ is a solution for any integer n. This nonuniqueness is a consequence of the large t limit used to obtain (6.98). Although it is possible to determine n from (6.94) and (6.95), we are interested solely in the question of synchronicity, and for this we only need to know the solution modulo integer multiples of 2π.

We have completed the derivation of the multi-scale approximation of the solution of (6.78)–(6.81) in the case where $\kappa_2 = \kappa_1 = \kappa$, and the coupling functions are given in (6.89) and (6.90). Of particular interest is the phase difference between the two oscillators, as given in (6.99). It shows that if $a < 0$, or if $c_0 = 0$, then $\phi \to 0$ as $t \to \infty$, otherwise $\phi \to \pm\pi$ (the sign depending on the sign of c_0). In other words, the weak coupling causes the two limit cycle solutions to become synchronous, being in-phase if $a < 0$

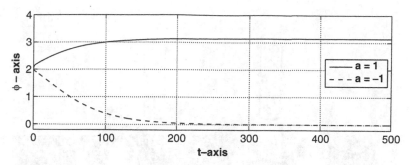

Figure 6.21 Numerically computed value of phase difference $\phi = \theta_2 - \theta_1$ in the two oscillators, showing that the sign of a determines whether $\phi \to \pi$ or $\phi \to 0$ as $t \to \infty$. In this calculation, $\varepsilon = 10^{-2}$

and antiphase if $a > 0$ (assuming $c_0 \neq 0$). The demonstration of this fact is shown in Fig. 6.21, which gives the value of $\phi(t)$ for $a = 1$ and for $a = -1$, as determined from the numerically computed solution of (6.82)–(6.85). It is also worth noting that this result differs from the linear oscillator problem, as expressed in Exercise 3.28, which only yields in-phase synchronicity.

The system (6.94)–(6.95) is an example of coupled phase equations. These play a central role in the more general theory of multiple weakly coupled oscillators. In such cases, the usual approach is to use what is known as Malkin's theorem to obtain the phase equations (Hoppensteadt and Izhikevich, 1997). This result requires that the uncoupled oscillators have an exponentially orbitally stable periodic solution. Although this assumption was not made here, as shown in (6.87), the solution does have exponential dependence as $r_{i0} \to \kappa_i$.

Exercises

6.37. This exercise investigates what happens when two van der Pol oscillators are weakly coupled. The problem to solve is

$$y_1'' - \varepsilon(1 - y_1^2)y_1' + y_1 = \varepsilon\alpha(-y_1 + y_2),$$
$$y_2'' - \varepsilon(1 - y_2^2)y_2' + y_2 = \varepsilon\alpha(y_1 - y_2).$$

As shown in Sect. 6.5, when $\alpha = 0$, both y_1 and y_2 evolve into limit cycles for $0 < \varepsilon \ll 1$.

(a) Carrying out a multiple-scale analysis, show that the first term for each oscillator has the form

$$y_{i0} = A_{i0}(t_2) \cos(t_1 + \phi_{i0}(t_2)),$$

Figure 6.22 Metronomes sitting on a moveable platform, as examined in Exercise 6.38 (Pantaleone, 2002)

where the A_{i0} are determined by solving

$$2A'_{10} = A_{10}\left(1 - \frac{1}{4}A^2_{10}\right) + \alpha A_{20}\sin(\phi),$$

$$2A'_{20} = A_{20}\left(1 - \frac{1}{4}A^2_{20}\right) - \alpha A_{10}\sin(\phi),$$

$$2\phi' = \alpha\left(\frac{A_{20}}{A_{10}} - \frac{A_{10}}{A_{20}}\right)\cos(\phi).$$

Also, explain how ϕ_{10} and ϕ_{20} are determined once ϕ is known.

(b) Assume that A_{10} and A_{20} approach positive constants as $t \to \infty$. Determine the possible values the phase difference ϕ approaches as $t \to \infty$.

(c) Show that the in-phase synchronized solution you found in part (b) is stable. As it turns out, the others are unstable (you do not need to show this).

6.38. A modified Huygens clock problem can be constructed by placing two metronomes on a movable platform (Fig. 6.22). The idea here is that the motion of the two pendulums can cause the platform to move back and forth, and this results in a weak coupling between the oscillators. A model for this system is (Pantaleone, 2002)

$$\theta''_1 + \sin\theta_1 + \varepsilon\left[\left(\frac{\theta_1}{\theta_0}\right)^2 - 1\right]\theta'_1 = \varepsilon\beta[\sin\theta_1 + \sin\theta_2]''\cos\theta_1,$$

$$\theta''_2 + \sin\theta_2 + \varepsilon\left[\left(\frac{\theta_2}{\theta_0}\right)^2 - 1\right]\theta'_2 = \varepsilon\beta[\sin\theta_1 + \sin\theta_2]''\cos\theta_2,$$

where $\theta_i(t)$ is the angular displacement of the ith pendulum. It is assumed that the angular deflections are small, so $\theta_0 = \varepsilon\rho_0$, where ρ_0 is independent of ε, and the expansions have the form

$$\theta_i \sim \varepsilon\theta_{i0}(t_1, t_2) + \varepsilon^2\theta_{i1}(t_1, t_2) + \cdots .$$

Also, both ρ_0 and β are positive constants.

(a) Carrying out a multiple-scale analysis shows that the first term has the form

$$\theta_{i0} = \rho_i(t_2)e^{i[t_1+\phi_i(t_2)]} + \rho_i(t_2)e^{-i[t_1+\phi_i(t_2)]},$$

where ρ_i and ϕ_i are solutions of a coupled first-order system. You are not expected to solve this system.

(b) Introducing the phase difference $\phi = \phi_2 - \phi_1$, show that the four equations for ρ_i and ϕ_i can be reduced to the following three equations:

$$2\rho_1' = \rho_1\left(1 - \frac{1}{\rho_0^2}\rho_1^2\right) - \beta\rho_2\sin\phi,$$

$$2\rho_2' = \rho_2\left(1 - \frac{1}{\rho_0^2}\rho_2^2\right) - \beta\rho_1\sin\phi,$$

$$2\phi' = \beta\left(\frac{\rho_1}{\rho_2} - \frac{\rho_2}{\rho_1}\right)\cos\phi.$$

(c) The steady-state solutions for the reduced system from part (b) are the possible synchronized solutions for this problem. Do any of these steady states correspond to an in-phase or an antiphase response?

(d) According to Pantaleone (2002), only in-phase synchronized solutions are observed. Is the in-phase steady state you found in part (c) linearly stable? What about the antiphase solution? Do your results show that the model is consistent with experimental observations?

6.39. This problem shows that an in-phase or antiphase response requires that the two oscillators have the same limit cycle solution.

(a) Solve (6.97) in the case where $\kappa_1 \neq \kappa_2$.

(b) Using your solution from part (a), describe what happens to the phase difference as $t \to \infty$. Explain why this shows that there is phase lock between the two oscillators, but they are not in-phase or antiphase.

6.40. This problem replaces the equal but opposite assumption with what might be called the equal but equal rule (i.e., $f_1 = f_2$ and $g_1 = g_2$). In particular, it is assumed that $f_1(y, v) = f_2(y, v) = ay + bv$ and $g_1(y, v) = g_2(y, v) = -by + av$, where a and b are nonzero. It is also assumed that $\kappa_1 = \kappa_2 = \kappa$.

(a) Solve (6.97) in the case where $\kappa_1 \neq \kappa_2$.

(b) What happens to the phase difference $t \to \infty$?

6.41. This problem considers some of the steps left out in the derivation of the reduced problem.

(a) From the initial conditions, in (6.87) show that $a_i(0) < 1$.

(b) Show that (6.91), when written out, gives rise to the following equations:

$$\partial_{t_1} r_{11} + \partial_{t_2} r_{10} = -g_1(t_1, t_2) r_{11} + a\left[r_{10} - r_{20}\cos(\phi)\right] - br_{20}\sin(\phi),$$
$$\partial_{t_1} r_{21} + \partial_{t_2} r_{20} = -g_2(t_1, t_2) r_{21} + a\left[r_{20} - r_{10}\cos(\phi)\right] + br_{10}\sin(\phi),$$

where

$$g_i(t_1, t_2) = \kappa_i^2 \frac{2 + a_i(t_2)\exp(-2\kappa_i^2 t_1)}{1 - a_i(t_2)\exp(-2\kappa_i^2 t_1)}.$$

(c) Solve the system in part (b), and use this solution to explain why it is not necessary to assume that a_i depends on t_2.

(d) From (6.92) and (6.93) show that

$$\partial_{t_1}\phi_1 + \phi' = b\left(\frac{r_{10}}{r_{20}} - \frac{r_{20}}{r_{10}}\right)\cos\phi + a\left(\frac{r_{20}}{r_{10}} + \frac{r_{10}}{r_{20}}\right)\sin\phi,$$

where $\phi_1 = \theta_{21} - \theta_{11}$.

6.8 An Example Involving a Nonlinear Partial Differential Equation

Many of the ideas discussed in previous sections can be applied to partial differential equations. One of the better-known examples of bifurcation involving a partial differential equation arises with what is known as buckling. If you take a meter stick and compress it along its axis, what you find is that the meter stick does essentially nothing until you press hard enough, in which case it "pops" into a bowed shape. This is called the first buckling mode and is an example of bifurcation from the zero state. This mode is very stable. As an experiment, you might try to hold the meter stick to prevent it from entering this bowed configuration and see if you can get it into the second buckling mode, which resembles the curve $\sin(2\pi x)$ for $0 \le x \le 1$. You should find that the second mode is fairly difficult to maintain, as a relatively small perturbation will cause the meter stick to pop back into the first mode.

For an introductory example, the equation for a meter stick is a little too cumbersome (Exercise 6.43). Therefore, to introduce the ideas underlying bifurcation for partial differential equations, we will consider the simpler problem

$$u_{tt} + u_t = u_{xx} + \lambda u + f(u) \quad \text{for } 0 < x < 1, \tag{6.100}$$

where

$$u = 0 \quad \text{for} \quad x = 0, 1. \tag{6.101}$$

The initial conditions will be discussed later. What we have is a (damped) nonlinear wave equation where the nonlinearity is contained in the function $f(u)$. To be able to make any headway in solving the problem, we need to be fairly clear on what form $f(u)$ can have. The type of nonlinearity appearing in Duffing's equation is a reasonable first example, so we will assume $f(0) = f'(0) = f''(0) = 0$ but $f'''(0) \neq 0$. This includes such functions as $f = \pm u^3$. Note that $f(0) = 0$ means $u = 0$ is a steady-state solution. This is a very common situation in physical problems and is therefore not an unreasonable assumption. The harder condition to explain is why we are assuming $f(u)$ does not include quadratic terms near $u = 0$. The short answer is that quadratic nonlinearities can take longer to produce secular terms (Exercise 3.3), although the ideas involved are essentially the same as in the case we will consider. This will be evident when working out Exercise 6.45. Anyway, we have a problem to solve, and that is what we now set out to do.

6.8.1 Steady State Solutions

To determine the steady states, we need to solve the problem

$$u_{xx} + \lambda u + f(u) = 0 \quad \text{for} \quad 0 < x < 1, \tag{6.102}$$

where

$$u = 0 \quad \text{for} \quad x = 0, 1. \tag{6.103}$$

This is a nonlinear eigenvalue problem where we are looking for values of λ that result in a nonzero solution. As in Sect. 6.3, we will look for the solutions that bifurcate from $u_s = 0$. This is done by letting

$$\varepsilon = \lambda - \lambda_b,$$

where λ_b is the value of λ where the bifurcation takes place. In this case, the appropriate expansion for small ε is

$$u \sim u_s + \varepsilon^\alpha u_1(x) + \varepsilon^\beta u_2(x) + \cdots.$$

We are interested in finding nonzero solutions of (6.102), (6.103), and so we will require that u_1 not be identically zero and $0 < \alpha < \beta$. Substituting this into (6.102) and using Taylor's theorem yields

$$\underset{①}{\varepsilon^\alpha (\partial_x^2 u_1 + \lambda_b u_1)} + \underset{②}{\varepsilon^\beta (\partial_x^2 u_2 + \lambda_b u_2)} + \underset{③}{\varepsilon^{1+\alpha} u_1 + \frac{1}{6} \varepsilon^{3\alpha} u_1^3 f'''(0)} + \cdots = 0. \tag{6.104}$$

This leads to the following problems:

$O(\varepsilon^{\alpha})$ $\partial_x^2 u_1 + \lambda_b u_1 = 0$, for $0 < x < 1$,
$u_1(0) = u_1(1) = 0$.

What we have here is a linear eigenvalue problem, and the bifurcation value λ_b is the eigenvalue. Solving this one finds

$$\lambda_b = (n\pi)^2 \quad \text{for} \quad n = 1, 2, 3, \cdots$$

and

$$u_1 = A_n \sin(n\pi x). \tag{6.105}$$

We have not yet completely determined u_1 since the coefficient A_n in (6.105) is arbitrary. Also, the value of α is not known. Both of these are determined from the next order problem. To determine what this problem is, note that in (6.104) the $O(\varepsilon^{\beta})$ term must deal with the contributions of u_1 found in either term ② or ③. The balancing of ① with ② (and ③ ≪ ①, ②) or ① with ③ (and ② ≪ ①, ③) leads to the conclusion that $A_n = 0$. Therefore, for a nonzero solution the terms ①, ②, and ③ must balance. From this we obtain that $\beta = 1 + \alpha = 3\alpha$. Thus, $\alpha = 1/2$ and $\beta = 3/2$.

$O(\varepsilon^{3/2})$ $\partial_x^2 u_2 + \lambda_b u_2 = -u_1 - \frac{1}{6} f'''(0) u_1^3$, for $0 < x < 1$,
$u_2(0) = u_2(1) = 0$.

This is an inhomogeneous version of the $O(\varepsilon^{\alpha})$ eigenvalue problem. We therefore have a situation similar to what we came across with the WKB method in Sect. 4.5. A solvability condition is going to come from this problem (as required by the Fredholm alternative theorem). To find this condition, multiply the differential equation by u_1, integrate both sides over the interval $0 < x < 1$, and then integrate by parts. The result is

$$\int_0^1 [u_1^2 + \frac{1}{6} f'''(0) u_1^4] dx = 0. \tag{6.106}$$

Using the solution in (6.105), we get from (6.106) that

$$A_n^2 = -\frac{8}{f'''(0)}. \tag{6.107}$$

From the preceding analysis we conclude that there is an infinite number of bifurcation points associated with the steady state $u_s = 0$. These are the eigenvalues associated with a linear boundary-value problem, and they are $\lambda_b = (n\pi)^2$. The functions bifurcating from these points are

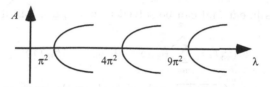

Figure 6.23 Bifurcation diagram for nonlinear wave problem in the case where $f'''(0) < 0$. Sketched here is $A = \varepsilon^{1/2}A_n$, where A_n is given in (6.107). In this case, there is a supercritical pitchfork bifurcation at each $\lambda_b = (n\pi)^2$ for $n = 1, 2, 3, \ldots$

$$u \sim \pm\sqrt{-\frac{8(\lambda - \lambda_b)}{f'''(0)}}\,\sin(n\pi x) \quad \text{for } n = 1, 2, 3, \ldots. \tag{6.108}$$

So there is supercritical pitchfork bifurcation if $f'''(0) < 0$ and subcritical pitchfork bifurcation if $f'''(0) > 0$ (Fig. 6.23). It should be emphasized that our approximation in (6.108) of the bifurcating solutions is valid only for λ close to λ_b.

6.8.2 Linearized Stability Analysis

The next step is to determine whether the steady-state solutions we have found are stable. As before, we will do this by solving the problem using the initial conditions

$$u(x, 0) = u_s(x) + \delta g(x) \quad \text{and} \quad \partial_t u(x, 0) = \delta h(x). \tag{6.109}$$

We are starting the solution out close to the equilibrium solution $u = u_s(x)$, and for this reason the parameter δ is small. In what follows, we will assume the value of λ does not correspond to a bifurcation value, that is, $\lambda \neq (n\pi)^2$. Also, the functions $g(x)$ and $h(x)$ in (6.109) are arbitrary (but smooth).

The appropriate expansion of the solution is

$$u(x, t) \sim u_s(x) + \delta v_1(x, t) + \cdots.$$

Substituting this into (6.100) and (6.101) yields

$$\partial_t^2 v_1 + \partial_t v_1 = \partial_x^2 v_1 + [\lambda + f'(u_s)]v_1 \quad \text{for } 0 < x < 1, \tag{6.110}$$

where

$$v_1 = 0 \quad \text{for } x = 0, 1 \tag{6.111}$$

and

$$v_1(x, 0) = g(x) \quad \text{and} \quad \partial_t v_1(x, 0) = h(x). \tag{6.112}$$

This problem is linear and can be solved using separation of variables. One finds that

$$v_1(x,t) = \sum_{m=1}^{\infty} (a_m e^{r_+ t} + b_m e^{r_- t}) F_m(x), \tag{6.113}$$

where $r_\pm = \frac{1}{2}(-1 \pm \sqrt{1 + 4\kappa_m})$ and κ_m is the mth separation constant (we are assuming here that $\kappa_m \neq -1/4$). The function $F_m(x)$ satisfies

$$F_m'' + [\lambda + f'(u_s) - \kappa_m] F_m = 0 \quad \text{for } 0 < x < 1, \tag{6.114}$$

where $F_m(0) = F_m(1) = 0$. It is interesting that this is another eigenvalue problem, where κ_m is the eigenvalue. To find κ_m, we must solve (6.114), and this requires us to specify which steady state we are considering. However, we are not particularly interested in the exact form of the solution but only whether $\text{Re}(r_+)$ is positive or negative. This is because $\text{Re}(r_-) \leq \text{Re}(r_+)$. Consequently, if $\text{Re}(r_+) < 0$, then $v_1 \to 0$ as $t \to \infty$, so u_s is asymptotically stable. On the other hand, if $\text{Re}(r_+) > 0$, then for any nonzero values of a_m, v_1 is unbounded as $t \to \infty$, and this means that u_s is unstable

What we have shown above is that this will happen if $\kappa_m < 0$, and it is this inequality we will concentrate on in the analysis to follow.

6.8.3 Stability of Zero Solution

We will first consider the particular case of $u_s = 0$. Since $f'(0) = 0$, the solution of (6.114) is

$$F_m(x) = \sin(m\pi x)$$

and

$$\kappa_m = \lambda - (m\pi)^2. \tag{6.115}$$

The coefficients a_m and b_m in (6.113) are determined by satisfying the initial conditions (6.112). Assuming

$$g(x) = \sum_{m=1}^{\infty} g_m \sin(m\pi x) \quad \text{and} \quad h(x) = \sum_{m=1}^{\infty} h_m \sin(m\pi x),$$

then

$$a_m + b_m = g_m \quad \text{and} \quad a_m r_+ + b_m r_- = h_m.$$

It is a simple matter to solve these equations for a_m and b_m.

The stability of the steady-state solution $u_s = 0$ depends on the exponents $r_\pm$ in (6.113). These depend on the mode number m; and using the result given in (6.115) we have the following cases:

1. If $\lambda < \pi^2$, then $r_- < r_+ < 0$. Therefore, $v_1 \to 0$ as $t \to \infty$, and this means the steady-state solution $u_s = 0$ is asymptotically stable.

2. If $\pi^2 < \lambda < 4\pi^2$, then $r_- < r_+ < 0$ if $m \geq 2$, but $r_- < 0 < r_+$ if $m = 1$. In this case, the steady state $u_s = 0$ is unstable. However, note that if the initial conditions are such that $h_1 = g_1 r_-$, then $v_1 \to 0$ as $t \to \infty$.

3. If $n^2 \pi^2 < \lambda < (n+1)^2 \pi^2$ for $n = 1, 2, 3, \ldots$, then the steady state $u_s = 0$ is unstable. In particular, it is stable only to initial disturbances that satisfy $h_m = g_m r_-$ for $m = 1, 2, \ldots, n-1$.

6.8.4 Stability of the Branches that Bifurcate from the Zero Solution

The next question is whether or not the solutions that bifurcate from $u_s = 0$ are stable. We do not have formulas for the nonzero steady states except right near the bifurcation point. It is therefore going to be necessary to incorporate our ε expansions into the stability argument. To do this, recall that we found that bifurcation occurs when $\lambda = (n\pi)^2$, and the expansion for the nonzero steady state is

$$u_s \sim \varepsilon^{1/2} A_n \sin(n\pi x).$$

We are going to find a first-term approximation to the solution of the linearized stability problem (6.110)–(6.112) for small ε. With this objective in mind, we first recall that $\lambda = \lambda_b + \varepsilon$ and $f'(0) = f''(0) = 0$, but $f'''(0) \neq 0$. Thus, from Taylor's theorem and (6.107),

$$\lambda + f'(u_s) \sim \lambda_b + \left[1 + \frac{1}{2}f'''(0)A_n^2 \sin^2(n\pi x)\right]\varepsilon$$

$$= \lambda_b + [1 - 4\sin^2(n\pi x)]\varepsilon.$$

Our objective is to solve (6.114), and the unknowns for this equation are F_m and κ_m. These quantities can depend on ε, and to account for this, we expand them as follows:

$$F_m \sim \overline{F}_0 + \varepsilon \overline{F}_1 + \cdots \quad \text{and} \quad \kappa_m \sim \overline{\kappa}_0 + \varepsilon \overline{\kappa}_1 + \cdots .$$

Substituting these into (6.114), the following problems appear:

$O(1)$ $\overline{F}_0'' + (\lambda_b - \bar{\kappa}_0)\overline{F}_0 = 0$ for $0 < x < 1$,
       $\overline{F}_0(0) = \overline{F}_0(1) = 0$.

Solving this eigenvalue problem yields $\overline{F}_0 = \sin(m\pi x)$ and $\bar{\kappa}_0 = (n^2 - m^2)\pi^2$. This information is sufficient to answer the question about stability except in the case where $n = m$. For this we need the next order problem.

$O(\varepsilon)$ $\overline{F}_1'' + (\lambda_b - \bar{\kappa}_0)\overline{F}_1 = -[1 - 4\sin^2(n\pi x) - \bar{\kappa}_1]\overline{F}_0$ for $0 < x < 1$.

As happened earlier, we are facing an inhomogeneous version of the eigenvalue problem. Multiplying by $\overline{F}_0$ and integrating yields

$$\bar{\kappa}_1 = \frac{\int_0^1 [1 - 4\sin^2(n\pi\xi)]\overline{F}_0^2 \, dx}{\int_0^1 \overline{F}_0^2 \, dx}$$

$$= \begin{cases} -1 & \text{if } m \neq n, \\ -2 & \text{if } m = n. \end{cases}$$

As stated previously, a bifurcating solution will be stable if $\kappa_m < 0$. To determine if this is the case, suppose, for the sake of argument, that we are considering the third bifurcation point, so $n = 3$. What we need to determine is whether or not it is true that $\kappa_m < 0$ for every value of m. The answer is no because we have found that, for $m \neq n$, $\kappa_m \sim (n^2 - m^2)\pi^2$. In particular, if $m = 2$ or if $m = 1$, then $\kappa_m > 0$. This conclusion holds for the other bifurcation points except, possibly, $n = 1$. For this value of n we have $\kappa_m \sim -2\varepsilon$ when $m = 1$ and $\kappa_m \sim -(m^2 - 1)\pi^2$ for $m \geq 2$. Hence this branch will be stable if $\varepsilon > 0$, that is, $\lambda > \lambda_b$.

Our expansions for the two branches are

$$u \sim \pm\sqrt{-\frac{8(\lambda - \lambda_b)}{f'''(0)}} \sin(\pi x).$$

From this we have that if $f'''(0) > 0$, then $\varepsilon < 0$ (i.e., it is a subcritical bifurcation), and if $f'''(0) < 0$, then $\varepsilon < 0$ (i.e., it is supercritical). Therefore, the solutions branching from the first bifurcation point ($n = 1$) are stable if $f'''(0) < 0$ and unstable if $f'''(0) > 0$.

To illustrate this situation, using the given boundary conditions, the solutions that bifurcate from the zero solution at $\lambda = \pi^2$ are unstable for the equation

$$u_{tt} + u_t = u_{xx} + \lambda u + u^3,$$

but they are stable for

$$u_{tt} + u_t = u_{xx} + \lambda u - u^3.$$

It should again be pointed out that our analysis is local. We can only say that the solutions that branch from $u_s = 0$ at $\lambda = \pi^2$ are stable if λ is close to π^2. For larger values of λ, we have no idea what is going on from what we have done. It is possible in certain cases to say something about the global structure of the bifurcating solutions and those who are interested should consult (Crandall and Rabinowitz, 1980). The approach used in global bifurcation is usually nonconstructive, that is, it does not actually determine the solution. About the only way this can be done is using numerical methods and for that aspect of the subject, the books by Seydel (2010) and by Krauskopf et al. (2007) are recommended.

Exercises

6.42. In the theory of Rayleigh–Bénard convection, one studies the motion of a thin fluid layer that is heated from below. A model for this is the Ginzburg–Landau equation

$$u_t = u_{xx} + \lambda u - u^3 \quad \text{for } 0 < x < \pi \text{ and } 0 < t,$$

where λ is associated with the Rayleigh number for the flow (DiPrima et al., 1971). The boundary conditions are $u = 0$ at $x = 0, \pi$.
(a) Find a first-term approximation of the steady-state solutions that bifurcate from $u_s = 0$.
(b) Carry out a linearized stability analysis of the steady states.

6.43. Consider the nonlinear diffusion problem

$$\partial_t u = \partial_x^2 u + \lambda u(1 - u) \quad \text{for } 0 < x < 1,$$

where $u = 0$ at $x = 0, 1$. This is known as Fisher's equation, or the Kolmogorov–Petrovsky–Piskunov (KPP) equation, and it has been used to model such phenomena as the spatial spread of certain genetic characteristics.

(a) Find a first-term approximation of the steady-state solutions that bifurcate from $u_s = 0$. (Hint: for n even, $\alpha = 1/2$, $\beta = 1$, and $\gamma = 3/2$.)
(b) For what values of λ is the steady state $u_s = 0$ stable or unstable? Assume λ is not a bifurcation value.

6.44. The equation for the transverse displacement $u(x,t)$ of a nonlinear beam subject to an axial load λ is (Woinowsky-Krieger, 1950; Eringen, 1952)

$$\partial_x^4 u + a(t)\partial_x^2 u + \partial_t^2 u + \partial_t u = 0 \quad \text{for } 0 < x < 1,$$

where

$$\alpha(t) = \lambda - \frac{1}{4} \int_0^1 u_x^2 dx.$$

The boundary conditions are $u = u_{xx} = 0$ at $x = 0, 1$.
(a) Find the steady-state solutions. The zero solution is called the unbuckled state and the others are buckled states. The values of λ where the buckled states appear are known as critical buckling loads.
(b) For what values of λ is each steady state stable or unstable?

6.45. This problem considers the effects of changing the nonlinearity in the wave equation (6.100). The first modification has a relatively minor effect but the second is more substantial.
(a) Explain how the results of this section change when $f(0) = f''(0) = 0$ but $f'(0) \neq 0$ and $f'''(0) \neq 0$. A simple example of this is $f(u) = u + u^3$.
(b) Carry out the analysis of this section when $f(0) = f'(0) = 0$ but $f''(0) \neq 0$. A simple example of this is $f(u) = \pm u^2$. [Hint: If n is odd, then $u \sim -3(\lambda - \lambda_n)\lambda_n \sin(\lambda_n x)/(2f''(0))$.]

6.46. The potential energy for the nonlinear string equation in (6.100) is

$$V(u) = \frac{1}{2} \int_0^1 [u_x^2 - \lambda u^2 - 2F(u)] dx,$$

where $F'(u) = f(u)$. In this problem take $f(u) = \kappa u^3$, where κ is a nonzero constant, and let $u_n(x)$ denote the steady state in (6.108). Sketch $V(u_n)$ for λ near λ_n (as is done for A in Fig. 6.23). On this basis, is it reasonable to conjecture that the stable solution has the smallest potential energy? Explain your reasoning.

6.47. The Kuramoto–Sivashinsky equation is

$$\partial_t u + \partial_x^4 u + \lambda \partial_x^2 u + \frac{1}{2}(\partial_x u)^2 = 0 \quad \text{for} \quad -\infty < x < \infty,$$

where $u(x, t)$ is required to be periodic in x. This equation arises in the study of the propagation of flame fronts, in the instabilities in reaction–diffusion systems, and in drift waves in plasmas (Mitani, 1984; Ishimura, 2001).
(a) Setting $u(x, t) = -\kappa t + v(x, t)$, where κ is a constant, find the problem satisfied by $v(x, t)$.
(b) For the problem in part (a), under what conditions on λ and κ is $v_s = 0$ an asymptotically stable steady-state solution?
(c) For the problem in part (a), determine the steady-state solution(s) that bifurcate from $v_s = 0$ at the point where v_s changes stability. To do this, take λ as the bifurcation parameter. Also note that the requirement of periodicity results in κ depending on λ.
(d) Determine if the steady-state solution(s) found in part (c) are asymptotically stable. In doing this, make sure to comment on the periodicity requirements of the initial conditions you use.

6.48. This problem reconsiders Fisher's equation (Exercise 6.43)

$$\partial_t u = \partial_x^2 u + \lambda u (1 - u) \quad \text{for} \quad -\infty < x < \infty.$$

We are now interested in traveling-wave solutions of the form $u(x,t) = f(x - \alpha t)$, where $-\infty < \alpha < \infty$.

(a) After finding the equation $f(z)$ satisfies, determine the stationary states for f and their stability.

(b) For physical reasons $u(x,t)$ must be nonnegative. Because of this, show that traveling waves of the form we have assumed will only propagate when $\alpha > 0$.

(c) Based on what you found in parts (a) and (b), can you say that these traveling waves are able to take a region near an unstable steady state to a stable state? For example, suppose the initial profile is $u(x,0) = \frac{1}{2}(1 + \tanh(kx))$, where k is a constant.

6.49. The Schlögl model in nonequilibrium phase transitions involves the reaction–diffusion equation (Schlogl, 1972)

$$\partial_t u = \partial_x^2 u + f(u, \lambda) \quad \text{for} \quad 0 < x < 1,$$

where $f(u, \lambda) = -u^3 + 2u^2 - u + \lambda$ and $u_x(0,t) = u_x(1,t) = 0$. The constant λ is positive. The solution $u(x,t)$ is the density of a chemical in solution and is therefore assumed to be nonnegative. Note that the nonlinearity here is cubic, which distinguishes it from the quadratic function appearing in Fisher's equation (Exercises 6.43 and 6.48). It also has different names, depending on the application area, and this includes the Zeldovich–Frank–Kamenetskii (ZFK) equation and the Nagumo equation (Idris and Biktashev, 2008).

(a) Find the homogeneous steady states (i.e., u_s constant) and determine their stability.

(b) For $0 < \lambda < \frac{4}{27}$ show that any steady state that bifurcates from one of the constant states is unstable.

(c) If $0 < \lambda < \frac{4}{27}$, then one can write $f(u, \lambda) = -(u - u_1)(u - u_2)(u - u_3)$, where the u_i depend on λ. Assuming $-\infty < x < \infty$, then find the traveling-wave solutions of the equation that have the form $u(x,t) = u_0 + \alpha \tanh[\beta(x - \chi t)]$.

(d) Using the results from parts (a) and (c), explain why certain of the traveling waves can take a region near an unstable steady state to one of the stable states. [Hint: Do this by examining the sign of χ and the behavior as $x \to \pm\infty$.]

6.50. In the study of phase transitions in nonlinear viscoelasticity, one comes across the following problem:

$$\partial_t^2 u - \partial_x \sigma(u_x) + \kappa u_{xxt} \quad \text{for} \quad 0 < x < 1 \text{ and } 0 < t,$$

where $u(0,t) = 0$ and $\sigma(u_x) + \kappa u_{xt} = \lambda$ at $x = 1$. Here $u(x,t)$ is the displacement of the material, $\sigma(u_x)$ is the elastic stress, $\kappa > 0$ is a constant associated with the viscoelastic properties of the material, and $\lambda \geq 0$ is a constant. A constitutive law is needed to specify σ, and in this problem we will let $\sigma(u_x) \equiv u_x(u_x^2 - 9u_x + 15)$.

(a) Sketch σ as a function of u_x and determine the steady states for this problem.

(b) Determine whether the steady states found in part (a) are stable or unstable.

6.51. Consider the nonlinear Klein–Gordon equation

$$u_{tt} = u_{xx} - \lambda g(u) \quad \text{for} \quad -\infty < x < \infty \text{ and } 0 < t,$$

where $u(x,t)$ is bounded as $x \to \pm\infty$ and $u(x, t+2\pi) = u(x,t)$. Also, assume $g(u)$ is smooth with $g(0) = 0$ and $g'(0) = 1$.

(a) One steady-state solution of this problem is $u_s = 0$. At what values of λ is it stable?

(b) To obtain the sine–Gordon equation, one takes $g(u) = \sin(u)$. In this case, show that

$$u_B(x,t) = 4\arctan\left(\frac{\beta \sin(nt)}{n \cosh(\beta x)}\right),$$

where n is a positive integer and $\beta = \sqrt{\lambda - n^2}$ is a solution of the problem (assuming $\lambda \geq n^2$). Sketch u_B as a function of x for $n = 1$ and $n = 10$ (in each case take $\beta = 1$). Because these solutions are localized in space and periodic in time, they are known as "breathers," and they arise in nonlinear optics and quantum mechanics. Note that they bifurcate from the trivial solution analyzed in part (a).

(c) To determine if breathers are stable, show that one must solve

$$v_{tt} = v_{xx} - \lambda g'(u_B)v \quad \text{for} \quad -\infty < x < \infty \text{ and } 0 < t,$$

where v is bounded as $x \to \pm\infty$ and $v(x, t + 2\pi) = v(x,t)$.

(d) For values of λ near where a breather bifurcates from $u_s = 0$ use the method of multiple scales to determine if the breather is stable or unstable.

6.9 Metastability

Metastability refers to the situation where you are watching something and it appears, for all intents and purposes, not to change. However, after a sufficiently long period of time it moves, or changes, and approaches another apparent steady state. This transition might occur very quickly, or proceed very slowly, and an example that illustrates both situations is

$$y(t) = \frac{2 - e^{\sigma(t)}}{2 + e^{\sigma(t)}}.$$

If $\sigma = (\varepsilon t - 1)/\varepsilon^3$, then $y \approx 1$ up until $t \approx 1/\varepsilon$, at which point it rather quickly switches and $y \approx -1$ for larger values of t. In contrast, consider the function $\sigma = e^{-1/\varepsilon}t$. If $\varepsilon = 10^{-2}$, then, using round numbers, $\sigma \approx 10^{-44}t$. In this case, even over the very long time interval $0 < t < 10^{40}$ the function is accurately approximated as $y \approx 1/3$, yet letting $t \to \infty$ one gets that $y = -1$. These are situations that give nightmares to those who use numerical solvers. The reason is that because of the minimal change in the solution over fairly long time intervals, it is easy to be fooled into thinking the solution has converged to a steady state.

From a perturbation viewpoint, metastability, as represented by the two examples described in the previous paragraph, can be handled using multiple scales in a straightforward manner. We are interested here in a complication that arises when transcendentally small terms must be included in the expansion to determine the behavior of the metastable solution. As illustrated in the examples in Sect. 2.4, such terms can often be ignored. However, there are situations where, over time, they are able to affect the solution sufficiently that they must be accounted for in the expansion. We saw this earlier, in Sect. 4.3, when using the WKB approximation to study the phenomenon of tunneling. Our goal here is to incorporate this into the multiple-scale approximation.

The example we will investigate that involves metastability is Burger's equation:

$$u_t + u u_x = \varepsilon u_{xx} \quad \text{for } 0 < x < 1 \text{ and } 0 < t, \tag{6.116}$$

where $u(0,t) = a$ and $u(1,t) = -a$, where a is a positive constant. The initial condition is

$$u(x,0) = \begin{cases} a & \text{for } 0 \le x < x_0, \\ -a & \text{for } x_0 < x \le 1, \end{cases} \tag{6.117}$$

where $0 < x_0 < \frac{1}{2}$.

A similar problem was considered in Sect. 2.7.6, where we found that the solution was a traveling wave that smoothly connected the left and right constant values appearing in the initial condition. This is shown in Fig. 2.32. It is not unreasonable to expect the solution of the preceding problem to behave in a similar manner. However, a very important difference is the spatial interval, which is infinite in Sect. 2.7.6 and is bounded in the preceding problem. Just how much this affects the solution will become evident as we attempt to construct an asymptotic approximation of the solution.

Given the bounded interval, one question to answer is whether the problem has a steady state. Letting $y(x)$ be the steady-state solution, then

$$yy' = \varepsilon y'' \quad \text{for } 0 < x < 1, \tag{6.118}$$

where $y(0) = a$ and $y(1) = -a$. As shown in Exercise 2.33,

$$y \sim a \, \frac{1 - e^{a(x-1/2)/\varepsilon}}{1 + e^{a(x-1/2)/\varepsilon}}. \tag{6.119}$$

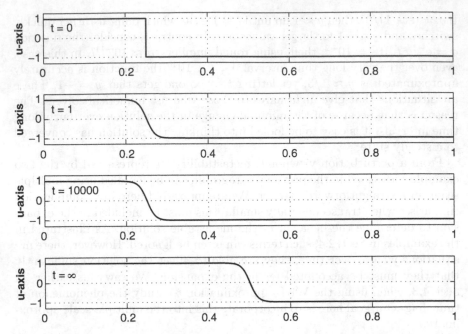

Figure 6.24 Numerical solution of Bruger's Eq. (6.116) on a finite interval starting with a jump located at $x_0 = \frac{1}{4}$ and $\varepsilon = 10^{-2}$. For these values the exponentially slow motion of the wave means that the transition layer does not reach $x = \frac{1}{2}$ until $t \approx 10^{20}$

In other words, the steady state consists of an interior layer, centered at $x = \frac{1}{2}$, that connects the two constant states on the left and right. This function is shown in Fig. 6.24 and corresponds to the solution for $t = \infty$.

Continuing our preliminary analysis of the problem, we turn to the asymptotic approximation of the time-dependent solution. It is not hard to show that the regular expansion $u(x,t) \sim u_0(x,t) + \cdots$ fails because u_0 ends up equaling the initial condition (6.117) for all time. This means that an interior-layer coordinate needs to be introduced to provide a smooth transition between the two constant states. Following the same argument used in Sect. 2.7.6, letting $\bar{x} = (x - s(t))/\varepsilon$ one finds that the first-term approximation in the transition layer is

$$U_0(\bar{x}, t) = a \frac{1 - B(t)e^{a\bar{x}}}{1 + B(t)e^{a\bar{x}}}, \tag{6.120}$$

where, from (2.135), $s(t) = x_0$. Also, $B(t)$ is an arbitrary positive function. Therefore, the layer analysis does provide a smooth transition but it does not move! This result is also seen in the numerical solution (Fig. 6.24), where the solution at $t = 10^4$ is indistinguishable from the solution at $t = 1$. However, there is no way that this result is a steady-state solution because the one and only steady state to this problem is given in (6.119) and it consists of a layer centered at $x = \frac{1}{2}$ and not one centered at $x = x_0$.

What we have here is an example of metastability. This is because when starting with the initial jump (6.117), the solution relatively quickly forms a smooth transition between $\pm a$ but does not move toward the known steady state (6.119), or, more precisely, it moves so slowly that our approximations simply consider the layer at $x = x_0$ to be stationary. To account for this very slow motion, we will introduce a scaled time coordinate. To explain how this will be done, note the similarities in the form of the steady-state solution (6.119) and the approximation in (6.120). What is happening is that the wave slowly moves from $x = x_0$ to $x = 1/2$, and we will change variables based on this observation.

The transition-layer coordinates will incorporate a time scale that reflects the location of the wave. The general form for the time variable is $\tau = q(\varepsilon, t)$. As in Sect. 3.6, q is assumed to be smooth with $q(\varepsilon, 0) = 0$ and satisfy $0 < \partial_t q \ll 1$. The corresponding layer coordinate is $\xi = (x - x_0 - \tau)/\varepsilon$. With this, the differential equation becomes

$$\varepsilon \partial_t q \partial_\tau u_t + (u - \partial_t q) \partial_\xi u = \partial_\xi^2 u. \tag{6.121}$$

The boundary conditions are

$$u|_{\xi = -(x_0 + \tau)/\varepsilon} = a, \quad u|_{\xi = (1 - x_0 - \tau)/\varepsilon} = -a. \tag{6.122}$$

In the examples in Chap. 2, the boundary conditions for the interior-layer problems were written as limits (e.g., $u \to a$ for $\xi \to -\infty$). This is not done here because of the sensitivity of the solution to the boundary conditions. In fact, it is this sensitivity that is responsible for the movement of the transition layer from $x = x_0$ over to $x = 1/2$.

Our approach for constructing the approximation will mimic what was done in Sect. 2.4. Given the various ε terms in (6.121), the appropriate expansion of the solution is

$$u \sim u_0(\xi, \tau) + u_1(\xi, \tau) + \cdots, \tag{6.123}$$

where $u_1 \ll u_0$. Substituting this into (6.121), we obtain the following problems:

$O(1)$ $u_0 \partial_\xi u_0 = \partial_\xi^2 u_0$.

Integrating this equation and then matching, one finds that $\partial_\xi u_0 = \frac{1}{2}(u_0^2 - a^2)$. Separating variables, it therefore follows that

$$u_0 = a \frac{1 - Be^{a\xi}}{1 + Be^{a\xi}}, \tag{6.124}$$

where $B(\tau)$ is a constant of integration. To produce a bounded solution that satisfies the matching conditions, B must be positive. Also, from balancing it follows that u_1 and $\partial_t q$ are of the same order.

$O(u_1) \ \partial_\xi^2 u_1 = u_0 \partial_\xi u_1 + u_1 \partial_\xi u_0 - \partial_t q \partial_\xi u_0$.

Writing the equation as $\partial_\xi^2 u_1 = \partial_\xi(u_0 u_1 - \partial_t q u_0)$ we can find the solution relatively easily, and the result is

$$u_1 = \partial_t q + \frac{e^{a\xi}}{(1 + B e^{a\xi})^2} \left[D + C(-e^{-a\xi} + 2aB\xi + B^2 e^{a\xi}) \right], \quad (6.125)$$

where $C(\tau, \varepsilon)$ and $D(\tau, \varepsilon)$ are constants of integration.

The matching of the second term requires some thought because of the importance of the exponentially small terms in the expansion. Starting with the right side, we are interested in what happens when you leave the transition layer and approach $x = 1$. The idea here is that $\xi = (1 - x_0 - q)/\varepsilon \to \infty$. From (6.124) and (6.125) we have that, in this limit,

$$u \sim -a(1 - 2B^{-1})e^{-a(1-x_0-q)/\varepsilon} + \partial_t q + C. \quad (6.126)$$

The right boundary condition is $u(1, t) = -a$, and so we conclude that

$$2aB^{-1}e^{-a(1-x_0-q)/\varepsilon} + \partial_t q + C = 0.$$

From the left side, where $\xi = -(x_0 + q)/\varepsilon \to -\infty$, one finds that

$$-2aB e^{-a(x_0+q)/\varepsilon} + \partial_t q - C = 0.$$

Adding the preceding two equations yields

$$\partial_t q = a\left(B e^{-a(x_0+q)/\varepsilon} - \frac{1}{B}e^{-a(1-x_0-q)/\varepsilon} \right).$$

Given that $\partial_t q = 0$ if $q = 1/2 - x_0$, it follows that $B = 1$. With this, the equation becomes

$$\frac{dq}{dt} = a\left(e^{-a(x_0+q)/\varepsilon} - e^{-a(1-x_0-q)/\varepsilon} \right). \quad (6.127)$$

Separating variables and integrating gives us that

$$q(\varepsilon, t) = \frac{1}{2} - x_0 + \frac{\varepsilon}{a} \ln\left[\frac{1 - \gamma e^{-\lambda t}}{1 + \gamma e^{-\lambda t}} \right], \quad (6.128)$$

where

$$\lambda = \frac{2a^2}{\varepsilon} e^{-a/(2\varepsilon)},$$

$$\gamma = \frac{1 - e^{-\kappa}}{1 + e^{-\kappa}},$$

and $\kappa = a(\frac{1}{2} - x_0)/\varepsilon$.

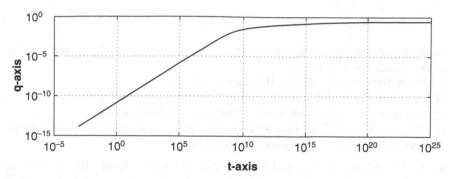

Figure 6.25 Plot of $q(\varepsilon, t)$, as given in (6.128), for $a = 1$, $x_0 = 0.25$, and $\varepsilon = 0.01$

This is interesting because it shows that the transition layer moves on an exponentially slow time scale, and this scale depends on the boundary conditions. To get an idea of the time the wave requires to move across the interval, the function q, corresponding to the example in Fig. 6.24, is plotted in Fig. 6.25. It shows that the wave reaches the half-way point between x_0 and $x = 1/2$ when $t \approx 2 \times 10^{14}$. The dependence of this result on the boundary conditions is dramatic. For example, doubling the boundary value, so $a = 2$, results in $t \approx 10^{30}$.

Exercises

6.52. This problem examines the solutions of (6.127) in more depth.
(a) Show that for large values of t, the solution in (6.128) satisfies
$$q \sim \frac{1}{2} - x_0 - \frac{2\varepsilon}{a} e^{-\lambda t}.$$

Similarly, for small values of t,
$$q \sim a e^{-ax_0/\varepsilon} t.$$

(b) Show that $q = \frac{1}{2} - x_0$ is a solution.
(c) Sketch the direction fields for (6.127) and use this to explain why the solution is constant, strictly monotonically increasing, or strictly monotonically decreasing. Explain how $q(\varepsilon, 0)$ determines which it is.
(d) Find the decreasing solution of (6.127).

6.53. In developing numerical approximations to solve Burger's equation it is common to test the procedure on known exact solutions. An often-used one is

$$u(x,t) = \frac{\mu + \alpha + (\mu - \alpha)e^{\eta}}{1 + e^{\eta}},$$

where $\eta = \alpha(x - \mu t - \gamma)/\varepsilon$, and α, μ, and γ are constants. Assume that α is positive and $0 < \gamma < 1$.

(a) Show that $u(x,t)$ satisfies Burger's Eq. (6.116).

(b) Describe the initial value $u(x,0)$ for $0 \le x \le 1$, assuming $\varepsilon \ll 1$.

(c) Assuming ε is small but fixed, describe what happens as $t \to \infty$.

(d) What is learned from this solution, if anything, as relates to the asymptotic approximation derived in this section?

6.54. One might try to avoid the complications of a layer that moves exponentially slowly by using an initial condition that does not have a jump discontinuity. This problem examines this by assuming that

$$u(x,0) = \begin{cases} a(1 - x/x_0) & \text{for } 0 \le x < x_0, \\ -a(x - x_0)/(1 - x_0) & \text{for } x_0 < x \le 1. \end{cases}$$

The important properties of this function are that it is continuous and monotonically decreasing. Continuity means that there is a point in the interval, $x = x_0$, where the function is zero. The monotonicity keeps the problem from becoming even more difficult than it already is. It is assumed in this problem that $0 < x_0 < \frac{1}{2}$.

(a) Consider the outer problem, so that $\varepsilon = 0$. Sketch the characteristics for $0 \le t \le x_0/a$, and use this to describe the solution for $0 \le x \le 1$ in this time interval. Explain why a shock appears in the solution at $t = x_0/a$.

(b) Continuing part (a), show that the equation for the shock is

$$x(t) = x_0 + \sqrt{1 - 2x_0}z - z^2$$

where $z = \sqrt{1 - x_0} - at$.

(c) Continuing part (b), show that the shock moves to the right until it reaches a point $x = x_s$, where $x_0 < x_s < \frac{1}{2}$, at which time it disappears. When this happens, a second shock appears. Explain why this second shock does not move.

(d) Use the result from part (c) to explain why using a continuous initial condition does not avoid the exponentially slowly transition layer, it just postpones its appearance.

6.55. This problem examines the initial behavior of the solution of Burger's equation.

(a) Show that the regular expansion $u(x,t) \sim u_0(x,t) + \cdots$ results in u_0 equaling the initial condition (6.117) for all time.

(b) Introducing the layer coordinates $\bar{x} = (x - x_0)/\varepsilon$ and $\bar{t} = t/\varepsilon$, find the first term in the layer expansion. The result from Exercise 2.56 will be helpful here.

(c) Show that the result from part (b) does not approach (6.119) when $\bar{t} \to \infty$. What function does it approach?

Appendix A
Taylor Series

A.1 Single Variable

(a) For x near a:

$$f(x) = f(a) + (x-a)f'(a) + \frac{1}{2}(x-a)^2 f''(a) + \cdots + \frac{1}{n!}(x-a)^n f^{(n)}(a) + \cdots .$$

(b) For h near 0:

$$f(x+h) = f(x) + hf'(x) + \frac{1}{2}h^2 f''(x) + \cdots + \frac{1}{n!}h^n f^{(n)}(x) + \cdots .$$

A.2 Two Variables

(a) For h and k near 0:

$$f(x+h, t+k) = f(x,t) + hf_x(x,t) + kf_t(x,t)$$
$$+ \frac{1}{2}h^2 f_{xx}(x,t) + hk f_{xt}(x,t) + \frac{1}{2}k^2 f_{tt}(x,t) + \cdots .$$

(b) For x near a and t near b:

$$f(x,t) = f(a,b) + (x-a)f_x(a,b) + (t-b)f_t(a,b)$$
$$+ \frac{1}{2}(x-a)^2 f_{xx}(a,b) + (x-a)(t-b)f_{xt}(a,b) + \frac{1}{2}(t-b)^2 f_{tt}(a,b)$$
$$+ \cdots .$$

M.H. Holmes, *Introduction to Perturbation Methods*, Texts in Applied
Mathematics 20, DOI 10.1007/978-1-4614-5477-9,
© Springer Science+Business Media New York 2013

(c) For h and k near 0:

$$f(x+h, t+k) = f(x,t) + Df(x,t) + \frac{1}{2}D^2 f(x,t) + \cdots + \frac{1}{n!}D^n f(x,t) + \cdots,$$

where

$$D = h\frac{\partial}{\partial x} + k\frac{\partial}{\partial t}.$$

A.3 Multivariable

(a) For $\mathbf{h}$ near $\mathbf{0}$:

$$f(\mathbf{x}+\mathbf{h}) = f(\mathbf{x}) + Df(\mathbf{x}) + \frac{1}{2}D^2 f(\mathbf{x}) + \cdots + \frac{1}{n!}D^n f(\mathbf{x}) + \cdots,$$

where $\mathbf{x} = (x_1, x_2, \ldots, x_k)$, $\mathbf{h} = (h_1, h_2, \ldots, h_k)$, and

$$D = \mathbf{h} \cdot \nabla$$
$$= h_1\frac{\partial}{\partial x_1} + h_2\frac{\partial}{\partial x_2} + \cdots + h_k\frac{\partial}{\partial x_k}.$$

(b) For $\mathbf{x}$ near $\mathbf{a}$:

$$f(\mathbf{x}) = f(\mathbf{a}) + (\mathbf{x}-\mathbf{a}) \cdot \nabla f(\mathbf{a}) + \frac{1}{2}(\mathbf{x}-\mathbf{a})^{\mathrm{T}}\mathbf{H}_a(\mathbf{x}-\mathbf{a}) + \cdots,$$

where $\mathbf{H}_a = \mathbf{H}(\mathbf{a})$, and $\mathbf{H}(\mathbf{x})$ is the Hessian defined as

$$\mathbf{H} = \begin{pmatrix} \dfrac{\partial^2 f}{\partial x_1^2} & \dfrac{\partial^2 f}{\partial x_2 \partial x_1} & \cdots & \dfrac{\partial^2 f}{\partial x_k \partial x_1} \\ \dfrac{\partial^2 f}{\partial x_1 \partial x_2} & \dfrac{\partial^2 f}{\partial x_2^2} & \cdots & \dfrac{\partial^2 f}{\partial x_k \partial x_2} \\ \vdots & \vdots & \ddots & \vdots \\ \dfrac{\partial^2 f}{\partial x_1 \partial x_k} & \dfrac{\partial^2 f}{\partial x_2 \partial x_k} & \cdots & \dfrac{\partial^2 f}{\partial x_k^2} \end{pmatrix}.$$

A.4 Useful Examples for x Near Zero

$$f(x) = f(0) + xf'(0) + \frac{1}{2}x^2 f''(0) + \frac{1}{6}x^3 f'''(0) + \cdots.$$

A.5 Power Functions

$$(a + x)^\gamma = a^\gamma + \gamma x a^{\gamma-1} + \frac{1}{2}\gamma(\gamma - 1)x^2 a^{\gamma-2} + \frac{1}{6}\gamma(\gamma - 1)(\gamma - 2)x^3 a^{\gamma-3} + \cdots,$$

$$\frac{1}{1 - x} = 1 + x + x^2 + x^3 + \cdots,$$

$$\frac{1}{(1 - x)^2} = 1 + 2x + 3x^2 + 4x^3 + \cdots,$$

$$\sqrt{1 + x} = 1 + \frac{1}{2}x - \frac{1}{8}x^2 + \frac{1}{16}x^3 + \cdots,$$

$$\frac{1}{\sqrt{1 - x}} = 1 + \frac{1}{2}x + \frac{3}{8}x^2 + \frac{5}{16}x^3 + \cdots.$$

A.6 Trig Functions

$$\sin(x) = x - \frac{1}{3!}x^3 + \frac{1}{5!}x^5 + \cdots,$$

$$\arcsin(x) = x + \frac{1}{6}x^3 + \frac{3}{40}x^5 + \cdots,$$

$$\cos(x) = 1 - \frac{1}{2}x^2 + \frac{1}{4!}x^4 + \cdots,$$

$$\arccos(x) = \frac{\pi}{2} - x - \frac{1}{6}x^3 - \frac{3}{40}x^5 + \cdots,$$

$$\tan(x) = x + \frac{1}{3}x^3 + \frac{2}{15}x^5 + \cdots,$$

$$\arctan(x) = x - \frac{1}{3}x^3 + \frac{1}{40}x^5 + \cdots,$$

$$\cot(x) = \frac{1}{x} - \frac{1}{3}x - \frac{1}{45}x^3 + \cdots,$$

$$\text{arccot}(x) = \frac{\pi}{2} - x + \frac{1}{3}x^3 - \frac{1}{5}x^5 + \cdots,$$

$$\sin(a + x) = \sin(a) + x \cos(a) - \frac{1}{2}x^2 \sin(a) + \cdots,$$

$$\cos(a + x) = \cos(a) - x \sin(a) - \frac{1}{2}x^2 \cos(a) + \cdots,$$

$$\tan(a + x) = \tan(a) + x \sec^2(a) + x^2 \tan(a) \sec^2(a).$$

A.7 Exponential and Log Functions

$$e^x = 1 + x + \frac{1}{2}x^2 + \frac{1}{6}x^3 + \cdots,$$

$$a^x = e^{x \ln(a)} = 1 + x \ln(a) + \frac{1}{2}[x \ln(a)]^2 + \frac{1}{6}[x \ln(a)]^3 + \cdots,$$

$$\ln(a + x) = \ln(a) + \frac{x}{a} - \frac{1}{2}\left(\frac{x}{a}\right)^2 + \frac{1}{3}\left(\frac{x}{a}\right)^3 + \cdots.$$

A.8 Hyperbolic Functions

$$\sinh(x) = x + \frac{1}{6}x^3 + \frac{1}{120}x^5 + \cdots,$$

$$\text{arcsinh}(x) = x - \frac{1}{6}x^3 + \frac{3}{40}x^5 + \cdots,$$

$$\cosh(x) = 1 + \frac{1}{2}x^2 + \frac{1}{24}x^4 + \cdots,$$

$$\text{arccosh}(x) = \sqrt{2x}\left(1 - \frac{1}{12}x + \frac{3}{160}x^2 + \cdots\right),$$

$$\tanh(x) = x - \frac{1}{3}x^3 + \frac{2}{15}x^5 + \cdots,$$

$$\text{arctanh}(x) = x + \frac{1}{3}x^3 + \frac{1}{5}x^5 + \cdots.$$

Appendix B
Solution and Properties of Transition Layer Equations

B.1 Airy Functions

This section concerns the solutions of Airy's equation. A more extensive presentation can be found in Vallée and Soares (2010), Abramowitz and Stegun (1972), and Olver et al. (2010).

B.1.1 Differential Equation

$$y'' = xy \quad \text{for} \quad -\infty < x < \infty.$$

B.1.2 General Solution

$$y(x) = \alpha_0 \text{Ai}(x) + \beta_0 \text{Bi}(x),$$

where

$$\text{Ai}(x) \equiv \frac{1}{3^{2/3}\pi} \sum_{k=0}^{\infty} \frac{1}{k!} \Gamma\left(\frac{k+1}{3}\right) \sin\left[\frac{2\pi}{3}(k+1)\right]\left(3^{1/3}x\right)^k$$

$$= \text{Ai}(0)\left(1 + \frac{1}{6}x^3 + \cdots\right) + \text{Ai}'(0)\left(x + \frac{1}{12}x^4 + \cdots\right)$$

and

$$\text{Bi}(x) \equiv e^{\pi i/6}\text{Ai}\left(xe^{2\pi i/3}\right) + e^{-\pi i/6}\text{Ai}\left(xe^{-2\pi i/3}\right)$$

$$= \text{Bi}(0)\left(1 + \frac{1}{6}x^3 + \cdots\right) + \text{Bi}'(0)\left(x + \frac{1}{12}x^4 + \cdots\right).$$

M.H. Holmes, *Introduction to Perturbation Methods*, Texts in Applied
Mathematics 20, DOI 10.1007/978-1-4614-5477-9,
© Springer Science+Business Media New York 2013

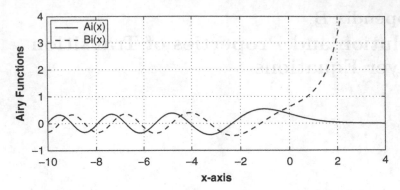

Figure B.1 Plot of the two Airy functions used in the development of the transition layer

B.1.3 Particular Values

$$\mathrm{Ai}(0) = \frac{\Gamma\left(\frac{1}{3}\right)}{2\pi 3^{1/6}}, \qquad \mathrm{Ai}'(0) = -\frac{3^{1/6}\Gamma\left(\frac{2}{3}\right)}{2\pi},$$

$$\mathrm{Bi}(0) = \sqrt{3}\,\mathrm{Ai}(0), \qquad \mathrm{Bi}'(0) = -\sqrt{3}\,\mathrm{Ai}'(0),$$

$$\int_0^\infty \mathrm{Ai}(x)\mathrm{d}x = \frac{1}{3}, \qquad \int_{-\infty}^0 \mathrm{Ai}(x)\mathrm{d}x = \frac{2}{3}, \qquad \int_{-\infty}^0 \mathrm{Bi}(x)\mathrm{d}x = 0,$$

$$\Gamma\left(\frac{1}{3}\right)\Gamma\left(\frac{2}{3}\right) = \frac{2\pi}{\sqrt{3}}.$$

B.1.4 Asymptotic Approximations

Setting $\zeta = \frac{2}{3}|x|^{3/2}$ and $\eta = \frac{5}{72\zeta}$, then

$$\mathrm{Ai}(x) \sim \begin{cases} \dfrac{1}{\sqrt{\pi}|x|^{1/4}}\left[\cos\left(\zeta - \dfrac{\pi}{4}\right) + \eta(x)\sin\left(\zeta - \dfrac{\pi}{4}\right)\right] & \text{if } x \to -\infty, \\[3mm] \dfrac{1}{2\sqrt{\pi}x^{1/4}}\,\mathrm{e}^{-\zeta}\left[1 - \eta(x)\right] & \text{if } x \to +\infty, \end{cases}$$

and

$$\mathrm{Bi}(x) \sim \begin{cases} \dfrac{1}{\sqrt{\pi}|x|^{1/4}}\left[\cos\left(\zeta + \dfrac{\pi}{4}\right) + \eta(x)\sin\left(\zeta + \dfrac{\pi}{4}\right)\right] & \text{if } x \to -\infty, \\[3mm] \dfrac{1}{\sqrt{\pi}x^{1/4}}\,\mathrm{e}^{\zeta}\left[1 + \eta(x)\right] & \text{if } x \to +\infty. \end{cases}$$

Also, setting $\nu = \frac{7}{72\zeta}$, then

$$\mathrm{Ai}'(x) \sim \begin{cases} -\frac{1}{\sqrt{\pi}} |x|^{1/4} \left[\cos\left(\zeta + \frac{\pi}{4}\right) - \nu(x) \sin\left(\zeta + \frac{\pi}{4}\right) \right] & \text{if } x \to -\infty, \\ -\frac{1}{2\sqrt{\pi}} x^{1/4} e^{-\zeta} [1 + \nu(x)] & \text{if } x \to +\infty, \end{cases}$$

and

$$\mathrm{Bi}'(x) \sim \begin{cases} -\frac{1}{\sqrt{\pi}} |x|^{1/4} \left[\cos\left(\zeta - \frac{\pi}{4}\right) - \nu(x) \sin\left(\zeta - \frac{\pi}{4}\right) \right] & \text{if } x \to -\infty, \\ -\frac{1}{2\sqrt{\pi}} x^{1/4} e^{\zeta} [1 - \nu(x)] & \text{if } x \to +\infty. \end{cases}$$

B.1.5 Connection with Bessel Functions

Setting $\zeta = \frac{2}{3}|x|^{3/2}$, then

$$\mathrm{Ai}(x) = \begin{cases} \sqrt{\frac{1}{3}|x|} \left[J_{\frac{1}{3}}(\zeta) + J_{-\frac{1}{3}}(\zeta) \right] & \text{if } x \le 0, \\ \sqrt{\frac{1}{3}x} \left[I_{-\frac{1}{3}}(\zeta) - I_{\frac{1}{3}}(\zeta) \right] & \text{if } x \ge 0, \end{cases}$$

and

$$\mathrm{Bi}(x) = \begin{cases} \sqrt{\frac{1}{3}|x|} \left[-J_{\frac{1}{3}}(\zeta) + J_{-\frac{1}{3}}(\zeta) \right] & \text{if } x \le 0, \\ \sqrt{\frac{1}{3}x} \left[I_{-\frac{1}{3}}(\zeta) + I_{\frac{1}{3}}(\zeta) \right] & \text{if } x \ge 0. \end{cases}$$

B.2 Kummer's Function

This section concerns the properties of the solutions of a differential equation that arises frequently when solving turning-point problems. It is related to the hypergeometric equation, and much of the material presented here is adopted from Slater (1960), Oldham et al. (2009), and Olver et al. (2010).

B.2.1 Differential Equation

$$y'' + \alpha x y' + \beta y = 0 \quad \text{for } -\infty < x < \infty,$$

where α and β are nonzero constants.

B.2.2 General Solution

$$y(x) = \alpha_0 M\left(\frac{\beta}{2\alpha}, \frac{1}{2}, -\frac{1}{2}\alpha x^2\right) + \beta_0 x M\left(\frac{\alpha+\beta}{2\alpha}, \frac{3}{2}, -\frac{1}{2}\alpha x^2\right),$$

where $M(a, b, z)$ is a confluent hypergeometric function and is known as Kummer's function [it is also denoted by $_1F_1(a, b, z)$]. Note $M(a, b, -\frac{1}{2}\alpha x^2)$ is an even function of x, so the foregoing solution consists of the sum of an even and an odd function. These functions are plotted for particular values of the coefficients in Fig. B.2.

The series definition of Kummer's function is

$$M(a, b, z) = \sum_{k=0}^{\infty} \frac{(a)_k}{(b)_k} \frac{1}{k!} z^k$$

$$= 1 + \frac{a}{b} z + \frac{a(a+1)}{2b(b+1)} z^2 + \cdots,$$

where $(a)_k = a(a+1)(a+2)\cdots(a+k-1)$ and $(a)_0 = 1$. This series is absolutely convergent for all values of a, b, and z except for $b = 0, -1, -2, -3, \ldots$. The latter values are assumed not to occur in the formulas that follow.

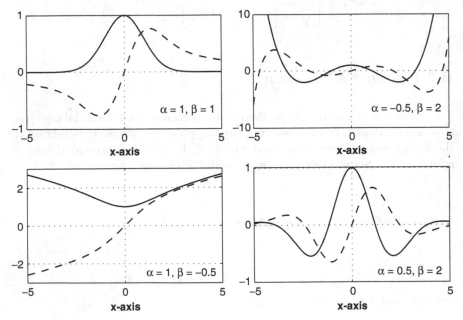

Figure B.2 Plots of $y_1 = M(\frac{\beta}{2\alpha}, \frac{1}{2}, -\frac{1}{2}\alpha x^2)$ (*solid curves*) and $y_2 = xM(\frac{\alpha+\beta}{2\alpha}, \frac{3}{2}, -\frac{1}{2}\alpha x^2)$ (*dashed curves*) for various values of α and β

B.2.3 Particular Values

$$M(a, b, 0) = 1, \quad \partial_z M(a, b, 0) = \frac{a}{b},$$
$$M(0, b, z) = 1, \quad M(a, a, z) = e^z.$$

B.2.4 Useful Formulas

The following formulas are useful for deriving some of the basic properties, and special cases, of Kummer's function:

$$M(a, b, z) = e^z M(b - a, b, -z),$$

$$M(a + 1, b, z) = \frac{2a - b + z}{a} M(a, b, z) + \frac{b - a}{a} M(a - 1, b, z),$$

$$M(a, b + 1, z) = \frac{b(b - 1 + z)}{(b - a)z} M(a, b, z) - \frac{b(b - 1)}{(b - a)z} M(a, b - 1, z),$$

$$\frac{d}{dz} M(a, b, z) = \frac{a}{b} M(a + 1, b + 1, z).$$

$$M(a, b, z) = \frac{\Gamma(b)}{\Gamma(b - a)\Gamma(a)} \int_0^1 e^{zt} t^{a-1}(1 - t)^{b-a-1} dt, \quad \text{for } 0 < a < b$$

B.2.5 Special Cases

If $\alpha = \beta$, then

$$M\left(\frac{1}{2}, \frac{1}{2}, -\frac{1}{2}\alpha x^2\right) = e^{-\alpha x^2/2}$$

and

$$M\left(1, \frac{3}{2}, -\frac{1}{2}\alpha x^2\right) = \frac{1}{x}\int_0^x e^{\alpha(s^2 - x^2)/2}\, ds.$$

If $\alpha = -\beta$, then

$$M\left(-\frac{1}{2}, \frac{1}{2}, -\frac{1}{2}\alpha x^2\right) = e^{-\alpha x^2/2} + \alpha x \int_0^x e^{-\alpha r^2/2}\, dr$$

and

$$M\left(0, \frac{3}{2}, -\frac{1}{2}\alpha x^2\right) = 1.$$

If $\beta = 2\alpha$, then

$$M\left(1, \frac{1}{2}, -\frac{1}{2}\alpha x^2\right) = 1 - \alpha x \int_0^x e^{\alpha(s^2 - x^2)/2}\, ds$$

and

$$M\left(\frac{3}{2}, \frac{3}{2}, -\frac{1}{2}\alpha x^2\right) = e^{-\alpha x^2/2}.$$

If $\alpha = 2\beta$, then

$$M\left(\frac{1}{4}, \frac{1}{2}, -\frac{1}{2}\alpha x^2\right) = \Gamma\left(\frac{3}{4}\right)\left(-\frac{1}{8}\alpha x^2\right)^{1/4} e^{-\alpha x^2/4} I_{-1/4}\left(-\frac{1}{4}\alpha x^2\right)$$

and

$$M\left(\frac{3}{4}, \frac{3}{2}, -\frac{1}{2}\alpha x^2\right) = \Gamma\left(\frac{5}{4}\right)\left(-\frac{1}{8}\alpha x^2\right)^{-1/4} e^{-\alpha x^2/4} I_{1/4}\left(-\frac{1}{4}\alpha x^2\right).$$

B.2.6 Polynomials

When the first argument of $M(a, b, z)$ is a nonpositive integer, the function reduces to an expression involving a Laguerre polynomial. In particular, if $n = 0, 1, 2, 3, \ldots$, then

$$M\left(-n, \frac{3}{2}, -\frac{1}{2}\alpha x^2\right) = \frac{\sqrt{\pi}\, n!}{2\Gamma\left(n + \frac{3}{2}\right)} \sum_{k=0}^{n} \binom{n + \frac{1}{2}}{n - k} \frac{1}{k!} \left(\frac{1}{2}\alpha x^2\right)^k$$

and

$$M\left(-n, \frac{1}{2}, -\frac{1}{2}\alpha x^2\right) = \frac{\sqrt{\pi}\, n!}{\Gamma\left(n + \frac{1}{2}\right)} \sum_{k=0}^{n} \binom{n - \frac{1}{2}}{n - k} \frac{1}{k!} \left(\frac{1}{2}\alpha x^2\right)^k.$$

B.2.7 Asymptotic Approximations

The asymptotic expansions for large $|x|$ depend on the sign of α and also on whether or not the first argument of the function is a nonpositive integer (see section *Polynomials*).

1. For $x^2 \to \infty$,

$$M\left(\frac{\alpha+\beta}{2\alpha}, \frac{3}{2}, -\frac{1}{2}\alpha x^2\right) \sim \begin{cases} \dfrac{\sqrt{\pi}}{2\Gamma\left(\frac{\alpha+\beta}{2\alpha}\right)}\left(-\dfrac{1}{2}\alpha x^2\right)^{\frac{\beta-2\alpha}{2\alpha}} e^{-\frac{1}{2}\alpha x^2} & \text{if } \alpha < 0, \\[3ex] \dfrac{\sqrt{\pi}}{2\Gamma\left(\frac{2\alpha-\beta}{2\alpha}\right)}\left(\dfrac{1}{2}\alpha x^2\right)^{-\frac{\alpha+\beta}{2\alpha}} & \text{if } \alpha > 0. \end{cases}$$

The approximation for $\alpha < 0$ does not hold when $\alpha + \beta = -2\alpha n$, where $n = 0, 1, 2, 3, \ldots$, and these cases are included in item 3 below. Also, the approximation for $\alpha > 0$ does not hold when $(2\alpha - \beta)/(2\alpha)$ is a nonpositive integer. The case where $\beta = 2\alpha$ is included in *Special Cases*, and Slater (1960) may be consulted for the others.

2. For $x^2 \to \infty$,

$$M\left(\frac{\beta}{2\alpha}, \frac{1}{2}, -\frac{1}{2}\alpha x^2\right) \sim \begin{cases} \dfrac{\sqrt{\pi}}{\Gamma\left(\frac{\beta}{2\alpha}\right)}\left(-\dfrac{1}{2}\alpha x^2\right)^{\frac{-\alpha+\beta}{2\alpha}} e^{-\frac{1}{2}\alpha x^2} & \text{if } \alpha < 0, \\[3ex] \dfrac{\sqrt{\pi}}{\Gamma\left(\frac{\alpha-\beta}{2\alpha}\right)}\left(\dfrac{1}{2}\alpha x^2\right)^{-\frac{\beta}{2\alpha}} & \text{if } \alpha > 0. \end{cases}$$

The approximation for $\alpha < 0$ does not hold when $\beta = -2\alpha n$, where $n = 1, 2, 3, \ldots$, and these cases are included in item 3 below. Also, the approximation for $\alpha > 0$ does not hold when $(\alpha - \beta)/(2\alpha)$ is a nonpositive integer. The case where $\beta = \alpha$ is included in *Special Cases*, and Slater (1960) may be consulted for the others.

3. The nonpositive-integer cases are

$$M\left(-n, \frac{3}{2}, -\frac{1}{2}\alpha x^2\right) \sim \frac{\sqrt{\pi}}{2\Gamma(\frac{3}{2}+n)}\left(\frac{1}{2}\alpha x^2\right)^n$$

and

$$M\left(-n, \frac{1}{2}, -\frac{1}{2}\alpha x^2\right) \sim \frac{\sqrt{\pi}}{\Gamma(\frac{1}{2}+n)}\left(\frac{1}{2}\alpha x^2\right)^n.$$

B.2.8 Related Special Functions

Kummer's differential equation reduces to Hermite's equation in the special case where $\alpha = -1$ and $\beta = 2n$, or $\alpha - 1$ and $\beta = n$, where n is a positive integer. Similarly, it can be transformed into Hermite's differential equation when $\alpha < 0$ and $\beta = -\alpha n$, where n is a positive integer. It is also possible to express the solution of Kummer's equation in terms of parabolic cylinder functions (Oldham et al., 2009).

B.3 Higher-Order Turning Points

Analysis of higher order-turning points can be found in Willner and Rubenfeld (1976).

B.3.1 Differential Equation

$$y'' = x^\gamma y \quad \text{for} \quad -\infty < x < \infty \text{ and } \gamma \geq 0.$$

B.3.2 General Solution

$$y(x) = \alpha_0 y_1(x) + \beta_0 y_2(x),$$

where, setting $\nu = 1/(\gamma + 2)$ and $\kappa = (\gamma + 2)/2$,

$$y_1(x) = \begin{cases} \sqrt{|x|} \left[J_\nu(2\nu|x|^\kappa) + J_{-\nu}(2\nu|x|^\kappa) \right] & \text{if } x \leq 0, \\ \sqrt{x} \left[I_{-\nu}(2\nu x^\kappa) - I_\nu(2\nu x^\kappa) \right] & \text{if } x \geq 0, \end{cases}$$

and

$$y_2(x) = \begin{cases} \sqrt{|x|} \left[-J_\nu(2\nu|x|^\kappa) + J_{-\nu}(2\nu|x|^\kappa) \right] & \text{if } x \leq 0, \\ \sqrt{x} \left[I_{-\nu}(2\nu x^\kappa) + I_\nu(2\nu x^\kappa) \right] & \text{if } x \geq 0. \end{cases}$$

B.3.3 Asymptotic Approximations

Setting $\xi = \frac{2}{\gamma+2}|x|^{\frac{2+\gamma}{2}}$, then

$$y_1(x) \sim \begin{cases} \dfrac{2\cos(\nu\pi/2)}{\sqrt{\pi\nu}} |x|^{-\gamma/4} \cos\left(\xi - \dfrac{\pi}{4}\right) & \text{if } x \to -\infty, \\ \dfrac{\sin(\nu\pi)}{\sqrt{\pi\nu}} x^{-\gamma/4} e^{-\xi} & \text{if } x \to +\infty, \end{cases}$$

and

$$y_2(x) \sim \begin{cases} \dfrac{2\sin(\nu\pi/2)}{\sqrt{\pi\nu}} |x|^{-\gamma/4} \cos\left(\xi + \dfrac{\pi}{4}\right) & \text{if } x \to -\infty, \\ \dfrac{1}{\sqrt{\pi\nu}} x^{-\gamma/4} e^{\xi} & \text{if } x \to +\infty. \end{cases}$$

Appendix C
Asymptotic Approximations of Integrals

C.1 Introduction

This appendix summarizes some formulas for approximating integrals. Specifically, approximations of integrals of the form

$$\int_a^b f(t)e^{-xg(t)}dt \quad \text{and} \quad \int_a^b f(t)e^{ixg(t)}dt$$

are given for the case of large x. Readers interested in a more extensive development of this material may consult Murray (1984) or Olver (1974).

In this appendix the following assumptions are made:

1. $-\infty \le a < b \le \infty$,

2. $f(t)$ and $g(t)$ are continuous and $g(t)$ is real-valued for $a < t < b$,

3. a, b, f, and g are independent of x.

The asymptotic behavior of the first integral depends on where $g(x)$ has a minimum value, while the second depends on where $g(x)$ has a stationary value.

C.2 Watson's Lemma

$$\int_a^b f(t)e^{-xt}dt$$

The exponential decay means that the value of the integral is determined by what happens at, or near, $t = a$. Thus, for $t \to a^+$, assume that

M.H. Holmes, *Introduction to Perturbation Methods*, Texts in Applied Mathematics 20, DOI 10.1007/978-1-4614-5477-9,
© Springer Science+Business Media New York 2013

$$f(t) \sim f_0(t-a)^\alpha + f_1(t-a)^\beta,$$

where $-1 < \alpha < \beta$. Also, assume that $-\infty < a$. In this case,

$$\int_a^b f(t)e^{-xt}dt \sim \left[\frac{f_0\Gamma(1+\alpha)}{x^{1+\alpha}} + \frac{f_1\Gamma(1+\beta)}{x^{1+\beta}}\right]e^{-ax} \quad \text{as } x \to \infty.$$

In particular, if $f(t) \sim f(a) + (t-a)f'(a) + \cdots$, where $f(a)$ and $f'(a)$ are nonzero, then

$$\int_a^b f(t)e^{-xt}dt \sim \left[\frac{f(a)}{x} + \frac{f'(a)}{x^2}\right]e^{-ax} \quad \text{as } x \to \infty.$$

C.3 Laplace's Approximation

$$\int_a^b f(t)e^{-xg(t)}dt.$$

The exponential decay means that the value of the integral is determined by what happens at, or near, points where $g(t)$ has a minimum value.

(a) The minimum of $g(t)$ for $a \leq t \leq b$ occurs only at $t = t_0$, where $a < t_0 < b$. Assuming that $f(t_0) \neq 0$ and $g(t)$ is a smooth function with $g''(t_0) > 0$, then

$$\int_a^b f(t)e^{-xg(t)}dt \sim f(t_0)\sqrt{\frac{2\pi}{xg''(t_0)}}\, e^{-xg(t_0)} \quad \text{as } x \to \infty.$$

More generally, for $t \to t_0$, assume that

$$f(t) \sim f_0(t-t_0)^\alpha \qquad \text{for } \alpha \geq 0 \text{ and } f_0 \neq 0,$$
$$g(t) \sim g_0 + g_1(t-t_0)^{2m} \quad \text{for } m \text{ a positive integer and } g_1 > 0.$$

In this case,

$$\int_a^b f(t)e^{-xg(t)}dt \sim \frac{f_0}{m}\Gamma(\kappa)\left(\frac{1}{xg_1}\right)^\kappa e^{-xg_0} \quad \text{as } x \to \infty,$$

where

$$\kappa = \frac{1+\alpha}{2m}.$$

(b) The minimum of $g(t)$ for $a \leq t \leq b$ occurs only at $t = a$ and $a > -\infty$. Assuming that $f(a) \neq 0$ and $g'(t) > 0$ for $a \leq t \leq b$, then

$$\int_a^b f(t)e^{-xg(t)}\,dt \sim \frac{f(a)}{g'(a)x}e^{-xg(a)} \quad \text{as } x \to \infty.$$

More generally, for $t \to a^+$, assume that

$$f(t) \sim f_0(t - a)^\alpha \qquad \text{for } \alpha > -1 \text{ and } f_0 \neq 0,$$
$$g(t) \sim g_0 + g_1(t - a)^\lambda \quad \text{for } \lambda > 1 + \alpha \text{ and } g_1 > 0.$$

In this case,

$$\int_a^b f(t)e^{-xg(t)}\,dt \sim \frac{f_0}{\lambda}\Gamma(\kappa)\left(\frac{1}{xg_1}\right)^\kappa e^{-xg_0} \quad \text{as } x \to \infty,$$

where $\kappa = (1 + \alpha)/\lambda$.

(c) The minimum of $g(t)$ for $a \leq t \leq b$ occurs only at $t = b$ and $b < \infty$. Assuming that $f(b) \neq 0$ and $g'(t) < 0$ for $a \leq t \leq b$, then

$$\int_a^b f(t)e^{-xg(t)}\,dt \sim -\frac{f(b)}{g'(b)x}e^{-xg(b)} \quad \text{as } x \to \infty.$$

C.4 Stationary Phase Approximation

$$\int_a^b f(t)e^{ixg(t)}\,dt.$$

The fast oscillations mean that the value of the integral is determined by what happens at, or near, points where $g(t)$ has a stationary point. It is assumed that $g(t)$ has only one stationary point. Specifically, in the interval $a \leq t \leq b$, $g'(t) = 0$ only at $t = t_0$, which is a finite point (i.e., $-\infty < t_0 < \infty$).

Assuming that $f(t_0) \neq 0$, $g''(t_0) \neq 0$, and $a < t_0 < b$, then

$$\int_a^b f(t)e^{ixg(t)}\,dt \sim f(t_0)\sqrt{\frac{2\pi}{x|g''(t_0)|}}\,e^{i\left(xg(t_0)\pm\frac{\pi}{4}\right)} \quad \text{as } x \to \infty,$$

where $+$ is used when $g''(t_0) > 0$ and $-$ when $g''(t_0) < 0$. If $t_0 = a$ or $t_0 = b$, then the approximation is

$$\int_a^b f(t)e^{ixg(t)}\,dt \sim f(t_0)\sqrt{\frac{\pi}{2x|g''(t_0)|}}\,e^{i\left(xg(t_0)\pm\frac{\pi}{4}\right)} \quad \text{as } x \to \infty,$$

where $+$ is used when $g''(t_0) > 0$ and $-$ when $g''(t_0) < 0$.

More generally, in the case where $t_0 = a$, for $t \to a^+$ assume that

$$f(t) \sim f_0 (t-a)^\alpha \qquad \text{for } \alpha \geq 0 \text{ and } f_0 \neq 0,$$
$$g(t) \sim g_0 + g_1 (t-a)^\lambda \quad \text{for } \lambda > 1 + \alpha.$$

In this case,

$$\int_a^b f(t) e^{ixg(t)} dt \sim \frac{f_0}{\lambda} \Gamma(\kappa) \left(\frac{1}{x|g_1|} \right)^\kappa e^{i(xg_0 \pm \frac{1}{2}\pi\kappa)} \quad \text{as } x \to \infty,$$

where $\kappa = (1+\alpha)/\lambda$. In the preceding expression, $+$ is used when $g_1 > 0$ and $-$ when $g_1 < 0$.

Appendix D
Second-Order Difference Equations

We will be interested mostly in second-order linear difference equations; an example is

$$y_{n+1} + a y_n + b y_{n-1} = f_n \quad \text{for } n = 1, 2, 3, \dots, \tag{D.1}$$

where $b \neq 0$. The theory and methods developed for such equations are very similar to what is found for second-order differential equations. For example, the general solution has the form $y_n = Y_n + Z_n$, where Y_n is the general solution of the associated homogeneous equation (where $f_n = 0$) and Z_n is a particular solution of the inhomogeneous equation. The specifics of this can be found in Elaydi (2005).

To determine Y_n, one assumes that $Y_n = r^n$. Substituting this into (D.1) and setting $f_n = 0$, the equation reduces to solving $r^2 + ar + b = 0$. The roots of this equation are

$$r_\pm = \frac{1}{2}\left[-a \pm \sqrt{a^2 - 4b}\right].$$

With this, the general solution of the associated homogeneous equation is

$$Y_n = \begin{cases} \alpha r_+^n + \beta r_-^n & \text{if } a^2 \neq 4b, \\[2mm] \alpha r^n + \beta n r^n & \text{if } a^2 = 4b, \end{cases} \tag{D.2}$$

where α and β are arbitrary constants and $r = -a/2$. In the special case where $a^2 < 4b$, Y_n can be written as

$$Y_n = A \rho^n \cos(n\theta + \phi),$$

where A and ϕ are arbitrary constants, $\rho = \sqrt{b}$, and $\cos\theta = -a/(?\rho)$. It is assumed that $0 < \theta < \pi$.

A particular solution can be found using a variety of methods, including reduction of order or the z-transform. However, writing $y_{n+1} = f_n - a y_n - b y_{n-1}$ and then using this to determine the first few terms it is evident that

M.H. Holmes, *Introduction to Perturbation Methods*, Texts in Applied Mathematics 20, DOI 10.1007/978-1-4614-5477-9,
© Springer Science+Business Media New York 2013

there is a particular solution of the form

$$Z_n = \sum_{i=1}^{n-1} f_{n-i} q_i \quad \text{for } n = 2, 3, 4, \ldots, \tag{D.3}$$

where $q_1 = 1$, $q_2 + aq_1 = 0$, and $q_i + aq_{i-1} + bq_{i-2} = 0$. The latter is the associated homogeneous equation, and so (D.2) applies. From this and the conditions on q_1 and q_2, one finds that

$$q_i = \begin{cases} \dfrac{r_+^i - r_-^i}{r_+ - r_-} & \text{if } a^2 \neq 4b, \\[3mm] ir^{i-1} & \text{if } a^2 = 4b. \end{cases}$$

Note that, by construction, the q_i satisfy the homogeneous equation, with $q_{-1} = -1/b$, $q_0 = 0$, and $q_1 = 1$. Also, in the case where $a^2 < 4b$, q_i can be written as

$$q_i = \frac{\rho^{i-1} \sin(i\theta)}{\sin(\theta)},$$

where $\rho = \sqrt{b}$ and $\cos\theta = -a/(2\rho)$. This is well defined because $0 < \theta < \pi$.

The complete solution depends on the particular problem being solved, and the most commonly studied are initial-value problems and boundary-value problems.

D.1 Initial-Value Problems

For an initial-value problem, (D.1) is to be satisfied, and y_0 and y_1 are pre-scribed. We have that the general solution has the form $y_n = Y_n + Z_n$, where $Z_0 = Z_1 = 0$. From the requirement that $Y_0 = y_0$ and $Y_1 = y_1$ it follows that the solution is

$$y_n = y_1 q_n - by_0 q_{n-1} + \sum_{i=1}^{n-1} f_{n-i} q_i \quad \text{for } n = 0, 1, 2, 3, \ldots.$$

The sum is understood to be zero in the case where $n = 0$ or $n = 1$.

In the special case where $a^2 < 4b$, note that

$$\sum_{i=1}^{n-1} f_{n-i} q_i = \frac{1}{\sin\theta} \sum_{i=1}^{n-1} f_{n-i} \rho^{i-1} \sin(i\theta).$$

In some cases it is possible to sum this series, although this usually requires knowing the right identities. Two of particular value are

$$\sum_{i=0}^{n-1} \sin(\alpha + i\beta) = \frac{\sin(n\beta/2)}{\sin(\beta/2)} \sin\left(\alpha + \frac{1}{2}(n-1)\beta\right)$$

and

$$\sum_{i=0}^{n-1} \cos(\alpha + i\beta) = \frac{\sin(n\beta/2)}{\sin(\beta/2)} \cos\left(\alpha + \frac{1}{2}(n-1)\beta\right),$$

where β is not an integer multiple of 2π. As an example, if $\rho = 1$ and $f_n = \cos(n\theta)$, then

$$\sum_{i=1}^{n-1} f_{n-i} q_i = \frac{1}{2}(n-1)\frac{\sin(n\theta)}{\sin\theta}.$$

Similarly, if $\rho = 1$ and $f_n = \sin(n\theta)$, then

$$\sum_{i=1}^{n-1} f_{n-i} q_i = \frac{1}{2\sin\theta}\left[-(n-1)\cos(n\theta) + \csc(\theta)\sin((n-1)\theta)\right].$$

D.2 Boundary-Value Problems

For a boundary-value problem, (D.1) is to be satisfied for $i = 1, 2, \ldots, N$, and y_0 and y_{N+1} are prescribed. In this case the solution is

$$y_n = y_0 Q_n + \sum_{i=1}^{n-1} f_{n-i} q_i + \frac{q_n}{q_{N+1}}\left[y_{N+1} - \sum_{i=1}^{N} f_{N+1-i} q_i\right]$$

$$\text{for } n = 0, 1, 2, \ldots, N+1, \quad \text{(D.4)}$$

where the first sum is understood to be zero in the case where $n = 0$ or $n = 1$. Also,

$$Q_n = \begin{cases} \dfrac{r_+^{N+1} r_-^n - r_+^n r_-^{N+1}}{r_+^{N+1} - r_-^{N+1}} & \text{if } a^2 \neq 4b, \\[3ex] \left(1 - \dfrac{n}{N+1}\right) r^n & \text{if } a^2 = 4b. \end{cases}$$

In the special case where $a^2 < 4b$, the preceding expressions can be written as

$$Q_n = \rho^n \frac{\sin(N+1-n)\theta}{\sin[(N+1)\theta]}$$

and

$$\frac{q_n}{q_{N+1}} = \rho^{n-N-1} \frac{\sin(n\theta)}{\sin[(N+1)\theta]},$$

where $\rho = \sqrt{b}$ and $\cos\theta = -a/(2\rho)$. This shows that the solution is well defined only so long as $(N+1)\theta \neq j\pi$ for $j = 1, 2, \cdots, N$.

Appendix E
Delay Equations

E.1 Differential Delay Equations

An example of a linear first-order delay equation is

$$y'(t) = ay(t) + by(t - \tau) \quad \text{for } t > 0, \tag{E.1}$$

where τ is a positive constant. It is assumed that $y(t) = \chi(t)$, for $-\tau \leq t \leq 0$, is known. This equation can be solved using the Laplace transform. To see this, note that

$$\int_0^\infty y(t - \tau)e^{-st}dt = e^{-\tau s}Y(s) + Z(s),$$

where $Y(s)$ is the Laplace transform of $y(t)$ and

$$Z(s) = e^{-\tau s} \int_{-\tau}^0 \chi(r)e^{-rs}dr.$$

With this, then from (E.1) we have that

$$Y(s) = \frac{y(0) + bZ(s)}{s - a - be^{-\tau s}}.$$

Using the definition of the inverse transform,

$$y(t) = \frac{1}{2\pi i} \int_{ic-\infty}^{ic+\infty} \frac{y(0) + bZ(s)}{s - a - be^{-\tau s}} e^{st}ds.$$

Assuming that $Z(s)$ is well behaved, from Cauchy's residue theorem we have that

$$y(t) = \sum_n \frac{y(0) + bZ(s_n)}{1 + \tau be^{-\tau s_n}} e^{s_n t}, \tag{E.2}$$

M.H. Holmes, *Introduction to Perturbation Methods*, Texts in Applied
Mathematics 20, DOI 10.1007/978-1-4614-5477-9,
© Springer Science+Business Media New York 2013

where the sum is over all roots of the equation $s = a + be^{-\tau s}$. This assumes the roots are simple. If not, then the terms in the preceding series must be modified using the formula for a higher-order pole. A more complete presentation of this method can be found in Pinney (1958).

The critical observation in the foregoing discussion is the form of the general solution in (E.2). This shows that to find the solution of a delay equation like that in (E.1), we simply assume the solution has the form $y = e^{\lambda t}$ and then use the delay equation to find what equation λ satisfies. For (E.1) this is $\lambda = a + be^{-\tau \lambda}$. We then sum over the roots of this equation to obtain a general solution of the form

$$y = \sum_{n} a_n e^{\lambda_n t}.$$

This assumes the roots are simple. If a root has order $m+1$, then its contribution to the preceding series will be of the form $(a_{n1} + a_{n2}t + \cdots + a_{nm}t^m)e^{\lambda_n t}$.

The preceding discussion can be extended to equations of the form

$$y''(x) = ay(x - h) + by(x) + cy(x + h),$$

where h is a positive constant. This is an example of what is called an advance-delay equation. Depending on what boundary or initial conditions are imposed, one can use the Laplace of Fourier transform to find the solution. However, the conclusion is the same as before. That is, the general solution can be obtained by simply assuming the solution has the form $y = e^{\lambda x}$ and then using the differential-difference equation to determine what equation λ satisfies. One then sums over the roots of this equation to obtain a general solution.

E.2 Integrodifferential Delay Equations

The equation of interest here has the general form

$$y'(t) = F(t, y(t), z(t)) \quad \text{for } t > 0, \tag{E.3}$$

where

$$z(t) = \int_{-\infty}^{t} K(t - \tau)y(\tau)\mathrm{d}\tau.$$

It is assumed that $y(t) = \chi(t)$, for $t \leq 0$, is known. Because of the improper integral, certain assumptions must be made about the kernel $K(t)$ and initial data $\chi(t)$. In particular, $K(t)$ is assumed to be smooth and $\int_0^\infty |K(s)|\mathrm{d}s$ is assumed to be finite. In addition, it is assumed that

$$z_0(t) = \int_{-\infty}^{0} K(t - \tau)\chi(\tau)\mathrm{d}\tau$$

is well defined.

An example is the Volterra delay equation that comes up in population modeling (Cushing, 1977):

$$y'(t) = ry(t)\left[Y - \int_{-\infty}^{t} K(t - \tau)y(\tau)\mathrm{d}\tau\right], \tag{E.4}$$

where Y is a positive constant and $K(t) = te^{-t}$. In this case, $F(t, y, z) = ry(Y - z)$.

These equations arise frequently in applications but are not usually studied in graduate mathematics programs. The objective here is to show that in some situations they can be written in more familiar terms, namely, as a system of differential equations. This does not necessarily make them easier to solve, but it does open up a number of possibilities on how to study such problems. For a more expansive discussion, including some of the theory, the book by Arino et al. (2006) may be consulted.

E.2.1 Basis Function Approach

The assumption is that it is possible to write

$$K(t) = k_1 f_1(t) + k_2 f_2(t) + \cdots + k_n f_n(t),$$

where $f_1(t), f_2(t), \ldots, f_n(t)$ are closed under differentiation. This means that given any $f_i(t)$, there are constants $a_{i1}, a_{i2}, \ldots, a_{in}$ such that

$$f_i'(t) = a_{i1} f_1(t) + a_{i2} f_2(t) + \cdots + a_{in} f_n(t).$$

As an example, if $K(t) = te^{-t}$, then we can take $f_1(t) = te^{-t}$ and $f_2(t) = e^{-t}$. Note that the closed-under-differentiation assumption is the same one as was made when using the method of undetermined coefficients to find a particular solution of a linear differential equation.

To write (E.3) as a first-order system, let

$$y_i(t) = \int_{0}^{t} f_i(t - \tau)y(\tau)\mathrm{d}\tau.$$

In this case,

$$y_i'(t) = f_i(0)y(t) + a_{i1}y_1(t) + a_{i2}y_2(t) + \cdots + a_{in}y_n(t).$$

Therefore, (E.3) can be written as

$$y' = F(t, y, k_1 y_1 + k_2 y_2 + \cdots + k_n y_n + z_0),$$
$$y_1' = f_1(0)y(t) + a_{11}y_1 + a_{12}y_2 + \cdots + a_{1n}y_n,$$
$$y_2' = f_2(0)y(t) + a_{21}y_1 + a_{22}y_2 + \cdots + a_{2n}y_n,$$
$$\vdots \qquad \vdots$$
$$y_n' = f_n(0)y(t) + a_{n1}y_1 + a_{n2}y_2 + \cdots + a_{nn}y_n.$$

The associated initial conditions are $y(0) = \chi(0)$ and $y_i(0) = 0$ for $i = 1, 2, \ldots, n$.

Applying this to (E.3), in the case where $K(t) = te^{-t}$, yields

$$y' = F(t, y, y_1 + z_0),$$
$$y_1' = -y_1 + y_2,$$
$$y_2' = y - y_2.$$

In the particular case of the linear integrodifferential equation where $F(t, y, z) = ay + bz$, for a and b constants, the preceding system can be written as

$$y' = Ay + g,$$

where

$$A = \begin{pmatrix} a & b & 0 \\ 0 & -1 & 1 \\ 1 & 0 & -1 \end{pmatrix} \quad \text{and} \quad g = \begin{pmatrix} bz_0 \\ 0 \\ 0 \end{pmatrix}.$$

This also shows that the general solution of a linear first-order integrodifferential equation can contain more than one arbitrary constant, and the exact number depends on the kernel.

E.2.2 Differential Equation Approach

It is possible to express the basis function approach in another form. The assumption is that $K(t)$ satisfies a constant-coefficient differential equation. The easiest way to explain this is to look at a couple of examples.

1. $K' + aK = 0$

   In this case,

$$z' = K_0 y - \frac{1}{a}z,$$

where $K_0 = K(0)$. With this, (E.3) can be written as

$$y' = F(t, y, z + z_0),$$
$$z' = K_0 y - \frac{1}{a} z.$$

As an example, if $K(t) = e^{-t}$, then $a = 1$, and the system is

$$y' = F(t, y, z + z_0),$$
$$z' = y - z.$$

2. $K'' + aK' + bK = 0$

In this case, letting

$$z_1 = \int_0^t K(t - \tau) y(\tau) d\tau,$$

and $z_2 = z_1'$, then (E.3) can be written as

$$y' = F(t, y, z_1 + z_0),$$
$$z_1' = z_2,$$
$$z_2' = (K_0' + aK_0)y - bz_1 - az_2 + K_0 F(t, y, z_1 + z_0),$$

where $K_0' = K'(0)$.

References

M. Abramowitz and I. A. Stegun. *Handbook of Mathematical Functions: With Formulas, Graphs, and Mathematical Tables.* Dover, New York, 1972.

R. C. Ackerberg and R. E. O'Malley. Boundary layer problems exhibiting resonance. *Studies Appl Math,* 49:277–295, 1970.

D. S. Ahluwalia and J. B. Keller. Exact and asymptotic representations of the sound field in a stratified ocean. In J. B. Keller and J. S. Papadakis, editors, *Wave Propagation and Underwater Acoustics,* pages 14–85, Berlin, 1977. Springer-Verlag.

G. Akay. Process intensification and miniaturisation, 2010. http://research.ncl.ac.uk/pim/resea.htm.

G. Allaire and R. Brizzi. A multiscale finite element method for numerical homogenization. *SIAM Multiscale Model Simul,* 4:790–812, 2005.

O. Arino, M. L. Hbid, and E. A. Dads, editors. *Delay Differential Equations and Applications.* Springer, Berlin, 2006.

T. W. Arnold and W. Case. Nonlinear effects in a simple mechanical system. *Am J Phys,* 50:220–224, 1982.

P. Bachmann. *Die Analytische Zahlentheorie.* Teubner, Leipzig, 1894.

A. Baggeroer and W. Munk. The Heard Island feasibility test. *Phys. Today,* Sept: 22–30, 1992.

A. A. Batista and J. M. Carlson. Bifurcations from steady sliding to stick slip in boundary lubrication. *Phys. Rev. E,* 57(5):4986–4996, May 1998.

C. M. Bender, K. Olaussen, and P. S. Wang. Numerological analysis of the WKB approximation in large order. *Phys Rev D,* 16:1740–1748, 1977.

V. S. Berman. On the asymptotic solution of a nonstationary problem on the propagation of a chemical reaction front. *Dokl Akad Nauk SSSR,* 242, 1978.

J. Bernoulli. Meditationes de chordis vibrantibus. *Comment. Acad. Sci. Imper. Petropol.,* 3:13–28, 1728.

F. W. Bessel. Untersuchung des thiels der planetarischen stroungen, welcher aus der bewegung der sonne entsteht. *Abh. Akad. Wiss. Berlin, math. Kl.,* pages 1–52, 1824.

W. Bleakney, D. K. Weimer, and C. H. Fletcher. The shock tube: a facility for investigations in fluid dynamics. *Rev Sci Instr,* 20:807–815, 1949.

N. Boccara. *Essentials of Mathematica: With Applications to Mathematics and Physics.* Springer, New York, 2007.

M. Born and F. Wolf. *Principles of Optics: Electromagnetic Theory of Propagation, Interference and Diffraction of Light.* Cambridge University Press, Cambridge, 7th edition, 1999.

S. Borowitz. *Fundamentals of Quantum Mechanics.* Benjamin, New York, 1967.

M.H. Holmes, *Introduction to Perturbation Methods,* Texts in Applied Mathematics 20, DOI 10.1007/978-1-4614-5477-9, 421
© Springer Science+Business Media New York 2013

M. Bouthier. Comparison of the matched asymptotic expansions method and the two-variable technique. *Q Appl Math*, 41:407–422, 1984.

M. Braun. *Differential Equations and Their Applications: An Introduction to Applied Mathematics*. Springer, New York, 4th edition, 1993.

P. A. Braun. WKB method for three-term recursion relations and quasienergies of an anharmonic oscillator. *Theor Math Phys*, 37:1070–1081, 1979.

H. Bremmer and S. W. Lee. Propagation of a geometrical optics field in an isotropic inhomogeneous medium. *Radio Sci*, 19:243–257, 1984.

D. Broutman, J. W. Rottman, and S. D. Eckermann. Ray methods for internal waves in the atmosphere and ocean. *Annu. Rev. Fluid Mech.*, 36:233–253, 2004.

D. L. Brown and J. Lorenz. A high-order method for stiff boundary value problems with turning points. *SIAM J Sci Stat Comput*, 8:790–805, 1987.

R. Burridge and J. B. Keller. Poroelasticity equations derived from microstructure. *J Acoust Soc Am*, 70:1140–1146, 1981.

R. Burridge and H. Weinberg. Horizontal rays and vertical modes. In J. B. Keller and J. S. Papadakis, editors, *Wave Propagation and Underwater Acoustics*, pages 86–152, Springer, Berlin, 1977.

I. A. Butt and J. A.D. Wattis. Asymptotic analysis of combined breather-kink modes in a Fermi-Pasta-Ulam chain. *Physica D*, 231(2):165–179, 2007.

R. Carles. *Semi-classical analysis for nonlinear Schrodinger equations*. World Scientific, Singapore, 2008.

G. F. Carrier, M. Krook, and C. E. Pearson. *Functions of a Complex Variable: Theory and Technique*. McGraw-Hill, New York, 1966.

E. R. Carson, C. Cobelli, and L. Finkelstein. *The Mathematical Modeling of Metabolic and Endocrine Systems*. Wiley, New York, 1983.

T. K. Caughey. Large amplitude whirling of an elastic string: a nonlinear eigenvalue problem. *SIAM J Applied Math*, 18:210–237, 1970.

V. Cerveny. *Seismic Ray Theory*. Cambridge University Press, Cambridge, 2001.

D. M. Christodoulou and R. Narayan. The stability of accretion tori. IV: Fission and fragamentation of slender self-gravitating annuli. *Astrophys J*, 388:451–466, 1992.

J. D. Cole. *Perturbation Methods in Applied Mathematics*. Blaisdell, Waltham, MA, 1968.

J. D. Cole and L. P. Cook. *Transonic Aerodynamics*. Elsevier, Amsterdam, 1986.

J. D. Cole and J. Kevorkian. Uniformly valid asymptotic approximations for certain differential equations. In J. P. LaSalle and S. Lefschetz, editors, *Nonlinear Differential Equations and Nonlinear Mechanics*, pages 113–120, Academic, New York, 1963.

J. B. Collings and D. J. Wollkind. A global analysis of a temperature-dependent model system for a mite predator-prey interaction. *SIAM J Appl Math*, 50:1348–1372, 1990.

J. M. Combes, P. Duclos, and R. Seiler. On the shape resonance. In L. S. Ferreira S. Albeverio and L. Streit, editors, *Resonances – Models and Phenomena*, pages 64–77, Springer, Berlin, 1984.

C. Comstock and G. C. Hsiao. Singular perturbations for difference equations. *Rocky Mtn J Math*, 6:561–567, 1976.

A. Comtet, A. D. Bandrauk, and D. K. Cambell. Exactness of semiclassical bound state energies for supersymmetric quantum mechanics. *Phys Lett B*, 150:159–162, 1985.

C. Conca and M. Vanninathan. Homogenization of periodic structures via bloch decomposition. *SIAM J Applied Math*, 57(6):pp. 1639–1659, 1997.

L. P. Cook and G. S. S. Ludford. The behavior as $\varepsilon \to 0^+$ of solutions to $\varepsilon \nabla^2 w = \partial w / \partial y$ in $|y| \leq 1$ for discontinuous boundary data. *SIAM J. Math. Anal.*, 2(4): 567–594, 1971.

L. P. Cook and G. S. S. Ludford. The behavior as $\varepsilon \to 0^+$ of solutions to $\varepsilon \nabla^2 w = (\partial/\partial y)w$ on the rectangle $0 \leq x \leq l, |y| \leq 1$. *SIAM J. Math. Anal.*, 4(1):161–184, 1973.

R. J. Cook. Quantum jumps. In E. Wolf, editor, *Progress in Optics, Vol XXVIII*, pages 361–416, Amsterdam, 1990. North-Holland.

O. Costin and R. Costin. Rigorous WKB for finite-order linear recurrence relations with smooth coefficients. *SIAM J. Math. Anal.*, 27(1):110–134, 1996.

M. G. Crandall and P. H. Rabinowitz. Mathematical theory of bifurcation. In C. Bardos and D. Bessis, editors, *Bifurcation Phenomena in Mathematical Physics and Related Topics*, pages 3–46, Boston, 1980. D. Reidel Pub Co.

J. M. Cushing. *Integrodifferential Equations and Delay Models in Population Dynamics*. Lecture Notes in Biomathematics. Springer, Berlin, 1977.

T. Dauxois, M. Peyrard, and C. R. Willis. Localized breather-like solution in a discrete Klein–Gordon model and application to DNA. *Physica D*, 57:267–282, 1992.

P. P. N. De Groen. The nature of resonance in a singular perturbation problem of turning point type. *SIAM J Math Anal*, 11:1–22, 1980.

E. de Micheli and G. Viano. The evanescent waves in geometrical optics and the mixed hyperbolic-elliptic type systems. *Appl Anal*, 85:181–204, 2006.

L. Debnath. *Nonlinear Partial Differential Equations for Scientists and Engineers*. Springer, New York, 3rd edition, 2012.

V. Denoel and E. Detournay. Multiple scales solution for a beam with a small bending stiffness. *J Eng Mech*, 136(1):69–77, 2010.

A. J. DeSanti. Nonmonotone interior layer theory for some singularly perturbed quasilinear boundary value problems with turning points. *SIAM J Math Anal*, 18: 321–331, 1987.

E. D'Hoker and R. Jackiw. Classical and quantal Liouville field theory. *Phys Rev D*, 26:3517–3542, 1982.

R. B. Dingle and G. J. Morgan. WKB methods for difference equations i. *Appl Sci Res*, 18:221–237, 1967a.

R. B. Dingle and G. J. Morgan. WKB methods for difference equations ii. *Appl Sci Res*, 18:238–245, 1967b.

R. J. DiPerna and A. Majda. The validity of nonlinear geometric optics for weak solutions of conservation laws. *Comm Math Phys*, 98:313–347, 1985.

R. C. DiPrima. Asymptotic methods for an infinitely long slider squeeze film bearing. *J Lub Tech*, 90:173–183, 1968.

R. C. DiPrima, W. Eckhaus, and L. A. Segel. Non-linear wave-number interaction in near-critical two-dimensional flows. *J Fluid Mech*, 49:705–744, 1971.

J. L. Dunham. The Wentzel-Brillouin-Kramers method of solving the wave equation. *Phys Rev*, 41:713–720, 1932.

T. M. Dunster. Asymptotic solutions of second-order linear differential equations having almost coalescent turning points, with an application to the incomplete gamma function. *Proc. R. Soc. Lond. A*, 452(1949):1331–1349, 1996.

B. D. Dushaw, P. F. Worcester, W. H. Munk, R. C. Spindel, J. A. Mercer, B. M. Howe, K. Metzger, T. G. Birdsall, R. K. Andrew, M. A. Dzieciuch, B. D. Cornuelle, and D. Menemenlis. A decade of acoustic thermometry in the north pacific ocean. *J. Geophys. Res.*, 114:C07021, 2009.

R. Dutt, A. Khare, and U. P. Sukhatme. Supersymmetry-inspired WKB approximation in quantum mechanics. *Am J Phys*, 59:723–727, 1991.

W. Eckhaus and E. M. d. Jager. Asymptotic solutions of singular perturbation problems for linear differential equations of elliptic type. *Arch Rational Mech Anal*, 23:26–86, 1966.

L. Edelstein-Keshet. *Mathematical Models in Biology*. Society for Industrial and Applied Mathematics, Philadelphia, 2005.

S. N. Elaydi. *An introduction to difference equations.* Springer, New York, 5th edition, 2005.

B. Engquist and P. E. Souganidis. Asymptotic and numerical homogenization. *Acta Numer,* 17:147–190, 2008.

U. Erdmann, W. Ebeling, and A. S. Mikhailov. Noise-induced transition from translational to rotational motion of swarms. *Phys Rev E,* 71(5):051904, May 2005.

A. C. Eringen. On the non-linear vibration of elastic bars. *Q Applied Math IX,* pages 361–369, 1952.

G. B. Ermentrout. Oscillator death in populations of "all to all" coupled nonlinear oscillators. *Physica D,* 41(2):219–231, 1990.

L. Euler. *Methodus inveniendi lineas curvas maximi minimive proprietate gaudentes, sive Solutio problematis isoperimetrici latissimo sensu accepti.* Apud Marcum-Michaelem Bousquet and Socios, Lausanne, 1774.

P. A. Farrell. Sufficient conditions for uniform convergence of a class of difference schemes for a singularly perturbed problem. *IMA J Num Anal,* 7:459–472, 1987.

M.V. Fedoryuk. Equations with rapidly oscillating solutions. In M.V. Fedoryuk, editor, *Partial Differential Equations V: Asymptotic Methods for Partial Differential Equations,* volume 34, pages 1–52, Berlin, 1999. Springer.

L. K. Forbes. Forced transverse oscillations in a simple spring-mass system. *SIAM J Appl Math,* 51:1380–1396, 1991.

W. B. Ford. *Studies on Divergent Series and Summability.* Macmillan, New York, 1916.

L. E. Fraenkel. On the method of matched asymptotic expansions. I: A matching principle. *Proc Cambridge Philos Soc,* 65:209–231, 1969.

K. O. Friedrichs. The mathematical structure of the boundary layer problem. In R. v. Mises and K. O. Friedrichs, editors, *Fluid Mechanics,* pages 171–174, Providence, RI, 1941. Brown University.

G. Frisk, D. Bradley, J. Caldwell, G. D'Spain, J. Gordon, M. Hastings, D. Hastings, J. Miller, D. L. Nelson, A. N. Popper, and D. Wartzok. *Ocean Noise and Marine Mammals.* National Academic Press, Washington, D.C., 2003.

Y. C. Fung. *Foundations of Solid Mechanics.* Prentice-Hall, Englewood Cliffs, NJ, 1965.

N. Ganapathisubramanian and K. Showalter. Relaxation behavior in a bistable chemical system near the critical point and hysteresis limits. *J Chem Phys,* 84:5427–5436, 1986.

R. Gans. Fortpflanzung des lichts durch ein inhomogenes medium. *Ann Phys (Lpz.),* 47:709–738, 1915.

C. Gao and D. Kuhlmann-Wilsdorf. On stick-slip and velocity dependence of friction at low speeds. *J Tribol,* 112:354–360, 1990.

M.J. Garlick, J.A. Powell, M.B. Hooten, and L.R. McFarlane. Homogenization of large-scale movement models in ecology. *Bull Math Biol,* 73(9):2088–2108, 2011.

J. F. Geer and J. B. Keller. Uniform asymptotic solutions for potential flow around a thin airfoil and the electrostatic potential about a thin conductor. *SIAM J Appl Math,* 16:75–101, 1968.

J. S. Geronimo and D. T. Smith. WKB (Liouville-Green) analysis of second order difference equations and applications. *J Approx Theory,* 69:269–301, 1992.

M. Gilli, M. Bonnin, and F. Corinto. Weakly connected oscillatory networks for dynamic pattern recognition. In R. A. Carmona and G. Linan-Cembrano, editors, *Bioengineered and Bioinspired Systems II. Proceedings of SPIE,* volume 5839, pages 274–285, 2005.

G. M. L. Gladwell. *Contact Problems in the Classical Theory of Elasticity.* Sijthoff and Noordhoff, Germantown, MD, 1980.

J. Goodman and Z. Xin. Viscous limits for piecewise smooth solutions to systems of conservation laws. *Arch Rational Mech Anal,* 121:235–265, 1992.

V. F. Goos and H. Hanchen. Ein neuer und fundamentaler versuch zur totalreflexion. *Annalen der Physik*, 436:333–346, 1947.

P. Gray and S. K. Scott. *Chemical Oscillations and Instabilities: Non-linear Chemical Kinetics*. Oxford University Press, Oxford, 1994.

P. Gray, S. K. Scott, and J. H. Merkin. The Brusselator model of oscillatory reactions. *J Chem Soc, Faraday Trans*, 84:993–1012, 1988.

R. M. Green. *Spherical Astronomy*. Cambridge University Press, Cambridge, 1985.

P. A. Gremaud and C. M. Kuster. Computational study of fast methods for the eikonal equation. *SIAM J. Sci. Comput.*, 27:1803–1816, 2006.

J. Guckenheimer and P. Holmes. *Nonlinear Oscillations, Dynamical Systems, and Bifurcations of Vector Fields*. Springer, New York, 1983.

R. B. Guenther and J. W. Lee. *Partial Differential Equations of Mathematical Physics and Integral Equations*. Dover, New York, 1996.

J. K. Hale and H. Kocak. *Dynamics and Bifurcations*. Springer, New York, 1991.

G. H. Hardy. *Orders of Infinity*. Cambridge University Press, Cambridge, 2nd edition, 1954.

S. Haszeldine. Diagenesis research at Edinburgh, 2010. http://www.geos.ed.ac.uk/research/subsurface/diagenesis/.

J. Heading. *An Introduction to Phase-Integral Methods*. Methuen, London, 1962.

J. H. Heinbockel and R. A. Struble. Resonant oscillations of an extensible pendulum. *ZAMP*, 14:262–269, 1963.

D. Hester. The nonlinear theory of a class of transistor oscillators. *IEEE Trans Circuit Theory*, 15:111–117, 1968.

M. H. Holmes. A mathematical model of the dynamics of the inner ear. *J Fluid Mech*, 116:59–75, 1982.

M. H. Holmes. A theoretical analysis for determining the nonlinear permeability of soft tissue from a permeation experiment. *Bull Math Biol*, 47:669–683, 1985.

M. H. Holmes. Finite deformation of soft tissue: analysis of a mixture model in uni-axial compression. *J Biomech Eng*, 108:372–381, 1986.

M. H. Holmes. Nonlinear ionic diffusion in polyelectrolyte gels: an analysis based on homogenization. *SIAM J Appl Math*, 50:839–852, 1990.

M. H. Holmes. *Introduction to Numerical Methods in Differential Equations*. Springer, Berlin, 2007.

M. H. Holmes. *Introduction to the Foundations of Applied Mathematics*. Springer, Berlin, 2009.

M. H. Holmes and F. M. Stein. Sturmian theory and transformations for the Riccati equation. *Port Math*, 35:65–73, 1976.

J. W. Hooker and W. T. Patula. Riccati type transformations for second-order linear difference equations. *J Math Anal Appl*, 82:451–462, 1981.

F. C. Hoppensteadt. *Mathematical Methods in Population Biology*. Cambridge University Press, Cambridge, 1982.

F. C. Hoppensteadt and E. M. Izhikevich. *Weakly connected neural networks*. Springer, Berlin, 1997.

F. C. Hoppensteadt and E. M. Izhikevich. Oscillatory neurocomputers with dynamic connectivity. *Phys Rev Lett*, 82(14):2983–2986, Apr 1999.

F. C. Hoppensteadt and W. L. Miranker. Multitime methods for systems of difference equations. *Stud Appl Math*, 56:273–289, 1977.

R. A. Horn and C. R. Johnson. *Matrix Analysis*. Cambridge University Press, Cambridge 1990.

P. Horwitz and W. Hill. *The Art of Electronics*. Cambridge University Press, Cambridge, 2nd edition, 1989.

F. A. Howes. *Boundary-interior layer interactions in nonlinear singular perturbation theory*. Memoirs of the American Mathematical Society. American Math Society, Providence, RI, 1978.

C. J. Howls. Exponential asymptotics and boundary-value problems: keeping both sides happy at all orders. *Proc. R. Soc. A*, 466(2121):2771–2794, 2010.

J. S. Hubert and E. Sanchez-Palencia. *Vibration and Coupling of Continuous Systems: Asymptotic methods*. Springer, Berlin, 1989.

G. E. Hutchinson. Circular causal systems in ecology. In R. W. Miner, editor, *Annals of the New York Academy of Science*, pages 221–246, New York, 1948. New York Academy of Science.

I. Idris and V. N. Biktashev. Analytical approach to initiation of propagating fronts. *Phys. Rev. Lett.*, 101(24):244101, Dec 2008.

A. M. Il'in. *Matching of Asymptotic Expansions of Solutions of Boundary Value Problems*. Translations of Mathematical Monographs v102. American Math Society, Providence, 1992.

T. Imbo and U. Sukhatme. Logarithmic perturbation expansions in nonrelativistic quantum mechanics. *Am J Phys*, 52:140–146, 1984.

G. Iooss and G. James. Localized waves in nonlinear oscillator chains. *Chaos*, 15(1): 015113, 2005.

N. Ishimura. On steady solutions of the Kuramoto-Sivashinsky equation. In R. Salvi, editor, *The Navier-Stokes Equations: Theory and Numerical Methods*, pages 45–52. Marcel Dekker, 2001.

A. K. Kapila. *Asymptotic Treatment of Chemically Reacting Systems*. Pitman, Boston, 1983.

D. R. Kassoy. Extremely rapid transient phenomena in combustion, ignition and explosion. In R. E. O'Malley, editor, *Asymptotic Methods and Singular Perturbations*, pages 61–72, Rhode Island, 1976. American Math Society.

T. Kato. *Perturbation Theory for Linear Operators*. Springer, Berlin, 1995.

J. P. Keener. Analog circuitry for the van der Pol and FitzHugh-Nagumo equations. *IEEE Trans Syst Man Cybern*, SMC-13:1010–1014, 1983.

J. B. Keller and S. Kogelman. Asymptotic solutions of initial value problems for nonlinear partial differential equations. *SIAM J Appl Math*, 18:748–758, 1970.

J. B. Keller and R. M. Lewis. Asymptotic methods for the partial differential equations: The reduced wave equation and Maxwell's equations. In J. B. Keller, D. W. McLaughlin, and G. C. Papanicolaou, editors, *Surveys in Applied Mathematics: Volume 1*, pages 1–82, New York, 1995. Plenum Press.

J. Kevorkian and J. D. Cole. *Perturbation Methods in Applied Mathematics*. Springer, New York, 1981.

J. R. King, M. G. Meere, and T. G. Rogers. Asymptotic analysis of a non-linear model for substitutional diffusion in semiconductors. *Z angew Math Phys*, 43: 505–525, 1992.

W. Klimesch, R. Freunberger, and P. Sauseng. Oscillatory mechanisms of process binding in memory. *Neurosci Biobehav Rev*, 34(7):1002–1014, 2010.

M. Kline. A note on the expansion coefficient of geometrical optics. *Comm Pure Appl Math*, XIV:473–479, 1961.

C. Knessl. The WKB approximation to the G/M/m queue. *SIAM J Appl Math*, 51: 1119–1133, 1991.

C. Knessl and J. B. Keller. Asymptotic properties of eigenvalues of integral equations. *SIAM J Appl Math*, 51:214–232, 1991a.

C. Knessl and J. B. Keller. Stirling number asymptotics from recursion equations using the ray method. *Stud Appl Math*, 84:43–56, 1991b.

L. P. Kollar and G. S. Springer. *Mechanics of Composite Structures*. Cambridge University Press, Cambridge, 2003.

A. Kolmogorov, I. Petrovsky, and N. Piskunov. Étude de l'équation de la diffusion avec croissance de la quantité de la matière et son application à un pròbleme biologique. *Bulletin Universitè d'Etat à Moscou*, A1:1–26, 1937.

B. Krauskopf, H. M. Osinga, and Jorge G.-V. *Numerical Continuation Methods for Dynamical Systems: Path Following and Boundary Value Problems.* Springer, 2007.

Y. A. Kravtsov. *Geometrical Optics in Engineering Physics.* Alpha Science International, 2005.

G. E. Kuzmak. Asymptotic solutions of nonlinear second order differential equations with variable coefficients. *J Appl Math Mech (PMM),* 23:730–744, 1959.

P. A. Lagerstrom. *Matched Asymptotic Expansions: Ideas and Techniques.* Springer, New York, 1988.

E. Landau. *Handbuch der Lehre von der Verteilung der Primzahlen.* Teubner, Berlin, 1909.

C. G. Lange and R. M. Miura. Singular perturbation analysis of boundary value problems for differential-difference equations. IV: A nonlinear example with layer behavior. *Stud Appl Math,* 84:231–273, 1991.

R. E. Langer. On the asymptotic solutions of ordinary differential equations, with an application to the Bessel functions of large order. *Trans Math Soc,* 33:23–64, 1931.

P.-S. Laplace. *Théorie Analytique des Probabilités.* Mme Ve Courcier, 1812.

R. W. Lardner. The formation of shock waves in Krylov-Bogoliubov solutions of hyperbolic partial differential equations. *J Sound Vib,* 39:489–502, 1975.

G. E. Latta. *Singular perturbation problems.* PhD thesis, California Institute of Technology, 1951.

C. F. Lee. Singular perturbation analysis of a reduced model for collective motion: a renormalization group approach. *Phys. Rev. E,* 83(3):031127, Mar 2011.

K. K. Lee. *Lectures on Dynamical Systems, Structural Stability and their Applications.* World Scientific, Singapore, 1992.

A. M. Legendre. *Traité des Fonctions Elliptiques.* Huzard-Courcier, 1825.

H. Leipholz. *Stability Theory.* Teubner, 2nd edition, 1987.

A. W. Leissa and A. M. Saad. Large amplitude vibrations of strings. *J Appl Mech,* 61:296–301, 1994.

U. Leonhardt and T. G. Philbin. Perfect imaging with positive refraction in three dimensions. *Phys. Rev. A,* 81:011804, Jan 2010.

M. Leontovich and V. Fock. Solution of the problem of electromagnetic wave propagation along the earth's surface by the method of parabolic equation. *J Phys USSR,* 10:13–24, 1946.

N. Levinson. The first boundary value problem for $\varepsilon\triangle u + a(x,y)u_x + b(x,y)u_y + c(x,y)u = d(x,y)$ for small ε. *Ann Math,* 51:428–445, 1950.

J. Lewis, J. M. W. Slack, and L. Wolpert. Thresholds in development. *J Theor Biol,* 65:579–590, 1977.

R. M. Lewis. Asymptotic theory of wave-propagation. *Arch Rational Mech Anal,* 20: 191–250, 1966.

G. C. Lie and J.-M. Yuan. Bistable and chaotic behavior in a damped driven Morse oscillator: A classical approach. *J Chem Phys,* 84:5486–5493, 1986.

M. J. Lighthill. A technique for rendering approximate solutions to physical problems uniformly valid. *Philos Mag,* XL:1179–1201, 1949.

M. J. Lighthill. Viscosity effects in sound waves of finite amplitude. In G. K. Batchelor and R. M. Davies, editors, *Surveys in Mechanics,* pages 250–351, Cambridge, 1956. Cambridge University Press.

M. J. Lighthill. Group velocity. *J Inst Math Appl,* 1:1–28, 1965.

A. Lindstedt. Über die integration einer für die störungstheorie wichtigen differentialgleichung. *Astron Nachr,* 103:211–220, 1882.

T. Linss, H.-G. Roos, and R. Vulanovic. Uniform pointwise convergence on shishkin-type meshes for quasi-linear convection-diffusion problems. *SIAM J. Numer. Anal.,* 38(3):897–912, 2000.

J. Lorentz. Nonlinear boundary value problems with turning points and properties of difference schemes. In W. Eckhaus and E. M. d. Jager, editors, *Theory and Applications of Singular Perturbations*, pages 150–169, Berlin, 1982. Springer.

D. Ludwig. Uniform asymptotic expansions at a caustic. *Comm Pure Appl Math*, 20:215–250, 1966.

J. C. Luke. A perturbation method for nonlinear dispersive wave problems. *Proc R Soc Lond A*, 292:403–412, 1966.

A. D. MacGillivray. Analytic description of the condensation phenomenon near the limit of infinite dilution based on the Poisson-Boltzmann equation. *J Chem Phys*, 56:83–85, 1972.

A. D. MacGillivray. Justification of matching with the transition expansion of van der Pol's equation. *SIAM J Math Anal*, 21:221–240, 1990.

A. D. MacGillivray. A method for incorporating transcendentally small terms into the method of matched asymptotic expansions. *Stud Appl Math*, 99(3):285–310, 1997.

G. M. Maggio, O. de Feo, and M. P. Kennedy. A general method to predict the amplitude of oscillation in nearly-sinusoidal oscillators. *IEEE Trans Circuits Syst*, 51:1586–1595, 2004.

P. A. Markowich, C. A. Ringhofer, and C. Schmeiser. *Semiconductor Equations*. Springer, New York, 1990.

B. E. McDonald and W. A. Kuperman. Time domain solution of the parabolic equation including nonlinearity. *Comp Math*, 11:843–851, 1985.

J. A. M. McHugh. An historical survey of ordinary linear differential equations with a large parameter and turning points. *Arch History Exact Sci*, 7:277–324, 1971.

C. C. Mei and B. Vernescu. *Homogenization Methods for Multiscale Mechanics*. World Scientific, Singapore, 2010.

MEIAF. Micro-environmental imaging and analysis facility, 2010. http://www.bren.ucsb.edu/facilities/meiaf/.

R. Merlin. Maxwell's fish-eye lens and the mirage of perfect imaging. *J Opt*, 13(2): 024017, 2011.

R. E. Mickens. *Difference Equations: Theory and Applications*. Van Nostrand Reinhold, New York, 2nd edition, 1990.

N. Minorsky. *Introduction to Non-linear Mechanics: Topological Methods, Analytical Methods, Non-linear Resonance, Relaxation Oscillations*. Edwards, Ann Arbor, MI, 1947.

T. Mitani. Stable solution of nonlinear flame shape equation. *Combustion Sci Tech*, 36:235–247, 1984.

M. P. Mortell and E. Varley. Finite amplitude waves in bounded media: nonlinear free vibrations of an elastic panel. *Proc R Soc Lond A*, 318:169–196, 1970.

T. J. Moser, G. Nolet, and R. Snieder. Ray bending revisited. *Bull Seism Soc Am*, 82:259–288, 1992.

W. H. Munk, R. C. Spindel, A. Baggeroer, and T. G. Birdsall. The Heard Island feasibility test. *J. Acoust. Soc. Am.*, 96:2330–2342, 1992.

J. A. Murdock. *Perturbations: Theory and Methods*. Classics in Applied Mathematics. SIAM, New York, 1999.

J. D. Murray. *Asymptotic Analysis*. Springer, New York, 1984.

J. D. Murray. *Mathematical Biology II: Spatial Models and Biomedical Applications*. Springer, Berlin, 3rd edition, 2003.

J. D. Murray and A. B. Tayler. An asymptotic solution of a class of nonlinear wave equations: a model for the human torso under impulsive stress. *SIAM J Appl Math*, 18:792–809, 1970.

C. J. Myerscough. A simple model of the growth of wind-induced oscillations in overhead lines. *J Sound Vib*, 28:699–713, 1973.

A. H. Nayfeh. *Perturbation Methods*. Wiley, New York, 1973.

Z. Neda, E. Ravasz, Y. Brechet, T. Vicsek, and A.-L. Barabasi. Self-organizing processes: the sound of many hands clapping. *Nature*, 403:849–850, 2000.

D. J. Ness. Small oscillations of a stabilized, inverted pendulum. *Am J Phys*, 35: 964–967, 1967.

A. M. Nobili and C. M. Will. The real value of Mercury's perihelion advance. *Nature*, 320:39–41, 1986.

K. B. Oldham, J. Myland, and J. Spanier. *An Atlas of Functions*. Springer, 2nd edition, 2009.

F. W. J. Olver. *Introduction to Asymptotics and Special Functions*. Academic, New York, 1974.

F. W. J. Olver, D. W. Lozier, R. F. Boisvert, and C. W. Clark. *NIST Handbook of Mathematical Functions*. Cambridge University Press, Cambridge, 2010.

R. E. O'Malley. Singular perturbation theory: a viscous flow out of Göttingen. In *Annu Rev Fluid Mech*, volume 42, pages 1–17, 2010.

C. H. Ou and R. Wong. On a two-point boundary-value problem with spurious solutions. *Stud Appl Math*, 111(4):377–408, 2003.

J. Pantaleone. Synchronization of metronomes. *Am J Phys*, 70(10):992–1000, 2002.

J. C. B. Papaloizou and J. E. Pringle. The dynamical stability of differentially rotating discs - iii. *Monthly Notices R Astron Soc*, 225:267–283, 1987.

M. Parang and M. C. Jischke. Adiabatic ignition of homogeneous systems. *AIAA J*, 13:405–408, 1975.

D. Park. *Classical Dynamics and its Quantum Analogues*. Springer, Berlin, 2nd edition, 1990.

W. T. Patula. Growth and oscillation properties of second order linear difference equations. *SIAM J Math Anal*, 10:55–61, 1979.

G. Pavliotis and A. M. Stuart. *Multiscale Methods: Averaging and Homogenization*. Texts in Applied Mathematics. Springer, Berline, 2008.

R. Penrose. A generalized inverse for matrices. *Proc Cambridge Philos Soc*, 51: 406–413, 1955.

V. Pereyra, Lee, W. H. K., and H. B. Keller. Solving two-point seismic-ray tracing problems in a heterogeneous medium. I: A general adaptive finite difference method. *Bull Seism Soc Am*, 70:79–99, 1980.

C. S. Peters and M. Mangel. New methods for the problem of collective ruin. *SIAM J Appl Math*, 50:1442–1456, 1990.

A. Pikovsky, M. Rosenblum, and J. Kurths. *Synchronization: A Universal Concept in Nonlinear Science*. Cambridge University Press, Cambridge, 2001.

E. Pinney. *Ordinary Difference-Differential Equations*. University of California Press, 1958.

P. Plaschko. Matched asymptotic approximations to solutions of a class of singular parabolic differential equations. *Z angew Math Mech*, 70:63–64, 1990.

H. Poincaré. Sur les intégrales irrégulières des équations linéaires. *Acta Math*, 8: 295–344, 1886.

K. Popp and P. Stelter. Stick-slip vibrations and chaos. *Philos Trans R Soc Lond A*, 332:89–105, 1990.

J. F. A. Poulet and C. H. Petersen. Internal brain state regulates membrane potential synchrony in barrel cortex of behaving mice. *Nature*, 454:881– 885, 2008.

L. Prandtl. Über Flüssigkeitsbewegung bei sehr kleiner Reibung. In A. Krazer, editor, *Verhandlungen des Dritten Internationalen Mathematiker-Kongresses, Heidelberg 1904*, pages 484–491, Leipzig, 1905. B. G. Teubner.

R. D. Rabbitt and M. H. Holmes. Three-dimensional acoustic waves in the ear canal and their interaction with the tympanic membrane. *J Acoust Soc Am*, 83:1064–1080, 1988.

M. Rafei and W. Van Horssen. On asymptotic approximations of first integrals for second order difference equations. *Nonlinear Dyn*, 61:535–551, 2010.

J. W. S. Rayleigh. On maintained vibrations. *Philos Mag xv*, pages 229–235, 1883.

Lord Rayleigh. On the instantaneous propagation of disturbance in a dispersive medium, exemplified by waves on water deep and shallow. *Phil. Mag.*, 18:1–6, 1909.

E. L. Reiss. On multivariable asymptotic expansions. *SIAM Rev*, 13:189–196, 1971.

D. Richards. *Advanced mathematical methods with Maple*. Cambridge University Press, Cambridge, 2002.

W. Rudin. *Principles of Mathematical Analysis*. McGraw-Hill, New York, 3rd edition, 1964.

E. Sanchez-Palencia. *Non-Homogeneous Media and Vibration Theory*. Lecture Notes in Physics 127. Springer, New York, 1980.

J. A. Sanders. The driven Josephson equation: an exercise in asymptotics. In F. Verhulst, editor, *Asymptotic Analysis II - Surveys and New Trends*, pages 288–318, New York, 1983. Springer.

A. S. Sangani. An application of an homogenization method to a model of diffusion in glassy polymers. *J Polymer Sci*, 24:563–575, 1986.

F. Santosa and W. W. Symes. A dispersive effective medium for wave propagation in periodic composites. *SIAM J Appl Math*, 51:984–1005, 1991.

A. E. Scheidegger. *The Physics of Flow Through Porous Media*. University of Toronto Press, Toronto, 3d edition, 1974.

A. Schlissel. The development of asymptotic solutions of linear ordinary differential equations, 1817–1920. *Arch History Exact Sci*, 16:307–378, 1977a.

A. Schlissel. The initial development of the WKB solutions of linear second order ordinary differential equations and their use in the connection problem. *Historia Math*, 4:183–204, 1977b.

F. Schlogl. Chemical reaction models for non-equilibrium phase transitions. *Z Phys*, 253:147, 1972.

J. Schnakenberg. Simple chemical reaction systems with limit cycle behaviour. *J theor Biol*, 3281:389–400, 1979.

D. Secrest, K. Cashion, and J. O. Hirschfelder. Power-series solutions for energy eigenvalues. *J Chem Phys*, 37:830–835, 1962.

E. E. Sel'kov. Self-oscillations in glycolysis. *Eur J Biochem*, 4(1):79–86, 1968.

J. A. Sethian and A. Vladimirsky. Fast methods for the eikonal and related Hamilton-Jacobi equations on unstructured meshes. *Proc. Natl. Acad. Sci.*, 97:5699–5703, 2000.

R. Seydel. *Practical Bifurcation and Stability Analysis*. Springer, Berlin, 3rd edition, 2010.

J. J. Shepherd. On the asymptotic solution of the Reynolds equation. *SIAM J Appl Math*, 34:774–791, 1978.

J. J. Shepherd and L. Stojkov. The logistic population model with slowly varying carrying capacity. In A. Stacey, B. Blyth, J. Shepherd, and A. J. Roberts, editors, *Proceedings of the 7th Biennial Engineering Mathematics and Applications Conference, EMAC-2005*, volume 47 of *ANZIAM J.*, pages C492–C506, 2007.

S.-D. Shih and R. B. Kellogg. Asymptotic analysis of a singular perturbation problem. *SIAM J Math Anal*, 18:1467–1511, 1987.

W. L. Siegmann, G. A. Kriegsmann, and D. Lee. A wide-angle three-dimensional parabolic wave equation. *J Acoust Soc Am*, 78:659–664, 1985.

A. Singer, D. Gillespie, J. Norbury, and R. S. Eisenberg. Singular perturbation analysis of the steady-state Poisson-Nernst-Planck system: Applications to ion channels. *Eur J Appl Math*, 19(5):541–569, 2008.

L. J. Slater. *Confluent Hypergeometric Functions*. Cambridge University Press, London, 1960.

D. R. Smith. *Singular-Perturbation Theory: An Introduction with Applications*. Cambridge University Press, Cambridge, 1985.

C. R. Steele. Application of the WKB method in solid mechanics. In S. Nemat-Nasser, editor, *Mechanics Today*, volume 3, pages 243–295, New York, 1976. Pergamon.

S. Stenholm. Quantum motion in a Paul trap. *J Modern Opt*, 39:279–290, 1992.

D. C. Stickler, J. Tavantzis, and E. Ammicht. Calculation of the second term in the geometrical acoustics expansion. *J Acoust Soc Am*, 75:1071–1076, 1984.

J. J. Stoker. *Nonlinear Vibrations in Mechanical and Electrical Systems*. Wiley-Interscience, New York, 1950.

G. G. Stokes. On some cases of fluid motion. *Trans Cambridge Philos Soc*, 8:105–165, 1843.

S. H. Strogatz. *Nonlinear Dynamics And Chaos: With Applications To Physics, Biology, Chemistry, And Engineering*. Westview, New York, 2001.

R. A. Struble and T. C. Harris. Motion of a relativistic damped oscillator. *J Math Phys*, 5:138–141, 1964.

J. T. Stuart. Stability problems in fluids. In W. H. Reid, editor, *Mathematical Problems in the Geophysical Sciences*, pages 139–155, Providence, RI, 1971. American Mathematical Society.

F. D. Tappert. The parabolic approximation method. In J. B. Keller and J. S. Papadakis, editors, *Wave Propagation and Underwater Acoustics*, pages 224–287, New York, 1977. Springer.

L. Tartar. *The General Theory of Homogenization: A Personalized Introduction*. Lecture Notes of the Unione Matematica Italiana. Springer, 2009.

H. C. Torng. Second-order non-linear difference equations containing small parameters. *J Franklin Inst*, 269:97–104, 1960.

A. N. Tychonov and A. A. Samarskii. *Partial Differential Equations of Mathematical Physics*. Holden-Day, 1970.

J. Um and C. Thurber. A fast algorithm for two-point seismic ray tracing. *Bull Seism Soc Am*, 77:972–986, 1987.

M. Ungarish. *Hydrodynamics of Suspensions: Fundamentals of Centrifugal and Gravity Separation*. Springer, Berlin, 1993.

O. Vallée and M. Soares. *Airy Functions and Applications to Physics*. Imperial College Press, London, 2nd edition, 2010.

B. van der Pol. On relaxation oscillations. *Philos Mag*, 2:978–992, 1926.

M. Van Dyke. *Perturbation Methods in Fluid Mechanics*. Parabolic Press, Stanford, CA, 1975.

A. van Harten. On an elliptic singular perturbation problem. In W. N. Everitt and B. D. Sleeman, editors, *Ordinary and Partial Differential Equations*, pages 485–495, Berlin, 1976. Springer.

W. van Horssen and M. ter Brake. On the multiple scales perturbation method for difference equations. *Nonlinear Dyn*, 55:401–418, 2009.

N. Voglis. Waves derived from galactic orbits. In *Galaxies and Chaos*, volume 626 of *Lecture Notes in Physics*, pages 56–74. Springer, Berlin Heidelberg, 2003.

M. Wang and D. R. Kassoy. Dynamic response of an inert gas to slow piston acceleration. *J Acoust Soc Am*, 87:1466–1471, 1990.

M. Wazewska-Czyzewska and A. Lasota. Mathematical problems of the dynamics of a system of red blood cells. *Math Stos, Seria III*, 6:23–40, 1976.

A. G. Webster. Acoustical impedance, and the theory of horns and the phonograph. *Proc Natl Acad Sci*, 5:275–282, 1919.

P. J. Westervelt. Parametric acoustic array. *J Acoust Soc Am*, 35:535–537, 1963.

G. B. Whitham. *Linear and Nonlinear Waves*. Wiley, New York, 1974.

J. H. Wilkinson. *Rounding Errors in Algebraic Processes*. Prentice-Hall, Englewood Cliffs, NJ, 1964.

B. Willner and L. A. Rubenfeld. Uniform asymptotic solutions for a linear ordinary differential equation with one m-th order turning point: analytic theory. *Comm Pure Appl Math*, XXIX:343–367, 1976.

P. Wilmott. A note on the WKB method for difference equations. *IMA J Appl Math*, 34:295–302, 1985.

S. Woinowsky-Krieger. The effect of an axial force on the vibration of hinged bars. *J. Appl. Mech.*, 17:35–36, March 1950.

D. J. Wollkind. Singular perturbation techniques: a comparison of the method of matched asymptotic expansions with that of multiple scales. *SIAM Rev*, 19:502–516, 1977.

H. Zhao. A fast sweeping method for eikonal equations. *Math. Comp.*, 74:603–627, 2005.

Index

M.H. Holmes, *Introduction to Perturbation Methods*, Texts in Applied
Mathematics 20, DOI 10.1007/978-1-4614-5477-9,
© Springer Science+Business Media New York 2013